T0328338

MICROPLASTIC CONTAMINATION IN AQUATIC ENVIRONMENTS

MICROPLASTIC CONTAMINATION IN AQUATIC ENVIRONMENTS

An Emerging Matter of Environmental Urgency

Edited by

EDDY Y. ZENG

Elsevier
Radarweg 29, PO Box 211, 1000 AE Amsterdam, Netherlands
The Boulevard, Langford Lane, Kidlington, Oxford OX5 1GB, United Kingdom
50 Hampshire Street, 5th Floor, Cambridge, MA 02139, United States

Library of Congress Cataloging-in-Publication Data
A catalog record for this book is available from the Library of Congress

British Library Cataloguing-in-Publication Data
A catalogue record for this book is available from the British Library

ISBN: 978-0-12-813747-5

Publisher: Candice Janco
Acquisition Editor: Louisa Hutchins
Editorial Project Manager: Hilary Carr
Production Project Manager: Bharatwaj Varatharajan
Cover Designer: Christian J. Bilbow

Typeset by SPi Global, India

CONTENTS

CONTRIBUTORS

Andrei Bagaev
Marine Hydrophysical Institute of Russian Academy of Sciences, Sevastopol, Russia

Margarita Bagaeva
Shirshov Institute of Oceanology of Russian Academy of Sciences, Moscow, Russia

Lian-Jun Bao
Jinan University, Guangzhou, China

Irina Chubarenko
Shirshov Institute of Oceanology of Russian Academy of Sciences, Moscow, Russia

Matthew Cole
Plymouth Marine Laboratory, Plymouth; University of Exeter, Exeter, United Kingdom

Rachel Coppock
Plymouth Marine Laboratory, Plymouth; University of Exeter, Exeter, United Kingdom

Natalia Demchenko
Shirshov Institute of Oceanology of Russian Academy of Sciences, Moscow, Russia

Isa Doverbratt
Lund University, Lund, Sweden

Rachid Dris
University of Bayreuth, Bayreuth, Germany

Dafne Eerkes-Medrano
Aberdeen, United Kingdom

Irina Efimova
Shirshov Institute of Oceanology of Russian Academy of Sciences, Moscow, Russia

Soeun Eo
Korean Institute of Ocean Science and Technology, Busan; Korea University of Science and Technology, Daejeon, South Korea

Elena Esiukova
Shirshov Institute of Oceanology of Russian Academy of Sciences, Moscow, Russia

Lei Gao
East China Normal University, Shanghai, China

Johnny Gasperi
UPEC, AgroParisTech, UPE, Paris, France

Lars-Anders Hansson
Lund University, Lund, Sweden

S. Michele Harmon
University of South Carolina Aiken, Aiken, SC, United States

Sang Hee Hong
Korean Institute of Ocean Science and Technology, Busan; Korean Institute of Ocean Science and Technology, Geoje; Korea University of Science and Technology, Daejeon, South Korea

Hannes K. Imhof
University of Bayreuth, Bayreuth, Germany

Igor Isachenko
Shirshov Institute of Oceanology of Russian Academy of Sciences, Moscow, Russia

Mi Jang
Korean Institute of Ocean Science and Technology, Geoje; Korea University of Science and Technology, Daejeon, South Korea

Simonne Jocic
Oak Crest Institute of Science, Monrovia, CA, United States

Lilia Khatmullina
Shirshov Institute of Oceanology of Russian Academy of Sciences, Moscow, Russia

Christian Laforsch
University of Bayreuth, Bayreuth, Germany

Maiju Lehtiniemi
Finnish Environment Institute, Helsinki, Finland

Daoji Li
East China Normal University, Shanghai, China

Wai Chin Li
Department of Science and Environmental Studies, The Education University of Hong Kong, Tai Po, Hong Kong, China

Martin G.J. Löder
University of Bayreuth, Bayreuth, Germany

Lei Mai
Jinan University, Guangzhou, China

Karin Mattsson
University of Gothenburg, Gothenburg; Lund University, Lund, Sweden

Huase Ou
Jinan University, Guangzhou, China

Outi Setälä
Finnish Environment Institute, Helsinki, Finland

Won Joon Shim
Korean Institute of Ocean Science and Technology, Busan; Korea University of Science and Technology, Daejeon; Korean Institute of Ocean Science and Technology, Geoje, South Korea

Bruno Tassin
LEESU, UMR MA 102, École des ponts, UPEC, AgroParisTech, UPE, Paris, France

Richard Thompson
University of Plymouth, Plymouth, United Kingdom

Fen Wang
Jinan University, Guangzhou, China

Fei Wang
Jinan University, Guangzhou, China

Charles S. Wong
Jinan University, Guangzhou, China; University of Winnipeg, Winnipeg, MB, Canada

Eddy Y. Zeng
Jinan University, Guangzhou, China

Shiye Zhao
East China Normal University, Shanghai, China

Lixin Zhu
East China Normal University, Shanghai, China

Mikhail Zobkov
Northern Water Problems Institute of the Karelian Research Centre of the Russian Academy of Sciences, Petrozavodsk, Russia

CHAPTER 1

Marine Microplastics: Abundance, Distribution, and Composition

Won Joon Shim, Sang Hee Hong, Soeun Eo
Korean Institute of Ocean Science and Technology, Busan, South Korea
Korea University of Science and Technology, Daejeon, South Korea

1.1 INTRODUCTION

The light weight, high durability, resistance to chemicals, plasticity, and high buoyancy of foamed and resin products and the cost-effectiveness of plastics (Thompson et al., 2009) make them so-called "essential" materials in our daily life. World plastic production of 1.7×10^6 ton in 1950 increased 189 times to 3.2×10^8 ton in 2015 (Plastics Europe, 2013, 2016). Moreover, the global production of polyethylene (PE) and polypropylene (PP) (the most common marine microplastics) grew at a rate of 8.7% per year from 1950 to 2012 (Andrady, 2017). However, it has been estimated that $4.8–12.7 \times 10^6$ ton of plastic waste entered the oceans from land-based sources in 192 coastal countries in 2010 (Jambeck et al., 2015), accounting for 1.8%–4.7% of the global plastic production in 2010. Although plastics can last for a long time in the environment, their surfaces become weathered, producing numerous micro- to nanosized fragments (Andrady, 2011; Koelman et al., 2015; Song et al., 2017).

Plastic is categorized in the plastic industry as one of five synthetic polymer groups that also include fibers, coatings, adhesives, and elastomers (Carraher, 2013) (Fig. 1.1). The term "plastic" in "microplastic" does refer not only to solid matter but also to any synthetic organic polymers that are not naturally derived. Although several definitions of microplastics have been proposed based on their size, the definition currently accepted by the scientific community is a particle size within 0.001–5 mm (GESAMP, 2015). Meanwhile, plastic particles <0.001 mm are classified as nanoplastics, those >5 mm are mesoplastics (5–25 mm), and those >25 mm are macroplastics (Lee et al., 2013a). In practical terms, microplastics can be defined as any solid synthetic organic polymers with a particle size in the range of 0.001–5 mm. Microplastics are further categorized as primary or secondary microplastics based on their origin (GESAMP, 2015). Primary microplastics are intentionally manufactured as small-sized particles for industrial purposes and include preproduction resin pellets, microbeads for abrasives in cosmetics, toothpaste and blasting, microsized powders for textile coatings, and drug delivery media (Fig. 1.2). Secondary microplastics are fragmented particles derived from any organic synthetic polymer products in use and in

Microplastic Contamination in Aquatic Environments
https://doi.org/10.1016/B978-0-12-813747-5.00001-1

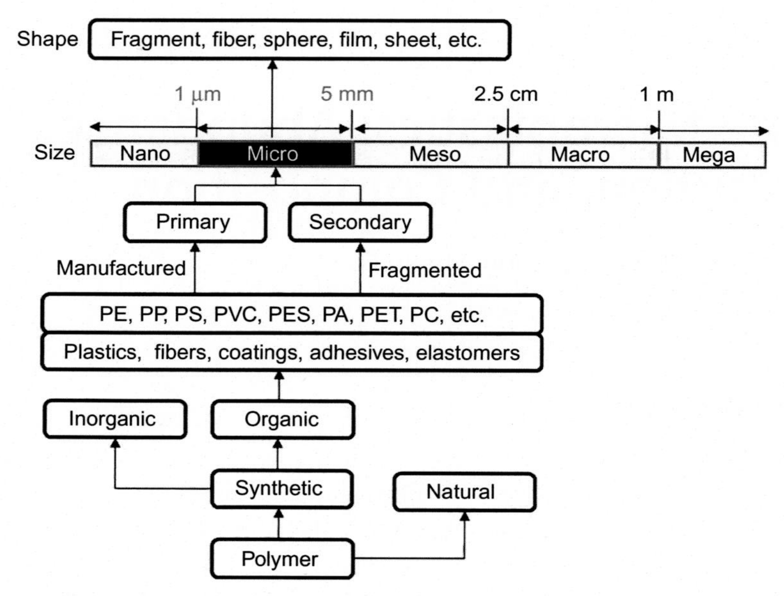

Fig. 1.1 Schematic diagram of microplastic definitions and classifications.

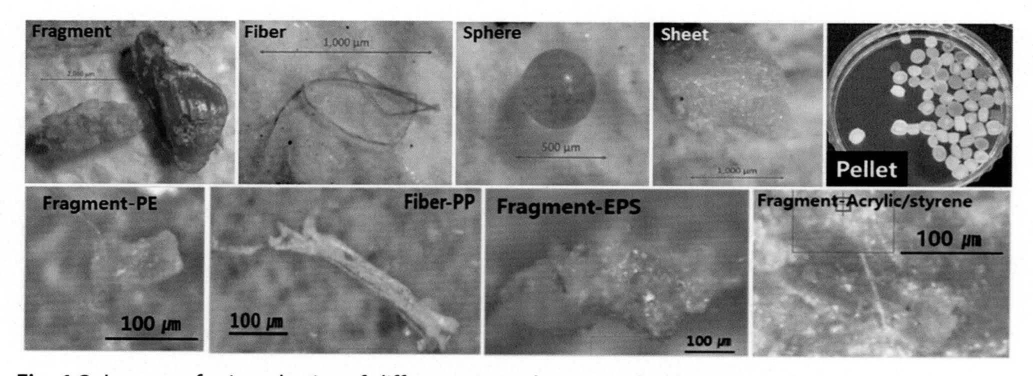

Fig. 1.2 Images of microplastics of different sizes, shapes, and polymer types.

the environment as litter and include solid plastic fragments, microfibers from fabric and rope, coatings that have peeled off, and debris from tire wear.

Microplastics that are less dense than seawater float on the sea surface and can travel globally via wind and oceanic currents (Maximenko et al., 2012). In addition, microplastics are ingestible to an increasing number of small aquatic organisms, and ingestion of small plastic particles may cause adverse biological effects (*see* Chapters 4 and 5 for biological effects (Wright et al., 2013). Their hydrophobic nature enables the accumulation of organic toxicants at concentrations up to a million times higher than in the surrounding water (Lee et al., 2014). Moreover, many chemicals, such as additives, unreacted monomers, catalysts, and by-products, may remain in microplastics (*see* Chapters 7 and 8 for associated chemicals) (Rani et al., 2017). Plastic additives such as plasticizers, antioxidants, ultraviolet and heat stabilizers, flame retardants, and pigments make up a variety of chemicals that exhibit a wide range of toxicity. Along with particle toxicity, adsorbed or additive chemicals desorbed or leached from ingested microplastics in the gastrointestinal tract of organisms can cause additional toxic effects (Rochman et al., 2013). High-trophic-level organisms can be exposed to microplastics via direct uptake from waterborne microplastics, depending on their feeding style and degree of dietary uptake of contaminated prey (Tanaka et al., 2015). Humans are not exempt from this process and can be exposed to microplastics via the consumption of tissues of microplastic-containing seafood, such as bivalves (Van Cauwenberghe and Janssen, 2014; Li et al., 2015; Rochman et al., 2015). Conversely, toxicity tests performed in the laboratory have shown that the particle toxicity of microplastic particles occurs at concentrations one-to-three orders of magnitude higher than those found in the environment (Lenz et al., 2016). In addition, the contribution of toxic chemical accumulation (except for additive chemicals) in marine organisms via microplastic ingestion has been predicted in modeling studies to be lower than other exposure routes (Herzke et al., 2016; Koelmans et al., 2016). In general, a large knowledge gap remains in terms of the exposure to and effects of microplastics in marine environments, and the ecological risk and risk to human health of exposure to microplastics have not yet been clarified.

The spatial and temporal distributions of microplastics, based on compositional data of size, shape, and polymer type in various environmental matrices, are essential elements in quantifying environmental exposure levels for further ecological risk assessment schemes, along with hazard identification (Shim and Thompson, 2015). After their presence in the marine environment was first revealed in the 1970s (Carpenter et al., 1972; Colton et al., 1974), microplastic pollution was reported sporadically from the 1980s to the early 2000s, and reports have increased exponentially since the mid-2000s (GESAMP, 2015). Reflecting the global use of plastics and transportation of plastic litter through ocean currents, microplastics are ubiquitous from coastal to open oceans (Eriksen et al., 2013; Suaria et al., 2016), from tropical to polar seas (Cincinelli et al., 2017; Kanhai et al., 2017), and from surface waters to deep-sea floors (Lattin et al., 2004; Van Cauwenberghe et al., 2013b; Woodall et al., 2014). This chapter summarizes the current abundance; composition in terms of size, shape, and polymer type; and spatial and temporal distribution of microplastics in the water and sediment of marine environments (*see* Chapter 3 for freshwater system, Chapter 12 for terrestrial environment, and Chapter 11 for biota). Further studies are recommended that reflect current knowledge and data gaps.

1.2 ABUNDANCE

1.2.1 Microplastics in Seawater

Microplastics that are less dense than seawater (e.g., PE and PP), including foamed plastics (e.g., expanded polystyrene (EPS) and polyurethane (PUR) foam), float on the sea surface. Since their presence was first revealed in the 1970s (Carpenter et al., 1972), microplastic abundance has been reported in estuarine, nearshore, offshore, and open oceans (Lusher, 2015). A total of 73 research papers and reports have been identified in the literature, among which 65 were published after 2010. These studies report the mean, median, and (or) range of microplastic abundances based on the number, mass, or both per unit area, volume, or weight of water. Because the data were in various formats and units, thereby preventing direct comparison, data reported as the mean number of items per cubic meter or means that could be converted from items per square kilometer into items per cubic meter were selected. When the abundance using neuston or manta trawl nets was reported as items per square kilometer or square meter with about top 20 cm sampling, the data were converted into a volumetric measurement (n/m^3) by adding a third dimension (i.e., conversion of items per square kilometer to items per square meter and multiplying by 0.20 m) (Lusher, 2015). Most water-monitoring studies targeted surface water where floating microplastics accumulated, using neuston or manta trawl nets. Floating microplastics accumulated on the sea surface were collected using nets with mesh sizes of 10–1000 μm or by bulk water filtering with different port sizes (Colton et al., 1974; Song et al., 2014; Zhao et al., 2014). However, several studies performed subsurface sampling with a vertical tow (Gorokhova, 2015), continuous plankton recorder (Thompson et al., 2004), multilayer net (Reisser et al., 2015), or underway sampling by pumping from the subsurface during ship movement (Desforges et al., 2014; Lusher et al., 2014). Among the 73 papers and reports, 61 studies with 70 mean values (multiple regions were surveyed in some studies) were chosen, and the results are summarized in Fig. 1.3.

The abundances of microplastics ranged from undetected at many stations in a number of studies to 102,550 n/m^3 in an industrial harbor in Stenungsund, Sweden (Norén, 2007). The abundances varied widely, not only among but also within studies. Thus, the mean value of each study or sampling region, in studies of multiple regions, was used to obtain a better understanding of concentrations worldwide. The mean abundance of microplastics in seawater reported worldwide ranged from $4.8 \times 10^{-6}\,n/m^3$ in eastern equatorial Pacific (Spear et al., 1995) to $8.6 \times 10^3\,n/m^3$ off the Swedish coast (Norén, 2007), except for a surface microlayer concentration of $1.6 \times 10^4\,n/m^3$ (Song et al., 2014; Fig. 1.3), showing a maximum difference of nine orders of magnitude. The median of the 70 mean values was $8.9 \times 10^{-2}\,n/m^3$, and 45% of studies reported mean abundances between 0.01 and 10 n/m^3.

Microplastic abundance tends to increase steeply with decreasing size (Cózar et al., 2014; Isobe et al., 2015). Thus, it is crucial to consider the lower bound of microplastic size for sampling and detection in any comparison of abundance. Table 1.1 divides the results by net mesh size in sampling or postsampling treatment of the 70 studies. Of the 70 cases, 47 collected samples with zooplankton nets with mesh sizes between

Fig. 1.3 Mean abundances of microplastics in seawater worldwide.

280 and 505 μm. Meanwhile, four cases used larger mesh sizes (900–1000 μm), and 11 cases used smaller mesh nets (120–250 μm). The other eight cases used fine phytoplankton nets with mesh sizes of 10–80 μm. Surface microlayer sampling involved direct filtration using 0.7 μm pore sizes (Song et al., 2014, 2015b). The mean (median) abundances of microplastics in surface water according to each net mesh size range were 2.4×10^3 (1.8×10^3) n/m^3 for 10–80 μm mesh, 93 (1.15) n/m^3 for 120–250 μm mesh, 9.6×10^{-1} (3.1×10^{-2}) n/m^3 for 280–350 μm mesh, 2.2×10^{-1} (1.5×10^{-2}) n/m^3 for

Table 1.1 Summary of microplastic abundance in water by sampling mesh size

Category	Abundance (n/m^3)				
Mesh (μm)	10–80	120–250	280–350	450–505	900–1000
No. of studies	8	11	38	9	4
Min-Max	0.17–8654	0.012–969	0.00028–7.68	0.00002–1.69	0.0000048–0.000341
Mean ± SD	2444 ± 2841	93 ± 291	0.96 ± 2.05	0.22 ± 0.55	0.00028 ± 0.00034
Median	1841	1.15	0.031	0.015	0.00018
75% percentile	2679	2.57	0.28	0.063	0.00044
25% percentile	599	0.49	0.01	0.012	0.00022

450–505 μm mesh, and 2.8×10^{-4} (1.8×10^{-4}) n/m^3 for 900–1000 μm mesh. Microplastic abundance demonstrated a negative relationship with net mesh size regardless of sampling region and time. Therefore, further comparison of the spatial distribution (*see* Section 1.3) was performed only for cases that used 300–350 μm mesh manta and neuston nets collecting surface water in marine environments. The microplastic abundance observed in surface water could also be influenced by the vertical distribution of microplastics, governed by the state of the ocean and turbulence according to wind speed (Kukulka et al., 2012; Reisser et al., 2015).

1.2.2 Microplastics in Sediment

Floating microplastics wash ashore due to a combination of landward wind and currents (Isobe et al., 2014). Selective removal of macrodebris during beach cleanup programs by local authorities or volunteers results in the accumulation of microplastics in intertidal and upper tidal zones and at the vegetation line (Turra et al., 2014; Lee et al., 2015; Moreira et al., 2016). Furthermore, direct exposure to ultraviolet radiation and high temperatures on beaches provides favorable conditions for weathering of macro- and microplastics, which can produce smaller micro- and nanosized particles (Lambert and Wagner, 2016; Song et al., 2017). Microplastics denser than seawater (e.g., polyester (PES) and polyvinyl chloride (PVC)) have relatively high likelihoods of settling on the seafloor, even though the size and shape (e.g., aspect ratio of dense microplastics) and surface tension on the sea surface may influence their sinking rate. Moreover, fouling by microorganisms on microplastics (Moret-Ferguson et al., 2010), interactions with plankton-forming aggregates (Long et al., 2015), and ejection of ingested microplastics in feces (Cole et al., 2016) facilitate the precipitation of lightweight microplastics into benthic environments. Conversely, the reduction of biofilm mass due to light limitations under the euphotic zone and grazing in the water column can cause the vertical oscillation of lightweight microplastics with different sinking depths and timescales according to their size and density (Kooi et al., 2017; Rummel et al., 2017).

An early microplastic monitoring study focused on large-sized microplastics (1–5 mm), such as preproduction resin pellets on sandy beaches (Gregory, 1978). Moreover, secondary small microplastics with sizes <1 mm on beaches and subtidal sediments (including deep-sea beds) have been intensively monitored since the 2000s (Thompson et al., 2004; Hidalgo-Ruz et al., 2012; Lee et al., 2013a; Van Cauwenberghe et al., 2013a). Microplastic abundances in inter- and subtidal sediments have been reported in 57 studies in the literature. As noted in the studies of microplastics in water (*see* Section 1.2.1), microplastic abundances in sediment have been reported in various formats and units. Sediment data were selected for further comparison when they were given as the mean value in units of items per square meter for beaches or items per kilogram for subtidal areas, and such units are relatively common. When studies provided information on quadrat size, sediment sampling depth, and volume of the sample for beaches, items per kilogram was converted into items per square meter. The conversion factor from weight to volume of sand sample (1 kg = 0.73 L) was based on Korean sand beach data ($n = 20$). Among the 57 studies published during 2004–17, 22 were chosen to represent beach environments and 11 for subtidal environments (Fig. 1.4).

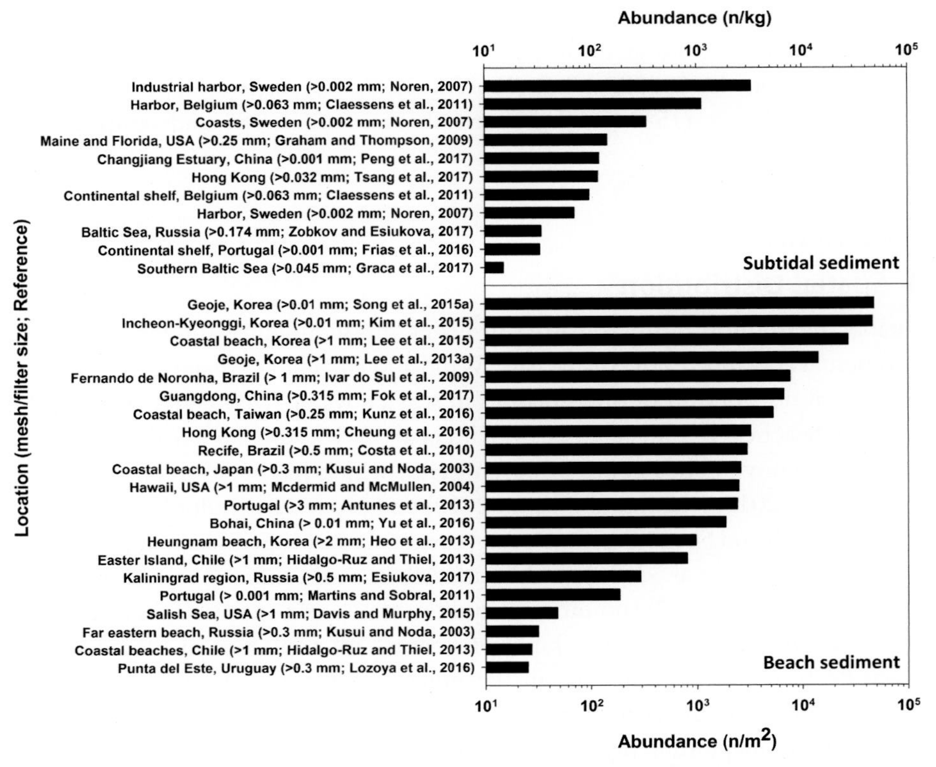

Fig. 1.4 Mean abundances of microplastics in beach and subtidal sediment worldwide.

The mean abundance of microplastics in sediments was in the range of 25–47,897 n/m^2 for beaches and 15–3320 n/kg for subtidal sediments. The maximum difference in mean abundance of microplastics in sediments among the studies was in the range of three orders of magnitude, which was much less than that of seawater. There was no negative relationship between the lower bound of microplastic size for sampling or detection and abundance in sediment. The narrow range of the mean abundance of microplastics in beach sediments might be explained by the sampling method, for example, sampling of only high strandlines or wreck lines on beaches. There were few transect (Heo et al., 2013) or multiline (Nel et al., 2017) surveys, and most studies collected a single (Wilber, 1987; Kunz et al., 2016) or multiple (Costa et al., 2010; Lee et al., 2013a; Fok et al., 2017) quadrat sample(s) only at the high strandline. High strandline sampling is easier, enables comparison with other similar studies, and may assume the worst-case scenario. However, it may frequently result in large overestimates and may be unrepresentative of microplastic abundance and exposure regime in organisms. The method of extracting microplastics from sediments, including the density of the separating solution such as seawater, NaCl, $ZnCl_2$, NaI, and lithium metatungstate, also influences the recovery rate and concentration of microplastics (Hidalgo-Ruz et al., 2012; Imhof et al., 2012).

Although there were limited microplastic mean abundance data in subtidal zones, the results showed a narrow range of distribution. Suspected hot spots, such as harbors, had high concentrations (340–3320 n/kg) (Norén, 2007; Claessens et al., 2011), and relatively low abundances (15–145 n/kg) were observed in coastal and continental shelf sediments.

1.3 DISTRIBUTION

1.3.1 Spatial Distribution

The spatial distribution of microplastics in water and sediment is influenced by various environmental and anthropogenic factors. Wind and current are major factors governing the horizontal distribution of microplastics in surface water (Law et al., 2010) and their stranding on beaches (Baztan et al., 2014; Kim et al., 2015). On large scales, microplastics accumulate in convergence zones of ocean gyres, which has been predicted by models (Lebreton et al., 2012; Maximenko et al., 2012) and proved by multiple in situ observations in the North and South Pacific Ocean (Law et al., 2014; Eriksen et al., 2013), North Atlantic Ocean (Law et al., 2010), and South Atlantic and Indian Oceans (Cózar et al., 2014). A 12-year (2001–12) monitoring study in the Northeast Pacific Ocean revealed that the concentration (156,800 n/km^2) of floating microplastics in the accumulating zone was 84 times higher than that in the nonaccumulating zone (1864 n/km^2) (Law et al., 2014). Each accumulation zone of the five ocean gyres showed microplastic abundance an order of magnitude higher than in nonaccumulation zones (Cózar et al., 2014). Even with increasing efforts toward observing floating microplastics on global ocean surfaces, large regions of open oceans, especially nonaccumulation zones, regional seas, and coastal zones, still lack data. Therefore, modeling is a useful approach for determining the global abundance of microplastics in oceans (Eriksen et al., 2014).

Recently, three different models were compared, and the total amount of floating microplastics was estimated in basins based on both a count and a mass basis (Van Sebille et al., 2015; Fig. 1.5). The three models yielded total microplastic particle abundances of 1.5×10^{13}–5.1×10^{13} and masses of 93–236 metric tons (Van Sebille et al., 2015). The predicted total microplastic counts per basin were in the order of Mediterranean > North Pacific > South Atlantic > North Atlantic ≅ South Pacific (Van Sebille et al., 2015), in good agreement with the distribution of the mean and median abundances observed in situ (Fig. 1.6). The mean abundances of microplastics derived from the literature (with lower particle sizes of 300–350 μm) differed by basin, although the difference was not significant due to the high variance among the studies (Kruskal-Wallis test, $P > 0.05$). However, the mean abundance differed significantly between nearshore (1.78 ± 2.63 items/m^3, $n=21$), gyre (0.46 ± 0.99 items/m^3, $n=10$), and offshore waters (0.11 ± 0.14 items/m^3, $n=5$) (Kruskal-Wallis test, $P < 0.001$).

Similarly, at a small scale, steep gradients of microplastic abundance and patchy distributions have been observed in coastal zones (Doyle et al., 2011; Desforges et al., 2014; Kang et al., 2015a; Pedrotti et al., 2016). A negative gradient of floating microplastics from nearshore to offshore waters was reported in the Mediterranean (Pedrotti et al., 2016) and off Vancouver Island, Canada (Desforges et al., 2014), while the reverse was found off the coast of California, the United States, and Geoje Island, Korea (Doyle et al., 2011; Kang et al., 2015a). The spatial distribution of microplastics in surface water in coastal zones is seasonally affected by the degree of riverine input of land-based sources (Lima et al., 2014). The spreading speed and range of river plumes near river mouths can influence short-term spatial distribution of microplastics and movement of flotsam patches. In addition, the location of sewage treatment plants can also affect the spatial distribution because sewage outfall is a source of microplastics (Horton et al., 2017).

Relatively, few studies have compared microplastics in sea surface water and in the water column. Microplastic concentrations decreased exponentially with increasing depth in the top 5 m (Reisser et al., 2015; Kooi et al., 2016). In southern California, offshore water, near the bottom at a depth of 30 m, showed much higher microplastic abundance than surface water and midcolumn water both before and after storm events (Lattin et al., 2004). Although a smaller mesh net (250 μm) was used for subsurface sampling compared with surface water sampling (333 μm), a higher mean abundance (2.68 n/m^3) was found in subsurface water than in surface water (0.34 n/m^3) in Arctic polar waters (Lusher et al., 2015). The vertical distribution of microplastics from surface water to bottom water was expected to be governed by complex interactions among density, size, shape, and attached biofilm mass of microplastics and the intensity of waves, turbulence, and seawater density profile (Rummel et al., 2017). Most studies focused on the sea surface, on which

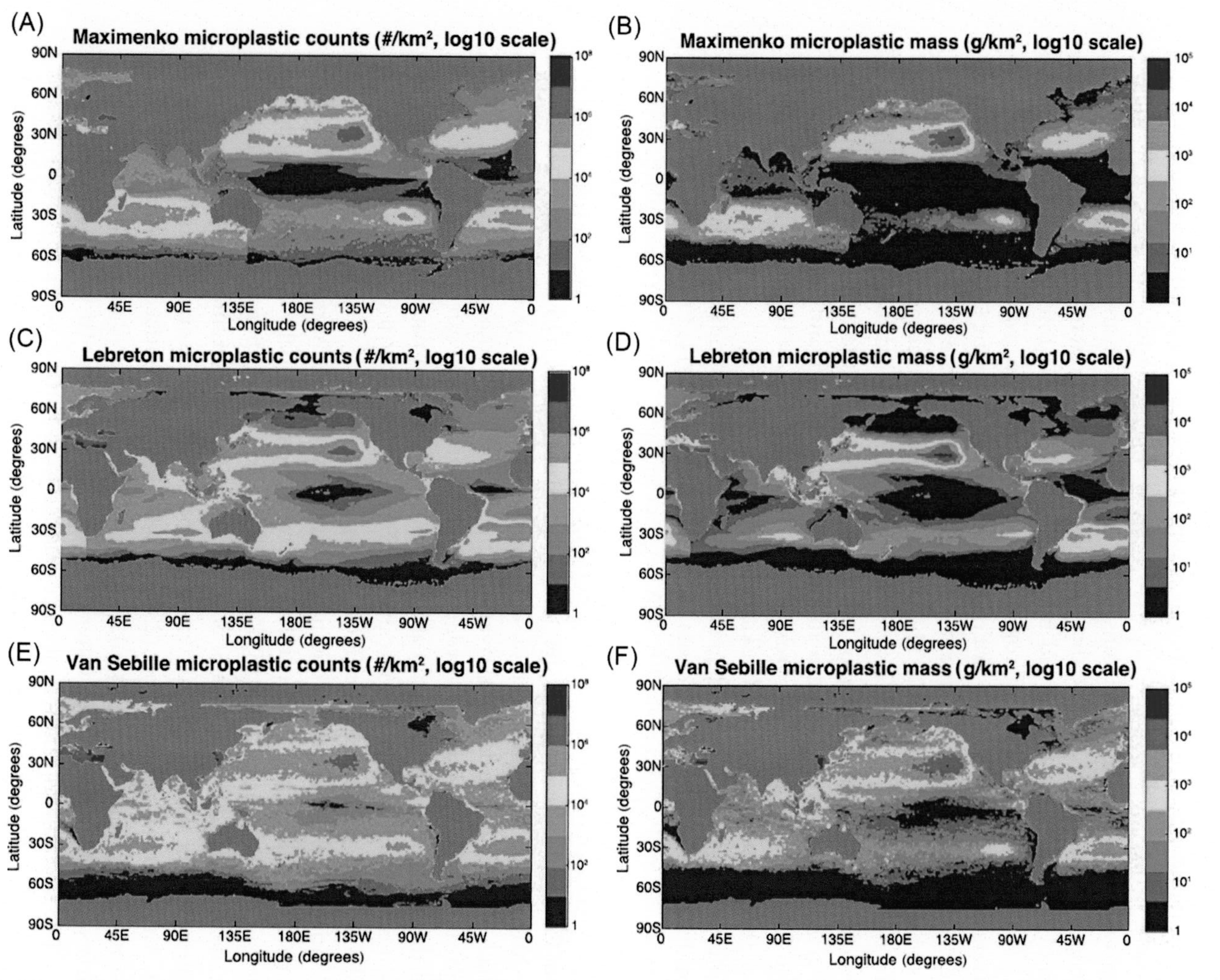

Fig. 1.5 Maps of microplastic count (left column) and mass (right column) distributions for three models. Because the fits were determined on a per-basin level, there are a few visible discontinuities (e.g., South of Tasmania in the Maximenko solution, panel (A)). Source: Van Sebille, E., Wilcox, C., Lebreton, L., Maximenko, N., Hardesty, B.D., van Franeker, J.A., Eriksen, M., Siegel, D., Galgani, F., Law, K.L., 2015. A global inventory of small floating plastic debris. Environ. Res. Lett. 10, 214006.

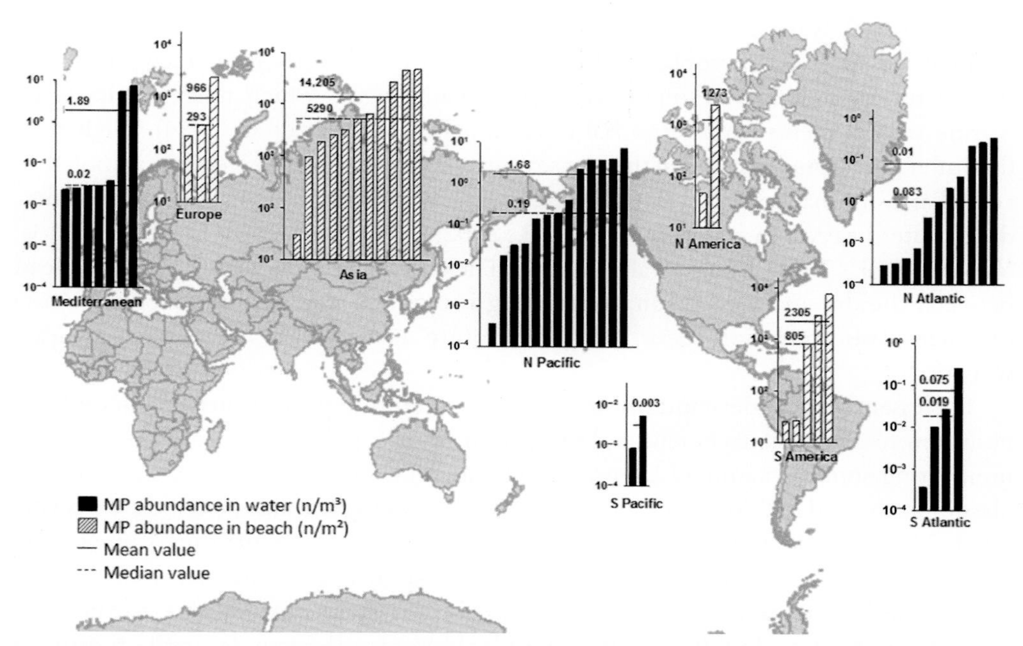

Fig. 1.6 Distribution of microplastic abundances in seawater by ocean basin, and on beaches by continent.

low-density microplastics accumulate in relatively high abundance, whereas most marine invertebrates and fish are likely to be exposed to suspended microplastics in the water column and overlying bottom water. Even though there are limited data on subsurface microplastics, marine organisms may be exposed to comparable microplastic levels in subsurface water (Lattin et al., 2004; Lusher et al., 2015). More extensive and systematic monitoring studies are required on subsurface water to elucidate more realistic exposure assessments of marine organisms to microplastics in their habitats.

Fig. 1.6 summarizes the microplastic abundance in beach sediments by continent according to the literature. The microplastic abundance in beaches in Asia was significantly (Kruskal-Wallis test, $P < 0.05$) higher than those in America and Europe. Relatively high abundances were recorded in East Asia, including China, Japan, Hong Kong, Korea, and Taiwan (1867–47,897 n/m^2). High abundances of microplastics on Asian beaches and in the North Pacific Ocean may reflect the relatively recent inputs of plastic from land-based sources in Asian countries (Jambeck et al., 2015).

1.3.2 Temporal Distribution

There are insufficient data in the literature to determine the proportions of microplastics that originate from various land- versus sea-based sources (Jang et al., 2014). Estuarine waters, coastal waters, and beaches where microplastics have washed ashore show seasonal variation in microplastic abundance. However, it has been recognized that surface runoff and riverine input are a major pathway by which microplastics reach marine environments from terrestrial environments (Lechner et al., 2014). High river flow after heavy precipitation in the wet season generally induces microplastic runoff into coastal areas (Lima et al., 2014), and there are many reports of high abundances immediately after heavy rain events (Moore et al., 2002; Lattin et al., 2004; Lee et al., 2013a; Yonkos et al., 2014; Kang et al., 2015b). These data indicate that seasonal, at least between the dry and wet season, monitoring is required to obtain representative data on microplastic pollution status and for exposure analyses that include a worst-case scenario.

It is essential to understand historical trends and future projections for microplastic pollution to allow management and mitigation of such pollution. Even though a few monitoring studies examined large microplastics and, to a lesser extent, small microplastics in the early 1970s, microplastic research is lacking for many periods, especially 1950–70 and 1980–2000, based on the number of papers in the literature survey. Fortunately, retrospective analyses of microplastic abundance in marine environments using archived samples and age-dated sediment cores are possible due to the high durability of plastics. Archived continuous plankton recorder samples in the Northeast Atlantic revealed that the microplastic abundance in the water column increased significantly from 1960–70 to 1980–90 (Thompson et al., 2004). Another increasing trend in microplastic concentrations was revealed in a nondisturbed, age-dated sediment core from the Belgian coast between 1993 and 2008 (Claessens et al., 2011). Meanwhile, long-term monitoring studies of floating microplastics in the Northwest Atlantic Ocean between 1986 and 2008 and in the Northwest Pacific Ocean between 2002 and 2012 showed no significant increases in abundance (Law et al., 2010, 2014). Global plastic demand and production have exponentially increased for more than half a century (Plastics Europe, 2016), and the input of plastic debris into marine environments is expected to increase in the future in the absence of counteracting measures. In addition, even if new inputs of plastic debris to the marine environment were stopped completely, weathering and fragmentation of previously accumulated large plastic debris in the marine environment would be expected to continue, producing large amounts of microplastics; therefore, the amount of microplastics is likely to increase in the long term. The standing stock of microplastics in the sea, on the shore, on the seafloor, and in organisms depends on both the input and removal rate. Therefore, an emission inventory for primary microplastics and measurements of the production rate of secondary microplastics due to weathering and natural removal processes (e.g., complete mineralization of microplastics and burial in deep-sea sediment) must be established to quantify and predict future microplastic pollution levels.

1.4 COMPOSITION BY SIZE, SHAPE, AND POLYMER TYPE

1.4.1 Size Distribution of Microplastics

Microplastics (0.001–5 mm) have a size range of three orders of magnitude, and the fate and biological effects of microplastics depend on their size. For example, the sinking rate and rising velocity are influenced by size (Reisser et al., 2015; Kowalski et al., 2016). In addition, the uptake rate by aquatic organisms, retention time in gastrointestinal tracts, and adverse biological effects on organisms are affected by size (Browne et al., 2008; Lee et al., 2013b). Moreover, the toxic effects depend on the target organism and the end point of toxicity, but there is a tendency toward increasing toxicity with decreasing particle size (Lee et al., 2013b; Jeong et al., 2016).

The size of microplastics in water and sediment has been determined by sequential sieving and by manual identification or image analysis under a microscope according to sampling time and pretreatment steps. The number of microplastic fragments produced in a laboratory by weathering experiment, by ultraviolet exposure, and by mechanical abrasion with sand increased exponentially down to a size of a few microns (Song et al., 2017). Similar size distribution patterns of microplastics were found in the sea-surface microlayer (Song et al., 2014; Chae et al., 2015; Song et al., 2015b), while the peak abundance was more often in the range of hundreds to thousands of microns in surface water (Lattin et al., 2004; Collignon et al., 2014; Cózar et al., 2014; Isobe et al., 2015). Because gradual fragmentation of plastics over time produces a larger number of smaller particles, there is a tendency toward increasing numbers of smaller microplastics. However, the mean microplastic size in water depends on the size range of the microplastics sampled and analyzed (Fig. 1.7). Water samples, collected using nets with mesh sizes of 200–1000 μm and analyzed mainly with microscopy (upper part of water data in Fig. 1.7), have shown a mean size of microplastics in the range of one to a few millimeters. Conversely, sea-surface microlayer and smaller net mesh size (50–63 μm) samples (lower part of water data in Fig. 1.7) analyzed by microspectroscopy have shown a mean size of <700 μm.

The difference in microplastic abundance at the lower end of the range of microplastic analyses is possibly caused by analytic artifacts and environmental factors. Analytic artifacts could result in lower recovery and identification rates of smaller particles versus larger particles. There is a greater chance to miss small microplastics during the extraction step (from environmental matrices) and a greater likelihood of failing to differentiate them from natural particles. In particular, using a microscopic identification method alone had a higher chance of missing small microplastics (Song et al., 2015a; Shim et al., 2016). This may in part explain why the mean microplastic size differs by the size range of different sampling and analytic methods. Environmental factors can also affect the efficacy or speed of selective removal of small microplastics. Four main routes to possible losses of small plastics have been proposed: shore deposition, nanofragmentation,

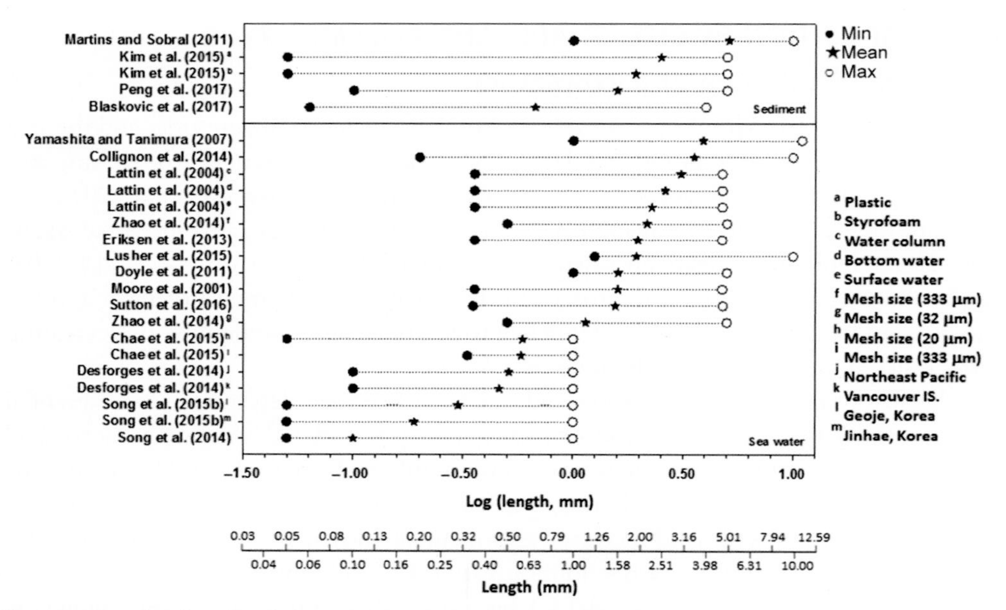

Fig. 1.7 Size distribution of microplastics in seawater and sediment.

biofouling, and ingestion (Cózar et al., 2014). However, Isobe et al. (2014) suggested that there was selective removal of mesoplastics (>5 mm) rather than microplastics in near-shore waters, resulting in an increasing proportion of microplastics over mesoplastics with increasing distance offshore. Small plastic particles have high surface-area-to-volume ratios, resulting in faster fragmentation down to the detection size limit (e.g., nanometer) (Song et al., 2017) and inducing more rapid biofouling, in turn enabling them to sink rapidly. In addition, smaller plastics are more bioavailable to a wider range of aquatic organisms, which could increase their removal rate via ingestion.

1.4.2 Microplastic Shapes

There is no standardized methodology for classifying microplastics by shape. In the literature, they are mainly categorized as fragments, solid (hard) plastics and foamed plastics, fibers, lines, filaments, spheres, films, sheets, and pellets. Microplastic shape can partly provide information on their origin, along with polymer type (*see* Section 1.4.3). Fragments are thought to originate mainly from hard plastics via fragmentation, even though microbeads used in cosmetics can be irregularly shaped. Fibers originate from fabrics, net, fishing line, and rope. Preproduction resin pellets and spheres can be considered as engineered primary plastics. Film and sheet microplastics might generally originate from plastic bags and packaging material. Floats, shock-absorbing packing material, heat-insulating boxes, and construction panels are the main sources of foamed plastics. Floating, sinking, and transportation behavior can be influenced by microplastic shape (Kowalski et al., 2016). In addition, the ingestion and removal rates of microplastics by aquatic organisms might be affected by their shape.

Among 23 studies in the literature providing numerical values of shape composition, fibers ($n=11$) and fragments ($n=9$) were dominant in seawater (Fig. 1.8A). Preproduction resin pellets and EPS (styrofoam) were predominant in Hong Kong (Tsang et al., 2017) and the Salish Sea, Alaska, and the United States (Davis and Murphy, 2015). Fibers ($n=6$), fragments ($n=4$), and styrofoam ($n=4$) were the dominant types in beach sediment, while fibers ($n=6$) were predominant in subtidal sediments. High proportions of

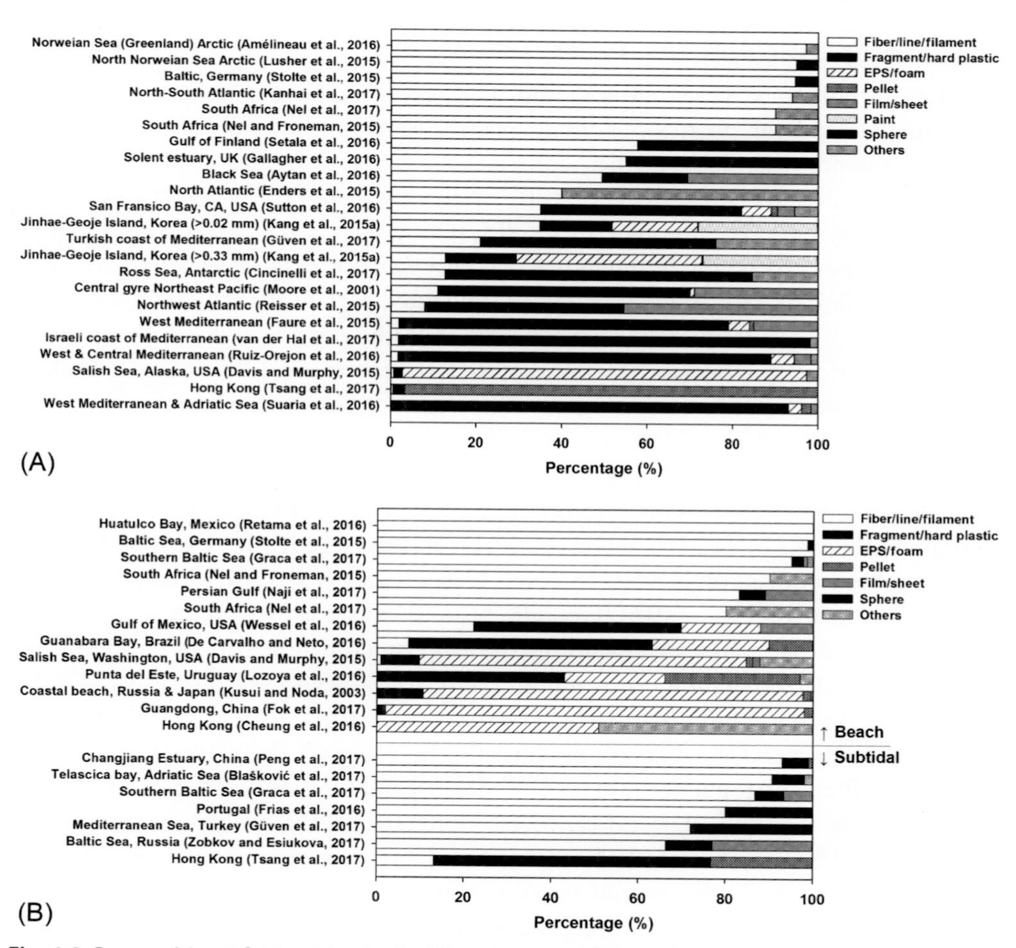

Fig. 1.8 Composition of microplastics in (A) seawater and (B) sediment by shape.

microplastic fibers and fragments in marine environments indicate that secondary microplastics contribute to microplastic abundance more so than do primary microplastics. Fibers originate from fabric, for example, via cloth washing (Browne et al., 2011; Napper and Thompson, 2016), and are a major land-based input source. Nets, ropes, and monofilaments used in fishery and aquaculture farms are additional sources of microfibers (Jang et al., 2014). Styrofoam floats are widely used in aquaculture farms and are a major input of microplastics in Asian countries (Kusui and Noda, 2003; Kang et al., 2015a; Cheung et al., 2016; Fok et al., 2017; Tsang et al., 2017). A high proportion of fibers consisting of polymers that are denser than seawater (e.g., PES and acrylic) has been found in subtidal sediments, indicative of their selective removal from surface water via sinking (Fig. 1.8B).

1.4.3 Microplastic Polymer Types

A variety of polymers are synthesized and used for domestic and industrial purposes. European demand for plastics, including plastic materials (thermoplastics and PUR) and other plastics (thermoset plastics, adhesives, coatings, and sealants) without fibers, is in the order of polyethylene (PE) > polypropylene (PP) > polyvinylchloride (PVC) > polystyrene (PS) > polyurethane (PUR) > polyester (PES) > other polymers (Plastics Europe, 2013). In the literature, PE ($n=23$), PP ($n=18$), PS ($n=13$), PES ($n=7$), PVC ($n=6$), polyamide (PA, $n=6$), polyvinyl acetate (PVA; $n=4$), and other polymers ($n<3$) have frequently been reported in both seawater and sediments. Based on the polymer composition in seawater reported in the literature, PE was the dominant polymer, followed by PP and PS (Fig. 1.9A), in agreement with the data on European demand for plastics. PVC and PA and acrylic PES blends were the second most abundant polymers in the surface water of Kyeonggi Bay, South Korea (Chae et al., 2015), and across the Atlantic Ocean (Kanhai et al., 2017), respectively. PE, PP, PS, and PES are major polymer types on beaches and in subtidal sediments (Fig. 1.9B). Expanded PS particles were predominant on beaches in South Korea (Chae et al., 2015; Lee et al., 2015) and Uruguay (Lozoya et al., 2016). Fused PS beads in expanded PS products can readily detach via weathering and are usually in the large microplastic size range (1–5 mm). Beach and subtidal sediment in the southern Baltic Sea showed a greater variety of polymer types than reported in other studies (Graca et al., 2017).

Different polymers have different densities (Hidalgo-Ruz et al., 2012). In the absence of inorganic fillers and fouling, PE (0.92–0.97 g/cm^3) and PP (0.90–0.91 g/cm^3) are generally less dense than seawater and can float. Conversely, other polymers, such as PS, PVC, PES, and PA, are heavier (1.02–2.3 g/cm^3) than seawater and sink, except when they are expanded, and contain void airspace. Less dense polymers, such as PE, PP, and expanded PS, are dominant in surface water and beaches, where floating plastics are washed ashore, while heavier polymers such as PES are often dominant in the water

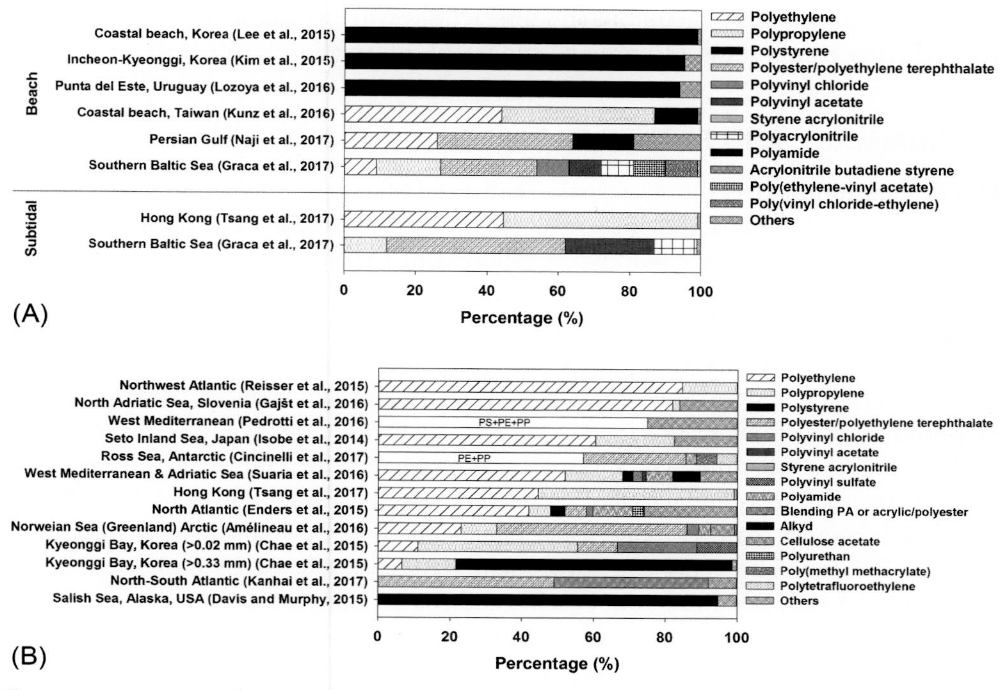

Fig. 1.9 Composition of microplastics in (A) seawater and (B) sediment by polymer type.

column and subtidal sediments. One study found that the fragmentation rate and size of fragments produced by ultraviolet exposure and subsequent mechanical abrasion differed among PE, PP, and expanded PS (Song et al., 2017). Sorption of hydrophobic contaminants showed a difference among polymers (Lee et al., 2014). Furthermore, the toxicity of additive chemicals in different types of plastic debris (and their leachates) differed among plastic products and polymer types (Lithner et al., 2011; Bejgarn et al., 2015; Rani et al., 2015, 2017). Biofilm formation and the microbial community also differed between PE and PP (Zettler et al., 2013), and heteroaggregation of microplastics and freshwater algae appeared to differ depending on the polymer type, that is, between PE and PP (Lagarde et al., 2016). Some polymer types provide information regarding their source. For example, alkyd and poly(acrylate/styrene) fragments originated from ship paint and fiber-reinforced plastics used in boats in Jinhae Bay, Korea (Song et al., 2014). Meanwhile, fabric is the main source of fiber-type PES and acrylic (Browne et al., 2010). Spherical- and granular-type PE is used for microbeads in cosmetics (Cheung and Fok, 2016; Isobe, 2016; Tanaka and Takada, 2016). Interestingly, a more diverse polymer composition was found in nearshore waters close to an urban source than in offshore waters (Song et al., 2015b). Each polymer type shows different environmental behaviors according to their distinct physicochemical characteristics and according to the weathering rate, additive chemicals, and interactions with chemicals and biota. In addition, polymer composition can provide additional information on their source.

1.5 SUMMARY AND FUTURE PROSPECTS

This chapter summarizes the abundance, spatial and temporal distributions, and composition of microplastics in abiotic matrices in marine environments. The mean abundance of microplastics in seawater worldwide reported in the literature ranged within nine orders of magnitude, while the range in microplastic abundance was narrower for beach and subtidal sediments. The microplastic abundance in water was relatively high in the enclosed Mediterranean Sea and the North Pacific Ocean, including Asian seas. Among continents, Asian beaches showed the highest microplastic abundance. These global distributions derived from in situ measurements matched well with the results from the modeled global predictions. The subtropical gyres accumulated more microplastics than other open-ocean areas, and nearshore waters contained higher abundances than offshore waters. On a short-term temporal scale, higher microplastic abundances tended to be reported after rainfall events and subsequent riverine discharge of freshwater in the wet season than dry season in both beaches and coastal waters. On a decadal scale, both significant increases and nonsignificant changes in microplastic abundances were observed depending on the study area. The abundance and mean particle size depended on the lower bound of the microplastic size sampled and analyzed, due to the increase in microplastic abundance according to decreasing size. Fibers, including lines and monofilaments, fragments, and foamed microplastics, considered as secondary microplastics, were the dominant forms of microplastics. Common polymer types included lightweight polymers and those produced in large amounts, such as PE, PP, and polystyrene (including EPS).

Although microplastic monitoring in abiotic matrices has rapidly increased since 2010, there are still limited data and information on both temporal and spatial scales. Therefore, further research and monitoring is recommended to support microplastic exposure analyses, which is a key component of ecological risk assessments of microplastics:

- *Harmonization of sampling and analytic methods*: Microplastic abundance is dependent on the size range of the microplastics sampled and analyzed. Furthermore, differences among samples in size range and reported units prevent direct comparison between global and regional assessments of microplastic pollution. Therefore, it is recommendable to use the most common sampling method worldwide so far (i.e., use of a neuston net with mesh size of 300–350 μm) in regular monitoring program to enable direct comparisons among studies. Furthermore, development and application of improved and novel sampling and analytic methods in monitoring studies are also required. Abundance data should be given on both a volume (n/m^3) and area (n/km^2) basis.
- *Small-sized microplastics*: There is a mismatch in the size of microplastics studied between in situ observations and laboratory toxicity tests (Shim and Thompson,

2015; Lenz et al., 2016). Even though microplastic toxicity depends on the target organisms and the measurement end point, smaller particles tend to be more bioavailable, bioaccumulative, and toxic. Many toxicity tests have been performed using microplastic particles that are orders of magnitude smaller than those found in the field to date. To fill this data gap, additional monitoring is required to assess environmental exposure of aquatic organisms to microplastics of sizes as small as a few microns or on submicron scale, along with regular monitoring of microplastics (>300 μm).

- *Data gaps by environmental compartment*: Most microplastic abundance data have been obtained in areas where microplastics accumulate, such as at the sea surface and on beaches. However, regarding the spatial scale of organisms exposed to microplastics in ecosystems and the ecological importance of habitat, microplastics in the water column and subtidal benthic environments should be assessed more intensively in the future for better assessments of environmental exposure levels.
- *Regional data gaps*: Microplastic monitoring is increasing in terms of spatial coverage. However, there are limited data for the Indian, South Pacific, and South Atlantic Oceans and Polar Seas on a global scale, as well as in highly productive shallow coastal zones on a local scale. To assess the status of microplastic pollution on global and national scales, national monitoring programs should be set up that include both nearshore and offshore waters and consider regional and global monitoring using harmonized sampling and analytic methods. Monitoring programs run by the National Oceanic and Atmospheric Administration in the United States and by the European Union Marine Strategy Framework Directive could be used as methodological baselines.
- *Temporal trends*: Historical and future pollution trends are crucial to understand and determine the severity of the microplastic problem in the context of the past, present, and future. In addition, temporal trends are important when evaluating the effectiveness of countermeasures. There are very limited data to date by which to describe temporal changes in microplastics over the past few decades on national, regional, and global scales. Most systematic monitoring and data collection efforts have been performed in the 21st century. Considering the increasing demand for plastic worldwide and the current degree of accumulated marine plastic debris in oceans, on seafloors, and on shores, microplastic pollution may present a long-term global environmental issue. Therefore, it is necessary to establish and maintain long-term monitoring programs at fixed locations using a standardized method to track changes in microplastic pollution levels and characteristics over time.

REFERENCES

Amélineau, F., Bonnet, D., Heitz, O., Mortreux, V., Harding, A.M.A., Karnovsky, N., Walkusz, W., Fort, J., Grémillet, D., 2016. Microplastic pollution in the Greenland Sea: background levels and selective contamination of planktivorous diving seabirds. Environ. Pollut. 219, 1131–1139.

Andrady, A.L., 2011. Microplastics in the marine environment. Mar. Pollut. Bull. 62, 1596–1605.

Andrady, A.L., 2017. The plastic in microplastics: a review. Mar. Pollut. Bull. 119, 12–22.

Antunes, J.C., Frias, J.G.L., Micaelo, A.C., Sobral, P., 2013. Resin pellets from beaches of the Portuguese coast and adsorbed persistent organic pollutants. Estuar. Coast. Shelf Sci. 130, 62–69.
Aytan, U., Valente, A., Senturk, Y., Usta, R., Sahin, F.B.E., Mazlum, R.E., Agirbas, E., 2016. First evaluation of neustonic microplastics in Black Sea waters. Mar. Environ. Res. 119, 22–30.
Baztan, J., Carrasco, A., Chouinard, O., Cleaud, M., Gabaldon, J.E., Huck, T., Jaffres, L., Jorgensen, B., Miguelez, A., Paillard, C., Vanderlinden, J., 2014. Protected areas in the Atlantic facing the hazards of micro-plastic pollution: first diagnosis of three islands in the Canary Current. Mar. Pollut. Bull. 80, 302–311.
Bejgarn, S., MacLeod, M., Bogdal, C., Breitholtz, M., 2015. Toxicity of leachate from weathering plastics: an exploratory screening study with *Nitocra spinipes*. Chemosphere 132, 114–119.
Blašković, A., Fastelli, P., Čižmek, H., Guerranti, C., Renzi, M., 2017. Plastic litter in sediments from the Croatian marine protected area of the natural park of Telaščica bay (Adriatic Sea). Mar. Pollut. Bull. 114, 583–586.
Browne, M.A., Dissanayake, A., Galloway, T.S., Lowe, D.M., Thompson, R.C., 2008. Ingested microscopic plastic translocates to the circulatory system of the mussel, *Mytilus edulis* (L.). Environ. Sci. Technol. 42, 5026–5031.
Browne, M.A., Galloway, T.S., Thompson, R.C., 2010. Spatial patterns of plastic debris along estuarine shorelines. Environ. Sci. Technol. 44, 3404–3409.
Browne, M.A., Crump, P., Niven, S.J., Teuten, E., Tonkin, A., Galloway, T., Thompson, R., 2011. Accumulation of microplastic on shorelines worldwide: sources and sinks. Environ. Sci. Technol. 45, 9175–9179.
Carpenter, E.J., Smith, K.L., 1972. Plastics on the Sargasso Sea surface. Science 175, 1240–1241.
Carpenter, E.J., Anderson, S.J., Harvey, G.R., Miklas, H.P., Peck, B.B., 1972. Polystyrene spherules in coastal waters. Science 178, 749–750.
Carraher Jr., C.E., 2013. Introduction to Polymer Chemistry. CRC Press, Boca Raton, FL.
Carson, H.S., Nerheim, M.S., Carroll, K.A., Eriksen, M., 2013. The plastic-associated microorganisms of the North Pacific Gyre. Mar. Pollut. Bull. 75, 126–132.
Castillo, A.B., Al-Maslamani, I., Obbard, J.P., 2016. Prevalence of microplastics in the marine waters of Qatar. Mar. Pollut. Bull. 111, 260–267.
Chae, D.-H., Kim, I.-S., Kim, S.-K., Song, Y.K., Shim, W.J., 2015. Abundance and distribution characteristics of microplastics in surface seawaters of the Incheon/Kyeonggi coastal region. Arch. Environ. Contam. Toxicol. 69, 269–278
Cheung, P.K., Fok, L., 2016. Evidence of microbeads from personal care product contaminating the sea. Mar. Pollut. Bull. 109, 582–585.
Cheung, P.K., Cheung, L.T.O., Fok, K., 2016. Seasonal variation in the abundance of marine plastic debris in the estuary of a subtropical macro-scale drainage basin in South China. Sci. Total Environ. 562, 658–665.
Cincinelli, A., Scopetani, C., Chelazzi, D., Lombardini, E., Martellini, T., Katsoyiannis, A., Fossi, M.C., Corsolini, S., 2017. Microplastic in the surface waters of the Ross Sea (Antarctica): occurrence, distribution and characterization by FTIR. Chemosphere 175, 391–400.
Claessens, M., De Meester, S., Van Landuyt, L., De Clerck, K., Janssen, C.R., 2011. Occurrence and distribution of microplastics in marine sediments along the Belgian coast. Mar. Pollut. Bull. 62, 2199–2204.
Cole, M., Webb, H., Lindeque, P.K., Fileman, E.S., Halsband, C., Galloway, T.S., 2014. Isolation of microplastics in biota-rich seawater samples and marine organisms. Sci. Rep. 4, 4528.
Cole, M., Lindeque, P.K., Fileman, E., Clark, J., Lewis, C., Halsband, C., Galloway, T.S., 2016. Microplastics alter the properties and sinking rates of zooplankton faecal pellets. Environ. Sci. Technol. 50, 3239–3246.
Collignon, A., Hecq, J.H., Galgani, F., Voisin, P., Collard, F., Goffart, A., 2012. Neustonic microplastic and zooplankton in the north western Mediterranean Sea. Mar. Pollut. Bull. 64, 861–864.
Collignon, A., Hecq, J.-H., Galgani, F., Collard, F., Goffart, A., 2014. Annual variation in neustonic micro- and meso-plastic particles and zooplankton in the Bay of Calvi (Mediterranean–Corsica). Mar. Pollut. Bull. 79, 293–298.

Colton, J.B., Knapp, F.D., Burns, B.R., 1974. Plastic particles in surface waters of the northwestern Atlantic. Science, 491–497.

Costa, M.F., do Sul, J.A.I., Silva-Cavalcanti, J.S., Araujo, M.C.B., Spengler, A., Tourinho, P.S., 2010. On the importance of size of plastic fragments and pellets on the strandline: a snapshot of a Brazilian beach. Environ. Monit. Assess. 168, 299–304.

Cózar, A., Echevarria, F., Gonzalez-Gordillo, J.I., Irigoien, X., Ubeda, B., Hernandez-Leon, S., Palma, A.T., Navarro, S., Garcia-de-Lomas, J., Ruiz, A., Fernandez-de-Puelles, M.L., Duarte, C.M., 2014. Plastic debris in the open ocean. Proc. Natl. Acad. Sci. 111, 10239–10244.

Davis III, W., Murphy, A.G., 2015. Plastic in surface waters of the Inside Passage and beaches of the Salish Sea in Washington State. Mar. Pollut. Bull. 97, 169–177.

Day, R.H., Shaw, D.G., Ignell, S.E., 1990. The quantitative distribution and characteristics of neuston plastic in the North Pacific Ocean, 1985–88. In R. S. Shomura and M. L. Godfrey (editors), Proceedings of the Second International Conference on Marine Debris, 2–7 April 1989, Honolulu, Hawaii. Memo. NHFS. NOM-TM-NMFS-SWFSC-154. 1990.

De Carvalho, D.G., Neto, J.A.B., 2016. Microplastic pollution of the beaches of Guanabara Bay, Southeast Brazil. Ocean Coast. Manag. 128, 10–17.

De Lucia, G.A., Caliani, I., Marra, S., Camedda, A., Coppa, S., Alcaro, L., Campani, T., Giannetti, M., Coppola, D., Cicero, A.M., Panti, C., Baini, M., Guerranti, C., Marsili, L., Massaro, G., Fossi, M.C., Matiddi, M., 2014. Amount and distribution of neustonic micro-plastic off the western Sardinian coast (Central-Western Mediterranean Sea). Mar. Environ. Res. 100, 10–16.

Desforges, J.W., Galbraith, M., Dangerfield, N., Ross, P.S., 2014. Widespread distribution of microplastics in subsurface seawater in the NE Pacific Ocean. Mar. Pollut. Bull. 79, 94–99.

Doyle, M.J., Watson, W., Bowlin, N.M., Sheavly, S.B., 2011. Plastic particles in coastal pelagic ecosystems of the northeast Pacific Ocean. Mar. Environ. Res. 71, 41–52.

Enders, K., Lenz, R., Stedmon, C.A., Nielsen, T.G., 2015. Abundance, size and polymer composition of marine microplastics $\geq$10 μm in the Atlantic Ocean and their modelled vertical distribution. Mar. Pollut. Bull. 100, 70–81.

Eriksen, M., Maximenko, N., Thiel, M., Cummins, A., Lattin, G., Wilson, S., Hafner, J., Zellers, A., Rifman, S., 2013. Plastic pollution in the South Pacific subtropical gyre. Mar. Pollut. Bull. 68, 71–76.

Eriksen, M., Lebreton, L.C.M., Carson, H.S., Thiel, M., Moore, C.J., Borerro, J.C., Galgani, F., Ryan, P.G., Reisser, J., 2014. Plastic pollution in the world's oceans: more than 5 trillion plastic pieces weighing over 250,000 tons afloat at sea. PLoS One 9, e111913.

Esiukova, E., 2017. Plastic pollution on the Baltic beaches of Kaliningrad region, Russia. Mar. Pollut. Bull. 114, 1072–1080.

Faure, F., Saini, C., Potter, G., Galgani, F., de Alencastro, L.F., Hagmann, P., 2015. An evaluation of surface micro- and mesoplastic pollution in pelagic ecosystems of the western Mediterranean Sea. Environ. Sci. Pollut. Res. 22, 12190–12197.

Fok, L., Cheung, P.K., Tang, G., Li, W.C., 2017. Size distribution of stranded small plastic debris on the coast of Guangdong, South China. Environ. Pollut. 220, 407–412.

Frias, J.P.G.L., Otero, V., Sobral, P., 2014. Evidence of microplastics in samples of zooplankton from Portuguese coastal waters. Mar. Environ. Res. 95, 89–95.

Frias, J.P.G.L., Gago, J., Otero, V., Sobral, P., 2016. Microplastics in coastal sediments from southern Portuguese shelf waters. Mar. Environ. Res. 114, 24–30.

Gago, J., Henry, M., Galgani, F., 2015. First observation on neustonic plastics in waters off NW Spain (spring 2013 and 2014). Mar. Environ. Res. 111, 27–33.

Gajšt, T., Bizjak, T., Palatinus, A., Liubartseva, S., Kržan, A., 2016. Sea surface microplastics in Slovenian part of the northern Adriatic. Mar. Pollut. Bull. 113, 392–399.

Gallagher, A., Rees, A., Rowe, R., Stevens, J., Wright, P., 2016. Microplastics in the Solent estuarine complex, UK: an initial assessment. Mar. Pollut. Bull. 102, 243–249.

GESAMP, 2015. Sources, fate and effects of microplastics in the marine environment: a global assessment (Kershaw, P. J., Ed.). IMO/FAO/UNESCO-IOC/UNIDO/WMO/IAEA/UN/UNEP/UNDP Joint Group of Experts on the Scientific Aspects of Marine Environmental Protection, Rep. Stud. GESAMP No. 90, 96 p.

Gorokhova, E., 2015. Screening for microplastic particles in plankton samples: how to integrate marine litter assessment into existing monitoring programs? Mar. Pollut. Bull. 99, 271–275.

Graca, B., Szewc, K., Zakrzewska, D., Dołęga, A., Szczerbowska-Boruchowska, M., 2017. Sources and fate of microplastics in marine and beach sediments of the southern Baltic Sea—a preliminary study. Environ. Sci. Pollut. Res. 24, 7650–7661.

Graham, E.R., Thompson, J.T., 2009. Deposit- and suspension-feeding sea cucumbers (Echinodermata) ingest plastic fragments. J. Exp. Mar. Biol. Ecol. 368, 22–29.

Gregory, M.R., 1978. Accumulation and distribution of virgin plastic granules on New Zealand beaches. N. Z. J. Mar. Freshw. Res. 12, 399–414.

Güven, O., Gökdağ, K., Jovanović, B., Kıdeyş, A.E., 2017. Microplastic litter composition of the Turkish territorial waters of the Mediterranean Sea, and its occurrence in the gastrointestinal tract of fish. Environ. Pollut. 223, 286–294.

Heo, N.W., Hong, S.H., Han, G.M., Hong, S., Lee, J., Song, Y.K., Jang, M., Shim, W.J., 2013. Distribution of small plastic debris in cross-section and high strandline on Heungnam beach, South Korea. Ocean Sci. J. 48, 225–233.

Herzke, D., Anker-Nilssen, T., Nøst, T.H., Götsch, A., Christensen-Dalsgaard, S., Langset, M., Fangel, K., Koelmans, A.A., 2016. Negligible impact of ingested microplastics on tissue concentrations of POPs in northern fulmars off coastal Norway. Environ. Sci. Technol. 50, 1924–1933.

Hidalgo-Ruz, V., Thiel, M., 2013. Distribution and abundance of small plastic debris on beaches in the SE Pacific (Chile): a study supported by a citizen science project. Mar. Environ. Res. 87–88, 12–18.

Hidalgo-Ruz, V., Gutow, L., Thompson, R.C., Thiel, M., 2012. Microplastics in the marine environment: a review of the methods used for identification and quantification. Environ. Sci. Technol. 46, 3060–3075.

Horton, A.A., Svendsen, C., Williams, R.J., Spurgeon, D.J., Lahive, E., 2017. Large microplastic particles in sediments of tributaries of the River Thames, UK—abundance, sources and methods for effective quantification. Mar. Pollut. Bull. 114, 218–226.

Imhof, H.K., Schmid, J., Niessner, R., Ivleva, N.P., Laforsch, C., 2012. A novel, highly efficient method for the separation and quantification of plastic particles in sediments of aquatic environments. Limnol. Oceanogr. Methods 10, 524–537.

Isobe, A., 2016. Percentage of microbeads in pelagic microplastics within Japanese coastal waters. Mar. Pollut. Bull. 110, 432–437.

Isobe, A., Kubo, K., Tamura, Y., Kako, S., Nakashima, E., Fujii, N., 2014. Selective transport of microplastics and mesoplastics by drifting in coastal waters. Mar. Pollut. Bull. 89, 324–330.

Isobe, A., Uchida, K., Tokai, T., Iwasaki, S., 2015. East Asian seas: a hot spot of pelagic microplastics. Mar. Pollut. Bull. 101, 618–623.

Isobe, A., Uchiyama-Matsumoto, K., Uchida, K., Tokai, T., 2017. Microplastics in the southern Ocean. Mar. Pollut. Bull. 114, 623–626.

Ivar do Sul, J.A., Spengler, A., Costa, M.F., 2009. Here, there and everywhere. Small plastic fragments and pellets on beaches of Fernando de Noronha (Equatorial Western Atlantic). Mar. Pollut. Bull. 58, 1236–1238.

Ivar do Sul, J.A., Costa, M.F., Barletta, M., Cysneiros, F.J.A., 2013. Pelagic microplastics around an archipelago of the equatorial Atlantic. Mar. Pollut. Bull. 75, 305–309.

Ivar do Sul, J.A., Costa, M.F., Fillmann, G., 2014. Microplastics in the pelagic environment around oceanic islands of the western tropical Atlantic Ocean. Water Air Soil Pollut. 225, 2004

Jambeck, J.R., Geyer, R., Wilcox, C., Siegler, T.R., Perryman, M., Andrady, A., Narayan, R., Law, K.L., 2015. Plastic waste inputs from land into the ocean. Science 347, 768–771.

Jang, Y.C., Lee, J., Hong, S., Lee, J.S., Shim, W.J., Song, Y.K., 2014. Sources of plastic marine debris on beaches of Korea: more from the ocean than the land. Ocean Sci. J. 49, 151–162.

Jeong, C.-B., Won, E.-J., Kang, H.-M., Lee, M.-C., Hwang, D.-S., Hwang, U.-K., Zhou, B., Souissi, S., Lee, S.-J., Lee, J.-S., 2016. Microplastic size-dependent toxicity, oxidative stress induction, and p-JNK and p-p38 activation in the monogonont rotifer (*Brachionus koreanus*). Environ. Sci. Technol. 50, 8849–8857.

Kang, J.-H., Kwon, O.Y., Lee, K.-W., Song, Y.K., Shim, W.J., 2015a. Marine neustonic microplastics around the southeastern coast of Korea. Mar. Pollut. Bull. 96, 304–312.

Kang, J.-H., Kwon, O.-Y., Shim, W.J., 2015b. Potential threat of microplastics to zooplanktivores in the surface waters of the southern sea of Korea. Arch. Environ. Contam. Toxicol. 69, 340–351.
Kanhai, L.D.K., Officer, R., Lyashevska, O., Thompson, R.C., O'Connor, I., 2017. Microplastic abundance, distribution and composition along a latitudinal gradient in the Atlantic Ocean. Mar. Pollut. Bull. 115, 307–314.
Kim, I.-S., Chae, D.-H., Kim, S.-K., Choi, S., Woo, S.B., 2015. Factors influencing the spatial variation of microplastics on high-tidal coastal beaches in Korea. Arch. Environ. Contam. Toxicol. 69, 299–309.
Koelman, A.A., Besseling, E., Shim, W.J., 2015. Nanoplastics in the aquatic environment: critical review. In: Bergmann, M., Gutow, L., Klages, M. (Eds.), Marine Anthropogenic Litter. Springer, New York, pp. 245–307.
Koelmans, A.A., Bakir, A., Burton, G.A., Janssen, C.R., 2016. Microplastic as a vector for chemicals in the aquatic environment: critical review and model-supported reinterpretation of empirical studies. Environ. Sci. Technol. 50, 3315–3362.
Kooi, M., Reisser, J., Slat, B., Ferrari, F.F., Schmid, M.S., Cunsolo, S., Brambini, R., Noble, K., Sirks, L.-A., Linders, T.E.W., Schoeneich-Argent, R.I., Koelmans, A.A., 2016. The effect of particle properties on the depth profile of buoyant plastics in the ocean. Sci. Rep. 633882.
Kooi, M., van Nes, E.H., Scheffer, M., Koelmans, A.A., 2017. Ups and downs in the ocean: effects of biofouling on vertical transport of microplastics. Environ. Sci. Technol. 51, 7963–7971.
Kowalski, N., Reichardt, A.M., Waniek, J.J., 2016. Sinking rates of microplastics and potential implications of their alteration by physical, biological, and chemical factors. Mar. Pollut. Bull. 109, 310–319.
Kukulka, T., Proskurowski, G., Morét-Ferguson, S., Meyer, D.W., Law, K.L., 2012. The effect of wind mixing on the vertical distribution of buoyant plastic debris. Geophys. Res. Lett. 39. L07601.
Kunz, A., Walther, B.A., Löwemark, L., Lee, Y.-C., 2016. Distribution and quantity of microplastic on sandy beaches along the northern coast of Taiwan. Mar. Pollut. Bull. 111, 126–135.
Kusui, T., Noda, M., 2003. International survey on the distribution of stranded and buried litter on beaches along the Sea of Japan. Mar. Pollut. Bull. 47, 175–179.
Lagarde, F., Olivier, O., Zanella, M., Daniel, P., Hiard, S., Caruso, A., 2016. Microplastic interactions with freshwater microalgae: heteroaggregation and changes in plastic density appear strongly dependent on polymer type. Environ. Pollut. 215, 331–339.
Lambert, S., Wagner, M., 2016. Formation of microscopic particles during the degradation of different polymers. Chemosphere 161, 510–517.
Lattin, G.L., Moore, C.J., Zellers, A.F., Moore, S.L., Weisberg, S.B., 2004. A comparison of neustonic plastic and zooplankton at different depths near the southern California shore. Mar. Pollut. Bull. 49, 291–294.
Law, K.L., Moret-Ferguson, S., Maximenko, N.A., Proskurowski, G., Peacock, E.E., Hafner, J., Reddy, C.M., 2010. Plastic accumulation in the north Atlantic subtropical gyre. Science 329, 1185–1188.
Law, K.L., Moret-Ferguson, S.E., Goodwin, D.S., Zettler, E.R., De Force, E., Kukulka, T., Proskurowski, G., 2014. Distribution of surface plastic debris in the eastern Pacific Ocean from an 11-year data set. Environ. Sci. Technol. 48, 4732–4738.
Lebreton, L.C.-M., Greer, S.D., Borrero, J.C., 2012. Numerical modelling of floating debris in the world's oceans. Mar. Pollut. Bull. 64, 653–661.
Lechner, A., Keckeis, H., Lumesberger-Loisl, F., Zens, B., Krusch, R., Tritthart, M., Glas, M., Schludermann, E., 2014. The Danube so colourful: a potpourri of plastic litter outnumbers fish larvae in Europe's second largest river. Environ. Pollut. 188, 177–181.
Lee, J., Hong, S., Song, Y.K., Hong, S.H., Jang, Y.C., Jang, M., Heo, N.W., Han, G.M., Lee, M.J., Kang, D., Shim, W.J., 2013a. Relationships among the abundances of plastic debris in different size classes on beaches in South Korea. Mar. Pollut. Bull. 77, 349–354.
Lee, K.W., Shim, W.J., Kwon, O.Y., Kang, J.H., 2013b. Size-dependent effects of micro polystyrene particles in the marine copepod *Tigriopus japonicus*. Environ. Sci. Technol. 47, 11278–11283.
Lee, H., Shim, W.J., Kwon, J.H., 2014. Sorption capacity of plastic debris for hydrophobic organic chemicals. Sci. Total Environ. 470, 1545–1552.

Lee, J., Lee, J.S., Jang, Y.C., Hong, S.Y., Shim, W.J., Song, Y.K., Hong, S.H., Jang, M., Han, G.M., Kang, D., Hong, S., 2015. Distribution and size relationships of plastic marine debris on beaches in South Korea. Arch. Environ. Contam. Toxicol. 69, 288–298.

Lenz, R., Enders, K., Nielsen, T.G., 2016. Microplastic exposure studies should be environmentally realistic. Proc. Natl. Acad. Sci. 113, E4121–E4122.

Li, J., Yang, D., Li, L., Jabeen, K., Shi, H., 2015. Microplastics in commercial bivalves from China. Environ. Pollut. 207, 190–195.

Lima, A.R.A., Costa, M.F., Barletta, M., 2014. Distribution patterns of microplastics with in the plankton of a tropical estuary. Environ. Res. 132, 146–155.

Lithner, D., Larsson, A., Dave, G., 2011. Environmental and health hazard ranking and assessment of plastic polymers based on chemical composition. Sci. Total Environ. 409, 3309–3324.

Long, M., Moriceau, B., Gallinari, M., Lambert, C., Huvet, A., Raffray, J., Soudant, P., 2015. Interactions between microplastics and phytoplankton aggregates: impact on their respective fates. Mar. Chem. 175, 39–46.

Lozoya, J.P., de Mello, F.T., Carrizo, D., Weinstein, F., Olivera, Y., Cedrés, F., Pereira, M., Fossati, M., 2016. Plastics and microplastics on recreational beaches in Punta del Este (Uruguay): unseen critical residents? Environ. Pollut. 218, 931–941.

Lusher, A.L., 2015. Microplastics in the marine environment: distribution, interaction, and effects. In: Bergmann, M., Gutow, L., Klages, M. (Eds.), Marine Anthropogenic Litter. Springer, New York, pp. 245–307.

Lusher, A.L., Burke, A., O'Connor, I., Officer, R., 2014. Microplastic pollution in the Northeast Atlantic Ocean: validated and opportunistic sampling. Mar. Pollut. Bull. 88 (1–2), 325–333.

Lusher, A.M., Tirelli, V., O'Connor, I., Officer, R., 2015. Microplastics in Arctic polar waters: the first reported values of particles in surface and sub-surface samples. Sci. Rep. 5, 14947.

Martins, J., Sobral, P., 2011. Plastic marine debris on the Portuguese coastline: a matter of size? Mar. Pollut. Bull. 62, 2649–2653.

Maximenko, N., Hafner, J., Niiler, P., 2012. Pathways of marine debris derived from trajectories of Lagrangian drifters. Mar. Pollut. Bull. 65, 51–62.

McDermid, K.J., McMullen, T.L., 2004. Quantitative analysis of small-plastic debris on beaches in the Hawaiian archipelago. Mar. Pollut. Bull. 48, 790–794.

Moore, C.J., Moore, S.L., Leecaster, M.K., Weisberg, S.B., 2001. A comparison of plastic and plankton in the north Pacific central gyre. Mar. Pollut. Bull. 42, 1297–1300.

Moore, C.J., Moore, S.L., Weisberg, S.B., Lattin, G.L., Zellers, A.F., 2002. A comparison of neustonic plastic and zooplankton abundance in southern California's coastal waters. Mar. Pollut. Bull. 44, 1035–1038.

Moreira, F.T., Balthazar-Silva, D., Barbosa, L., Turra, A., 2016. Revealing accumulation zones of plastic pellets in sandy beaches. Environ. Pollut. 218, 313–321.

Moret-Ferguson, S., Law, K.L., Proskurowski, G., Murphy, E.K., Peacock, E.E., Reddy, C.M., 2010. The size, mass, and composition of plastic debris in the western north Atlantic Ocean. Mar. Pollut. Bull. 60, 1873–1978.

Morris, R.J., 1980. Plastic debris in the surface waters of the south Atlantic. Mar. Pollut. Bull. 11, 164–166.

Naji, A., Esmaili, Z., Khan, F.R., 2017. Plastic debris and microplastics along the beaches of the Strait of Hormuz, Persian Gulf. Mar. Pollut. Bull. 114, 1057–1062.

Napper, I.E., Thompson, R.C., 2016. Release of synthetic microplastic plastic fibres from domestic washing machines: effects of fabric type and washing conditions. Mar. Pollut. Bull. 112, 39–45.

Nel, H.A., Froneman, P.W., 2015. A quantitative analysis of microplastic pollution along the south-eastern coastline of South Africa. Mar. Pollut. Bull. 101, 274–279.

Nel, H.A., Hean, J.W., Noundou, X.S., Froneman, P.W., 2017. Do microplastic loads reflect the population demographics along the southern African coastline? Mar. Pollut. Bull. 115, 115–119.

Norén, F., 2007. Small plastic particles in coastal Swedish waters. N-Research report, commissioned by KIMO, Sweden, 11 pp.

Pedrotti, M.L., Petit, S., Elineau, A., Bruzaud, S., Crebassa, J., Dumontet, B., Martí, E., Gorsky, G., Cózar, A., 2016. Changes in the floating plastic pollution of the Mediterranean Sea in relation to the distance to land. PLoS One 11, e0161581.

Peng, G., Zhu, B., Yang, D., Su, L., Shi, H., Li, D., 2017. Microplastics in sediments of the Changjiang estuary, China. Environ. Pollut. 225, 283–290.

Plastics Europe, 2013. Plastics – The Facts 2013: An Analysis of European Latest Plastics Production, Demand and Waste Data. Plastics Europe, Belgium.

Plastics Europe, 2016. Plastics – The Facts 2016: An Analysis of European Plastics Production, Demand and Waste Data. Plastics Europe, Belgium.

Rani, M., Shim, W.J., Han, G.M., Jang, M., Najat, A.A., Song, Y.K., Hong, S.H., 2015. Qualitative analysis of additives in plastic marine debris and its new products. Arch. Environ. Contam. Toxicol. 69, 352–366.

Rani, M., Shim, W.J., Han, G.M., Jang, M., Song, Y.K., Hong, S.H., 2017. Benzotriazole-type ultraviolet stabilizers and antioxidants in plastic marine debris and their new products. Sci. Total Environ. 579, 745–754.

Reisser, J., Shaw, J., Wilcox, C., Hardesty, B.D., Proietti, M., Thums, M., Pattiaratchi, C., 2013. Marine plastic pollution in waters around Australia: characteristics, concentrations, and pathways. PLoS One 8, e80466.

Reisser, J., Slat, B., Noble, K., du Plessis, K., Epp, M., Proietti, M., de Sonneville, J., Becker, T., Pattiaratchi, C., 2015. The vertical distribution of buoyant plastics at sea: an observational study in the north Atlantic gyre. Biogeosciences 12, 1249–1256.

Retama, I., Jonathan, M.P., Shruti, V.C., Velumani, S., Sarkar, S.K., Roy, P.D., Rodríguez-Espinosa, P.F., 2016. Microplastics in tourist beaches of Huatulco Bay, Pacific coast of southern Mexico. Mar. Pollut. Bull. 113, 530–535.

Rochman, C.M., Hoh, E., Kurobe, T., Teh, S.J., 2013. Ingested plastic transfers hazardous chemicals to fish and induces hepatic stress. Sci. Rep. 3, 3263.

Rochman, C.M., Tahir, A., Williams, S.L., Baxa, D.V., Lam, R., Miller, J.T., Teh, F.-C., Werorilangi, S., Teh, S.J., 2015. Anthropogenic debris in seafood: plastic debris and fibers from textiles in fish and bivalves sold for human consumption. Sci. Rep. 5, 14340.

Ruiz-Orejón, L.F., Sardá, R., Ramis-Pujol, J., 2016. Floating plastic debris in the central and western Mediterranean Sea. Mar. Environ. Res. 120, 136–144.

Rummel, C.D., Jahnke, A., Gorokhova, E., Kühnel, D., Schmitt-Jansen, M., 2017. Impacts of biofilm formation on the fate and potential effects of microplastic in the aquatic environment. Environ. Sci. Technol. Lett. 4, 258–267.

Ryan, P.G., 1988. The characteristics and distribution of plastic particles at the sea-surface off the southwestern Cape Province, South Africa. Mar. Environ. Res. 25, 249–273.

Setälä, O., Magnusson, K., Lehtiniemi, M., Norén, F., 2016. Distribution and abundance of surface water microlitter in the Baltic Sea: a comparison of two sampling methods. Mar. Pollut. Bull. 110, 177–183.

Shim, W.J., Thompson, R.C., 2015. Microplastics in the ocean. Arch. Environ. Contam. Toxicol. 69, 265–268.

Shim, W.J., Song, Y.K., Hong, S.H., Jang, M., 2016. Identification and quantification of microplastics using Nile Red staining. Mar. Pollut. Bull. 113, 469–476.

Song, Y.K., Hong, S.H., Jang, M., Kang, J.H., Kwon, O.Y., Han, G.M., Shim, W.J., 2014. Large accumulation of micro-sized synthetic polymer particles in the sea surface microlayer. Environ. Sci. Technol. 48, 9014–9021.

Song, Y.K., Hong, S.H., Jang, M., Han, G.M., Rani, M., Lee, J., Shim, W.J., 2015a. A comparison of microscopic and spectroscopic identification methods for analysis of microplastics in environmental samples. Mar. Pollut. Bull. 93, 202–209.

Song, Y.K., Hong, S.H., Jang, M., Han, G.M., Shim, W.J., 2015b. Occurrence and distribution of microplastics in the sea surface microlayer in Jinhae Bay, South Korea. Arch. Environ. Contam. Toxicol. 69, 279–287.

Song, Y.K., Hong, S.H., Jang, M., Han, G.M., Jung, S.W., Shim, W.J., 2017. Combined effects of UV exposure duration and mechanical abrasion on microplastic fragmentation by polymer type. Environ. Sci. Technol. 51, 4368–4376.

Spear, L.B., Ainley, D.G., Ribic, C.A., 1995. Incidence of plastic in seabirds from the tropical pacific, 1984–1991: relation with distribution of species, sex, age, season, year and body weight. Mar. Environ. Res. 40, 123–146.

Stolte, A., Forster, S., Gerdts, G., Schubert, H., 2015. Microplastic concentrations in beach sediments along the German Baltic coast. Mar. Pollut. Bull. 99, 216–229.

Suaria, G., Avio, C.G., Mineo, A., Lattin, G.L., Magaldi, M.G., Belmonte, G., Moore, C.J., Regoli, F., Aliani, S., 2016. The Mediterranean plastic soup: synthetic polymers in Mediterranean surface waters. Sci. Rep. 6, 37551.

Sutton, R., Mason, S.A., Stanek, S.K., Willis-Norton, E., Wren, I.F., Box, C., 2016. Microplastic contamination in the San Francisco Bay, California, USA. Mar. Pollut. Bull. 109, 230–235.

Tanaka, K., Takada, H., 2016. Microplastic fragments and microbeads in digestive tracts of planktivorous fish from urban coastal waters. Sci. Rep. 6, 34351.

Tanaka, K., Takada, H., Yamashita, R., Mizukawa, K., Fukuwaka, M., Watanuki, Y., 2015. Facilitated leaching of additive-derived PBDEs from plastic by seabirds' stomach oil and accumulation in tissues. Environ. Sci. Technol. 49, 11799–11807.

Thompson, R.C., Olsen, Y., Mitchell, R.P., Davis, A., Rowland, S.J., John, A.W.G., McGonigle, D., Russell, A.E., 2004. Lost at sea: where is all the plastic? Science 304, 838.

Thompson, R.C., Moore, C.J., vom Saal, F.S., Swan, S.H., 2009. Plastics, the environment and human health: current consensus and future trends. Philos. Trans. Royal Soc. B Biol. Sci. 364, 2153–2166.

Tsang, Y.Y., Mak, C.W., Liebich, C., Lam, S.W., Sze, E.T.-P., Chan, K.M., 2017. Microplastic pollution in the marine waters and sediments of Hong Kong. Mar. Pollut. Bull. 115, 20–28.

Turra, A., Manzano, A.B., Dias, R.J.S., Mahiques, M.M., Barbosa, L., Balthazar-Silva, D., Moreira, F.T., 2014. Three-dimensional distribution of plastic pellets in sandy beaches: shifting paradigms. Sci. Rep. 4, 4435.

Van Cauwenberghe, L., Janssen, C.R., 2014. Microplastics in bivalves cultured for human consumption. Environ. Pollut. 193, 65–70.

Van Cauwenberghe, L., Claessens, M., Vandegehuchte, M.B., Mees, J., Janssen, C.R., 2013a. Assessment of marine debris on the Belgian continental shelf. Mar. Pollut. Bull. 73, 161–169.

Van Cauwenberghe, L., Vanreusel, A., Mees, J., Janssen, C.R., 2013b. Microplastic pollution in deep-sea sediments. Environ. Pollut. 182, 495–499.

Van der Hal, N., Ariel, A., Angel, D.L., 2017. Exceptionally high abundances of microplastics in the oligotrophic Israeli Mediterranean coastal waters. Mar. Pollut. Bull. 116, 151–155.

Van Sebille, E., Wilcox, C., Lebreton, L., Maximenko, N., Hardesty, B.D., van Franeker, J.A., Eriksen, M., Siegel, D., Galgani, F., Law, K.L., 2015. A global inventory of small floating plastic debris. Environ. Res. Lett. 10, 214006.

Wessel, C.C., Lockridge, G.R., Battiste, D., Cebrian, J., 2016. Abundance and characteristics of microplastics in beach sediments: insights into microplastic accumulation in northern Gulf of Mexico estuaries. Mar. Pollut. Bull. 109, 178–183.

Wilber, R.J., 1987. Plastic in the north Atlantic. Oceanus 30, 61–68.

Woodall, L.C., Sanchez-Vidal, A., Canals, M., Paterson, G.L.J., Coppock, R., Sleight, V., Calafat, A., Rogers, A.D., Narayanaswamy, B.E., Thompson, R.C., 2014. The deep sea is a major sink for microplastic debris. Royal Soc. Open Sci. 1, 140317.

Wright, S.L., Thompson, R.C., Galloway, T.S., 2013. The physical impacts of microplastics on marine organisms: a review. Environ. Pollut. 178, 483–492.

Yamashita, R., Tanimura, A., 2007. Floating plastic in the Kuroshio current area, western north Pacific Ocean. Mar. Pollut. Bull. 54, 485–488.

Yonkos, L.T., Friedel, E.A., Perez-Reyes, A.C., Ghosal, S., Arthur, C.D., 2014. Microplastics in four estuarine rivers in the Chesapeake Bay, U.S.A. Environ. Sci. Technol. 48, 14195–14202.

Yu, X., Peng, J., Wang, J., Wang, K., Bao, S., 2016. Occurrence of microplastics in the beach sand of the Chinese inner sea: the Bohai Sea. Environ. Pollut. 214, 722–730.

Zettler, E.R., Mincer, T.J., Amaral-Zettler, L.A., 2013. Life in the "Plastisphere": microbial communities on plastic marine debris. Environ. Sci. Technol. 47, 7137–7146.

Zhao, S., Zhu, L., Wang, T., Li, D., 2014. Suspended microplastics in the surface water of the Yangtze estuary system, China: first observations on occurrence, distribution. Mar. Pollut. Bull. 86, 562–568.

Zobkov, M., Esiukova, E., 2017. Microplastics in Baltic bottom sediments: quantification procedures and first results. Mar. Pollut. Bull. 114, 724–732.

CHAPTER 2

Limitations for Microplastic Quantification in the Ocean and Recommendations for Improvement and Standardization

Shiye Zhao, Lixin Zhu, Lei Gao, Daoji Li
East China Normal University, Shanghai, China

2.1 INTRODUCTION

Annual global plastic production has increased exponentially since the mid-20th century, reaching 322 million tons in 2015 (Plastics Europe, 2015). As a result of deliberate disposal and accidental release, 10% of all plastics will end up as wastes in marine environments (Jambeck et al., 2015). By 2050, plastics are expected to be more abundant than fish (by weight) in the oceans (Neufeld et al., 2016). Marine plastic debris is presumably capable of posing a global-scale threat (Jahnke et al., 2017).

In the past decade, microsized plastic debris, often referred to as microplastics (MPs) (<5 mm) (Thompson et al., 2004; Arthur et al., 2009), have become a growing global concern. MPs may originate from weathered remnants of larger plastic pieces (secondary MPs) and engineered micron-sized plastic products (primary MPs), such as plastic nurdles, cosmetics, cleaning agents, and industrial shot-blasting abrasives. MPs are widely distributed in the environment and have been recovered throughout the world's oceans from the Arctic to the Antarctic (Sul et al., 2013; Cózar et al., 2014; Obbard et al., 2014), and MPs have been found in samples ranging from the sea surface and shoreline to the seafloor (Cauwenberghe et al., 2013; Fischer et al., 2015; Woodall et al., 2014; Zhao et al., 2017). Due to their ubiquitous nature and minute sizes, MPs are readily consumed by organisms in marine environments. The consumption of plastic particles by marine biota can cause sublethal physical injury and accumulation of chemical contaminants (including both additives and reversibly sorbed waterborne pollutants). Moreover, the toxicity of MPs can be transferred to the upper trophic level (including humans) (Rochman et al., 2013; Rochman, 2015). Additionally, plastics can carry pathogens, which are damaging to organisms that ingest MPs (Zettler et al., 2013; Mincer et al., 2016). To date, numerous mechanisms by which MPs interact with individual organisms and render adverse impacts have been identified in laboratory dose-response experiments

Microplastic Contamination in Aquatic Environments
https://doi.org/10.1016/B978-0-12-813747-5.00002-3

(Rochman et al., 2016a). Laboratory ecotoxicity experiments have become common and have grown to include a wide diversity of taxa exposed to a wide array of MPs (Besseling et al., 2017; Browne et al., 2008; Cole et al., 2015; Nadal et al., 2016; Rochman et al., 2016a; Sussarellu et al., 2016; von Moos et al., 2012). However, most laboratory studies are not environmentally realistic. The experimental concentrations of MPs are far above the levels reported from field studies (Lenz et al., 2016). Additionally, the types (i.e., sizes and shapes) of MPs tested are different from those reported in the field (Cole, 2016). These discrepancies are partly attributable to the inconsistency of sampling methods and greatly hinder our understanding of the deleterious effects of MPs on the ecosystem (Rochman et al., 2016b; Sussarellu et al., 2016). Risk assessments of ecological impacts associated with MPs are of critical importance to assist decision-makers in cultivating legislation to mitigate marine plastic debris accumulation (Rochman et al., 2016a,b).

The lack of cohesive and standardized approaches for collecting, fractionating, characterizing, and quantifying MPs in various marine matrices (e.g., coastlines, sea surface, water column, seafloor, and biota) hampers spatial and temporal comparisons of MP levels between and across these marine reservoirs (Law, 2017). Some attempts have been made to harmonize the approaches for evaluating plastic wastes (including MPs) in various marine compartments. A pilot standardization of protocols for detecting, quantifying, and analyzing fulmars that mistake plastics for prey in the North Sea has been proposed by Franeker et al. (2005). Since then, numerous studies have been conducted to improve the standard operation protocols for plastic ingestion in seabirds (Franeker, 2011; van Franeker et al., 2011; Averygomm et al., 2016; Provencher et al., 2017). Although these studies did not exclusively target MPs, the protocols are applicable to MPs. In 2009, the United Nations Environment Programme published a set of guidelines on standardizing beach plastic litter survey methods (Cheshire et al., 2009). Based on critical reviews and a case study at Meijendel beach (the Netherlands), Besley et al. (2017) proposed a standard operating procedure for beach MP studies. The Technical Subgroup on Marine Litter under the program of Good Environmental Status established a series of general methods to monitor MP pollution across Europe (Galgani et al., 2013). However, these studies basically focused on MPs in one marine reservoir. Some prerequisites for the analytic techniques and methods mentioned in the literature are not normally available in chemical, biological, and environmental laboratories, for instance, employing an ultra-cleanroom to minimize airborne microplastic contamination (Zhao et al., 2016). An easily achievable protocol that is applicable to multimedia settings (i.e., water, sediment, and biota) is urgently needed.

This chapter summarizes and compares the existing analytic approaches for assessing MPs in marine environments (Fig. 2.1). Similar to the general procedure for quantifying MPs in different marine matrices, this chapter is structured in four parts: (1) sample collection, (2) separation, (3) analysis, and (4) quality assurance/quality control. Ultimately, we offer a set of cost-effective recommendations on method standardization and

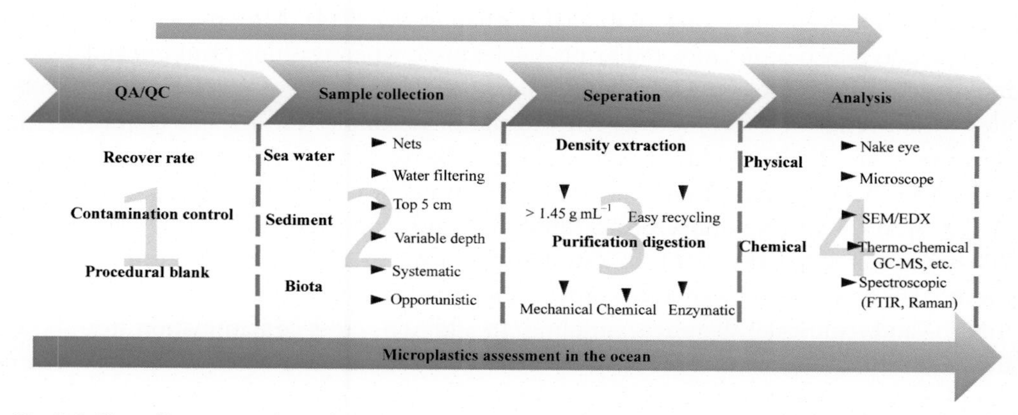

Fig. 2.1 Flow diagram to show the procedure and currently used analytic techniques for microplastic investigation.

harmonization for measuring MPs in the oceans. Additionally, investigations of MPs in freshwater systems are equally important as those in marine systems (Horton et al., 2017). While this chapter focuses on methodologies for measuring MPs in marine ecosystems, these methods are applicable to other aquatic environments, such as freshwater and estuarine environments.

2.2 SAMPLE COLLECTION

A variety of approaches have been used to collect MPs in different marine matrices, such as sea surface, water column, sediment, and biota (Lusher et al., 2014). Analytic methods in literature are generally based on techniques applied in limnology/biology. The following sections present the most common methods used in MP research, the limitations of these methods, and our recommendations.

2.2.1 Water Samples

The most common sampling method for MPs in marine waters is the use of plankton-sampling nets, such as neuston nets for sea surface samples and zooplankton nets for water column samples. Nets with a mesh size of 333 or 335 μm are most commonly employed, although mesh sizes ranging from 50 to 3000 μm have been used (Hidalgo-Ruz et al., 2012). Net-based methods are capable of sampling large volumes of water, which allows for reasonable interstudy comparisons and understanding of large-scale distribution of MPs (Barrows et al., 2017). The size limit of plastics captured by plankton nets is dependent upon the mesh size. Thus, these nets cannot sample MP particles smaller than 300 μm, resulting in the omission of lower size fractions and underestimation of MP abundances in the oceans. As expected, a smaller mesh size can retain considerably more plastics than a large one (Song et al., 2014; Zhao et al., 2014). Moreover, samples captured in nets may be contaminated by plastic net mesh when washing plankton and debris into the cod end of the net. The nets must be thoroughly rinsed; otherwise, residual samples would result in a carryover of MPs to the next samples (Löder and Gerdts, 2015; Barrows et al., 2017). Besides nets sampling, bulk water sampling was also conducted to collect MP samples from the water column, sea surface, and its microlayer (1–1000 μm thick) (Ng and Obbard, 2006; Dubaish and Liebezeit, 2013; Song et al., 2014; Zhao et al., 2015, 2016). In a study by Barrows et al. (2017), MP concentrations captured by this method were three orders of magnitude greater and had a lower size limit (micro- and nanoscale) than nets sampling. In addition, cross contamination from both field and laboratory procedures can be conveniently minimized when using bulk water sampling. In contrast to net samples, bulk samples with small volumes could result in a highly heterogeneous distribution of MPs on a small spatial scale. Large sample volumes by bulk sampling are highly recommended to assess MPs in waters.

2.2.2 Sediment Samples

Sediment has been considered a major deposit for MPs (Cauwenberghe et al., 2013; Cózar et al., 2014; Woodall et al., 2014; Fischer et al., 2015; Zhao et al., 2017). Depending on the sedimentary environments (i.e., sandy beaches, subtidal coastal habitats, and the deep-sea floor), different sampling protocols should be applied.

Beaches are easier to assess compared with other sediment environments; therefore, recent studies determining MP densities mostly focused on sandy beaches. Transects perpendicular or parallel to the water line are the most common sampling areas of beaches. Different sized quadrats, selected at random or at regular intervals, are usually utilized along these transects (Hanvey et al., 2017). It is well known that recent flotsam and plastic debris mostly accumulates along drift lines. Moreover, most researchers assumed that MPs on beaches distribute homogeneously within a small spatial range (Dekiff et al., 2014). Therefore, monitoring MP contamination at the strandline could potentially yield biased results. A specific beach zone for sampling of MPs should stretch from the wrack line to the berm (the vegetation line). Sampling depth is another parameter that potentially impacts the loads of plastic litter. Sampling depths reported in the literature range from 1 to 200 cm (Hidalgo-Ruz et al., 2012; Turra et al., 2014). Comparing samples taken from top 1, 2, 5, and 10 cm, Besley et al. (2017) found the MP concentrations within the top 1–5 cm of sandy samples differed insignificantly, which were higher than those in 2 and 10 cm. Over 50% of previous studies have sampled the top 5 cm of sand, which is beneficial for comparison across these specific studies. Carson et al. (2011) took sediment core samples and indicated that more than 50% of plastic debris occurred in the top 5 cm, with nearly 95% in the top 15 cm. Turra et al. (2014) obtained a three-dimensional distribution pattern of plastic pellets in beaches and suggested that

$<10\%$ of the total MPs occurred in the top layers. Sampling depth apparently impacts the quantification of plastic loadings on beaches. Ineffective sampling will result in underestimated plastic loading measurements. Hence, a method sampling the top 5 cm of sand and variable depth layers selectively should be employed in beach sediment sampling.

2.2.3 Biota Samples

Approximately 557 marine species throughout the food web have been affected by plastic debris, either by entanglement or ingestion (Kühn et al., 2015). Sampling methods of MPs in marine species vary depending on the target organisms (invertebrates and vertebrates) and their habitats, that is, water column, sediments, and estuaries (Hidalgo-Ruz et al., 2012). Planktonic and nektonic biota samples (e.g., zooplankton) are traditionally collected by Bongo, neuston, and manta nets (Frias et al., 2014; Desforges et al., 2015; Sun et al., 2016). Fish residing at the surface or in midwater and benthic habitats are usually sampled by trawls (Neves et al., 2015; Lusher et al., 2015b). Some organisms (e.g., bivalves and crabs) are commonly collected by hand (Wójcik-Fudalewska et al., 2016; Santana et al., 2016). Opportunistic sampling of organisms has also been employed, largely for sea birds and whales (Lusher et al., 2013; Besseling et al., 2015; Lusher et al., 2015a; Zhao et al., 2017; Kedzierski et al., 2017). However, opportunistic sample collection is likely affected by sampling biases, large amounts of samples (i.e., hundreds or thousands of beach-washed seabirds), and scarcity of organisms (i.e., whales and turtles). If possible, both systematic (long-term monitoring of animals) and opportunistic methods should be utilized in field sampling. In addition to standardized sampling methods, Provencher et al. (2017) reviewed the improved methods of quantifying ingested plastics by wildlife.

2.3 SEPARATION

2.3.1 Density-Based Extraction

Density-based separation was identified as a fundamental step to extract lighter MP particles (Table 2.1) from complex matrices containing dense constituents, such as sediment grains and the digestive tracts of marine organisms (Hidalgo-Ruz et al., 2012; Löder and Gerdts, 2015). The use of higher-density solvents causes plastic particles within water samples or supernatant to float and heavier materials to settle to the bottom of a sampling vessel. Researchers have employed various solutions with specific densities. Concentrated sodium chloride (NaCl; $1.2\,g\,cm^{-3}$), initially used by Thompson et al. (2004), has been widely applied to isolate MPs from different environmental samples (Claessens et al., 2011). However, this approach could lose plastics with higher densities (e.g., polyvinyl chloride (PVC), polyethylene terephthalates (PET), and polyurethane

Table 2.1 Densities of virgin plastic polymers commonly detected in oceans

Polymer type	Abbreviation	Density ($g\,cm^{-3}$)
Polyethylene	PE	0.91–0.94
Polystyrene	PS	1.04–1.1
Expanded polystyrene	EPS	0.01–0.04
Polypropylene	PP	0.83–0.85
Polyvinyl chloride	PVC	1.16–1.58
Polyethylene terephthalate	PET	0.96–1.45
Polyamide 6/6	PA6/6	1.02–1.16
Polymethyl methacrylate (acrylic)	PMMA	1.09–1.20
Polyester	PES	1.24–2.30
Polytetrafluoroethylene	PTFE	2.10–2.30
Cellulose acetate	CA	1.29
Polyurethane	PU	1.2

(PU)), resulting in an underestimation of MPs. To solve this issue, other denser salt solutions have been used to increase the recovery efficiency. Saturated zinc chloride (1.81 $g\,cm^{-3}$) has been utilized to extract MPs from sediments (Imhof et al., 2012; Liebezeit and Dubaish, 2012; Imhof et al., 2013). Corcoran et al. (2009) employed sodium polytungstate solution (1.4 $g\,cm^{-3}$) to separate plastic litter from beach sand. Dense sodium iodide (NaI; 1.84 $g\,cm^{-3}$) was also used to separate plastics (Cauwenberghe et al., 2013; Zhao et al., 2017). Employing these higher-density salt solutions allows for the separation of denser plastics (>1.2 $g\,cm^{-3}$). On the other hand, these salts are considerably more expensive than NaCl, more hazardous to work with, and not environmentally friendly. Preextraction methods based on air-induced overflow (Nuelle et al., 2014) and fluidization (Claessens et al., 2013) in a NaCl solution were proposed to decrease the volume of NaI consumed in the subsequent isolation step, which also somewhat reduced the separation cost. However, it is still an issue when large volumes of samples are processed. Recently, the recyclability of NaI was evaluated by monitoring the density, mass, pH, and chemical reactions of NaI during storage. The recycling and reusing methods significantly decrease the cost (Kedzierski et al., 2017). In order to recover all commonly encountered plastic polymers, salt solutions with densities >1.45 $g\,cm^{-3}$ are recommended.

2.3.2 Purification Digestion

Before the physical and chemical characterization of MPs, a step to eliminate matrix interferences and purify plastic particles needs to be conducted. These impurities can be inorganic chemicals adsorbed onto the plastics or organic materials, such as hydrophobic adherents, biogenic residues in the form of tissue detritus or secretion from biotic samples, and microbial communities colonized on the surface of plastics (Löder and Gerdts, 2015; Zhao et al., 2017). Interfering materials can cover the plastics and impede the identification of MPs. Both mechanical and chemical removal methods have been employed in earlier investigations. Mechanical methods, for example, ultrasonic bath, were used to remove extraneous material excess, such as loose debris, sand, calcium carbonate, NaCl, and other residues (Löder and Gerdts, 2015; Zhao et al., 2017). Treatment with ultrasonic cleaning, which might artificially introduce secondary MPs into samples by breaking weathered plastics and lead to overestimation of MP loads (Löder and Gerdts, 2015), must be considered with caution. A range of digestion agents have been proposed for the removal of chemical materials in various environmental samples, including oxidizing agents such as hydrogen peroxide (H_2O_2) (Löder and Gerdts, 2015); alkali agents such as potassium hydroxide (KOH) (Foekema et al., 2013; Lusher et al., 2015b) and sodium hydroxide (NaOH) (Nuelle et al., 2014; Cole et al., 2014; Dehaut et al., 2016); acids such as nitric acid (HNO_3) (Claessens et al., 2013; Wright et al., 2013; Devriese et al., 2015; Davidson and Dudas, 2016), hydrochloric acid (HCl) (Cole et al., 2014; Zhao et al., 2017), perchloric acid ($HClO_4$) (Zoeter Vanpoucke, 2015), and formic acid (Hall et al., 2015); sodium hypochlorite (NaClO) (Collard et al., 2015; Enders et al., 2016); a mixture of H_2O_2 and sulfuric acid (Klein et al., 2015); a mixture of HNO_3 and $HClO_4$ (De et al., 2014; Devriese et al., 2015); a mixture of peroxodisulfate potassium ($K_2S_2O_8$) and NaOH (Dehaut et al., 2016); sodium hypochlorite and HNO_3 (Collard et al., 2015); KOH and NaClO (Enders et al., 2016); and enzymes (Cole et al., 2014; Löder and Gerdts, 2015; Catarino et al., 2016). However, some of these techniques can alter the physical properties (color, shape and size, etc.) of samples or destroy pH-sensitive polymers present in samples. This results in inaccurate qualification and quantification of MPs. Color loss and softening were observed for acrylonitrile butadiene styrene, polymethyl methacrylate, PVC, polycarbonate (PC), expanded and solid polystyrene (EPS, PS), and PET after submersion in a mixture of acids recommended by the International Council for the Exploration of the Sea (ICES) (Cole et al., 2014; Löder and Gerdts, 2015; Catarino et al., 2016). A solution of 40% NaOH resulted in the deformation of polyamide (PA), yellowing of PVC granules, and fusing of polyethylene (PE) particles (Cole et al., 2014). Nuelle et al. (2014) tested a solution of 35% hydrogen peroxide for 7 days on several polymer types and observed discoloration and apparent size losses. Modification of morphological characteristics (size, color, and shape) could result in misidentification during visual or microscopic identification of MPs, which would limit the comparability across studies. Despite the possible changes in polymer morphology, polymeric composition of these particles is still identifiable using spectroscopy. However, the loss of materials and merging of plastics by chemical digestion may lead to underestimation of MP quantification. A solution of 35% HNO_3 melded some PET, high-density PE (HDPE), and polystyrene (PS) particles and dissolved all PA 6/6 fibers (Avio et al., 2015; Catarino et al., 2016; Dehaut et al., 2016). The complete dissolution of PA, PU, and a tire rubber elastomer in the acid mixture (HNO_3 and $HClO_4$) were observed as well (Enders et al., 2016). Although some chemical digestion approaches did not affect plastic size, shape, and chemical constituents (Avio et al., 2015; Enders et al., 2016; Nuelle et al., 2014; Zhao et al., 2017), environmental plastics may be subjected to more adverse effects, as they have poor structural integrity and resistance to chemicals in comparison with virgin plastics used in some tests (Enders et al., 2016; Lusher et al., 2017). These caustic digestive reagents in natural samples should be used with caution. Various enzymatic digestion techniques have been suggested to remove biological materials in field samples. Cole et al. (2014) first developed a sequential enzymatic digestion protocol to extract MPs in biota-rich seawater samples and marine zooplankton. Enzymes have been successfully used to purify other MPs from river water samples (Nuelle et al., 2014), mussel tissues (Catarino et al., 2016), and the digestive tracts of turtles (Duncan et al., 2017). Those results imply that enzymatic purification ensures MP integrity, meaning there is no loss, degradation, or surface modification during digestion. Additionally, enzymatic purification is a more environmentally friendly technique and should be considered by the scientific community for monitoring MP accumulation in natural samples.

2.4 ANALYSIS

Conventionally, the procedures for MP analysis include physical characterization of plastic-like particles and chemical characterization for the identification of specific polymer types. Physical properties of MP wastes, such as shape, size, color, and aging state, are helpful for predicting the source (e.g., fishery, clothing, and industrial processes), origin (primary or secondary wastes) and the relevant fragmentation processes as well as the potential physical and/or physiological hazards to exposed organisms. The confirmation of chemical characteristics, such as chemical structures, compositions, and additives, is necessary for accurate quantification of MP loads and provides important clues in identifying the source of MPs, especially for polymers with limited applications (Käppler et al., 2016; Shim et al., 2016). Analytic techniques employed for characterizing MPs include visual identification (e.g., the naked eye and microscope) (Hidalgo-Ruz et al., 2012; Song et al. 2015), thermal analysis (e.g., pyrolysis gas chromatography coupled to mass spectrometry, Pyro-GC/MS) (Fries et al., 2013; Dekiff et al., 2014), thermogravimetric analysis (TGA) coupled with differential scanning calorimetry (DSC) or GC/MS, TGA-solid-phase extraction (SPE)-thermal desorption (TDS)-GC/MS (Dümichen et al., 2015), vibrational spectroscopic techniques (e.g., Fourier transform infrared spectroscopy, FTIR) (Harrison et al., 2012; Käppler et al., 2015; Tagg et al., 2015; Löder et al., 2015), Raman spectroscopy (Imhof et al., 2012; Lenz et al., 2015; Zhao et al., 2017), and scanning electron microscopy with energy-dispersive X-ray spectroscopy (SEM/EDX) (Corcoran et al., 2009; Eriksen et al., 2013; Fries et al., 2013; Vianello et al., 2013).

2.4.1 Visual Identification

Small particles of plastic (1–5 mm) in marine habitats were first investigated and sorted into different categories, based on polymer type, color, size, and morphology, by visual inspection (Carpenter and Smith, 1972; Coltan et al., 1974; Gregory, 1977; Mcdermid and Mcmullen, 2004; Morétferguson et al., 2010). Large MP particles (>1 mm) were sorted and identified with the naked eye (Heo et al., 2013). When large MPs are identified in a study, visual sorting and identification is an easy, quick, and economical approach. However, interfering (nonplastic) materials from complex environmental samples or ambiguous plastics (resembling nonplastics) make it difficult to accurately confirm and classify MPs (<1 mm) visually. Smaller MP particles (<1 mm) are generally identified with the aid of a microscope, following the proposed morphological criteria (Doyle et al., 2011; Hidalgo-Ruz et al., 2012; Löder and Gerdts, 2015; Norén, 2007). Microscope-aided visual inspection can acquire relatively detailed morphological and structural information of objects, which is essential for separating putative plastic particles from other materials (Shim et al., 2016). To prevent the high risk of misidentification, visual inspection is only applicable for confirming MP particles larger than 500 μm (Rochman et al., 2015) or 1 mm in some cases (Rochman et al., 2015). Regardless, this approach often results in overestimation since differentiating between MPs and other extracted particles (chitin fragments, animal parts, dried algae, sand, tar, glass, etc.) within a similar size range is very difficult. The misidentification rate of visual sorting reported in previous studies ranges from 20% to 70%. Misidentification is especially common for transparent and fibrous particles (Hidalgo-Ruz et al., 2012; Eriksen et al., 2013; Song et al., 2015) and increases considerably with decreasing particle size. Additionally, this approach is also time-consuming and labor-intensive. Therefore, complementary techniques are needed to facilitate the accurate identification of MPs.

2.4.2 Scanning Electron Microscopy With Energy-Dispersive X-Ray Spectroscopy

SEM can offer a clear, high-resolution images of the size and surface texture of plastic-like particles, allowing researchers to discriminate MPs from interfering particles in the environment matrices (Cooper and Corcoran, 2010; Remy et al., 2015). In addition, SEM or environmental SEM equipped with EDX can determine the main atomic composition of putative plastic particles, which is useful for identifying carbon-dominant plastics from inorganic particles (Eriksen et al., 2013; Vianello et al., 2013). Inorganic plastic additives, such as titanium dioxide nanoparticles, barium, sulfur, and zinc, were identified by SEM/EDX (Fries et al., 2013). However, particle colors cannot be determined by SEM. SEM/ESEM-EDX instrumentation and sample preparation (e.g., precious coating materials) are expensive and not conducive to processing large sample quantities. SEM/EDX analysis should be applied on selective particles based on the results of other identification methods such as vibrational spectroscopy.

2.4.3 Thermo-Chemical Analysis

Three thermoanalytic techniques have been introduced for the identification of MPs (Fries et al., 2013; Dümichen et al., 2015; Majewsky et al., 2016). Pyro-GC/MS has been successfully employed to identify the polymer types from laboratory and field samples in previous studies (Fries et al., 2013; Dekiff et al., 2014; Nuelle et al., 2014; Fischer and Scholz-Böttcher, 2017). By analyzing thermal decomposition products of polymers, pyro-GC/MS analysis can confirm the identity of plastic polymers and organic plastic additives simultaneously without background contamination. Fries et al. (2013) reported that the disadvantage of pyro-GC/MS was the size limitation, meaning particles below a certain size cannot be manually transferred to the pyrolysis tube. Moreover, only one particle can be analyzed per run, which limits the number of samples handled. However, these disadvantages have been overcome by Fischer and Scholz-Böttcher (2017), who filtered plastic particles through an Anodisc filter, milled the filtered samples, and finally transferred them to the pyrolysis tube. TGA coupled with SPE, followed by TDS-GC/MS, is another thermal analysis method, which was first used to test PE-spiked environmental samples (Dümichen et al., 2015). Recently, Dümichen et al. (2017) demonstrated that this approach is capable of fast identification of MPs in real aquatic samples (three rivers and a waste water treatment plant) and terrestrial systems (a biogas plant). The polymer nature of MPs can be identified by TGA-DSC (Majewsky et al., 2016). Unfortunately, both TGA-SPE-TDS-GC-MS and TGA-DSC are not routinely applicable. Furthermore, all thermal decomposition approaches are destructive to the analyzed particles, and the results are reported in terms of the mass of plastics per sample. Thermal analysis with mass-related results should be complementary to specific particles that have not been thoroughly characterized by established spectroscopic methods.

2.4.4 Spectroscopic Methods

Vibrational spectroscopic methods, including FTIR and Raman spectroscopy, are the most common techniques to determine the chemical structure of plastics from sediments (Harrison et al., 2012; Imhof et al., 2012), waters (Lenz et al., 2015; Song et al., 2015), and marine organisms (Collard et al., 2015; Neves et al., 2015). When compared with a polymer spectrum library, the compound-specific spectral fingerprints allow for the identification of specific polymer compositions and differentiation of plastics from other organic and inorganic materials (Löder and Gerdts, 2015). If coupled with a microscope, spectroscopic methods are able to achieve high lateral resolution. Furthermore, both spectroscopic approaches are nondestructive and straightforward characterization techniques, which are highly recommended for identifying polymer nature for particles smaller than 1 mm in size (GESAMP, 2015).

With respect to FTIR spectroscopy, there are three measuring modes, including attenuated total reflection (ATR), transmission, and reflectance, available for MP analysis. By probing the surface region of a sample, ATR-FTIR can produce stable spectra from irregular MP surfaces. Although the micro-FTIR can theoretically detect the characteristic bands of particles at a minimum of 10 μm (Lenz et al., 2015), it is more suitable for analyzing large MP particles in practice. It is difficult to acquire clear and reliable spectra from particles smaller than 50 μm (Löder et al., 2015). The disadvantages of micro-ATR-FTIR are as follows: (1) Aged and brittle MPs can be destroyed by the pressure of the ATR probe, and (2) the ATR crystal is prone damage by hard inorganic particle remnants. Micro-FTIR analysis in reflectance mode may result in distorted and noninterpretable spectra when measuring irregularly shaped particles. However, reflectance mode has the advantage of obtaining FTIR spectra of thick and opaque samples (Ojeda et al., 2009; Harrison et al., 2012; Löder et al., 2015). Transmittance micro-FTIR spectroscopy gives high-quality spectra but requires infrared-transparent filter substrates, such as silicon filters (Käppler et al., 2015) and aluminum oxide membrane filters (Löder et al., 2015). In addition, transmission mode is subjected to total absorption patterns and limited by a certain thickness of the measured samples. Besides FTIR spectroscopy, Raman spectroscopy has been widely used to detect MPs in various environmental samples. Raman does offer some advantages over FTIR, including a higher spatial resolution (down to 500 nm in size) (Ojeda et al., 2009; Harrison et al., 2012; Löder et al., 2015), noncontact analysis (Imhof et al., 2013), the ability to analyze wet samples (Tagg et al., 2015), and simultaneous identification of fillers or pigments as well as the polymeric chemical structure (Tagg et al., 2015). In some cases, however, undesired fluorescence absorption from additives (e.g., filler and colorants) or environmental organic matters (e.g., biofilm and algae) overlaps with the Raman spectra of polymer matrices and hampers particle identification. Until now, most attempts to employ spectroscopic (FTIR and Raman) techniques for the detection of MPs have been preceded by presorting and marking potential plastic particles on the filters with the help of the microscopy inspection (Thompson et al., 2004; Claessens et al., 2011; Sun et al., 2016). However, this protocol for the spectroscopic analysis of MPs is both very time-consuming and inaccurate.

To counteract these limitations, semiautomatic spectral mapping without visual presorting of potential plastic particles is needed for analyzing complex field samples. Focal plane array (FPA)-based FTIR (transmission) and Raman imaging allow for high throughput and thorough analysis of total MP particles on a sample filter (Käppler et al., 2015; Löder and Gerdts, 2015; Tagg et al., 2015; Imhof et al., 2016; Käppler et al., 2016; Löder et al., 2015). In this case, Raman and FPA-FTIR spectroscopic imaging are complementary techniques and should be applied sequentially (Käppler et al., 2016). The disadvantages of one method can be circumvented by the other method during the semiautomated process. For example, transmission FTIR imaging can lead to total absorption in the IR transmission spectra when analyzing thick particles (50–100 μm). However, Raman imaging does not suffer from particle thickness limitations (Käppler et al., 2015). Additionally, the combination of FTIR imaging, with a lower size limit >20 μm, and Raman imaging, with lateral resolution >1 μm, can result in a detailed and comprehensive MP analysis of the smaller size fraction. This procedure seems to be recommended, especially when considering that the abundance of MPs in marine and freshwater samples increases with the decrease of size.

2.5 QUALITY ASSURANCE/QUALITY CONTROL

2.5.1 Recovery

Due to overestimation and underestimation of MP pollution level readily occurring at the MP separation stage in the laboratory, method validation is imperative to generate reliable data. Recently, information on recovery rates of various extraction methods has expanded. In a study by Claessens et al. (2011), who mixed plastics (fibers and granules) with clean sandy sediments to validate the method modified from Thompson et al. (2004), recovery efficiencies ranged from 68.8% to 97.5%. Hereafter, numerous validation studies were conducted by spiking different types of chemicals and sizes of polymers into diverse matrices, such as sediments, waters, and organisms. In Table 2.2, we summarize the recovery rates of MPs across a range of environmental compartments. Extraction efficiencies varied upon the size (Imhof et al., 2012; Quinn et al., 2017; Zhao et al., 2017), color (Lavers and Bond, 2015; Stolte et al., 2015; Lavers et al., 2016), density (Nuelle et al., 2014; Fuller and Gautam, 2016), and morphology (Claessens et al., 2013; Hanvey et al., 2017). Spiked plastics with a continuous size spectrum, various colors, and densities need to be taken into consideration when conducting recovery tests.

2.5.2 Minimizing Cross Contamination

Cross contamination of MP samples is known to occur from field sampling, laboratory processing, and analysis (Lusher et al., 2013, 2017). There are many potential sources of contamination, such as rasps or shavings from plastic sampling gear, nets, experimental instruments, airborne plastic particles, instruments containing polymers, and fibers from the clothes (Hidalgo-Ruz et al., 2012; Filella, 2015). Utmost precautions should be performed to prevent sample contamination. In the field, samples for MP analysis are often collected and carried by way of plastic nets, containers, or traps (Hidalgo-Ruz et al., 2012). In addition, the plastic gear, colored plaques, and ropes of the scientific vessels may contaminate the samples. However, few investigations implemented preventive measures in the field. Woodall et al. (2014) took simple precautions (cotton coats, headwear, and latex gloves) during sample collection. Reference samples of the sampling gear, such as PE bottles (Woodall et al., 2014), PE bags (Imhof et al., 2013), plastic nets (Lusher et al., 2013), and polypropylene (PP) Falcon tubes (Zhao et al., 2017), were retained and checked for possible contamination. In contrast to laboratory conditions, reducing

Table 2.2 Summary of the literature reporting recovery rates of quantifying microplastic

Medium	Recovery rate %	Polymer type	Shape	Size range	Extraction method	References
Sediments	68.8–97.5	–	Fibers and granules/spheres	–	NaCl flotation	Claessens et al. (2011)
	100, 93, 98	PVC, PE, and natural fibers	Rasps	–	Elutriation + NaI solution	Claessens et al. (2013)
	95.5–100	PC, PVC, PC, PA, PP, PE, PA, PET	Pellet, plastic rasps	40 μm–5 mm	MPSS + ZnCI solution	Imhof et al. (2012)
	68–99	PE, PP, PVC, PET, PS, EPS, and PUR	Plastic rasps	<1 mm, 1–5 mm	AIO + NaI solution	Nuelle et al. (2014)
	>90	PA6, PVC	Sharply angular and spherical powder	63–1000 μm	Granulometric approach	Kedzierski et al. (2016)
	>90	PP, HDPE, PS, PVC, PET, polyamide 6/6	Rasps, spheres, fiber	180–1000 μm	NaI, $ZnBr_2$	Quinn et al. (2017)
	80–100	PE	Pellet rasps	200 μm	NaCl	Fries et al. (2013)
	54.9–71.8	PE	Sphere	500–600 μm	NaCl	Nor and Obbard (2014)
	97.25	–	–		Elutriation + aeration	Wessel et al. (2016)
	50	PE	–	100–1000 μm	$CaCl_2$ separation, air venting, settle overnight, filtration and digestion	Stolte et al. (2015)

Continued

Table 2.2 Summary of the literature reporting recovery rates of quantifying microplastic—cont'd

Medium	Recovery rate %	Polymer type	Shape	Size range	Extraction method	References
Municipal solid waste	85–94	HDPE, PS, PVC, PET, PP	Powder	50 μm, <1 mm	Pressurized fluid extraction	Fuller and Gautam (2016)
Marine snow	90–98	PE	Sphere	63–75, 212–250, 500–600 μm	Twofold density (NaI and MeOH)	Zhao et al. (2016)
Waste water	98.33	PE, PP, PS, or PVC		–	Filtration	Tagg et al. (2015)
Fish tissue	80–95	PP, PS	Powder	10 μm–5 mm	NaCl + digestion	Avio et al. (2015)
Mussels tissue	94, 98, >98	PS, polyamide 6/6	Sphere, fiber	10, 30 μm (dimensions: 100 × 400, 30 × 200 μm)	Digestion + filtration	Claessens et al. (2013)
Personal-care products	92–96	PP	Powder		Two-step density	Hintersteiner et al. (2015)

contamination in the field is more complicated and should be given the priority. Protective measures, such as the use of systematic sampling equipment, thoroughly cleaned containers, suitable personal precautions and controls, and minimal exposure to air of samples, should be employed in the future studies.

Airborne plastic fibers are ubiquitous in our environment and are easily transported by air (Dris et al., 2017). Not surprisingly, airborne fibers are the main cause of contamination problems in the laboratory. The use of a common set of compatible instruments and procedures has been mentioned in published studies. Filtering all liquid reagents and media used in sample collection and processing has been adopted in several studies (Dris et al., 2017). The use of glassware that is rinsed with Milli-Q water and ethanol or acetone is recommended as much as possible throughout the entire procedure (De et al., 2014; Zhao et al., 2014; Avio et al., 2015; Collard et al., 2015; Zhao et al., 2015; Sussarellu et al., 2016). Glassware should be covered with combusted tinfoil whenever possible. Zhao et al. (2017) further suggested burning all glassware, including beakers, filtration system, glass filters, Petri dishes, and Pasteur pipettes, at 450°C for 8h to minimize contamination from any organic materials. Steel tweezers should be cleaned via flame prior to use. Personal protective devices, such as nitrile gloves, cotton headwear, and lab coats, should be worn during the entire experimental procedure (Lusher et al., 2014). A laminar flow cabinet (Cole et al., 2014) or fume hood (Santana et al., 2016; Zhao et al., 2016) has been employed by some researches to reduce microfiber contamination. However, these hoods would not be deemed sufficient by Woodall et al. (2015), who adopted aseptic forensic techniques, including a designated clean lab separated by air locks. Similarly, a cleanroom rated at class 2 k (maximum 2000 particles of 0.5 μm per cubic foot of air) was utilized for processing MP samples in a study by Wagner et al. (2017). However, cleanroom protocols are expensive and not always achievable in normal marine science laboratories. Instead, Zhao et al. (2017) successfully applied a polymerase chain reaction (PCR) hood with a high efficiency particulate air (HEPA) filtration system to avoid contamination from the air. Michele et al. (2016) drastically reduced airborne microfibers by employing some hermetic enclosure devices, including microscope covers and a glove box for sample processing. Containers and other apparatuses used in the study, whenever possible, should be made of glass and sufficiently cleaned prior to use. All liquids or medium should be filtered. Simple and cost-effective measurements, such as the use of HEPA filtered air curtains or glove boxes, should be implemented for preventing sample contamination.

2.5.3 Procedural Blank

Procedural blanks have to be present at every step of the MP evaluation process to ensure appropriate extraction, minimize the sample contamination, and achieve a higher recovery rate. In the laboratory, controls have been used to verify potential airborne particle

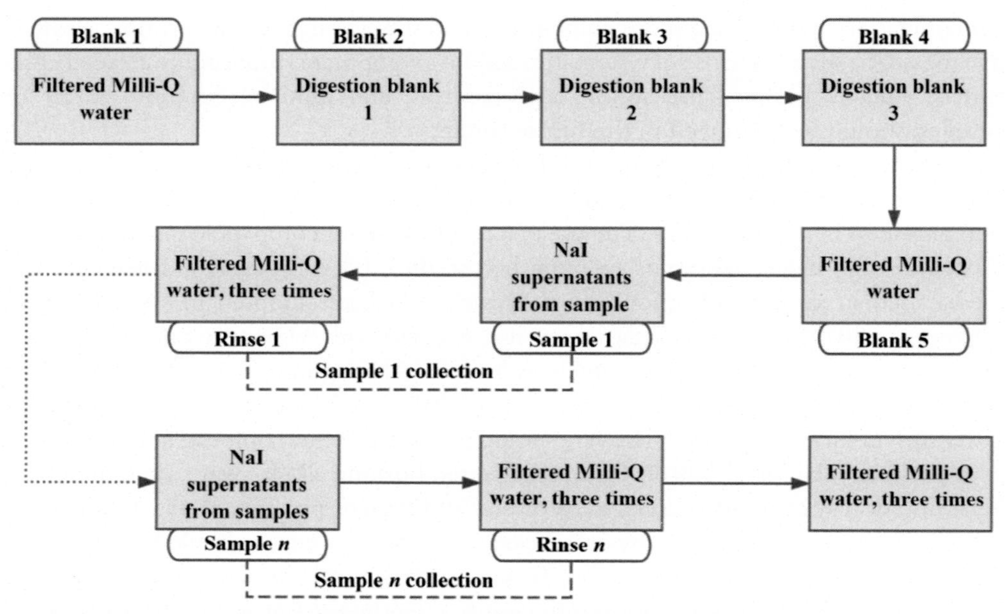

Fig. 2.2 Flow diagram of microplastic extraction protocol. *Reproduced from Zhao, S., Danley, M., Ward, J.E., Li, D., Mincer, T.J., 2017. An approach for extraction, characterization and quantitation of microplastic in natural marine snow using Raman microscopy. Anal. Methods 9, 1470–1478.*

contamination, especially by fibers. These controls include vacuum-filtering negative control samples, laboratory air samples, and filter samples that have been exposed to air (Nuelle et al., 2014; Woodall et al., 2015; Lusher et al., 2017; Ziajahromi et al., 2017). Additionally, forensic tape lifting methods and clean filter wipes were also used before and after sample processing to determine the possible fiber and particle contamination in the laboratory surroundings and on lab coats (Fries et al., 2013; Murphy et al., 2016; Quinn et al., 2017; Zhao et al., 2017; Ziajahromi et al., 2017). The use of procedural blanks in the field was rarely mentioned in published studies (Lusher et al., 2014). Procedural blanks (Fig. 2.2) should be in place after each sample filtration to collect MPs lost in the interior wall of the separatory funnel, avoid cross contamination, and increase the recovery rates (Zhao et al., 2017). To monitor potential contaminants and improve the recovery, systematic blanks should be run at all stages of the research process, including blanks for field sampling, laboratory environments, atmospheric depositions, sample filtration, and negative control samples during the digestion and separation.

2.6 CONCLUSIONS

Although there has been growing interest in investigating marine MP pollution since 2004, reliable and cohesive approaches to quantify MPs in various environmental compartments have remained scarce. This is partly because research on MPs remains an evolving field. The lack of standardized protocols greatly hampers data comparison across studies from specific regions, risk assessment of ecological and human health impacts, and the development of appropriate management and mitigation policies. To achieve a systematic understanding of MP contamination in the world's oceans, improving the consistency of analytic techniques should be the first priority.

ACKNOWLEDGMENTS

This study was supported by National Key Research and Development project (2016YFC1402205 and 2016YFC1402201), National Natural Science Fund of China (41676190), and the State Key Laboratory of Estuarine and Coastal Research of China (2017RCDW05). We thank two anonymous reviewers and the editorial team for their constructive comments and suggestions, which contributed to considerable improvements in the manuscript.

REFERENCES

Arthur, C., Baker, J., Bamford, H., 2009. Proceedings of the International Research Workshop on the Occurrence, Effects, and Fate of Microplastic Marine Debris. Tech. Memo. NOS-OR&R-30, Natl. Ocean. Atmos. Adm, Washington, DC.

Averygomm, S., Valliant, M., Schacter, C.R., Robbins, K.F., Liboiron, M., Daoust, P.Y., Rios, L.M., Jones, I.L., 2016. A study of wrecked Dovekies (Alle alle) in the western North Atlantic highlights the importance of using standardized methods to quantify plastic ingestion. Mar. Pollut. Bull. 113, 75–80.

Avio, C.G., Gorbi, S., Regoli, F., 2015. Experimental development of a new protocol for extraction and characterization of microplastics in fish tissues: first observations in commercial species from Adriatic Sea. Mar. Environ. Res. 111, 18–26.

Barrows, A.P.W., Neumann, C.A., Berger, M.L., Shaw, S.D., 2017. Grab vs. neuston tow net: a microplastic sampling performance comparison and possible advances in the field. Anal. Methods 9, 1446–1453.

Besley, A., Vijver, M.G., Behrens, P., Bosker, T., 2017. A standardized method for sampling and extraction methods for quantifying microplastics in beach sand. Mar. Pollut. Bull. 114, 77–83.

Besseling, E., Foekema, E.M., Van Franeker, J.A., Leopold, M.F., Kühn, S., Bravo Rebolledo, E.L., Heße, E., Mielke, L., Ijzer, J., Kamminga, P., 2015. Microplastic in a macro filter feeder: humpback whale *Megaptera novaeangliae*. Mar. Pollut. Bull. 95, 248–252.

Besseling, E., Foekema, E.M., van den Heuvel-Greve, M.J., Koelmans, A.A., 2017. The effect of microplastic on the uptake of chemicals by the lugworm *Arenicola marina* (L.) under environmentally relevant exposure conditions. Environ. Sci. Technol. 51, 8795–8804.

Browne, M.A., Dissanayake, A., Galloway, T.S., Lowe, D.M., Thompson, R.C., 2008. Ingested microscopic plastic translocates to the circulatory system of the mussel, *Mytilus edulis* (L). Environ. Sci. Technol. 42, 5026–5031.

Carpenter, E.J., Smith, K.L., 1972. Plastics on the Sargasso sea surface. Science 175, 1240–1241.

Carson, H.S., Colbert, S.L., Kaylor, M.J., Mcdermid, K.J., 2011. Small plastic debris changes water movement and heat transfer through beach sediments. Mar. Pollut. Bull. 62, 1708–1713.

Catarino, A.I., Thompson, R., Sanderson, W., Henry, T.B., 2016. Development and optimisation of a standard method for extraction of microplastics in mussels by enzyme digestion of soft tissues. Environ. Toxicol. Chem. 36, 947–951.

Cauwenberghe, L.V., Vanreusel, A., Mees, J., Janssen, C.R., 2013. Microplastic pollution in deep-sea sediments. Environ. Pollut. 182, 495–499.

Cheshire A, Adler E, Barbière J, Cohen Y, Evans S, Jarayabhand S, Jeftic L, Jung RT, Kinsey S, Kusui ET (2009). UNEP/IOC guidelines on survey and monitoring of marine litter. Unep Regional Seas Reports & Studies.
Claessens, M., De, M.S., Van, L.L., De, C.K., Janssen, C.R., 2011. Occurrence and distribution of microplastics in marine sediments along the Belgian coast. Mar. Pollut. Bull. 62, 2199–2204.
Claessens, M., Van, C.L., Vandegehuchte, M.B., Janssen, C.R., 2013. New techniques for the detection of microplastics in sediments and field collected organisms. Mar. Pollut. Bull. 70, 227–233.
Cole, M., 2016. A novel method for preparing microplastic fibers. Sci. Rep. 6, 34519.
Cole, M., Webb, H., Lindeque, P.K., Fileman, E.S., Halsband, C., Galloway, T.S., 2014. Isolation of microplastics in biota-rich seawater samples and marine organisms. Sci. Rep. 4, 4528.
Cole, M., Lindeque, P., Fileman, E., Halsband, C., Galloway, T.S., 2015. The impact of polystyrene microplastics on feeding, function and fecundity in the marine copepod *Calanus helgolandicus*. Environ. Sci. Technol. 49, 1130–1137.
Collard, F., Gilbert, B., Eppe, G., Parmentier, E., Das, K., 2015. Detection of anthropogenic particles in fish stomachs: an isolation method adapted to identification by Raman spectroscopy. Arch. Environ. Contam. Toxicol. 69, 331–339.
Coltan Jr., J.B., Burns, B.R., Knapp, F.D., 1974. Plastic particles in surface waters of the Northwestern Atlantic. Science 185, 491–497.
Cooper, D.A., Corcoran, P.L., 2010. Effects of mechanical and chemical processes on the degradation of plastic beach debris on the island of Kauai, Hawaii. Mar. Pollut. Bull. 60, 650–654.
Corcoran, P.L., Biesinger, M.C., Grifi, M., 2009. Plastics and beaches: a degrading relationship. Mar. Pollut. Bull. 58, 80–84.
Cózar, A., Echevarría, F., Gonzálezgordillo, J.I., Irigoien, X., Ubeda, B., Hernándezleón, S., Palma, A.T., Navarro, S., Garcíadelomas, J., Ruiz, A., 2014. Plastic debris in the open ocean. Proc. Natl. Acad. Sci. U. S. A. 111, 10239–10244.
Davidson, K., Dudas, S.E., 2016. Microplastic ingestion by wild and cultured manila clams (*Venerupis philippinarum*) from Baynes Sound, British Columbia. Arch. Environ. Contam. Toxicol. 71, 147–156.
De, W.B., Devriese, L., Bekaert, K., Hoffman, S., Vandermeersch, G., Cooreman, K., Robbens, J., 2014. Quality assessment of the blue mussel (*Mytilus edulis*): comparison between commercial and wild types. Mar. Pollut. Bull. 85, 146–155.
Dehaut, A., Cassone, A.L., Frère, L., Hermabessiere, L., Himber, C., Rinnert, E., Rivière, G., Lambert, C., Soudant, P., Huvet, A., 2016. Microplastics in seafood: benchmark protocol for their extraction and characterization. Environ. Pollut. 215, 223–233.
Dekiff, J.H., Remy, D., Klasmeier, J., Fries, E., 2014. Occurrence and spatial distribution of microplastics in sediments from Norderney. Environ. Pollut. 186, 248–256.
Desforges, J.P.W., Galbraith, M., Ross, P.S., 2015. Ingestion of microplastics by zooplankton in the Northeast Pacific Ocean. Arch. Environ. Contam. Toxicol. 69, 320–330.
Devriese, L.I., Meulen, M.D.V.D., Maes, T., Bekaert, K., Paul-Pont, I., Frère, L., Robbens, J., Vethaak, A.D., 2015. Microplastic contamination in brown shrimp (Crangon crangon, Linnaeus 1758) from coastal waters of the Southern North Sea and Channel area. Mar. Pollut. Bull. 98, 179–187.
Doyle, M.J., Watson, W., Bowlin, N.M., Sheavly, S.B., 2011. Plastic particles in coastal pelagic ecosystems of the Northeast Pacific ocean. Mar. Environ. Res. 71, 41–52.
Dris, R., Gasperi, J., Mirande, C., Mandin, C., Guerrouache, M., Langlois, V., Tassin, B., 2017. A first overview of textile fibers, including microplastics, in indoor and outdoor environments. Environ. Pollut. 221, 453–458.
Dubaish, F., Liebezeit, G., 2013. Suspended microplastics and black carbon particles in the Jade System, Southern North Sea. Water Air Soil Pollut. 224, 1352–1358.
Dümichen, E., Barthel, A.K., Braun, U., Bannick, C.G., Brand, K., Jekel, M., Senz, R., 2015. Analysis of polyethylene microplastics in environmental samples, using a thermal decomposition method. Water Res. 85, 451–457.
Dümichen, E., Eisentraut, P., Bannick, C.G., Barthel, A.-K., Senz, R., Braun, U., 2017. Fast identification of microplastics in complex environmental samples by a thermal degradation method. Chemosphere 174, 572–584.
Duncan, E., Broderick, A., Galloway, T., Lindeque, P., Godley, B., 2017. Investigating the presence and effects of microplastics in sea turtles. In: Fate and Impact of Microplastics in Marine Ecosystems, Elsevier, pp. 33–34.

Enders, K., Lenz, R., Beer, S., Stedmon, C.A., 2016. Extraction of microplastic from biota: recommended acidic digestion destroys common plastic polymers. ICES J. Mar. Sci. 74, 326–331.
Eriksen, M., Mason, S., Wilson, S., Box, C., Zellers, A., Edwards, W., Farley, H., Amato, S., 2013. Microplastic pollution in the surface waters of the Laurentian Great Lakes. Mar. Pollut. Bull. 77, 177–182.
Filella, M., 2015. Questions of size and numbers in environmental research on microplastics: methodological and conceptual aspects. Environ. Chem. 12, 527–538.
Fischer, M., Scholz-Böttcher, B., 2017. Simultaneous trace identification and quantification of common types of microplastics in environmental samples by pyrolysis-gas chromatography-mass spectrometry. Environ. Sci. Technol. 51, 5025–5060.
Fischer, V., Elsner, N.O., Brenke, N., Schwabe, E., Brandt, A., 2015. Plastic pollution of the Kuril–Kamchatka Trench area (NW pacific). Deep Sea Res. Part II Top. Stud. Oceanogr. 111, 399–405.
Foekema, E.M., Gruijter, C.D., Mergia, M.T., Franeker, J.A.V., Murk, A.T.J., Koelmans, A.A., 2013. Plastic in North Sea fish. Environ. Sci. Technol. 47, 8818–8824.
Franeker, J.A.V., 2011. In: A standard protocol for monitoring marine debris using seabird stomach contents: the Fulmar EcoQO approach from the North Sea.Fifth International Marine Debris Conference, pp. 114–120.
Franeker JAV, Heubeck M, Fairclough K, Turner DM, Grantham M, Stienen EWM, Guse N, Pedersen J, Olsen KO, Andersson PJ (2005). 'Save the North Sea' Fulmar Study 2002–2004: a regional pilot project for the Fulmar-Litter-EcoQO in the OSPAR area.
Frias, J.P.G.L., Otero, V., Sobral, P., 2014. Evidence of microplastics in samples of zooplankton from Portuguese coastal waters. Mar. Environ. Res. 95, 89–95.
Fries, E., Dekiff, J.H., Willmeyer, J., Nuelle, M.T., Ebert, M., Remy, D., 2013. Identification of polymer types and additives in marine microplastic particles using pyrolysis-GC/MS and scanning electron microscopy. Environ. Sci. Process. Impacts 15, 1949–1956.
Fuller, S., Gautam, A., 2016. A procedure for measuring microplastics using pressurized fluid extraction. Environ. Sci. Technol. 50, 5774–5780.
Galgani, F., Hanke, G., Werner, S., Oosterbaan, L., Nilsson, P., Fleet, D., Kinsey, S., Thompson, R.C., Van, F.J., Th, V., 2013. Guidance on Monitoring of Marine Litter in European Seas. European Commission. Available online: http://hdl.handle.net/10508/1649.
GESAMP 2015. Sources, fate and effects of microplastics in the marine environment: a global assessment, Kershaw, P.J. (Ed.). IMO/FAO/UNESCO-IOC/UNIDO/WMO/IA EA/UN/UNEP/UNDP Joint Group of Experts on the Scientific Aspects of Marine Environmental Protection, Rep. Stud. GESAMP No. 90, p. 96.
Gregory, M.R., 1977. Plastic pellets on New Zealand beaches. Mar. Pollut. Bull. 8, 82–84.
Hall, N.M., Berry, K.L.E., Rintoul, L., Hoogenboom, M.O., 2015. Microplastic ingestion by scleractinian corals. Mar. Biol. 162, 725–732.
Hanvey, J.S., Lewis, P.J., Lavers, J.L., Crosbie, N.D., Pozo, K., Clarke, B.O., 2017. A review of analytical techniques for quantifying microplastics in sediments. Anal. Methods 9, 1369–1383.
Harrison, J.P., Ojeda, J.J., Romero-González, M.E., 2012. The applicability of reflectance micro-Fourier-transform infrared spectroscopy for the detection of synthetic microplastics in marine sediments. Sci. Total Environ. 416, 455–463.
Heo, N.W., Hong, S.H., Han, G.M., Hong, S., Lee, J., Song, Y.K., Mi, J., Shim, W.J., 2013. Distribution of small plastic debris in cross-section and high strandline on Heungnam beach, South Korea. Ocean Sci. J. 48, 225–233.
Hidalgo-Ruz, V., Gutow, L., Thompson, R.C., Thiel, M., 2012. Microplastics in the marine environment: a review of the methods used for identification and quantification. Environ. Sci. Technol. 46, 3060–3075.
Hintersteiner, I., Himmelsbach, M., Buchberger, W.W., 2015. Characterization and quantitation of polyolefin microplastics in personal-care products using high-temperature gel-permeation chromatography. Anal. Bioanal. Chem. 407, 1253–1259.
Horton, A.A., Walton, A., Spurgeon, D.J., Lahive, E., Svendsen, C., 2017. Microplastics in freshwater and terrestrial environments: evaluating the current understanding to identify the knowledge gaps and future research priorities. Sci. Total Environ. 586, 127–141.

Imhof, H.K., Schmid, J., Niessner, R., Ivleva, N.P., Laforsch, C., 2012. A novel, highly efficient method for the separation and quantification of plastic particles in sediments of aquatic environments. Limnol. Oceanogr. Methods 10, 524–537.

Imhof, H.K., Ivleva, N.P., Schmid, J., Niessner, R., Laforsch, C., 2013. Contamination of beach sediments of a subalpine lake with microplastic particles. Curr. Biol. 23, 867–868.

Imhof, H.K., Laforsch, C., Wiesheu, A.C., Schmid, J., Anger, P.M., Niessner, R., Ivleva, N.P., 2016. Pigments and plastic in limnetic ecosystems: a qualitative and quantitative study on microparticles of different size classes. Water Res. 98, 64–74.

Jahnke, A., Arp, H.P.H., Escher, B.I., Gewert, B., Gorokhova, E., Kühnel, D., Ogonowski, M., Potthoff, A., Rummel, C., Schmitt-Jansen, M., Toorman, E., MacLeod, M., 2017. Reducing uncertainty and confronting ignorance about the possible impacts of weathering plastic in the marine environment. Environ. Sci. Technol. Lett. 32, 32–48.

Jambeck, J.R., Geyer, R., Wilcox, C., Siegler, T.R., Perryman, M., Andrady, A., Narayan, R., Law, K.L., 2015. Marine pollution. Plastic waste inputs from land into the ocean. Science 347, 768–771.

Käppler, A., Windrich, F., Löder, M.G., Malanin, M., Fischer, D., Labrenz, M., Eichhorn, K.J., Voit, B., 2015. Identification of microplastics by FTIR and Raman microscopy: a novel silicon filter substrate opens the important spectral range below 1300 cm(−1) for FTIR transmission measurements. Anal. Bioanal. Chem. 407, 6791–6801.

Käppler, A., Fischer, D., Oberbeckmann, S., Schernewski, G., Labrenz, M., Eichhorn, K.J., Voit, B., 2016. Analysis of environmental microplastics by vibrational microspectroscopy: FTIR, Raman or both? Anal. Bioanal. Chem. 408, 1–15.

Kedzierski, M., Le, T.V., Bourseau, P., Bellegou, H., César, G., Sire, O., Bruzaud, S., 2016. Microplastics elutriation from sandy sediments: a granulometric approach. Mar. Pollut. Bull. 107, 315–323.

Kedzierski, M., Tilly, V.L., César, G., Sire, O., Bruzaud, S., 2017. Efficient microplastics extraction from sand. A cost effective methodology based on sodium iodide recycling. Mar. Pollut. Bull. 115, 120–129.

Klein, S., Worch, E., Knepper, T.P., 2015. Occurrence and spatial distribution of microplastics in river shore sediments of the Rhine-Main area in Germany. Environ. Sci. Technol. 49, 6070–6076.

Kühn, S., Rebolledo, E.L.B., Franeker, J.A.V., 2015. Deleterious effects of litter on marine life. In: Marine Anthropogenic Litter. vol. 4. Springer International Publishing, Berlin, pp. 75–116.

Lavers, J.L., Bond, A.L., 2015. Selectivity of flesh-footed shearwaters for plastic colour: evidence for differential provisioning in adults and fledglings. Mar. Environ. Res. 113, 1–6.

Lavers, J.L., Oppel, S., Bond, A.L., 2016. Factors influencing the detection of beach plastic debris. Mar. Environ. Res. 119, 245–251.

Law, K.L., 2017. Plastics in the marine environment. Annu. Rev. Mar. Sci. 9, 205–229.

Lenz, R., Enders, K., Stedmon, C.A., Mackenzie, D.M.A., Nielsen, T.G., 2015. A critical assessment of visual identification of marine microplastic using Raman spectroscopy for analysis improvement. Mar. Pollut. Bull. 100, 82–91.

Lenz, R., Enders, K., Stedmon, C.A., Nielsen, T.G., 2016. Microplastic exposure studies should be environmentally realistic. Proc. Natl. Acad. Sci. U. S. A. 113, E4121–E4122.

Liebezeit, G., Dubaish, F., 2012. Microplastics in beaches of the East Frisian Islands Spiekeroog and Kachelotplate. Bull. Environ. Contam. Toxicol. 89, 213–217.

Löder, M.G.J., Gerdts, G., 2015. Methodology used for the detection and identification of microplastics—a critical appraisal. In: Marine Anthropogenic Litter. vol. 4. Springer International Publishing, Cham, pp. 201–227.

Löder, M.G.J., Kuczera, M., Mintenig, S., Lorenz, C., Gerdts, G., 2015. Focal plane array detector-based micro-Fourier-transform infrared imaging for the analysis of microplastics in environmental samples. Environ. Chem. 12, 563–581.

Lusher, A.L., Mchugh, M., Thompson, R.C., 2013. Occurrence of microplastics in the gastrointestinal tract of pelagic and demersal fish from the English Channel. Mar. Pollut. Bull. 67, 94–99.

Lusher, A.L., Burke, A., O'Connor, I., Officer, R., 2014. Microplastic pollution in the Northeast Atlantic Ocean: validated and opportunistic sampling. Mar. Pollut. Bull. 88, 325–333.

Lusher, A.L., Hernandez-Milian, G., O'Brien, J., Berrow, S., O'Connor, I., Officer, R., 2015a. Microplastic and macroplastic ingestion by a deep diving, oceanic cetacean: the True's beaked whale *Mesoplodon mirus*. Environ. Pollut. 199, 185–191.

Lusher, A.L., O'Donnell, C., Officer, R., O'Connor, I., 2015b. Microplastic interactions with North Atlantic mesopelagic fish. ICES J. Mar. Sci. 73, 1214–1225.
Lusher, A.L., Welden, N.A., Sobral, P., Cole, M., 2017. Sampling, isolating and identifying microplastics ingested by fish and invertebrates. Anal. Methods 9, 1346–1360.
Majewsky, M., Bitter, H., Eiche, E., Horn, H., 2016. Determination of microplastic polyethylene (PE) and polypropylene (PP) in environmental samples using thermal analysis (TGA-DSC). Sci. Total Environ. 568, 507–511.
Mcdermid, K.J., Mcmullen, T.L., 2004. Quantitative analysis of small-plastic debris on beaches in the Hawaiian Archipelago. Mar. Pollut. Bull. 48, 790–794.
Michele, T., Digka, N., Anastasopoulou, A., Tsangaris, C., Mytilineou, C., 2016. Anthropogenic microfibers pollution in marine biota. A new and simple methodology to minimize airborne contamination. Mar. Pollut. Bull. 113, 55–61.
Mincer, T.J., Zettler, E.R., Amaral-Zettler, L.A., 2016. Biofilms on plastic debris and their influence on marine nutrient cycling, productivity, and hazardous chemical mobility. In: The Handbook of Environmental Chemistry. Springer International Publishing, Berlin. https://doi.org/10.1007/698_2016_12.
Morétferguson, S., Law, K.L., Proskurowski, G., Murphy, E.K., Peacock, E.E., Reddy, C.M., 2010. The size, mass, and composition of plastic debris in the western North Atlantic Ocean. Mar. Pollut. Bull. 60, 1873–1878.
Murphy, F., Ewins, C., Carbonnier, F., Quinn, B., 2016. Wastewater treatment works (WwTW) as a source of microplastics in the aquatic environment. Environ. Sci. Technol. 50, 5800–5808.
Nadal, M.A., Alomar, C., Deudero, S., 2016. High levels of microplastic ingestion by the semipelagic fish bogue *Boops boops* (L.) around the Balearic Islands. Environ. Pollut. 214, 517–523.
Neufeld, L, Stassen, F, Sheppard, R, Gilman, T, (2016). The new plastics economy: rethinking the future of plastics. World Economic Forum.
Neves, D., Sobral, P., Ferreira, J.L., Pereira, T., 2015. Ingestion of microplastics by commercial fish off the Portuguese coast. Mar. Pollut. Bull. 101, 119–126.
Ng, K.L., Obbard, J.P., 2006. Prevalence of microplastics in Singapore's coastal marine environment. Mar. Pollut. Bull. 52, 761–767.
Nor, N.H.M., Obbard, J.P., 2014. Microplastics in Singapore's coastal mangrove ecosystems. Mar. Pollut. Bull. 79, 278–283.
Norén, F., 2007. Small Plastic Particles in Coastal Swedish Waters. KIMO Sweden, N-Research, Lysekil, Sweden.
Nuelle, M.T., Dekiff, J.H., Remy, D., Fries, E., 2014. A new analytical approach for monitoring microplastics in marine sediments. Environ. Pollut. 184, 161–169.
Obbard, R.W., Sadri, S., Wong, Y.Q., Khitun, A.A., Baker, I., Thompson, R.C., 2014. Global warming releases microplastic legacy frozen in Arctic Sea ice. Earths Future 2, 315–320.
Ojeda, J.J., Romerogonzález, M.E., Banwart, S.A., 2009. Analysis of bacteria on steel surfaces using reflectance micro-Fourier transform infrared spectroscopy. Anal. Chem. 81, 6467–6473.
Plastics Europe, 2015. Plastics—The Facts 2015. An Analysis of European Plastics Production, Demand and Waste Data Plastics Europe. Brussels, Belgium.
Provencher, J.F., Bond, A.L., Avery-Gomm, S., Borrelle, S.B., Rebolledo, E.L.B., Hammer, S., Kühn, S., Lavers, J.F., Mallory, M.L., Trevail, A., van Franeker, J.A., 2017. Quantifying ingested debris in marine megafauna: a review and recommendations for standardization. Anal. Methods 9, 1454–1469.
Quinn, B., Murphy, F., Ewins, C., 2017. Validation of density separation for the rapid recovery of microplastics from sediment. Anal. Methods 9, 1491–1498.
Remy, F., Collard, F., Gilbert, B., Compère, P., Eppe, G., Lepoint, G., 2015. When microplastic is not plastic: the ingestion of artificial cellulose fibers by macrofauna living in seagrass macro-phytodetritus. Environ. Sci. Technol. 49, 11158–11166.
Rochman, C.M., 2015. The complex mixture, fate and toxicity of chemicals associated with plastic debris in the marine environment. In: Marine Anthropogenic Litter. vol. 4. Springer International Publishing, Cham, pp. 117–140.
Rochman, C.M., Browne, M.A., Halpern, B.S., Hentschel, B.T., Hoh, E., Karapanagioti, H.K., Riosmendoza, L.M., Takada, H., Teh, S., Thompson, R.C., 2013. Policy: classify plastic waste as hazardous. Nature 494, 169–171.

Rochman, C.M., Tahir, A., Williams, S.L., Baxa, D.V., Lam, R., Miller, J.T., Teh, F.C., Werorilangi, S., Teh, S.J., 2015. Anthropogenic debris in seafood: plastic debris and fibers from textiles in fish and bivalves sold for human consumption. Sci. Rep. 5, 14340.

Rochman, C.M., Browne, M.A., Underwood, A.J., Van Franeker, J.A., Thompson, R.C., Amaral Zettler, L.A., 2016a. The ecological impacts of marine debris: unraveling the demonstrated evidence from what is perceived. Ecology 97, 302–312.

Rochman, C.M., Cook, A.M., Koelmans, A.A., 2016b. Plastic debris and policy: using current scientific understanding to invoke positive change. Environ. Toxicol. Chem. 35, 1617–1626.

Santana, M.F., Ascer, L.G., Custódio, M.R., Moreira, F.T., Turra, A., 2016. Microplastic contamination in natural mussel beds from a Brazilian urbanized coastal region: rapid evaluation through bioassessment. Mar. Pollut. Bull. 106, 183–189.

Shim, W.J., Hong, S.H., Eo, S., 2016. Identification methods in microplastic analysis: a review. Anal. Methods 9, 1384–1391.

Song, Y.K., Hong, S.H., Mi, J., Kang, J.H., Kwon, O.Y., Han, G.M., Shim, W.J., 2014. Large accumulation of micro-sized synthetic polymer particles in the sea surface microlayer. Environ. Sci. Technol. 48, 9014–9021.

Song, Y.K., Hong, S.H., Jang, M., Han, G.M., Rani, M., Lee, J., Shim, W.J., 2015. A comparison of microscopic and spectroscopic identification methods for analysis of microplastics in environmental samples. Mar. Pollut. Bull. 93, 202–209.

Stolte, A., Forster, S., Gerdts, G., Schubert, H., 2015. Microplastic concentrations in beach sediments along the German Baltic coast. Mar. Pollut. Bull. 99, 216–229.

Sul, J.A.I.D., Costa, M.F., Barletta, M., 2013. Pelagic microplastics around an archipelago of the Equatorial Atlantic. Mar. Pollut. Bull. 75, 305–309.

Sun, X., Li, Q., Zhu, M., Liang, J., Zheng, S., Zhao, Y., 2016. Ingestion of microplastics by natural zooplankton groups in the northern South China Sea. Mar. Pollut. Bull. 115, 217–224.

Sussarellu, R., Suquet, M., Thomas, Y., Lambert, C., Fabioux, C., Pernet, M.E., Le, G.N., Quillien, V., Mingant, C., Epelboin, Y., 2016. Oyster reproduction is affected by exposure to polystyrene microplastics. Proc. Natl. Acad. Sci. U. S. A. 113. 201519019.

Tagg, A.S., Sapp, M., Harrison, J.P., Ojeda, J.J., 2015. Identification and quantification of microplastics in wastewater using focal plane array-based reflectance micro-FT-IR imaging. Anal. Chem. 87, 6032–6040.

Thompson, R.C., Olsen, Y., Mitchell, R.P., Davis, A., Rowland, S.J., John, A.W., Mcgonigle, D., Russell, A.E., 2004. Lost at sea: where is all the plastic? Science 304, 838.

Turra, A., Manzano, A.B., Dias, R.J., Mahiques, M.M., Barbosa, L., Balthazarsilva, D., Moreira, F.T., 2014. Three-dimensional distribution of plastic pellets in sandy beaches: shifting paradigms. Sci. Rep. 4, 4435.

van Franeker, J.A., Blaize, C., Danielsen, J., Fairclough, K., Gollan, J., Guse, N., Hansen, P.L., Heubeck, M., Jensen, J.K., Le, G.G., 2011. Monitoring plastic ingestion by the northern fulmar Fulmarus glacialis in the North Sea. Environ. Pollut. 159, 2609–2615.

Vianello, A., Boldrin, A., Guerriero, P., Moschino, V., Rella, R., Sturaro, A., Ros, L.D., 2013. Microplastic particles in sediments of Lagoon of Venice, Italy: first observations on occurrence, spatial patterns and identification. Estuar. Coastal Shelf Sci. 130, 54–61.

von Moos, N., Burkhardt-Holm, P., Koehler, A., 2012. Uptake and effects of microplastics on cells and tissue of the blue mussel *Mytilus edulis* L. after an experimental exposure. Environ. Sci. Technol. 46, 11327–11335.

Wagner, J., Wang, Z.M., Ghosal, S., Rochman, C., Gassel, M., Wall, S., 2017. Novel method for the extraction and identification of microplastics in ocean trawl and fish gut matrices. Anal. Methods 9, 1479–1490.

Wessel, C.C., Lockridge, G.R., Battiste, D., Cebrian, J., 2016. Abundance and characteristics of microplastics in beach sediments: insights into microplastic accumulation in northern Gulf of Mexico estuaries. Mar. Pollut. Bull. 109, 178–183.

Wójcik-Fudalewska, D., Normant-Saremba, M., Anastácio, P., 2016. Occurrence of plastic debris in the stomach of the invasive crab *Eriocheir sinensis*. Mar. Pollut. Bull. 113, 306–311.

Woodall, L.C., Sanchezvidal, A., Canals, M., Paterson, G.L., Coppock, R., Sleight, V., Calafat, A., Rogers, A.D., Narayanaswamy, B.E., Thompson, R.C., 2014. The deep sea is a major sink for microplastic debris. Royal Soc. Open Sci. 1, 140317.

Woodall, L.C., Gwinnett, C., Packer, M., Thompson, R.C., Robinson, L.F., Paterson, G.L., 2015. Using a forensic science approach to minimize environmental contamination and to identify microfibers in marine sediments. Mar. Pollut. Bull. 185, 40–46.
Wright, S.L., Rowe, D., Thompson, R.C., Galloway, T.S., 2013. Microplastic ingestion decreases energy reserves in marine worms. Curr. Biol. 23, 1031–1033.
Zettler, E.R., Mincer, T.J., Amaralzettler, L.A., 2013. Life in the "plastisphere": microbial communities on plastic marine debris. Environ. Sci. Technol. 47, 7137–7146.
Zhao, S., Zhu, L., Wang, T., Li, D., 2014. Suspended microplastics in the surface water of the Yangtze Estuary System, China: first observations on occurrence, distribution. Mar. Pollut. Bull. 86, 562–568.
Zhao, S., Zhu, L., Li, D., 2015. Microplastic in three urban estuaries, China. Environ. Pollut. 206, 597–604.
Zhao, S., Zhu, L., Li, D., 2016. Microscopic anthropogenic litter in terrestrial birds from Shanghai, China: not only plastics but also natural fibers. Sci. Total Environ. 550, 1110–1115.
Zhao, S., Danley, M., Ward, J.E., Li, D., Mincer, T.J., 2017. An approach for extraction, characterization and quantitation of microplastic in natural marine snow using Raman microscopy. Anal. Methods 9, 1470–1478.
Ziajahromi, S., Neale, P.A., Rintoul, L., Leusch, F.D.L., 2017. Wastewater treatment plants as a pathway for microplastics: development of a new approach to sample wastewater-based microplastics. Water Res., 93–99.
Zoeter Vanpoucke, M., 2015. Impact of Microplastic Uptake: Contaminationin Sprat and Microplastic-Mediated Uptake of PAHs by European Shore Crab (Master thesis). Institute for Agricultural and Fisheries Research (ILVO), Ghent.

FURTHER READING

Andrady, A.L., 2011. Microplastics in the marine environment. Mar. Pollut. Bull. 62, 1596–1605.
Ashton, K., Holmes, L., Turner, A., 2010. Association of metals with plastic production pellets in the marine environment. Mar. Pollut. Bull. 60, 2050–2055.
Browne, M.A., Crump, P., Niven, S.J., Teuten, E., Tonkin, A., Galloway, T., Thompson, R., 2011. Accumulation of microplastic on shorelines woldwide: sources and sinks. Environ. Sci. Technol. 45, 9175–9179.
Cole, M., Lindeque, P., Halsband, C., Galloway, T.S., 2011. Microplastics as contaminants in the marine environment: a review. Mar. Pollut. Bull. 62, 2588–2597.
Frère, L., Paul-Pont, I., Moreau, J., Soudant, P., Lambert, C., Huvet, A., Rinnert, E., 2016. A semi-automated Raman micro-spectroscopy method for morphological and chemical characterizations of microplastic litter. Mar. Pollut. Bull. 113, 461–468.
Frias, J.P., Gago, J., Otero, V., Sobral, P., 2015. Microplastics in coastal sediments from Southern Portuguese shelf waters. Mar. Environ. Res. 114, 24–30.
Koelmans, A.A., Gouin, T., Thompson, R., Wallace, N., Arthur, C., 2014. Plastics in the marine environment. Eerdmans 33, 5–10.
Li, J., Yang, D., Li, L., Jabeen, K., Shi, H., 2015. Microplastics in commercial bivalves from China. Environ. Pollut. 207, 190–195.
Mani, T., Hauk, A., Walter, U., Burkhardtholm, P., 2015. Microplastics profile along the Rhine River. Sci. Rep. 5, 17988.
Mathalon, A., Hill, P., 2014. Microplastic fibers in the intertidal ecosystem surrounding Halifax Harbor, Nova Scotia. Mar. Pollut. Bull. 81, 69–79.
Ryan, P.G., Moore, C.J., van Franeker, J.A., Moloney, C.L., 2009. Monitoring the abundance of plastic debris in the marine environment. Philos. Trans. Royal Soc. B Biol. Sci. 364, 1999–2012.
Zbyszewski, M., Corcoran, P.L., 2011. Distribution and degradation of fresh water plastic particles along the beaches of Lake Huron, Canada. Water Air Soil Pollut. 220, 365–372.

CHAPTER 3

Microplastic Contamination in Freshwater Systems: Methodological Challenges, Occurrence and Sources

Rachid Dris*, Hannes K. Imhof*, Martin G.J. Löder*, Johnny Gasperi†, Christian Laforsch*, Bruno Tassin†
*University of Bayreuth, Bayreuth, Germany
†LEESU, UMR MA 102, École des ponts, UPEC, AgroParisTech, UPE, Paris, France

3.1 INTRODUCTION

Over the past decade, microplastics have received considerable attention by both the scientific community and the global media. Their ubiquity in the marine environments has been widely documented, and their possible impact was highly investigated. It has been shown that a wide range of organisms ingests microplastics (Ivar do Sul and Costa, 2014). This might cause physical harm for the organisms, most often related to a disruption of the digestive system (blockage of intestinal tract, false sensation of satiation, etc.) (Farrell and Nelson, 2013; Tourinho et al., 2010; Derraik, 2002; Carr et al., 2012; Cole et al., 2013). The extents of this impact are not entirely known. A second category of ecological risks, which is controversially discussed (Koelmans et al., 2016), pertains to the fact that these particles can carry a toxic "cocktail of chemicals." These chemicals can be transported over long distances by microplastics or released inside an organism after being ingested with the microplastics (Cole et al., 2011). Such chemicals are incorporated directly into the plastic polymers as additives during the production or may adsorb to microplastics once they are introduced into the environment (Rochman and Browne, 2013). A further anticipated risk is that microplastics may act as a vector of harmful microorganisms, which adhere to the surface of these particles. However, the complex nature of microplastics consisting of different synthetic polymers present in different shapes, sizes, and states of degradation, possessing different blends of additives and a different potential to adsorb pollutants, hampers at present general conclusions on possible ecological impacts.

Knowledge on the sources, fate, occurrences, and dynamics of microplastics, as well as their interactions with biota and pollutants, has been constantly expanding, and opinions of the scientific community have often changed. Nonetheless, scientific work related to plastic pollution (both macro- and microplastics) is mainly oriented toward marine environments. Surprisingly, the focus on plastic pollution in freshwater environments has

Microplastic Contamination in Aquatic Environments
https://doi.org/10.1016/B978-0-12-813747-5.00003-5

been very limited. The dynamics of microplastics in continental aquatic environments are practically unknown; their fate, transfer routes, and destination in continental waters are yet to be determined in detail. Moreover, while different methodologies for the determination of the environmental pollution with microplastics have evolved tremendously during the past decade, there is a lack of homogeneous and standardized methods. This fact is hindering comparability between studies (Ivleva et al., 2017; Hidalgo-Ruz et al., 2012; Löder and Gerdts, 2015; Dris et al., 2015a; Wagner et al., 2014).

In this chapter, the topic of microplastics in freshwater is reviewed. We discuss the different methodologies for microplastic sampling, sample preparation and analysis, the occurrence of these particles in limnetic systems, and the different potential sources and input pathways of microplastics into freshwater.

3.2 METHODOLOGICAL RECOMMENDATIONS AND CHALLENGES

3.2.1 Prevention of Microplastic Contamination

Synthetic polymers are omnipresent in our daily life. Thus, it is of utmost importance to prevent that samples are contaminated during microplastic sampling, preparation, and analysis by, for example, plastic sampling or laboratory equipment, fibers from synthetic clothing, or airborne plastic particles that have a high potential to contaminate samples (Hidalgo-Ruz et al., 2012; Nuelle et al., 2014; Norén, 2007). Several points should be considered to avoid the contamination of samples:

- All fluids used for sample preparation should be filtered, and all laboratory equipment and material should be thoroughly rinsed with prefiltered deionized water or 35% ethanol before and during all working steps.
- To prevent airborne contamination, all materials and the samples should be covered with aluminum foil or glass lids during phases with no treatment.
- To avoid potential sources of contamination, all plastic equipment should be replaced by glass or metal equipment (including lids; be aware of the fact that seals may consist of synthetic polymers).
- Within the laboratory, only nonsynthetic polymer-based clothing and 100% cotton lab coats should be worn.
- For the monitoring of possible sample contaminations, it is of utmost importance to apply blank samples that undergo the same treatment as environmental samples.
- It is mandatory to work only in clean environments (clean the laboratory from the presence of any dust, work—if possible—in a laminar flow hood, etc.).

3.2.2 Sampling of Microplastics

3.2.2.1 Water Samples

Large sample volumes are generally required to obtain representative samples when collecting microplastics, due to their concentrations. One possibility to concentrate samples is the use of nets applied in different mesh sizes during studies. The water surface in studies on the marine environment is mostly sampled by using manta trawls (Doyle et al., 2011; Eriksen et al., 2013a) or neuston nets (Carpenter and Smith, 1972; Colton et al., 1974; Morét-Ferguson et al., 2010). The same methods are frequently applied in sampling campaigns in freshwater systems (McCormick et al., 2014; Moore et al., 2011; Yonkos et al., 2014; Faure et al., 2015). Given the different dimensions between the marine and the freshwater environment, smaller vessels and smaller manta trawls are generally more suitable for sampling the water surface in limnetic ecosystems. Flowmeters, which record the water volume that passes through the net, enable the normalization of microplastic numbers to filtered water volume. By multiplying the trawl distance by the horizontal width of the net opening, concentrations can be also related to the sampled area. During trawling or vertical net hauling, the sample is concentrated in the cod end of the net. After sampling, the net should be carefully rinsed from the exterior to assure that all collected material is washed into the cod end (Doyle et al., 2011) and that the net is clean for the next sampling to prevent a carryover of microplastics to the next sample. The material in the cod end is then transferred to a sample vessel and conserved with plastic-friendly fixatives (e.g., formalin) or stored frozen until further analysis.

Trawling speed is highly dependent on the on-site conditions, for example, flow velocity of a river or wave action in a lake, but usually lies between 1 and 5 knots. If the flow velocity of a river is higher, it can also be necessary to sample with stationary drift nets to prevent a high ram pressure. Generally, the trawling time lies between a few minutes and half an hour. The time of sampling is however limited by the potential clogging of the net, which is dependent on the levels of suspended matter in the water.

The used mesh size limits the size of the retained particles and also the filterable volume. Nets are usually 2–4.5 m long, and a mesh size of 300 μm is commonly used. Those nets only sample microplastic particles >300 μm representatively. However, since microplastic particles <300 μm may occur in colloidal aggregations or adhere to larger organic fragments, some particles in this size fraction are also retained. These nets allow also sampling of relatively large volumes of water. Depending on the suspended matter concentration, a few hundred liters to several cubic meters can be filtered until a net gets clogged. Seasons with high seston loads during high water phases or plankton blooms are generally unfavorable for sampling large volumes of water. Currently, the sampling techniques are not yet standardized, preventing a complete comparability of data. Different mesh sizes are often used depending on the required volumes and the targeted size of plastic. A 80 μm-mesh-size net collects thin fibers more efficiently (Dris et al., 2015b), while a 500 μm net enables the collection of larger volumes of sample without clogging, and therefore, bigger and scarcer debris are sampled more representatively (Lechner et al., 2014).

Further, devices sometimes applied in studies on the plastic contamination of rivers are streambed samplers and hand nets (Moore et al., 2011), eel fyke nets (Morritt et al., 2014), and stationary drift nets. Alternative approaches for the microplastics down to 1 μm involve the pumping of the water sample over filters. The disadvantage of the pump approach is a lower water volume that can be sampled reducing the representability (Wang et al., 2017a; Zhao et al., 2015).

3.2.2.2 Sediment Samples

Beaches

The sampling of lake beaches or riverbanks is similar to marine beach sampling and suffers from the same difficulties. Sampling is relatively easily feasible and basically requires a frame or a corer to define the sampling area, a nonplastic sampling tool (tablespoon, trowel, or small shovel) and a nonplastic sampling vessel. A great advantage of sampling sediments is the possibility to achieve quantitative bulk sampling that is only limited by the subsequent sample processing and particle identification method. A sample quantity between <500 g to up to 10 kg of beach sediment samples is reported in the literature (Hidalgo-Ruz et al., 2012).

The recent finding of a high temporal and a high small-scale spatial variability of microplastic abundance in beach sediments (Imhof et al., 2017) indicates that snapshot sampling might deliver biased results. The microplastic distribution on a beach is highly dynamic and depends on environmental factors like wind and wave action; thus, the representative location of a sample on the beach is still under debate (Imhof et al., 2017; Browne et al., 2015; Imhof et al., 2018). Until now, no clear distribution pattern of plastic debris on marine beaches has been identified at different sampling locations (Hidalgo-Ruz and Thiel, 2013; Mathalon and Hill, 2014). The same is true for freshwater beaches.

In addition to this fact, a comprehensive comparison of studies is hampered by the varying experimental designs (reviewed in Browne et al., 2015). Most studies use natural accumulation zones like the high-tide line or the drift line (flotsam accumulation zone) for sampling (Browne et al., 2010). Sampling of the drift line was also suggested by a recent study comparing microplastic distributions in beach sediments of one lake (Imhof et al., 2018). However, further examinations of the accumulation of plastic particles on beach sediments are necessary to confirm the drift line as appropriate sampling area. It should still be kept in mind that the appropriate sampling area is related to the aims and objectives of the study.

Sampling strategies used consist of, for example, random sampling, transect sampling perpendicular or parallel to the water, or sampling in single squares. Samples of a whole beach can be pooled for an integrated estimate of the microplastic contamination. Commonly, the top 5 cm are sampled; however, one study sampled down to a depth of 0.3 m (Claessens et al., 2011). The use of corers facilitates the subdivision of the sample in different depth layers, and microplastic concentrations can be related to depth in stratified sediments.

Microplastic concentrations can be normalized to sampling area, sediment weight, or volume. Thus, similar to microplastics in the water column, the comparability of data on microplastics in beach sediments is limited as studies differ in the reference units they use.

The technical subgroup on marine litter for the Marine Strategy Framework Directive has made a suggestion for a first step toward a standardized sampling for microplastics at beaches for the EU (MSFD Technical Subgroup on Marine Litter, 2013). These suggestions could also be applied to freshwater beaches as the configuration and challenges are similar. It suggests sampling microplastics at sandy beaches at the strandline where a minimum of five replicate samples should be taken, separated by at least 5 m. Two size categories should be distinguished: small microplastics (20 μm–1 mm) and large microplastics (1–5 mm). Large microplastics should be sampled from the top 5 cm by sieving the sample over a 1 mm sieve directly at the beach. This can lead to a strong reduction of several kilograms of sediment sample. Small microplastics should be also sampled from the top 5 cm with a metal spoon, and by combining several scoops at arm length in an arc-shaped area at the strandline, ~250 g of sediment should be collected. These suggestions could also be applied to freshwater beaches as the conditions and challenges are similar, and future monitoring programs should definitely consider spatial and temporal variations of plastic abundance.

Bottom Sediments

Bottom sediment samples can be obtained with grabs, for example, Van Veen or Ekman grabs or corers of different design, by sampling from a boat. The sediment samples obtained by grabs are usually disturbed, and grab samples are thus suited for surface (e.g., top 5 cm) or bulk sampling. Core sampling enables the simultaneous and undisturbed sampling of surface and depth layers; however, the overall amount of sample is limited. The instrument size strongly depends on the water depth of the sampling location. To our knowledge, studies that sampled river bottom sediments used different types of grabs, while lake bottom studies applied either grabs or corers (Tables 3.2 and 3.4).

3.2.3 Laboratory Preparation of Samples

3.2.3.1 Extraction of Microplastics

The density of common consumer plastic polymers lies between 0.8 (silicone) and 1.4 g/cm^3 (e.g., polyethylene terephthalate (PET) and polyvinyl chloride (PVC)). In contrast to original polymers, the density of the corresponding expanded plastic foams is merely a fraction of the initial densities (e.g., expanded polystyrene <0.05 g/cm^3). Matrices with higher densities, like sand (2.65 g/cm^3) or other sediments, can thus be separated from microplastics by density separation with saturated salt solutions of high density. Therefore, the sediment sample is often dried and then mixed with concentrated salt solution. After agitation (e.g., stirring, shaking, and aeration) for a defined period of time, microplastics and other lightweight particles float to the surface or stay suspended, while heavy sediment particles settle quickly. By removing the light material in the supernatant, the microplastics are recovered. The higher the density of the solution, the more polymers also of heavier types can be extracted. Many studies use saturated sodium chloride solution for the extraction of microplastics (Browne et al., 2010; Claessens et al., 2011; Browne et al., 2011; Ng and Obbard, 2006; Thompson et al., 2004). Because of the relatively low density of this solution (~1.2 g/cm^3), this inexpensive and environmentally friendly approach does not extract all common polymers (e.g., PVC, PET, polycarbonate, and polyurethane). Other solutions for density separation range from sodium polytungstate solution (Corcoran et al., 2009) (1.4 g/cm^3, theoretically up to 3.1 g/cm^3) and zinc chloride solution (1.5–1.7 g/cm^3) (Imhof et al., 2013, 2016) to sodium iodine solution (1.8 g/cm^3) (Nuelle et al., 2014). Principally, all these solutions are suitable for extracting the common consumer plastics. However, for financial reasons, zinc chloride is often applied, while the recycling of the solution by filtration is recommended.

Combined with the different high-density solutions, the applied extraction techniques range from simple density separation with stirring in a jar or bucket to apparatuses developed specifically for this purpose, such as the "Munich Plastic Sediment Separator" (MPSS) (Imhof et al., 2012) or different elutriation/fluidization setups with subsequent flotation (Nuelle et al., 2014; Claessens et al., 2013; Kedzierski et al., 2016). The latter approaches reach high recovery rates of 68%–99% (Nuelle et al., 2014), 96%–100% (Imhof et al., 2012), and 98%–100% (Claessens et al., 2013). Repeated extraction steps are recommended in order to maximize the recovery especially for particles <500 μm (Nuelle et al., 2014; Browne et al., 2011; Claessens et al., 2013). In contrast to that, the MPSS showed a recovery rate of 96% of small microplastics within the first extraction in experiments with spiked sediments (Imhof et al., 2012).

3.2.3.2 Size Fractionation

The fractionation of microplastic samples into (at least) two size classes is useful for the subsequent laboratory processing; however, this is highly related to the identification method that will be used afterward. A standardized size fractionation would facilitate an intercomparison between different studies, even if one of the fractions would not be of interest (Hidalgo-Ruz et al., 2012). A separation into fractions of 1–5 mm and 20 μm–1 mm was recently suggested (MSFD Technical Subgroup on Marine Litter, 2013). Most samples (water, sediment, and biota) can be fractionated easily by sieving over stainless steel sieves. Here, the size fractionation of sediment samples can be better performed after the density separation. A cascade of different mesh sizes allows for a simultaneous size separation within one step and for a quantification of several size classes of microplastics (McDermid and McMullen, 2004). If large amounts of biological matrix bear the risk of clogging the sieve, a purification step prior to sieving is recommended.

We suggest the separation step at 500 μm, as the visual sorting out of potential microplastic particles >500 μm under a stereomicroscope is a reasonable step prior to any analysis (visually, spectroscopically, and through other techniques). Particles from 500 to 20 μm can be analyzed by micro-FTIR (Fourier-transform infrared) spectroscopy (Löder et al., 2015). Where relevant, the microplastic size fraction below 20 μm can be analyzed by Raman microspectroscopy. Here, an additional size fractioning at 20 μm is recommended to reduce the amount of sample.

3.2.3.3 Sample Purification

The purification of microplastic samples facilitates the analysis and is a mandatory step prior to a spectroscopic characterization (FTIR/Raman microspectroscopy and pyrolysis-gas chromatography combined with mass spectrometry (GC/MS)). For a proper spectroscopic identification, biofilms and other organic and inorganic adherents have to be removed from the microplastics, and the sample matrix has to be reduced to a minimum to minimize the nonplastic filter residue when filters have to be generated for spectroscopic measurements. Ultrasonic cleaning (Cooper and Corcoran, 2010) for larger microplastics should be considered with caution, as this technique bears a considerable risk to artificial generation of secondary microplastics from aged and brittle plastic material. A gentle way to clean larger plastic particles is stirring and rinsing with filtered freshwater (McDermid and McMullen, 2004) or ethanol (~30%). Organic material in microplastic samples is usually reduced by strong acidic (Imhof et al., 2016; Claessens et al., 2013; Cole et al., 2014) or alkaline solutions (Cole et al., 2014; Dehaut et al., 2016), oxidation agents like hydrogen peroxide (Nuelle et al., 2014; Collard et al., 2015), or a combination of these agents (Dehaut et al., 2016). All these purification approaches are, however, limited, because sensitive synthetic polymers can be lost during the treatment as a cause of their chemical susceptibility (Claessens et al., 2013; Cole et al., 2014; Dehaut et al., 2016).

An alternative plastic-friendly approach is the purification of environmental samples with enzymes (Cole et al., 2014; Catarino et al., 2017). This has been first suggested by Cole et al. (2014) who used proteinase-K to isolate microplastics from seawater samples with a high content of planktonic organisms. A significant drawback of the approach is the high cost for the enzyme. Meanwhile, the purification of plastic particles from mussel tissue with inexpensive enzymes has been reported (Catarino et al., 2017), and a universal enzymatic digestion protocol involving the sequential application of different technical enzymes and oxidation agents was recently developed (Löder et al., 2017).

3.2.4 Identification of Microplastics

3.2.4.1 Visual Identification

Visual sorting is often used to separate microplastics from the sample matrix and for their identification (Hidalgo-Ruz et al., 2012). While smaller microplastic particles should generally be sorted out using a dissection microscope (Doyle et al., 2011), large microplastics can be (>1 mm) identified by the naked eye (Morét-Ferguson et al., 2010). The use of sorting chambers (e.g., Bogorov counting chambers) can be helpful when sorting aqueous samples. The visual identification of particles according to standardized criteria in tandem with a strict and conservative examination to reduce misidentification was suggested (Norén, 2007).

However, especially for smaller microplastic particles, it is highly recommended to analyze potential microplastics with reliable techniques (e.g., spectroscopic approaches) for a proper identification of synthetic polymer origin (Dekiff et al., 2014; Song et al., 2015). The cause for this claim is that the quality of the identification by visual sorting is highly subjective and depends on (1) the experience of the identifying person, (2) the magnification and optical quality of the microscope, and (3) the sample matrix itself (e.g., plankton, sediment, and biota samples). A second fundamental drawback is the downward size limitation of visual sorting, as particles smaller than a certain size cannot be discriminated visually from other sample material or are unmanageable with forceps due to their minuteness. Depending on the sample amount, visual sorting is additionally extremely time-consuming. In summary, even with experience, it is extremely difficult to separate and identify all potential microplastic particles unambiguously from the sample matrix.

3.2.4.2 Identification of Microplastics by Their Chemical Composition

Density Separation With Subsequent C:H:N Analysis

The elemental composition of polymers determined by C:H:N analysis in combination with the determination of the specific density of the particles has been used for a rough polymer classification of potential plastic particles (Morét-Ferguson et al., 2010). The results of each particle was compared with the densities and C:H:N ratios of virgin polymers and thus assigned to a group of potential polymers. This approach represents not a rigorous chemical analysis and only an approximation by narrowing the search for the potential polymer type. The technique is furthermore not applicable to smaller particles and involves a high time effort that hampers a high sample throughput.

Pyrolysis-GC/MS

Pyrolysis-GC/MS can be used to obtain informations about the chemical composition of potential microplastic particles in environmental samples by analyzing their thermal degradation products (Fries et al., 2013). The pyrolysis products of plastic polymers result in characteristic pyrograms, which facilitate a proper identification of different polymer types via comparison with reference pyrograms of known virgin polymer samples. This approach was already used after extraction and visual sorting of microplastics from sediments (Nuelle et al., 2014; Fries et al., 2013). To date, it has not been applied to freshwater samples. Plastic additives can be determined simultaneously during pyrolysis-GC/MS analysis if a thermal desorption step precedes the actual pyrolysis. Standard pyrolysis-GC/MS facilitates the identification of potential microplastics; however, it involves manual handling of particles that have to be inserted in pyrolysis tubes, and this results in a lower size limitation of particles that can be analyzed. As only one particle per run can be analyzed, the technique is thus not suitable for processing large amounts of samples. Latest research suggested the concentration of a whole environmental sample on filters after purification and the simultaneous pyrolysis-GC/MS analysis of microplastic particles in one run that relativize the abovementioned restrictions (Fischer and Scholz-Böttcher, 2017). In a similar GC-MS technique, the sample is subjected to a complete thermal decomposition. The degradation products of synthetic polymers are adsorbed on a solid-phase adsorber and subsequently analyzed by thermal extraction and desorption-gas chromatography-mass spectrometry (TED-GC/MS) (Dümichen et al., 2017). This method requires no preceding purification of the sample, as the solid-phase adsorber is loaded after the organic material is decomposed. However, only 20 mg of dry sample can be analyzed during one TED-GC/MS run; thus, a combination of several sequential runs is required.

Raman Spectroscopy

Raman spectroscopy has been successfully used to identify microplastic particles in different environmental sample matrices with high reliability (Cole et al., 2013; Imhof et al., 2012, 2013, 2016; Van Cauwenberghe et al., 2013; Murray and Cowie, 2011). During Raman analysis, the sample is irradiated with a monochromatic laser source, and the interaction of the molecules and atoms (vibrational, rotational, and other low-frequency interactions) with the laser light results in differences in the frequency of the backscattered light compared with the initial laser frequency. This "Raman shift" can be measured and leads to substance-specific Raman spectra. Due to the highly characteristic Raman spectra of synthetic polymers, the technique facilitates the identification of plastic polymers by comparison with reference spectra. Raman spectroscopy is a "surface-reflectance" technique and thus facilitates the measurement of a wide size range of samples. Raman spectroscopy can also be coupled with microscopy, and this so-called Raman microspectroscopy allows for the identification of very small plastic particles of sizes below 1 μm (Cole et al., 2013; Käppler et al., 2016). Raman microscopy combined with Raman chemical imaging theoretically allows for the chemical analysis of whole membrane filters at a spatial resolution below 1 μm, if time is not a limiting factor. Raman chemical imaging can also be combined with confocal laser scanning microscopy, which facilitates the determination of the location of polymer particles within biological tissues with high precision (Cole et al., 2013). One drawback of Raman spectroscopy is the fluorescence of the samples excited by the laser, which is often the case if residues of biological origin are measured. Such samples cannot be analyzed easily as the strong fluorescence signal overlies the weak Raman signal and prevents the generation of interpretable Raman spectra. The fluorescence can be minimized by optimizing acquisition time and laser energy for each particle measurement or by using lasers with higher wavelengths (>1000 nm), which excite less fluorescence than high energy lasers, with the drawback that the lower energy of the laser results also in a lower signal of the sample (Käppler et al., 2016). Thus, it is recommended to purify microplastic samples from organic residues prior to measurements to prevent fluorescence for a clear identification of microplastic particles. Moreover, Raman spectroscopy at a high spatial resolution is still very time-consuming compared with micro-FTIR spectroscopy when the analysis of an entire filter is required.

Infrared Spectroscopy

Infrared (IR) or FTIR spectroscopy is a complementary technique to Raman spectroscopy and also offers the possibility of proper identification of plastic polymer particles according to their characteristic IR spectra. IR radiation excites wavelength specific molecular vibrations in a substance, and the excitable vibrations depend on the composition and molecular structure of this substance. The energy of the IR radiation exiting a specific vibration is absorbed to a certain amount. This absorption can be measured and results in characteristic IR spectra. Due to the fact that plastic polymers possess highly specific IR spectra, IR spectroscopy is an optimal technique for the identification of microplastics (Hidalgo-Ruz et al., 2012), but the comparison with reference spectra is necessary for polymer identification. FTIR spectroscopy can also be used to detect the intensity of surface oxidation and to provide information on physicochemical weathering of microplastics (Corcoran et al., 2009).

"Attenuated total reflectance" (ATR) FTIR spectroscopy facilitates the fast and easy measurement and analysis of the IR spectrum of larger particles (>500 μm) with high reliability. The combination of IR spectroscopy with an IR microscope is a step forward with respect to the characterization of small sized particles down to 10 μm in size concentrated on filters (Löder and Gerdts, 2015). The use of two measuring modes, reflectance and transmittance, is feasible. However, the reflectance mode may result in noninterpretable spectra as measurements of irregularly shaped microplastic particles often lead to refractive error (Harrison et al., 2012). This problem does not exist in the transmittance mode. This mode, in turn, needs IR transparent filters (e.g., aluminum oxide (Löder et al., 2015) or silica (Käppler et al., 2015)) and is limited to a certain thickness of the microplastic sample due to the total absorption of samples that are too thick. Combining transmittance measurements with micro-ATR measurements of larger particles that show total absorption can be a solution. Micro-FTIR mapping, that is, the sequential measurement of IR spectra at spatially separated points on the sample filter surface with a single element detector, has been successfully applied for microplastic identification. However, when targeting a large area at a high spatial resolution, this technique is still extremely time-consuming (Harrison et al., 2012; Vianello et al., 2013). In contrast, focal plane array (FPA)-based FTIR imaging—that is, the simultaneous recording of several thousand spectra within an area with a single measurement with a detector array—allows for a detailed and unbiased high-throughput analysis of microplastics. By sequentially measuring FPA fields on an area of interest, whole sample filters can be analyzed via FTIR imaging in a reasonable time (Löder and Gerdts, 2015; Fries et al., 2013). One drawback is that black particles are difficult to analyze, as these often lead to total absorption during measurement. The lateral resolution of micro-FTIR spectroscopy is diffraction limited (e.g., 10 μm at 1000 cm^{-1}) and thus lower compared with Raman spectroscopy. Furthermore, samples must be dried prior to measurement as water strongly absorbs IR radiation. Because many substances with either organic or inorganic origin possess characteristic IR spectra, the purification of samples to reduce the general amount of particles present in a sample is mandatory for a proper identification of microplastic particles via IR spectroscopy.

3.2.4.3 Data Reporting

Currently, the abundance of macro- and microplastic particles in water samples are often reported normalized to water volume (particle per liter or particle per cubic meter) or water surface (particle per square meter or particle per square kilometer). The latter is more common for lake or ocean surface samples. Sediment samples are often reported as particles per surface (particles per square meter), particles per sediment volume (particles per liter), or particles per sediment dry weight (particles per kilogram dry weight). Reporting microplastic abundances in a certain unit as mentioned above is generally a result of the scientific question of a study and the investigated water body. However, to facilitate a global comparison, it would be favorable to provide more than one reference unit or to allow for a conversion of the reported values to other reference units by supplying all necessary data like sampled volume, sampled surface, sampled sediment dry weight, and volume.

3.3 REVIEW OF MICROPLASTICS IN FRESHWATER

Since the first evidence for the existence of macro- and microplastics in lakes in Switzerland (Faure et al., 2012) and in a subalpine lake in northern Italy (Imhof et al., 2013), a variety of studies concerning the contamination of lakes, rivers, and estuaries with macro- and microplastics were published. However, the quantification was performed with a variety of different methods. This involves different sampling strategies, sample processing methods, and the subsequent analysis of the microplastics and leads to the incomparability of a large portion of the available data. Because of these methodological differences, the analyzed microplastic size classes are often different, and often, studies even miss to provide a clear indication of the upper and lower size limits. Especially in early studies on microplastics in the environment, the identification of the particles was often performed by visual discrimination, which was shown to be questionable

(Löder and Gerdts, 2015; Lenz et al., 2015). Although they are of higher relevance for risk assessment due to their high bioavailability, especially the lower size classes down to 1 μm are often not quantified with suitable methods (Tables 3.1–3.4).

However, the current trend is toward the use of reliable identification methods, which are described above. More and more studies perform the identification of the potential plastic particles with spectroscopic or equivalent methods for all or at least for a small subset of the particles found. Nevertheless, often, it is not mentioned if the total particle number was then corrected with the findings of the subset analysis. An attempt to compare the results of various studies with different analytic reliability is presented in this review. Nonetheless, this comparison is limited and requires keeping in mind the high heterogeneity of the underlying data.

Our extensive search resulted in 46 studies reporting on microplastic contamination of lakes and streams. Concerning the problems mentioned earlier, only 39 studies were considered as suitable for a global comparison. However, the number of studies that identified microplastics by reliable methods are still rare (lakes, 4 out of 19, and rivers, 3 out of 20) at this point of time. Therefore, this review includes studies that performed visual identification, studies that did a reliable identification of only a (small) subset of the particles, and studies that identified all particles by reliable methods. Even if this fact needs to be considered, this compilation of different studies from all over the world gives a first assessment of the global contamination of limnetic water bodies with microplastics. A few studies were not included, since newer data from the same area, analyzed with more reliable methods, were available. A detailed list of the studies including the respective identification methods is given in Tables 3.1–3.4.

3.3.1 Microplastics in Lakes

The contamination of lakes with microplastic debris was assessed by 19 studies (Tables 3.1 and 3.2). This resulted in comparable datasets of the surface water of 28 lakes, beach sediments of 17 lakes, and lake bottom sediments of 2 lakes. Currently, no studies exist that examined the microplastic contamination of the water column as for a few studies in the marine system (Doyle et al., 2011; Di Mauro et al., 2017).

3.3.1.1 Surface Water

Microplastic concentration of lake surface water ranged from 0.01 to 5.00 microplastic particles/m^2 (median, 0.17 microplastic particles/m^2). When considering particles per cubic meter as a reference unit, water lake surface water contained 0.06–15,000 particles/m^3 (median, 832 microplastic particles/m^3). Macroplastic was often reported in early studies in rather low numbers (0.0003–0.008 macroplastic debris/m^2 and median, 0.001 macroplastic debris/m^2). In recent studies, the abundance of macroplastic was often not assessed.

Table 3.1 Overview over the available studies assessing microplastic contamination in lake surface water

Publication	Location	Compartment	Field method (sampling)	Processing	Size class	Identification	Reference unit
Anderson et al. (2017)	Lake Winnipeg (CA)	Water surface	Manta net (333 μm)	Subsample, wet peroxide oxidation	333–5000 μm	Visual, subsample SEM/EDX	Particle/km^2
Eriksen et al. (2013b)	Lakes Superior, Huron, and Erie (the United States and CA)	Water surface	Manta net (333 μm)	Sieved, subsamples treated with hydrochloric acid prior SEM/EDS	>4.75 mm, 1.00–4.749 mm, 0.355–0.999 mm	Subsamples SEM/EDX	Particle/km^2
Faure et al. (2012)	Lake Geneva (CH and FR)	Water surface	Manta net (300 μm)	Sieved	Macroplastic: >5 mm Microplastic: <5 mm	Stereomicroscope (visual)	Particle/km^2
Faure et al. (2015)	Lake Geneva, Lake Constance, Lake Maggiore, Lake Neuchatel (CH)	Water surface	Manta net (300 μm)	Sieved, wet peroxide oxidation 1–300 μm	>5 mm, 1–5 mm, 1 mm–300 μm, 300–1 μm	Stereomicroscope (visual), subsample of the particles >1 mm (10%) with ATR-FTIR	Particle/km^2
Fischer et al. (2016)	Lake Chiusi, Lake Bolsena (IT)	Water surface	Manta net (300 μm)	Sieved, density separation (NaCl, 1.2 g/cm^3), hot digestion of organic material with hydrochloric acid, staining with nile red	Macroplastic: >5 mm (Macroplastic) Microplastic: 5–1 mm, 1–0.5 mm, 0.5–0.3 mm, <0.3 mm	Fluorescence microscopy under UV light (visual), SEM for the differentiation of polymer fibers and cotton	Particle/m^2 Particle/m^3

Continued

Table 3.1 Overview over the available studies assessing microplastic contamination in lake surface water—cont'd

Publication	Location	Compartment	Field method (sampling)	Processing	Size class	Identification	Reference unit
Free et al. (2014)	Lake Hovsgol (Mongolia)	Water surface	Manta net (333 μm)	Density separation (salt water, 1.6 g/cm^3), wet peroxide oxidation	>4.75 mm, 1.00–4.749 mm, 0.355–0.999 mm	Stereomicroscope (visual), subsample with DSC	Particle/km^2
Mason et al. (2016b)	Lake Michigan (the United States)	Water surface	Manta net (333 μm)	Sieved, wet peroxide oxidation	Microplastic: <5 mm	>0.999 μm: visual 0.355–0.999 mm: randomly selected sites (20%, 11 out of 52) by SEM/EDS >4.75: random number 72 out of 122 (59%) dissolved in dichlorobenzene and analyzed by FT-IT (destructive)	Particle/km^2
Su et al. (2016)	Taihu Lake (CN)	Water surface	Plankton net (333 μm)	Hydrogen peroxide (30%)	Microplastic: 5 mm–333 μm	Visual, subset by micro-FIR or SEM/EDS	Particle/km^2
Wang et al. (2017a)	12 Lakes in the Wuhan City area (CN)	Water surface	Water volume of 20 L pumped over 50 μm sieve	Hydrogen peroxide (30%)	Microplastic <5 mm, 50–500 μm, 500 μm–1 mm, 1–2 mm, 2–3 mm, 3–4 mm, 4–5 mm	Visual, subsamples SEM	Particles/m^3

Table 3.2 Overview over the available studies assessing microplastic contamination in lake beach and lake bottom sediment

Publication	Location	Compartment	Field method (sampling)	Processing	Size class	Identification	Reference unit
Ballent et al. (2016)	Lake Ontario (CA)	Beach sediment, bottom sediment	Beach sediment: split-spoon corer Bottom sediment: Glew gravity corer (PVC core 6.5 cm), Shipek grab sampler (20 cm), sediment traps	Sieved, <63 μm: density separation—sodium polytungstate solution (1.5 g/cm³)	<5 mm limit of visual identification of plastics is estimated to be ~0.25 mm	Stereomicroscope (visual), random particles by Raman and X-ray fluorescence spectroscopy (90 out of 6331 particles analyzed by Raman and a small number by XRF)	Particle/kg dw
[a]Corcoran et al. (2015)	Lake Ontario (CA)	Bottom sediment	Two cores (depth 30 cm)	Density separation (sodium polytungstate, 1.5 g/cm³)	<0.5 mm, 0.5–0.71 mm, 0.71–0.85 mm, 0.85–1 mm, >1 mm	ATR-FTIR	Number of particles
Corcoran et al. (2015)	Lake Ontario (CA)	Beach sediment,	Visual identifiable plastic	–	>5 cm, 1–5 cm, <1 cm	Visual, subsample with Raman spectroscopy	Particle/m²
[a]Faure et al. (2012)	Lake Geneva (CH, FR)	Beach sediment	Sampled defined surface, direct sampling of visual identifiable plastics	Sieved, density separation (water)	>5 mm, >2 mm	Stereomicroscope (visual)	Particle/L
Faure et al. (2015)	Lake Geneva, Lake Constance, Lake Maggiore, Lake Neuchatel (CH)	Beach sediment	Quadrats at drift line (depth 5 cm)	Density separation (saturated NaCl), sieved (300 μm), wet peroxide oxidation for particles 1–300 μm	>5 mm, 1–5 mm, 1 mm–300 μm, 300–1 μm	Stereomicroscope (visual), subsample of the particles >1 mm (10%) with ATR-FTIR	Particle/m²

Continued

Table 3.2 Overview over the available studies assessing microplastic contamination in lake beach and lake bottom sediment—cont'd

Publication	Location	Compartment	Field method (sampling)	Processing	Size class	Identification	Reference unit
Fischer et al. (2016)	Lake Chiusi, Lake Bolsena (IT)	Beach sediment	Quadrats along transects (depth 3 cm)	Sieved, density separation (NaCl, 1.2 g/cm^3), hot digestion of organic material with hydrochloric acid, staining with nile red	>5 mm macroplastic, 5–1 mm, 1–0.5 mm, 0.5–0.3 mm, <0.3 mm	Fluorescence microscopy under UV light (visual), SEM for the differentiation of polymer fibers and cotton	$Particle/m^2$ $Particle/m^3$ $Particle/kg^3$
Imhof et al. (2013)	Lake Garda (IT)	Beach sediment	Random distributed quadrats (depth 5 cm)	Density separation (zinc chloride, 1.6–1.8 g/mL)	Macroplastic: >5 mm Microplastic: <5 mm	Raman microspectroscopy	$Particle/m^2$
Imhof et al. (2016)	Lake Garda (IT)	Beach sediment	Sediment cores along transects parallel to the shoreline (depth 5 cm)	Density separation (MPSS, zinc chloride, 1.6–1.8 g/mL)	Macroplastic: >5 mm Microplastic (<5 mm): 5 mm–500 μm, 500–1 μm	Raman microspectroscopy	$Particle/m^2$
[a]Sruthy and Ramasamy (2017)	Vembanad Lake (IN)	Bottom sediment	Van Veen grab ($25 cm^2$)	Sieved, wet peroxide oxidation, density separation (NaCl, density 1.3 g/cm^3)	Microplastic: <5 mm	Raman microspectroscopy	$Particle/m^2$

Su et al. (2016)	Lake Taihu (CN)	Bottom sediment	Sediments collected with three grabs (Peterson grab) and pooled into steel mesh, rinsed in situ with lake water	Density separation with saturated sodium chloride solution, hydrogen peroxide (30%)	Microplastic: <5 mm	Visual, subset by micro-FTIR or SEM/EDS	Particle/kg
Zbyszewski and Corcoran (2011)	Lake Huron (CA and the United States)	Beach sediment	Direct sampling of visual identifiable plastic within quadrats and along transects	Ultrasonic bath for 4 min	>5 mm, <5 mm	Visual, subsample with FTIR	Particle/m^2
Zbyszewski et al. (2014)	Lakes Erie and St Clair, Lake Huron (the United States and CA)	Beach sediment	Direct sampling of visual identifiable plastic along transects	Ultrasonic bath for 4 min, dried	>10 cm, 10–2 cm, <2 cm	Visual, subsample with FTIR	Particle/m^2
Zhang et al. (2016)	Tibet Plateau (CN): Geren Co, Wuru Co, Mujiu Co, Siling Co	Beach sediment	Quadrats (20 × 20 cm) along lakeshore between water and drift line, top 2 cm of sediments	Sieved (1 mm), <1 mm density separation (potassium formate, 1.5 g/cm^3)	Microplastic: 5–1 mm, 1–0.5 mm, <0.5 mm	Visual selection, all particles identified by Raman microspectroscopy	Particle/m^2

[a]Not included in Fig. 3.3.

Table 3.3 Overview over the available studies assessing microplastic and macroplastic contamination in river surface water

Publication	Location	Compartment	Field method (sampling)	Processing	Size class	Identification	Reference unit
Baldwin et al. (2016)	29 Great Lakes tributaries (the United States)	Water surface	Manta net (300 μm)	Sieved (125 μm), wet peroxide oxidation	>125 μm	Visual	Particle/m^3
Dris et al. (2015b)	Seine and Marne (FR)	Water surface	Manta net (330 μm) and plankton net (80 μm)	Enzymatic digestion according to Löder et al. (2017)	1–5 mm, 1 mm–500 μm, 500–100 μm	Stereomicroscope (visual)	Particle/m^3 Particle/m^2
Faure et al. (2015)	Rhone, Aubonne, Venoge, Vuachere (CH)	Water surface	Manta net (300 μm)	Density separation (NaCl), wet peroxide oxidation	>5 mm, 5–1 mm, 1 mm–300 μm	Stereomicroscope (visual), subsample with ATR-FTIR	Particle/m^3 Particle/h mg/h mg/m^3
[a]Gasperi et al. (2014)	Seine (FR)	Water surface	Floating booms	–	Macroplastic >5 mm	ATR-FTIR	Weight percentage of polymer types
Lechner et al. (2014)	Danube (DE, AT)	Water column (below water surface)	Stationary drift net (~500 μm)	Density separation (water)	Mesodebris: 2–20 mm Microdebris: <2 mm	Visual	Particle/1000 m^3
[a]Leslie et al. (2017)	Urban canals in the Amsterdam area	Water surface	2 L surface water samples were collected in glass bottles	Subsampled, filtered on 0.7 μm glass filters, visually examined	Microplastic: 5000–300 μm, <300 μm	Subsample (6% of the particles) by FTIR spectroscopy	Particle/kg dw

Mani et al. (2015)	Rhine (CH, DE, NL)	Water surface	Manta net (300 μm)	Enzymatic digestion according to Löder et al. (2017) , density separation (salt water, 1.16 g/mL)	>5 mm, 5–1 mm, 1 mm–300 μm	Stereomicroscope (visual, in Bogorov chamber), subsample with ATR-FTIR (0.45%)	Particle/km^2
McCormick et al. (2014)	North Shore Channel, Chicago (the United States)	Water surface	Neuston net (333 μm)	Density separation (NaCl), wet peroxide oxidation	2 mm–330 μm	Stereomicroscope (visual)	Particle/m^3 Particle/km^2
McCormick et al. (2016)	10 Rivers in Illinois and Indiana (the United States)	Water surface	Neuston net (333 μm)	Sieved (125 μm), wet peroxide oxidation	Microplastic: <5 mm	Visual, subsample by pyrolysis-GC/MS	Particles/m^3 Particles/km^2
Moore et al. (2011)	Los Angeles and San Gabriel River (the United States)	Water column (surface and bottom)	Manta net (333 μm), water bottom with hand net (800 and 500 μm)	Sieved	>4.75 mm, 4.75–2.8 mm, 2.8–1 mm	Stereomicroscope (visual)	Weight/m^3 Particle/m^{-3}
[a]Morritt et al. (2014)	Thames River (GB)	Water column (bottom)	Eel traps at the water bottom	–	Macroplastic: >5 mm	Visual	Absolute number of particles found

Continued

Table 3.3 Overview over the available studies assessing microplastic and macroplastic contamination in river surface water—cont'd

Publication	Location	Compartment	Field method (sampling)	Processing	Size class	Identification	Reference unit
van der Wal et al. (2015)	Po (IT), Danube (RO), Rhine (NL), Dalålven (SE)	Water surface	Manta net (330 μm, depth 0.1 m), 5000 L pumped through manta net from a depth of 0.3 m, waste-free water sampler (NOAA, 3–5 and 20–70 cm below the water surface)	Sieved, rinsed with water and ethanol	>5 mm, <5 mm	Subsamples with NIR spectroscopy and ATR-FTIR	Particle/m^3 Particle/km^2
Wang et al. (2017a)	Yangtze and Hanjing River in the area of Wuhan (CN)	Water surface	Water volume of 20 L pumped over 50 μm sieve	Hydrogen peroxide (30%)	Microplastic <5 mm: 50–500 μm, 500 μm–1 mm, 1–2 mm, 2–3mm, 3–4 mm, 4–5 mm	Visual, subsamples SEM	Particle/m^3

Zhang et al. (2015)	Yangtze River above the three gorges dam and four tributaries (Qinggan, Yuanshui, Tongzhuang, Xiangxi)	Water surface	Manta net (112 μm)	Sieved (1.6 mm), >1.6 mm visually screened, <1.6 in a separating funnel for 7 days, and floating potential plastic particles extracted and dried (60°C)	Microplastic: 1.6–5 mm, 500 μm–1.6 mm, 300–500 μm, 112–300 μm	Stereomicroscope (visual), subsample of 50–100 random particles of each sample by ATR-FTIR	Particle/km^2
Zhang et al. (2017)	Xiangxi River (largest Yangtze tributary, CN)	Water surface	Manta net (112 μm)	Sieved (1.6 mm), separated liquid collected and floating particles separated, retained material screened visually, material <1.6 mm, density separation (potassium formate, 1.5 g/cm^3)	Microplastic: <5 mm	Raman microspectroscopy	Particle/m^3

[a]Not included in Fig. 3.4.

Table 3.4 Overview over the available studies assessing microplastic and macroplastic contamination in river beach sediment and river bottom sediment

Publication	Location	Compartment	Field method (sampling)	Processing	Size class	Identification	Reference unit
Ballent et al. (2016)	Tributaries of Lake Ontario	Bottom sediment	Petite Ponar grab	Sieved, <63 μm: density separation (sodium polytungstate solution, 1.5 g/cm^3)	<5 mm, limit of visual identification of plastics is estimated to be ~0.25 mm	Stereomicroscope (visual), random particles by Raman and X-ray fluorescence spectroscopy (90 out of 6331 particles analyzed by Raman and a small number by XRF)	Particle/kg dw
Castañeda et al. (2014)	Saint Lawrence River (CA)	Beach sediment	Petite Ponar grab and Peterson grab	Sieved (500 μm)	>500 μm	Stereomicroscope (visual), subsample with spectroscopy, thermal degradation analysis (DSC)	Particle/m^2 Particle/L
Horton et al. (2016)	Thames River tributaries (the United Kingdom)	Beach sediment	Four sediment samples along a 3 m transect, each collected with a spoon to a depth of 10 cm to fill a 1 L glass Kilner jar	Visual inspection of sieved sediment, flotation, and visual inspection of sediments post-flotation and overflowed particles, density separation (ZnCl 1.7–1.8 kg/L)	Microplastic: 1–2 mm and 2–4 mm	Subsamples (20% of the particles) by Raman	Particle/kg Particle/L

Klein et al. (2015)	Rhine, Rhine-Main area (DE)	Beach sediment	Quadrats (30 cm^2, depth 2–3 cm, 3–4 kg sediment)	Density separation (saturated NaCl), digestion with a mixture of hydrogen peroxide and concentrated sulfuric acid, sieved	630–5000 μm, 200–630 μm, 63–200 μm	>630 μm: ATR-FTIR <630 μm: microscope (visual)	Particle/kg
Leslie et al. (2017)	Urban canals in the Amsterdam area (NL)	Bottom sediment	Two or more Van Veen grab samples taken and pooled for a single sample (1 L)	Homogenized, subsamples (20 g), density separation (NaCl, 1.2 g/mL), filtered over 0.7 μm filters, visually examined	Microplastic: 5000–300 μm, <300 μm	Subsample (6% of the particles) by FTIR	Particle/kg dw Particle/L
Rech et al. (2014)	Elqui, Maipo, Maule und Bio Bio (CL)	Beach sediment	Direct sampling of visual identifiable plastic >1.5 cm	–	>1.5 cm	Visual	Particle/m^2

Continued

Table 3.4 Overview over the available studies assessing microplastic and macroplastic contamination in river beach sediment and river bottom sediment—cont'd

Publication	Location	Compartment	Field method (sampling)	Processing	Size class	Identification	Reference unit
Wang et al. (2017b)	Beijiang River (CN)	Beach sediment	20 cm quadrats, randomly in the littoral zone, depth 2 cm	Dried, investigated in 30 g triplicates, density separation (saturated NaCl)	Microplastic: <5 mm	Visual (digital handheld microscope)	Particle/kg
Zhang et al. (2017)	Xiangxi River (largest Yangtze tributary, CN)	Bottom sediment	From the midstream channel with a Petersen grab sampler (0.0635 m^2)	Sieved (1.6 mm), separated liquid collected and floating particles separated, retained material screened visually, material <1.6 mm density separation (potassium formate, 1.5 g/cm^3)	Microplastic: <5 mm	Raman microspectroscopy	Particle/m^2

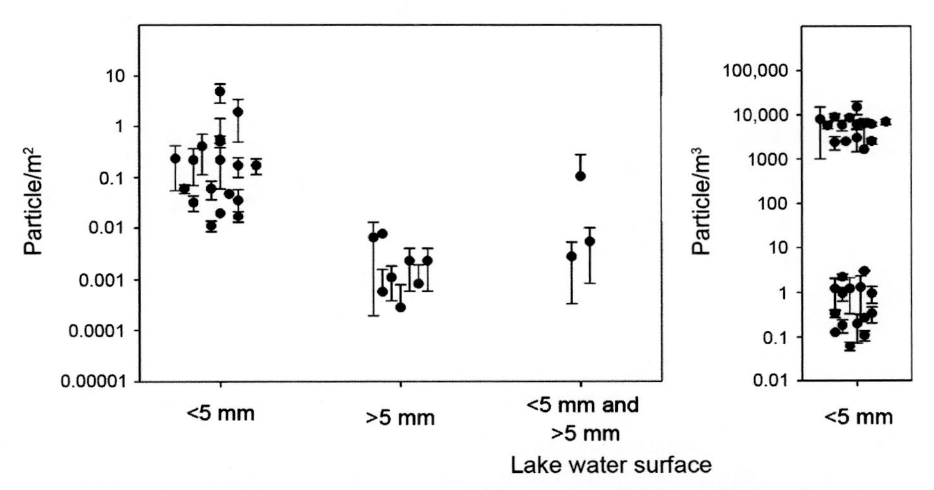

Fig. 3.1 Macro- and microplastic contamination of lake water surface considering the water surface (particle per square meter) or water volume (particle per cubic meter). Every data point corresponds to one lake. If studies provide multiple data points of single lakes, these are drawn as mean values with standard deviation. Mean values were calculated when they were not provided. For some lakes, it was not possible to differentiate between macro- and microplastic. When possible, the number of particles per square meter was calculated out of particles per cubic meter and vice versa.

The highest microplastic abundances on a lake surface water were detected in China with concentrations ranging between 1660 and 15,000 particles/m^3 (median, 6162 microplastic particles/m^3) (Wang et al., 2017a; Su et al., 2016). In contrast to that, the concentrations in North America, Europe, and Mongolia were much lower (0.06–3.02 microplastic particles/m^3 and median, 0.34 particles/m^3). Major differences of the two studies on Chinese lakes were that one study worked with a pump system and analyzed only 20 L of water (Wang et al., 2017a) and the other study used a plankton net instead of a dedicated surface/neuston net (Su et al., 2016). Additionally, both studies analyzed only a subsample of the potential particles with spectroscopic methods. This might explain the comparable high numbers by an overestimation of the microplastic concentration due to the methodological approaches (Fig. 3.1).

3.3.1.2 Beach and Bottom Sediment

Compared with the reported values of microplastics in surface water samples, the contamination of beach and bottom sediments is considerably high. This is obviously in coherence with the fact, that both, the bottom sediment and the drift line, which were often sampled to estimate microplastic concentration on beaches, are thought to be strong accumulation zones for microplastics (Browne et al., 2015; Imhof et al., 2018;

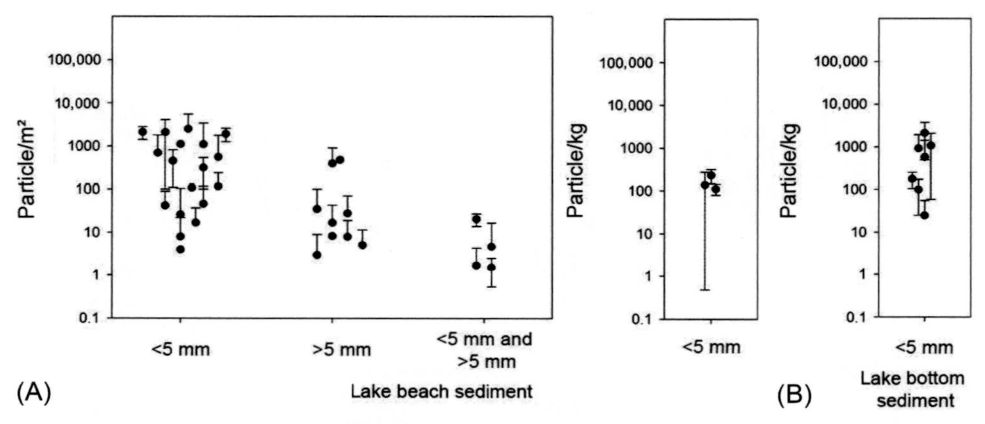

Fig. 3.2 Macro- and microplastic contamination of lake beach sediment (A) and lake bottom sediment (B) considering the sediment surface (particle per square meter) or sediment dry weight (particle per kilogram dry weight). Every data point corresponds to one lake. If studies provide multiple data points of single lakes, these are drawn as mean values with standard deviation. Mean values were calculated when they were not provided. For some lakes, it was not possible to differentiate between macro- and microplastic.

Hidalgo-Ruz and Thiel, 2013; Moreira et al., 2016; Van Cauwenberghe et al., 2015). In beach sediments, a microplastic concentration of 4–2500 particles/m^2 or 112–234 particles/kg dw was detected (median, 390 particles/m^2 or 139 particles/kg dw). For most studies, only one reference unit, particles/m^2 or particles/kg dw, was reported, and only a few studies provided both values. One additional study provided particles per liter of sediment and could therefore not be included in the comparison. Comparable to the water surface, the macroplastic concentration of beach sediments was much lower than the concentration of microplastics (3–483 macroplastic particles/m^2, and median, 17 macroplastic particles/m^2). Lake bottom sediments were analyzed in four studies and showed a microplastic concentration of 25–2128 particles/kg dw (median, 929 particles/kg dw) (Fig. 3.2).

3.3.1.3 Polymer Distribution

Only four out of the 17 studies on the abundance of macro- and microplastics in lakes identified all particles with spectroscopic methods and thus provided data on the complete polymer composition. All four studies investigated sediments (three beach sediments and one bottom sediment). Eleven lake studies performed visual identification accompanied by the spectroscopic characterization of a random subset of particles, and two studies relied solely on visual identification.

However, the polymer types with a high market share (PE, polypropylene (PP), and PS) were commonly detected on the water surface and in the sediments throughout all studies that performed a reliable polymer identification. Among the studies, which identified all particles in beach and bottom sediments, also, other polymer types (e.g., PVC, PET, and PA) were found but generally in a lower abundance (Imhof et al., 2013; Zhang et al., 2016). These microplastics were mainly found in the fraction smaller than 500 μm. The small particles comprise additional paint particles (Imhof et al., 2016; Zhang et al., 2017; Wang et al., 2017b). The high abundance of raw pellets detected in the North American Great Lakes (Corcoran et al., 2015; Eriksen et al., 2013b; Zbyszewski et al., 2014) was not observed in European studies that found raw pellets only in small amounts.

3.3.2 Microplastics in Rivers

Currently, 22 studies on microplastics in rivers are available. However, some rivers were sampled multiple times within one study, and others were sampled simultaneously within different studies. In summary, the current studies analyzed the water surface of 63 rivers, the beach sediment of 27 rivers, and the bottom sediment of 6 rivers (Tables 3.3 and 3.4).

3.3.2.1 Surface Water

The amount of microplastics reported on the water surface of rivers with respect to the filtered volume is 0.10–2933 particles/m^3 (median, 2.6 particles/m^3) and therefore higher than the values reported for lake water surfaces. In studies where the microplastic abundance is normalized to the water surface, the microplastic concentration reported ranges between 0.02 and 24.6 particles/m^2 (median, 0.8 particles/m^2). Similar to lake water surface, two Chinese rivers show the highest microplastic concentration at the water surface (Yangtze, 2517 ± 912, and Hanjing, 2933 ± 306 particles/m^3) (Wang et al., 2017a). It is unclear if this high contamination is a result of the vicinity of the city Wuhan far below the three gorges dam, the low amount of water sampled from the water surface (20 L, mesh size 50 μm) and the resulting extrapolation, or the different mesh sizes of the sampling equipment used in the studies. The microplastic concentration above the three gorges dam, sampled with a manta net (mesh size 112 μm), was 16.9 ± 10.3 microplastic particles/m^3 (Zhang et al., 2015). Nevertheless, further downstream in the Yangtze estuary, comparable high microplastic concentrations between 680 ± 285 and 4137 ± 2462 particles/m^3 were detected by sampling microplastics with neuston nets (mesh size 333) (Zhao et al., 2014) (Fig. 3.3).

3.3.2.2 Beach and Bottom Sediments

Microplastic concentrations detected in river beach sediments ranged between 7 and 243 particles/m^2 (median, 57 particles/m^2) or 178–3763 particles/kg dw (median, 333 particles/kg dw). One outlier with a concentration of $136{,}926 \pm 83{,}947$ microplastic particles/m^2 corresponds to beach sediment samples from the effluent canal of the Gentilly-2 nuclear power plant (Castañeda et al., 2014). This dataset was excluded from the presented range and the median calculation due to the exceptional location and the

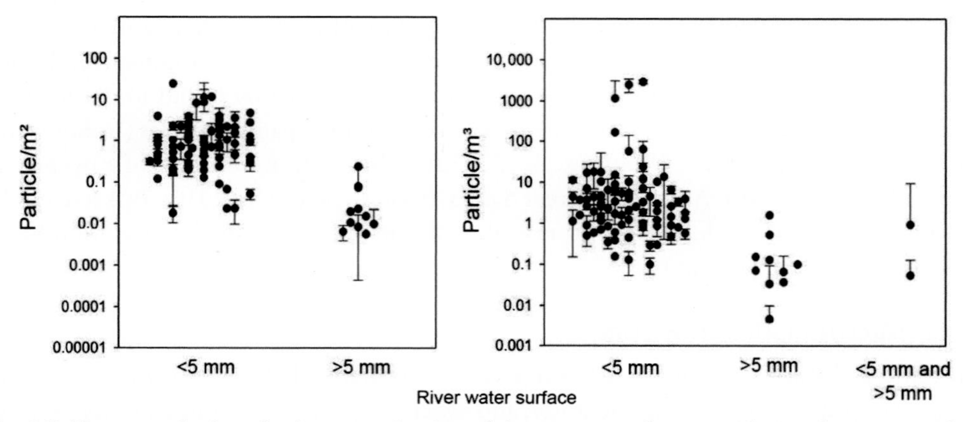

Fig. 3.3 Macro- and microplastic contamination of river water surface considering the water surface (particle per square meter) or the water volume (particle per cubic meter). Whereas the lake data were summarized for each lake, this was not possible due to the heterogeneity of the rivers along their course. Therefore, single values or regional means for highly detailed studies are presented here. When possible, the number of particles per square meter was calculated out of particles per cubic meter and vice versa.

extraordinary high concentration. Compared with the microplastic concentration, the macroplastic concentration of river beach sediments is rather low (0.03–2.11 macroplastic debris/m^2, and median, 0.29), as observed for lake beach sediments (Fig. 3.4).

River bottom sediments were analyzed in three studies. Two provided particles per kilogram dry weight and one particles per square meter sediment surface. The concentrations ranged from 28 to 10,500 particles/kg dw (median, 480) in tributaries of Lake

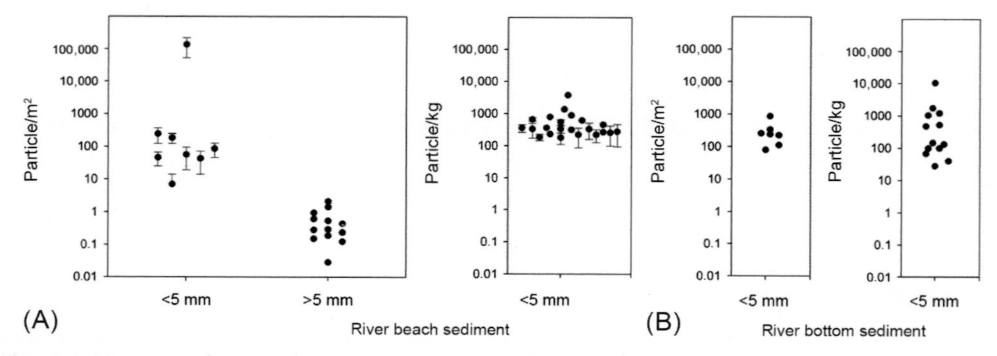

Fig. 3.4 Macro- and microplastic contamination of river beach sediment (A) and river bottom sediment (B) considering the sediment surface (particle per square meter) or the sediment dry weight (particle per kilogram dry weight). Every data point corresponds to one sample location from one river. Whereas the lake data were summarized for each lake, this was not possible due to the heterogeneity of the rivers along their course. Therefore, single values or regional means for highly detailed studies are presented here.

Ontario (CA and the United States) and an urban channel in Amsterdam (NL) and from 80 to 864 microplastic particles/m^2 (median, 240) at seven locations in the Xiangxi River (CN).

3.3.2.3 *Polymer Distribution*

Comparable with the lake studies, only three studies identified all particles with reliable methods. From these, two are not included into the graphs in this chapter, as one provided only the absolute microplastic number not normalized to a unit and the other collected larger debris in floating booms. Consistently, PE and PP were the most common polymers followed by PS (expanded and nonexpanded) and polyester in river surface water and sediments of the Xiangxi River (CN) (Zhang et al., 2017). A similar pattern was found in studies analyzing only a subset of the particles, for example, in the water surface of the Rhine (29.7% PS and 16.9% PP) (Mani et al., 2015) and in Rhine beach sediment samples in the Rhine-Main area (Klein et al., 2015). The macro debris from the Seine surface collected by floating booms consisted mainly of PP, PE, PET, and PS (according to their abundance) (Gasperi et al., 2014).

3.4 SOURCES OF MICROPLASTICS IN FRESHWATER

Through the past years, knowledge on the occurrence of microplastics in marine environments and to a lesser extent also freshwater ecosystems has been increasing. Currently, we have exact data on the amount of the production of plastics on one hand and also first estimations of the amount of microplastics that end up in the environment on the other hand. However, data on the relevance of different potential entry pathways and the amount of microplastics that enter the environment via different routes are lacking almost completely at present. Additionally, practically nothing is known about either the behavior of plastics in different environmental source compartments or the dynamics between them. Hence, the need for further investigations is obvious.

For mitigation measures to reduce the input of microplastics into the aquatic environment, it is of utmost importance to understand and identify the origin of the microplastics encountered in freshwater and consider all the important entry routes. The main potential pathways are depicted in Fig. 3.5 and discussed later.

Plastic originating from human activities will reach all three environmental compartments, namely, water, atmosphere, and soil.

Here, the aquatic compartment is often reached directly by plastics due to illegal dumping and littering. Macroplastics that reach freshwater environments by this pathway can fragment over time and thus be a source for microplastic particles in the freshwater itself. In contrast to microplastics stemming from macroplastic degradation in freshwater, microplastics (either primary or secondary) can also be introduced directly to freshwater already on the microscopic scale.

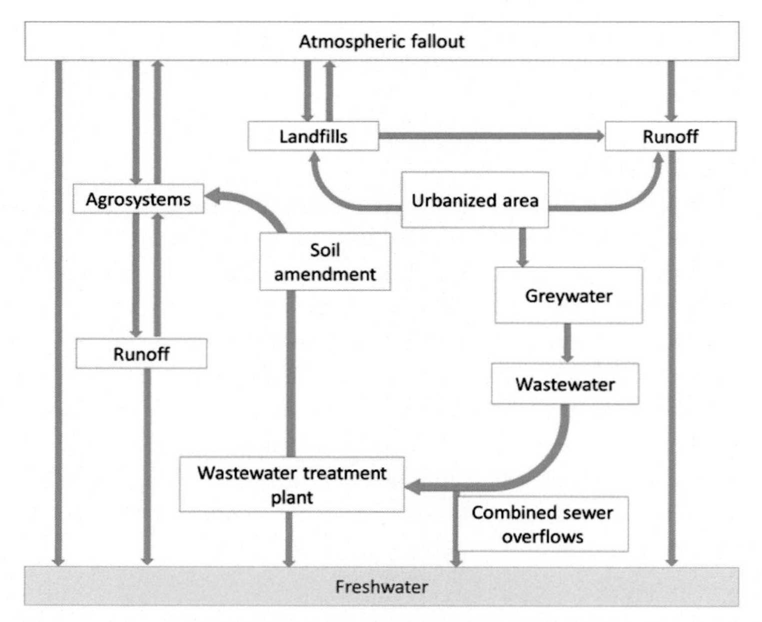

Fig. 3.5 A schematic diagram showing potential microplastic pathways and entry routes of microplastics into freshwater ecosystems. Macroplastics may also be present in most of the compartments. They can degrade to microplastics within the different compartments, during their transfer from one compartment to the other or once they reached freshwater ecosystems.

Another source, the atmosphere, is itself fueled by microplastics stemming from industrial emissions, landfill, particle resuspension, and other anthropogenic causes (littering, traffic, building activities, urban infrastructure, etc.). Additionally, the textiles worn by the population might also emit fibrous microplastics into the atmospheric compartment. These particles can be subsequently deposited as atmospheric fallout, contaminating urban surfaces, agricultural soils, and water surfaces.

On the other hand, urban water systems serve as a potential major point of entry for plastics (macro- and microplastics) into the environment. In this context, wastewater treatment plant (WWTP) disposals are a potential input source for microplastics as these particles are often not completely retained in the sewage sludge during the wastewater treatment, in addition to combined sewer overflows (CSOs) where microplastics in wastewater and in runoff bypass the WWTP completely during heavy rain events.

Lastly, runoff in general can transport not only a lot of microplastics but also macroplastics into aquatic systems. Urban runoff transports microplastics deposited from atmospheric fallout (including all the above mentioned sources), stemming from litter degradation or tire wear. Rural runoffs could also, in addition to the previous sources, carry microplastics originating from plasticulture, including larger plastic items used for temperature regulation and crop protection, or from soil amendment with WWTP sewage sludge or industrial compost.

3.4.1 Microplastics Stemming From Macroplastic Degradation

Estimating the amount of microplastics directly stemming from the degradation of macroplastics that have been introduced in the freshwater environments is not an easy task. It relies on the availability of two important data: the macroplastics present in freshwater (their occurrence, type, volume, etc.) on one hand and the degradation processes, mechanisms, and times on the other.

As the fragmentation of large plastic debris is one of the known sources of microplastics, the entry pathways of these larger debris in freshwater ecosystems have to be studied. A study estimated the amount of plastic debris flowing from cities to the oceans (Jambeck et al., 2015). This estimation was performed by linking worldwide data on solid waste, population density, and economic conditions. It was estimated that 4.8–12.7 million metric tons entered the ocean annually. To derive this estimation, a number of hypotheses had to be adopted. First, only the population living a distance of <50 km from the coast was considered. The paper then estimated the amounts of waste produced by this population and hypothesized on the fraction of this waste that gets mismanaged. The proportion of plastic materials among this waste also had to be approximated. Some of these approximations contain high uncertainties, and their accuracy is still impossible to determine. A second study estimated that between 1.15 and 2.41 million metric tons are conveyed worldwide from the rivers into the oceans per year (Lebreton et al., 2017). Both studies present estimations in the same order of magnitude reinforcing their potential accuracy. The latter estimation is carried out with the use of a model based on population density, rates of mismanaged plastic waste production, monthly catchment runoff, and the presence of artificial barriers like dams.

In order to support such global and theoretical estimations, local field studies are required. These are, however, completely absent. It can be supposed that informal dumping and littering is the main source of macroplastics in the environment. The extent of informal dumping is not easy to estimate and was never, to our knowledge, investigated by scientists. As a consequence, the easier way is to provide estimations on macroplastic inputs from urban environments, and, thus, macroplastic fluxes and amounts have been investigated directly on rivers impacted by large cities by a few studies (see Section 3.2) (Morritt et al., 2014; Gasperi et al., 2014).

After the assessment of the inputs and amounts of macroplastics in the environment, their degradation rate is required in order to estimate the microplastic production by degradation of macroplastics. While the degradation processes were widely examined experimentally(Cooper and Corcoran, 2010; Barnes et al., 2009), to our knowledge, there is no study linking both, abundance of macroplastics and their degradation, to estimate the production of secondary microplastics in the environment. This is due to the fact that degradation rates in the real environment depend on a variety of factors and are hard to determine.

3.4.2 Atmospheric Fallout

Very limited data exist on the input of microplastic into environmental compartments via atmospheric fallout. The role of atmospheric transport of microplastics was suspected when these particles were observed in remote areas like Lake Hovsgol in Mongolia (Free et al., 2014). To our knowledge, only one single study investigated atmospheric fallout at an urban and a suburban site (Dris et al., 2016). A continuous annual monitoring was carried out. Fibers accounted for almost all of the microplastics collected, and an atmospheric fallout between 2 and 355 fibers/m^2/day was reported. The lower size limit of the analysis was 50 μm, and it could be expected that smaller particles could be more abundant. The registered fluxes were systematically higher at the urban than at the suburban site. The chemical characterization of a small subsample allowed to estimate that 29% of the collected fibers were synthetic (made with petrochemicals) or made from a mixture of natural and synthetic material. Extrapolation using weight and volume assumptions of the collected fibers allowed the rough estimation that between 3 and 10 tons of fibers are deposited by atmospheric fallout at the scale of the Paris Metropolitan Area every year (2500 km^2) (Dris et al., 2016).

The presence of fibers in the atmospheric fallout suggested their presence in the air. Therefore, concentrations of fibers in indoor and outdoor air were also estimated in the same study. Two private apartments and one office were considered for indoor air analysis. The outdoor air was sampled in one urban site (with a pump filtering the air, thus obtaining concentrations and not atmospheric fallout) (Dris et al., 2017). Overall, indoor concentrations ranged between 1.0 and 60.0 fibers/m^3. Outdoor concentrations were significantly lower and ranged between 0.3 and 1.5 fibers/m^3. Regarding fiber type, 67% of the analyzed fibers in the indoor environments were made of natural material, primary cellulose, while the remaining 33% fibers were of synthetic origin with PP being predominant.

3.4.3 Waste Water Treatment Plants

Several studies have considered WWTP disposal as a source of microplastics in aquatic ecosystems. It has first been shown that laundry washing machines potentially discharge a large amount of plastic fibers into wastewater, with one study estimating that a single garment releases up to 1900 fibers in a single wash (Browne et al., 2011). These fibers will be subsequently found in wastewater. The same study found an average of 1 fiber/L in the effluent discharged from a tertiary-level WWTP.

A mere 3 years later, the first studies assessing microplastics in wastewater and the potential input of microplastics in the environment through WWTPs were published (Table 3.5). Among the 12 studies listed, four focused only on the disposal of the WWTPs, while the remaining studies included raw wastewater and in some cases various treatment stages in order to estimate removal efficiencies. Although the units reported were always the same (particles per liter), the comparison is however still hindered due to very heterogeneous methodologies used for the determination of the microplastic amount. For instance, several studies carried out an in situ size fractioning of the samples using nets, single sieves, or a stack of sieves (Talvitie et al., 2015, 2017; Murphy et al., 2016; Carr et al., 2016; Mason et al., 2016a; Michielssen et al., 2016; Ziajahromi et al., 2016; Magnusson and Norén, 2014). The lowest mesh size used for collecting microplastic however varied from study to study (between 20 and 300 μm, Table 3.5). Additionally, the fact that the targeted size range of particles was often smaller compared with the size range of sampled particles in freshwater studies hinders the establishment of a link between WWTP disposals and found freshwater concentrations.

The reported microplastic removal during the wastewater treatment is between 40% and 99.9% (Table 3.5). The highest and the lowest removal by wastewater treatment were both reported in the same study (Talvitie et al., 2017). The highest removal value was recorded in a WWTP using a membrane bioreactor, while the lowest value was found for a WWTP using disc filters. Most of the studies report removal efficiencies higher than 95%. It was shown that 80% of the removed microplastics end in the dried sludge (Talvitie et al., 2017). Two studies agree on the fact that the grease removal is the most important step for reducing microplastic concentrations (of lightweight polymers) between the influent and the effluent (Murphy et al., 2016; Carr et al., 2016).

The concentrations in the treated water vary between 0.005 particles/L in Finland (Talvitie et al., 2017) and 91 particles/L in the Netherlands (Leslie et al., 2017). The reasons behind the reported variations are not easy to identify due to various factors that potentially play a role: the different treatment facilities and efficiencies in different WWTPs, the size of the respective WWTPs, the density of the population that is connected to the WWTP, the temporal factors often not reported in the studies (seasonal changes, daily changes, rainy vs dry weather, etc.), and last but not least the different methodologies for sampling and processing the samples. A study used effluent concentrations to estimate that daily discharges of microplastics range from approx. 50,000 up to nearly 15 million particles per day (Mason et al., 2016a). Daily estimations are important and should be provided as it helps in estimating the contribution and importance of WWTP as a source of microplastics in freshwater. In fact, various WWTPs discharge different volumes of waters daily, and sole concentrations are therefore not meaningful.

We only found one study reporting results on microplastic abundances in CSOs. The study found between 190 and 1046 fibers/L and between 35 and 3100 fragments/L

Table 3.5 Overview over the available studies assessing microplastics in the influents and/or effluents of the WWTPs

Publication	Location	Studied WWTPs		Influent (number/L)	Effluent (number/L)	Removal (%)	Sampling	Processing	Assessed size classes	Identification
		#	Type							
Browne et al. (2011)	Australia	2	Tertiary	–	1 particle/L	–	Bulk sampling in 750 mL glass bottles	Vacuum filtration	<1 mm	ATR-FTIR spectroscopy
Talvitie et al. (2015)	Finland	1	Tertiary	180 particles/L 430 fragments/L	5 fibers/L 9 fragments/L	Fibers: 97% Fragments 98%	Pump system with in situ fractioning filtration	No further processing	20 μm–5 mm	Microscope (visual)
Talvitie et al. (2017)	Finland	4	Membrane bioreactor Rapid sand filtered Dissolved air flotation Disc filter	6.9 particles/L 0.7 particles/L 2 particles/L 0.5–2 particles/L	0.005 particles/L 0.02 particles/L 0.1 particles/L 0.03–0.3 particles/L	99.90% 97.00% 95.00% 40%–98.5%	Pump system with in situ fractioning filtration + automated 24 h composite samplers	No further processing	20 μm–5 mm	Microscope (visual) and FTIR analysis for preselected particles
Dris et al. (2015b)	France	1	Secondary with biofilters	260–320 fibers/L	14–50 fibers/L	81%–96%	Automated 24 h composite samplers	Vacuum filtration	50 μm–5 mm	Microscope (visual)
Murphy et al. (2016)	Ireland	1	Secondary	15.7 particles/L	0.25 particles/L	98.41%	30–50 L filtered through 65 μm sieves	Vacuum filtration	>65 μm	Microscope (visual) followed by FTIR validation
Carr et al. (2016)	The United States	12	Different facilities	–	–	95%–99%	Stack of 8 sieves from 400 to 20 μm +surface skimming	Pipetting of representative 5 mL aliquots	20 μm–5 mm	Microscope (visual) subsamples with FTIR
Mason et al. (2016a)	The United States	17	Different facilities	–	0.05 ± 0.024 particles/L Daily discharge 50,000 to 15 million particles	–	Up to 24 h filtration through a 125 and a 355 μm mesh	30% hydrogen peroxide with Fenton's reagent	125–355 μm >355 μm	Microscope (visual)

Michielssen et al. (2016)	The United States	3	Secondary Tertiary Membrane bioreactor system	133 particles/L 93 particles/L 83 particles/L	5.9 particles/L 2.6 particles/L 0.5 particles/L	95.60% 97.20% 99.4%	Grab samples from 1 L (raw wastewater) to 38 L (treated effluent)	Stack of sieves from 4.75 mm to 20 μm	20 μm–4.75 mm	Microscope (visual)
Leslie et al. (2017)	The Netherlands	7	Different facilities	20–910 particles/L	9–91 particles/L	72%	Bulk 2 L sampling	Density separation with NaCl (1.2 kg/L)	10–300 μm 300 μm–5 mm	Microscope (visual) with FTIR confirmation for preselected particles
Mintenig et al. (2017)	Germany	12	8 secondary 4 tertiary	–	0.08–9 particles/L	–	Pump system (0.39–1 m^3), stainless steel filter candle (10 μm)	Enzymatic digestion	5000–500 μm 500–20 μm	FPA-based FTIR microspectroscopy (<500 μm) ATR-FTIR spectroscopy (<500 μm)
Magnusson and Norén (2014)	Sweden	1	Secondary	15.1 ± 0.89 particles/L	8.25 ± 0.85 particles/L	99.90%	Influent Ruttner sampler, effluent pump system, both filtered over 300 μm	No further processing	≥300 μm	Microscope (visual), subsample with ATR-FTIR
Ziajahromi et al. (2016))	Australia	3	Primary Secondary Tertiary	–	1.5 particles/L 0.48 particles/L 0.28 particles/L	–	3–200 L filtered through a stack of sieves from 25 to 500 μm	30% hydrogen peroxide treatment Density separation on NaI (1.49 kg/L)	25 μm–5 mm	Microscope (visual) with FTIR confirmation for preselected particles

(Dris et al., 2018). While these concentrations are higher than concentrations in WWTP discharge, CSOs represent often only punctual events. A normalization of the found concentrations with annual volumes discharged by CSOs and WWTPs on the same area will allow for the evaluation of the importance of both sources.

3.4.4 Runoff

To the best of our knowledge, no study has focused on microplastics in runoff, and the contribution of this potential microplastic source has not been estimated.

In addition to the atmospheric microplastics washed out directly by rain and the runoff of the dry deposition of microplastics on surfaces, a study has suggested that additional sources for microplastics in runoff are automobile tire wear, building materials and road paints (Norwegian Environment Agency, 2014), or littering. The assumptions of this study remain profound speculation; however, microplastics have not been investigated in runoff directly.

A study analyzed the presence of microplastics in urban dust on a street and reported a load of 88–605 particles per 30 g of dry dust. The particles had a size ranging from 250 to 500 μm (Dehghani et al., 2017). Although the identification was only visual, the study provided a first evidence of microplastics in urban dust, which will be transported by urban runoff.

Car tire wear is often suspected as a major part of the microplastic pollution in urban runoff (Kole et al., 2017). A report estimated that between 60,000 and 111,000 metric tons of microplastics are released annually from car tires in Germany and between 375.000 and 693.750 metric tons in Europe (Essel et al., 2015). The lack of data on car tire wear in urban runoff may be related to the challenge to analyze such black particles with the common spectroscopic identification methods.

Besides the urban runoff, the runoff from agrosystems is frequently discussed as an input pathway for microplastics into aquatic ecosystems (Hussain and Hamid, 2003; Horton et al., 2016). However, currently, no direct evidence for microplastics in agricultural runoff exists. A reason might be the high effort required for the purification and analysis of soil samples.

Nevertheless, in addition to microplastics introduced into agricultural land by atmospheric fallout and wind (Dris et al., 2016), several farming methods are the cause for introducing large plastic fragments into agrarian soil. The agricultural industry provides various plastic products, which are used in order to prolong growth periods, improve soil quality, etc. (Hussain and Hamid, 2003; Steinmetz et al., 2016). A high proportion of agricultural plastic products are foils, for example, plastic films as mulch films, greenhouse coverings, labels, and silage wraps (Lament, 1993). During their use, agricultural foils undergo weathering and fragmentation and are consequently incorporated in the soils by mulching (Hussain and Hamid, 2003; Steinmetz et al.,

2016; Huerta Lwanga et al., 2017; Brodhagen et al., 2017). In the long run, the fragments become smaller and smaller, and microplastics are formed.

Other sources on how microplastics enter the terrestrial ecosystem are compost and sewage sludge applied for fertilizing agricultural soils (Habib et al., 1998; Zubris and Richards, 2005). Due to the longevity of plastic particles and fibers stemming from these sources, microplastics potentially accumulate in soils (Horton et al., 2017) and are thus, for example, found even 15 years after the last sewage sludge application (Zubris and Richards, 2005).

In conclusion, agricultural soils are potentially partly highly contaminated with microplastics, and the transfer of these particles with runoff into aquatic ecosystems needs to be investigated. However, the extent of the contribution of this source for microplastics in freshwater is completely unknown.

3.5 CONCLUSIONS

After the first evidences of microplastics in streams and lakes, the contamination of freshwater ecosystems with microplastics is of increasing concern in the scientific community and the global media due to their potential negative environmental impacts.

The studies we evaluated during our literature search show the ubiquitous presence of microplastic particles in freshwater ecosystems from the sediment to the water surface. The microplastic concentration of the analyzed compartments was often in a narrow range, although the difference of the used methods and the reported units largely hamper the comparability of the data.

Although first evidences that microplastics can have negative impacts on single aquatic species have been reported, at the moment, we are far away from a detailed picture on the environmental impacts of microplastics. However, the underlying baselines for a risk assessment are, however, reliable abundance data. Only when relevant environmental microplastic abundances are considered in experiments on the effects of microplastics on organisms, we are really able to determine environmental risks. This is especially important for the lower micron range of microplastic particles as these particles are presumed to have the highest impact. Those particles not only are bioavailable but also are able to pass cell barriers. However, the currently performed methods to analyze microplastics either are often not reliable or involve a very time-consuming effort to analyze particles down to 1 μm. Both facts impede large-scale routine monitoring programs that would be necessary for local and global comparative studies.

A prerequisite would be the use of reliable identification methods to reduce one of the major uncertainties within the currently available data. Methods such as FTIR spectroscopy, Raman spectroscopy, or weight-based methods like thermogravimetry in combination with mass spectroscopy ensure the production of reliable data. A pure visual identification alone, especially of microplastics smaller than 1 mm, is not suitable and should not be performed in future approaches. Furthermore, standard operation protocols for sampling, sample processing, and analyses are urgently necessary for producing comparable data. Nevertheless, it is still a long way until routine monitoring approaches, spanning from macro to the micron scale, will be established. Simultaneously, to the increasing data on the occurrence of microplastics, a deeper understanding of their dynamics in freshwater is required. This involves the importance of the different sources of microplastics and their dynamics in the environment. Additionally, the fate and potential impacts of different microplastic types have to be considered, as they can be highly different and depend on their size, shape, polymer composition, and state of degradation. Future modeling approaches, which consider as much of these parameters as possible in tandem with realistic experiments on the effects of microplastics on biota, will be helpful to understand the complex interactions of microplastics and the environment and will allow for a reliable risk assessment.

REFERENCES

Anderson, P.J., et al., 2017. Microplastic contamination in Lake Winnipeg, Canada. Environ. Pollut. 225, 223–231.

Baldwin, A.K., Corsi, S.R., Mason, S.A., 2016. Plastic debris in 29 Great Lakes tributaries: relations to watershed attributes and hydrology. Environ. Sci. Technol. 50, 10377–10385.

Ballent, A., Corcoran, P.L., Madden, O., Helm, P.A., Longstaffe, F.J., 2016. Sources and sinks of microplastics in Canadian Lake Ontario nearshore, tributary and beach sediments. Mar. Pollut. Bull. 110, 383–395.

Barnes, D.K.A., Galgani, F., Thompson, R.C., Barlaz, M., 2009. Accumulation and fragmentation of plastic debris in global environments. Philos. Trans. R Soc. B Biol. Sci. 364, 1985–1998.

Brodhagen, M., et al., 2017. Policy considerations for limiting unintended residual plastic in agricultural soils. Environ. Sci. Pol. 69, 81–84.

Browne, M.A., Galloway, T.S., Thompson, R.C., 2010. Spatial patterns of plastic debris along estuarine shorelines. Environ. Sci. Technol. 44, 3404–3409.

Browne, M.A., et al., 2011. Accumulation of microplastic on shorelines woldwide: sources and sinks. Environ. Sci. Technol. 45, 9175–9179.

Browne, M.A., et al., 2015. Spatial and temporal patterns of stranded intertidal marine debris: is there a picture of global change? Environ. Sci. Technol. 49, 7082–7094.

Carpenter, E.J., Smith, K.L., 1972. Plastics on the Sargasso sea surface. Science 175, 1240–1241.

Carr, K.E., Smyth, S.H., McCullough, M.T., Morris, J.F., Moyes, S.M., 2012. Morphological aspects of interactions between microparticles and mammalian cells: intestinal uptake and onward movement. Prog. Histochem. Cytochem. 46, 185–252.

Carr, S.A., Liu, J., Tesoro, A.G., 2016. Transport and fate of microplastic particles in wastewater treatment plants. Water Res. 91, 174–182.

Castañeda, R.A., Avlijas, S., Simard, M.A., Ricciardi, A., Smith, R., 2014. Microplastic pollution in St. Lawrence River sediments. Can. J. Fish. Aquat. Sci. 71, 1767–1771.

Catarino, A.I., Thompson, R., Sanderson, W., Henry, T.B., 2017. Development and optimization of a standard method for extraction of microplastics in mussels by enzyme digestion of soft tissues: standard method for microplastic extraction from mussels. Environ. Toxicol. Chem. 36, 947–951.

Claessens, M., Meester, S.D., Landuyt, L.V., Clerck, K.D., Janssen, C.R., 2011. Occurrence and distribution of microplastics in marine sediments along the Belgian coast. Mar. Pollut. Bull. 62, 2199–2204.

Claessens, M., Van Cauwenberghe, L., Vandegehuchte, M.B., Janssen, C.R., 2013. New techniques for the detection of microplastics in sediments and field collected organisms. Mar. Pollut. Bull. 70, 227–233.

Cole, M., Lindeque, P., Halsband, C., Galloway, T.S., 2011. Microplastics as contaminants in the marine environment: a review. Mar. Pollut. Bull. 62, 2588–2597.
Cole, M., et al., 2013. Microplastic ingestion by zooplankton. Environ. Sci. Technol. 130606145528005, https://doi.org/10.1021/es400663f.
Cole, M., et al., 2014. Isolation of microplastics in biota-rich seawater samples and marine organisms. Sci. Rep. 4, Article No.: 4528.
Collard, F., Gilbert, B., Eppe, G., Parmentier, E., Das, K., 2015. Detection of anthropogenic particles in fish stomachs: an isolation method adapted to identification by Raman spectroscopy. Arch. Environ. Contam. Toxicol. 69, 331–339.
Colton Jr., J.B., Burns, B.R., Knapp, F.D., 1974. Plastic particles in surface waters of the northwestern Atlantic. Science 185, 491–497.
Cooper, D.A., Corcoran, P.L., 2010. Effects of mechanical and chemical processes on the degradation of plastic beach debris on the island of Kauai, Hawaii. Mar. Pollut. Bull. 60, 650–654.
Corcoran, P.L., Biesinger, M.C., Grifi, M., 2009. Plastics and beaches: a degrading relationship. Mar. Pollut. Bull. 58, 80–84.
Corcoran, P.L., et al., 2015. Hidden plastics of Lake Ontario, Canada and their potential preservation in the sediment record. Environ. Pollut. 204, 17–25.
Dehaut, A., et al., 2016. Microplastics in seafood: benchmark protocol for their extraction and characterization. Environ. Pollut. 215, 223–233.
Dehghani, S., Moore, F., Akhbarizadeh, R., 2017. Microplastic pollution in deposited urban dust, Tehran metropolis, Iran. Environ. Sci. Pollut. Res. Int. 24, 20360–20371.
Dekiff, J.H., Remy, D., Klasmeier, J., Fries, E., 2014. Occurrence and spatial distribution of microplastics in sediments from Norderney. Environ. Pollut. 186, 248–256.
Derraik, J.G., 2002. The pollution of the marine environment by plastic debris: a review. Mar. Pollut. Bull. 44, 842–852.
Di Mauro, R., Kupchik, M.J., Benfield, M.C., 2017. Abundant plankton-sized microplastic particles in shelf waters of the northern Gulf of Mexico. Environ. Pollut. (Barking Essex: 1987) 230, 798–809.
Doyle, M.J., Watson, W., Bowlin, N.M., Sheavly, S.B., 2011. Plastic particles in coastal pelagic ecosystems of the Northeast Pacific ocean. Mar. Environ. Res. 71, 41–52.
Dris, R., et al., 2015a. Beyond the ocean: contamination of freshwater ecosystems with (micro-) plastic particles. Environ. Chem. 12, 539–550.
Dris, R., et al., 2015b. Microplastic contamination in an urban area: a case study in Greater Paris. Environ. Chem. 12, 592–599.
Dris, R., Gasperi, J., Saad, M., Mirande, C., Tassin, B., 2016. Synthetic fibers in atmospheric fallout: a source of microplastics in the environment? Mar. Pollut. Bull. 104, 290–293.
Dris, R., et al., 2017. A first overview of textile fibers, including microplastics, in indoor and outdoor environments. Environ. Pollut. 221, 453–458.
Dris, R., Gasperi, J., Tassin, B., 2018. Sources and fate of microplastics in urban areas: a focus on Paris Megacity. In: Wagner, M., Lambert, S. (Eds.), Freshwater Microplastics. In: vol. 58. Springer International Publishing, Berlin, pp. 69–83.
Dümichen, E., et al., 2017. Fast identification of microplastics in complex environmental samples by a thermal degradation method. Chemosphere 174, 572–584.
Eriksen, M., et al., 2013a. Plastic pollution in the South Pacific subtropical gyre. Mar. Pollut. Bull. 68 (1–2), 71–76.
Eriksen, M., et al., 2013b. Microplastic pollution in the surface waters of the Laurentian Great Lakes. Mar. Pollut. Bull. 77, 177–182.
Essel, R., Engel, L. & Carus, M. Quellen für Mikroplastik mit Relevanz für den Meeresschutz in Deutschland. In: Umweltbundesam, editor. UBA Texte. (2015).
Farrell, P., Nelson, K., 2013. Trophic level transfer of microplastic: *Mytilus edulis* (L.) to *Carcinus maenas* (L.). Environ. Pollut. 177, 1–3.
Faure, F., Corbaz, M., Baecher, H., de Alencastro, L., 2012. Pollution due to plastics and microplastics in Lake Geneva and in the Mediterranean Sea. Arch. Sci. 65, 157–164.

Faure, F., Demars, C., Wieser, O., Kunz, M., de Alencastro, L.F., 2015. Plastic pollution in Swiss surface waters: nature and concentrations, interaction with pollutants. Environ. Chem. 12, 582.

Fischer, M., Scholz-Böttcher, B.M., 2017. Simultaneous trace identification and quantification of common types of microplastics in environmental samples by pyrolysis-gas chromatography–mass spectrometry. Environ. Sci. Technol. 51, 5052–5060.

Fischer, E.K., Paglialonga, L., Czech, E., Tamminga, M., 2016. Microplastic pollution in lakes and lake shoreline sediments – a case study on Lake Bolsena and Lake Chiusi (central Italy). Environ. Pollut. 213, 648–657.

Free, C.M., et al., 2014. High-levels of microplastic pollution in a large, remote, mountain lake. Mar. Pollut. Bull. 85, 156–163.

Fries, E., et al., 2013. Identification of polymer types and additives in marine microplastic particles using pyrolysis-GC/MS and scanning electron microscopy. Environ. Sci. Process. Impacts 15, 1949.

Gasperi, J., Dris, R., Bonin, T., Rocher, V., Tassin, B., 2014. Assessment of floating plastic debris in surface water along the Seine River. Environ. Pollut. 195, 163–166.

Habib, D., Locke, D.C., Cannone, L.J., 1998. Synthetic fibers as indicators of municipal sewage sludge, sludge products, and sewage treatment plant effluents. Water Air Soil Pollut. 103, 1–8.

Harrison, J.P., Ojeda, J.J., Romero-González, M.E., 2012. The applicability of reflectance micro-Fourier-transform infrared spectroscopy for the detection of synthetic microplastics in marine sediments. Sci. Total Environ. 416, 455–463.

Hidalgo-Ruz, V., Thiel, M., 2013. Distribution and abundance of small plastic debris on beaches in the SE Pacific (Chile): a study supported by a citizen science project. Mar. Environ. Res. 87–88, 12–18.

Hidalgo-Ruz, V., Gutow, L., Thompson, R.C., Thiel, M., 2012. Microplastics in the marine environment: a review of the methods used for identification and quantification. Environ. Sci. Technol. 46, 3060–3075.

Horton, A.A., Svendsen, C., Williams, R.J., Spurgeon, D.J., Lahive, E., 2016. Large microplastic particles in sediments of tributaries of the River Thames, UK – abundance, sources and methods for effective quantification. Mar. Pollut. Bull. https://doi.org/10.1016/j.marpolbul.2016.09.004.

Horton, A.A., Walton, A., Spurgeon, D.J., Lahive, E., Svendsen, C., 2017. Microplastics in freshwater and terrestrial environments: evaluating the current understanding to identify the knowledge gaps and future research priorities. Sci. Total Environ. 586, 127–141.

Huerta Lwanga, E., et al., 2017. Incorporation of microplastics from litter into burrows of Lumbricus terrestris. Environ. Pollut. (Barking Essex: 1987) 220, 523–531.

Hussain, I., Hamid, H., 2003. Plastics in agriculture. In: Andrady, A.L. (Ed.), Plastics and the Environment. John Wiley & Sons, Inc., Hoboken, NJ, pp. 185–209. https://doi.org/10.1002/0471721557.ch5

Imhof, H.K., Schmid, J., Niessner, R., Ivleva, N.P., Laforsch, C., 2012. A novel, highly efficient method for the separation and quantification of plastic particles in sediments of aquatic environments. Limnol. Oceanogr. Methods 10, 524–537.

Imhof, H.K., Ivleva, N.P., Schmid, J., Niessner, R., Laforsch, C., 2013. Contamination of beach sediments of a subalpine lake with microplastic particles. Curr. Biol. 23, R867–R868.

Imhof, H.K., et al., 2016. Pigments and plastic in limnetic ecosystems: a qualitative and quantitative study on microparticles of different size classes. Water Res. 98, 64–74.

Imhof, H.K., et al., 2017. Spatial and temporal variation of macro-, meso- and microplastic abundance on a remote coral island of the Maldives, Indian Ocean. Mar. Pollut. Bull. 116, 340–347.

Imhof, H.K., et al., 2018. Variation in plastic abundance at different lake beach zones—a case study. Sci. Total Environ. 613–614, 530–537.

Ivar do Sul, J.A., Costa, M.F., 2014. The present and future of microplastic pollution in the marine environment. Environ. Pollut. 185, 352–364.

Ivleva, N.P., Wiesheu, A.C., Niessner, R., 2017. Microplastic in aquatic ecosystems. Angew. Chem. Int. Ed. 56, 1720–1739.

Jambeck, J.R., et al., 2015. Plastic waste inputs from land into the ocean. Science 347, 768–771.

Käppler, A., et al., 2015. Identification of microplastics by FTIR and Raman microscopy: a novel silicon filter substrate opens the important spectral range below 1300 cm(−1) for FTIR transmission measurements. Anal. Bioanal. Chem. 407, 6791–6801.

Käppler, A., et al., 2016. Analysis of environmental microplastics by vibrational microspectroscopy: FTIR, Raman or both? Anal. Bioanal. Chem., 408, 8377–8391.

Kedzierski, M., et al., 2016. Microplastics elutriation from sandy sediments: a granulometric approach. Mar. Pollut. Bull. 107, 315–323.

Klein, S., Worch, E., Knepper, T.P., 2015. Occurrence and spatial distribution of microplastics in river shore sediments of the Rhine-Main area in Germany. Environ. Sci. Technol. 49, 6070–6076.

Koelmans, A.A., Bakir, A., Burton, G.A., Janssen, C.R., 2016. Microplastic as a vector for chemicals in the aquatic environment: critical review and model-supported reinterpretation of empirical studies. Environ. Sci. Technol. 50, 3315–3326.

Kole, P.J., Löhr, A.J., Van Belleghem, F.G.A.J., Ragas, A.M.J., 2017. Wear and tear of tyres: a stealthy source of microplastics in the environment. Int. J. Environ. Res. Public Health 14, Article No.: 1265.

Lament, W.J., 1993. Plastic mulches for the production of vegetable crops. HortTechnology 3, 35–39.

Lebreton, L.C.M., et al., 2017. River plastic emissions to the world's oceans. Nat. Commun. 8, ncomms15611.

Lechner, A., et al., 2014. The Danube so colourful: a potpourri of plastic litter outnumbers fish larvae in Europe's second largest river. Environ. Pollut. 188, 177–181.

Lenz, R., Enders, K., Stedmon, C.A., Mackenzie, D.M.A., Nielsen, T.G., 2015. A critical assessment of visual identification of marine microplastic using Raman spectroscopy for analysis improvement. Mar. Pollut. Bull. 100, 82–91.

Leslie, H.A., Brandsma, S.H., van Velzen, M.J.M., Vethaak, A.D., 2017. Microplastics en route: field measurements in the Dutch river delta and Amsterdam canals, wastewater treatment plants, North Sea sediments and biota. Environ. Int. 101, 133–142.

Löder, M.G.J., Gerdts, G., 2015. Methodology used for the detection and identification of microplastics—a critical appraisal. In: Bergmann, M., Gutow, L., Klages, M. (Eds.), Marine Anthropogenic Litter. Springer International Publishing, Berlin, pp. 201–227. https://doi.org/10.1007/978-3-319-16510-3_8.

Löder, M.G.J., Kuczera, M., Mintenig, S., Lorenz, C., Gerdts, G., 2015. Focal plane array detector-based micro-Fourier-transform infrared imaging for the analysis of microplastics in environmental samples. Environ. Chem. 12, 563–581.

Löder, M.G.J., et al., 2017. Enzymatic purification of microplastics in environmental samples. Environ. Sci. Technol 51 (24), 14283–14292. https://doi.org/10.1021/acs.est.7b03055.

Magnusson, K., Norén, F., 2014. Screening of microplastic particles in and down-stream a wastewater treatment plant. Available at: http://www.diva-portal.org/smash/get/diva2:773505/FULLTEXT01.pdf.

Mani, T., Hauk, A., Walter, U., Burkhardt-Holm, P., 2015. Microplastics profile along the Rhine River. Sci. Rep. 5, 17988.

Mason, S.A., et al., 2016a. Microplastic pollution is widely detected in US municipal wastewater treatment plant effluent. Environ. Pollut. 218, 1045–1054.

Mason, S.A., et al., 2016b. Pelagic plastic pollution within the surface waters of Lake Michigan, USA. J. Great Lakes Res. 42, 753–759.

Mathalon, A., Hill, P., 2014. Microplastic fibers in the intertidal ecosystem surrounding Halifax Harbor, Nova Scotia. Mar. Pollut. Bull. 81, 69–79.

McCormick, A., Hoellein, T.J., Mason, S.A., Schluep, J., Kelly, J.J., 2014. Microplastic is an abundant and distinct microbial habitat in an urban river. Environ. Sci. Technol. 48, 11863–11871.

McCormick, A.R., et al., 2016. Microplastic in surface waters of urban rivers: concentration, sources, and associated bacterial assemblages. Ecosphere 7.

McDermid, K.J., McMullen, T.L., 2004. Quantitative analysis of small-plastic debris on beaches in the Hawaiian archipelago. Mar. Pollut. Bull. 48, 790–794.

Michielssen, M.R., Michielssen, E.R., Ni, J., Duhaime, M.B., 2016. Fate of microplastics and other small anthropogenic litter (SAL) in wastewater treatment plants depends on unit processes employed. Environ. Sci. Water Res. Technol. 2, 1064–1073.

Mintenig, S.M., Int-Veen, I., Löder, M.G.J., Primpke, S., Gerdts, G., 2017. Identification of microplastic in effluents of waste water treatment plants using focal plane array-based micro-Fourier-transform infrared imaging. Water Res. 108, 365–372.

Moore, C.J., Lattin, G.L., Zellers, A.F., 2011. Quantity and type of plastic debris flowing from two urban rivers to coastal waters and beaches of southern. J. Integr. Coast. Zone Manage 11, 65–73.
Moreira, F.T., et al., 2016. Small-scale temporal and spatial variability in the abundance of plastic pellets on sandy beaches: methodological considerations for estimating the input of microplastics. Mar. Pollut. Bull. 102, 114–121.
Morét-Ferguson, S., et al., 2010. The size, mass, and composition of plastic debris in the western North Atlantic Ocean. Mar. Pollut. Bull. 60, 1873–1878.
Morritt, D., Stefanoudis, P.V., Pearce, D., Crimmen, O.A., Clark, P.F., 2014. Plastic in the Thames: a river runs through it. Mar. Pollut. Bull. 78, 196–200.
MSFD Technical Subgroup on Marine Litter, 2013. Guidance on Monitoring of Marine Litter in European Seas. Available at: http://publications.jrc.ec.europa.eu/repository/bitstream/JRC83985/lb-na-26113-en-n.pdf.
Murphy, F., Ewins, C., Carbonnier, F., Quinn, B., 2016. Wastewater treatment works (WwTW) as a source of microplastics in the aquatic environment. Environ. Sci. Technol. 50, 5800–5808.
Murray, F., Cowie, P.R., 2011. Plastic contamination in the decapod crustacean Nephrops norvegicus (Linnaeus, 1758). Mar. Pollut. Bull. 62, 1207–1217.
Ng, K.L., Obbard, J.P., 2006. Prevalence of microplastics in Singapore's coastal marine environment. Mar. Pollut. Bull. 52, 761–767.
Norén, F. Small plastic particles in Coastal Swedish waters—Kimo reports. (2007).
Norwegian Environment Agency, 2014. Sources of Microplastic Pollution to the Marine Environment. Norwegian Environment Agency. Available at: http://www.miljodirektoratet.no/Documents/publikasjoner/M321/M321.pdf.
Nuelle, M.-T., Dekiff, J.H., Remy, D., Fries, E., 2014. A new analytical approach for monitoring microplastics in marine sediments. Environ. Pollut. 184, 161–169.
Rech, S., et al., 2014. Rivers as a source of marine litter – a study from the SE Pacific. Mar. Pollut. Bull. 82, 66–75.
Rochman, C.M., Browne, M.A., 2013. Classify plastic waste as hazardous. Nature 494, 169–171.
Song, Y.K., et al., 2015. A comparison of microscopic and spectroscopic identification methods for analysis of microplastics in environmental samples. Mar. Pollut. Bull. https://doi.org/10.1016/j.marpolbul.2015.01.015.
Sruthy, S., Ramasamy, E.V., 2017. Microplastic pollution in Vembanad Lake, Kerala, India: the first report of microplastics in lake and estuarine sediments in India. Environ. Pollut. 222, 315–322.
Steinmetz, Z., et al., 2016. Plastic mulching in agriculture. Trading short-term agronomic benefits for long-term soil degradation? Sci. Total Environ. 550, 690–705.
Su, L., et al., 2016. Microplastics in Taihu Lake, China. Environ. Pollut. 216, 711–719.
Talvitie, J., et al., 2015. Do wastewater treatment plants act as a potential point source of microplastics? Preliminary study in the coastal Gulf of Finland, Baltic Sea. Water Sci. Technol. 72, 1495–1504.
Talvitie, J., Mikola, A., Setälä, O., Heinonen, M., Koistinen, A., 2017. How well is microlitter purified from wastewater? – A detailed study on the stepwise removal of microlitter in a tertiary level wastewater treatment plant. Water Res. 109, 164–172.
Thompson, R.C., et al., 2004. Lost at sea: where is all the plastic? Science 304, 838.
Tourinho, P.S., Ivar do Sul, J.A., Fillmann, G., 2010. Is marine debris ingestion still a problem for the coastal marine biota of southern Brazil? Mar. Pollut. Bull. 60, 396–401.
Van Cauwenberghe, L., Vanreusel, A., Mees, J., Janssen, C.R., 2013. Microplastic pollution in deep-sea sediments. Environ. Pollut. 182, 495–499.
Van Cauwenberghe, L., Devriese, L., Galgani, F., Robbens, J., Janssen, C.R., 2015. Microplastics in sediments: a review of techniques, occurrence and effects. Mar. Environ. Res. 111, 5–17.
van der Wal, M. et al. Final report on Identification and assessment of riverine input of (marine) litter. (Eunomia Research & Consulting Ltd, 2015).
Vianello, A., et al., 2013. Microplastic particles in sediments of Lagoon of Venice, Italy: first observations on occurrence, spatial patterns and identification. Estuar. Coast. Shelf Sci. 130, 54–61.
Wagner, M., et al., 2014. Microplastics in freshwater ecosystems: what we know and what we need to know. Environ. Sci. Eur. 26, 1–9.
Wang, W., Ndungu, A.W., Li, Z., Wang, J., 2017a. Microplastics pollution in inland freshwaters of China: a case study in urban surface waters of Wuhan, China. Sci. Total Environ. 575, 1369–1374.

Wang, J., et al., 2017b. Microplastics in the surface sediments from the Beijiang River littoral zone: composition, abundance, surface textures and interaction with heavy metals. Chemosphere 171, 248–258.
Yonkos, L.T., Friedel, E.A., Perez-Reyes, A.C., Ghosal, S., Arthur, C.D., 2014. Microplastics in four estuarine rivers in the Chesapeake Bay, U.S.A. Environ. Sci. Technol. 48, 14195–14202.
Zbyszewski, M., Corcoran, P.L., 2011. Distribution and degradation of fresh water plastic particles along the beaches of Lake Huron, Canada. Water Air Soil Pollut. 220, 365–372.
Zbyszewski, M., Corcoran, P.L., Hockin, A., 2014. Comparison of the distribution and degradation of plastic debris along shorelines of the Great Lakes, North America. J. Great Lakes Res. 40, 288–299.
Zhang, K., Gong, W., Lv, J., Xiong, X., Wu, C., 2015. Accumulation of floating microplastics behind the Three Gorges Dam. Environ. Pollut. 204, 117–123.
Zhang, K., et al., 2016. Microplastic pollution of lakeshore sediments from remote lakes in Tibet plateau, China. Environ. Pollut. 219, 450–455.
Zhang, K., et al., 2017. Occurrence and characteristics of microplastic pollution in Xiangxi Bay of Three Gorges Reservoir, China. Environ. Sci. Technol. 51, 3794–3801.
Zhao, S., Zhu, L., Wang, T., Li, D., 2014. Suspended microplastics in the surface water of the Yangtze Estuary System, China: first observations on occurrence, distribution. Mar. Pollut. Bull. 86, 562–568.
Zhao, S., Zhu, L., Li, D., 2015. Microplastic in three urban estuaries, China. Environ. Pollut. 206, 597–604.
Ziajahromi, S., Neale, P.A., Leusch, F.D.L., 2016. Wastewater treatment plant effluent as a source of microplastics: review of the fate, chemical interactions and potential risks to aquatic organisms. Water Sci. Technol. 74, 2253–2269.
Zubris, K.A.V., Richards, B.K., 2005. Synthetic fibers as an indicator of land application of sludge. Environ. Pollut. 138, 201–211.

CHAPTER 4

Occurrence, Fate, and Effect of Microplastics in Freshwater Systems

Dafne Eerkes-Medrano*, Richard Thompson†
*Aberdeen, United Kingdom
†University of Plymouth, Plymouth, United Kingdom

4.1 INTRODUCTION

The accumulation of small plastic items (<5 mm) in aquatic environments was first reported in marine settings in the 1970s. These included polyethylene and polypropylene pellets in New Zealand (Gregory, 1977); polystyrene spherules in the Northwest Atlantic (Carpenter et al., 1972); and high-density polyethylene, polymethyl methacrylate, polystyrene, and polyethylene pellets in the Mediterranean Sea (Shiber, 1979). At this time, the term "microplastic" had not been introduced, though plastics observed were <5 mm and of the size range now termed *microplastic*. In 2004, a paper by Thompson et al. described the accumulation of microscopic pieces of plastic and fiber around the United Kingdom. Using archived plankton samples, they showed increases in the abundance of these microscopic pieces from the 1960s to the 1990s. This study used the term *microplastic* to describe the tiny plastic pieces, and since then, there has been an exponential increase in the number of scientific papers on the topic of microplastics (Fig. 4.1).

In the literature on freshwater habitats, few studies on microplastics existed prior to the 21st century; Hays and Cormons (1974) and Moore et al. (2011) documented plastic particles (<5 mm) in the rivers of North America, while Faure et al. (2012) reported on microplastics in Lake Geneva. Since these early publications, the number of studies has increased rapidly, and microplastics are now documented in freshwater systems around the globe including lakes and rivers. This chapter discusses the latest research on the occurrence, fate, and effects of microplastics in freshwater systems and draws examples from the marine field where research of microplastics is more advanced.

4.2 SOURCES OF MICROPLASTICS

Microplastics are a very heterogeneous mixture of particles of differing shapes, sizes, colors, and polymers that originate from diverse sources (Fig. 4.2). Microplastics can be described as originating from *primary* sources if they are manufactured or used as small particles <5 mm in diameter. These include preproduction resin pellets, industrial

Microplastic Contamination in Aquatic Environments
https://doi.org/10.1016/B978-0-12-813747-5.00004-7

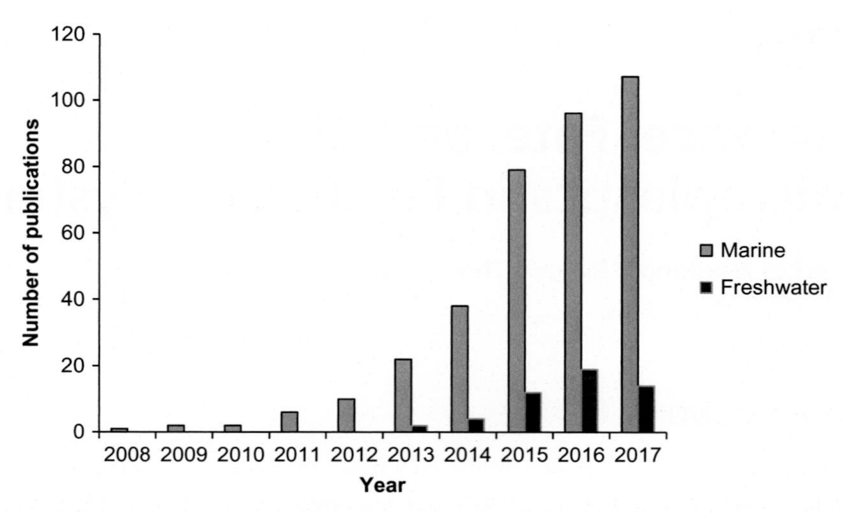

Fig. 4.1 Web of Science citation search on microplastic studies in marine (*light bars*) and freshwater systems (*dark bars*). Search terms used for marine studies were (marine) and (microplastic). Search terms used for freshwater studies were (freshwater) and (microplastic). Search was for all years covered by the database and search was conducted on 5 August 2017.

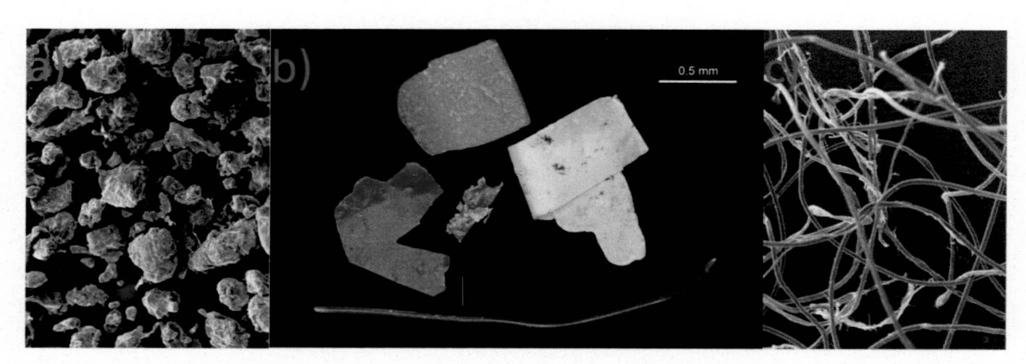

Fig. 4.2 Microplastics: (A) particles from a cosmetic product; (B) fragments of microplastics collected from a shoreline near to Plymouth, the United Kingdom; (C) polyester fibers released from clothing during laundering. *((A,C) Courtesy Napper and Thompson, Plymouth University Electron Microscopy Suite; (B) Courtesy Thompson, Plymouth University.)*

scrubbers for abrasive blast cleaning, and capsules and microbeads in personal care products (Fig. 4.2A). *Secondary* microplastics arise from the disintegration and fragmentation of larger plastics after they have entered the environment (Fig. 4.2B). Examples of secondary microplastics include fragments arising from plastic bottles, bags, and other packaging. A further category of microplastics is those generated as a consequence of product wear during a plastic item's service life, for example, fiber from textiles (Fig. 4.2C) or particles generated from tire wear.

The spectrum of uses and compositions of primary microplastics is varied; for example, they are used as raw material for the plastic industry (Lechner et al., 2014; Zbyszewski et al., 2014); abrasives in cleaning, personal care, and cosmetic products (Zitko and Hanlon, 1991; Gregory, 1996; Fendall and Sewell, 2009); and air-blasting media (Gregory, 1996). Paralleling the varied range of uses a wide spectrum of polymers are reported as primary microplastics. Plastic powders, pellets, or granules serving as raw materials for the plastic industry may consist of polyethylene, commonly used for making plastic packaging, or polyvinyl chloride and polypropylene, used to fabricate building and construction supplies and automotive parts (Vasile, 2000; PlasticsEurope, 2016). Microplastic beads used as abrasives in personal care and cosmetic products like shower gels, face wash, and liquid soaps are commonly made of polyethylene (UNEP, 2015; Gouin et al., 2015).

Plastic microbeads, used for personal care and cosmetic products, deserve discussion due to their widespread use (Fendall and Sewell, 2009; Gouin et al., 2015; Chang, 2015), numerical abundance (Napper et al., 2015) and their presence in freshwater and marine environments (Eriksen et al., 2013; Isobe, 2016). When plastic particles were first proposed for use in skin cleansers, they were suggested as an alternative to natural abrasives because they would be less harsh on skin, easier to flush away due to their lighter density, and would cause less wear to the plastic and metal containers that they were dispensed from (Beach, 1967). A recent study, based on an industry survey of major cosmetic companies operating in the European Union, reported that the typical quantity of polyethylene microbeads included in cosmetic products ranged from 0.05% to 12% with 70% of microbeads in a size range >450 μm (Gouin et al., 2015). A similar size range (250–420 μm) exists in US patented cleansers (Beach, 1967), though some products contain smaller particles (60 μm; Chang, 2015). A single 150 mL container of cosmetic product can contain around 3 million microbeads (Napper et al., 2015), and based on a Cosmetics Europe survey, the total volume of liquid soap used in 2012 by countries within the North Sea watershed was 6.88×10^8 L, with per capita consumption values of microplastic beads estimated at 17.5 mg day^{-1} (Gouin et al., 2015). Based on concerns over quantities of microbeads released into the aquatic environment, legislation has been introduced in Australia, Canada, and the United States, and the European Union is considering a ban (Rochman et al., 2016). Evidence of microplastic particles in the Great Lakes (e.g., Eriksen et al., 2013) contributed to many US states banning the sale and in some cases manufacture of products containing microbeads (Schroeck, 2016).

Marine literature noting *primary* source plastics in the environment suggests that primary microplastics enter aquatic systems in more than one way. Polyethylene, polypropylene, and polystyrene particles in cleaning and cosmetic products enter aquatic systems through household sewage discharge (Zitko and Hanlon, 1991; Gregory, 1996; Fendall and Sewell, 2009). Early investigations of industrial plastic pellets (polyethylene and polypropylene) on beaches suggested that likely sources were from plastic manufacturing through losses or spillage during shipment, handling, and unloading at ports (Gregory, 1978; Shiber, 1982). These early studies also suggested that pathways for primary microplastics to beaches were from spills, at inland factories, that made their way to coastlines via rivers, streams, and storm-water drains (Gregory, 1978). Similarly, in recent freshwater literature, great accumulations of virgin plastics have been found proximal to plastic production and processing plants along the Rhine River (Mani et al., 2015). In the Danube River, where industrial plastics at times comprised large proportions of the river load, plastic production sites adjacent to the river have been identified as potential sources of primary microplastics (Lechner et al., 2014).

Secondary microplastics arise from the disintegration and fragmentation of larger plastics once they have entered the environment. Examples of secondary microplastics include fragments arising from litter, for example, bottles, bags, and packaging. In the environment, plastics are exposed to environmental conditions that affect their properties in some of the following ways: (1) Air causes oxidation, (2) sunlight leads to UV degradation, (3) physical or mechanical forces can lead to fragmentation, and (4) thermal or chemical stress from the environment can lead to disintegration (Andrady et al., 1998, 2003; Vasile, 2000; Cooper and Corcoran, 2010; Lambert, 2013). A large plastic item in the environment, such as a discarded drinking bottle, fishing gear, or plastic film, can disintegrate into multiple smaller items. The slow degradation of plastics (e.g., years for carrier bags; O'Brine and Thompson, 2010) means that secondary source microplastics are likely to accumulate over long timescales, and even if all inputs of plastic to aquatic environments were to cease with immediate effect, an increase in microplastic particles would still occur as a consequence of the fragmentation of legacy items of large plastic already present. A 2017 study that sampled microplastics and modeled their distributions and transport in Lake Erie concluded that as a result of slow degradation rates of plastics and the lake's long hydraulic residence time, some microplastics in existence at Lake Erie are likely fragments from some of the very first plastic products that entered the consumer market (Cable et al., 2017).

Further descriptors of microplastic sources are those generated as a consequence of wear during a product's lifetime in service, as opposed to microplastics generated from discarded products. Examples are fibers from textile wear or particles from the use of tires. Some publications describe particles and fibers released from synthetic textiles as secondary microplastics (e.g., Duis and Coors, 2016). The release of these particles and fibers may occur before the microplastics enter the aquatic environment such as while washing clothes (Browne et al., 2011) or while textiles are used in the terrestrial environment (Dris et al., 2017). Microplastics that come from washing clothes are mainly polyester, acrylic, and polyamide and can reach >100 fibers L^{-1} of effluent (Habib et al., 1998; Browne et al., 2011). Other studies report maximum fiber loads released from washing of clothes as >700,000 fibers from a 6 kg wash of acrylic clothes (Napper and Thompson, 2016) and 7360 fibers m^{-2} in one wash of polyester fleece clothes (Åström, 2016). The main factor influencing fiber release appears to be the type of textile used in garments with those made from acrylic releasing more than other types (Napper and Thompson, 2016). Other factors that may affect fiber release are the use of detergent and conditioner (Åström, 2016; Napper and Thompson, 2016). Clothes receiving an abrasion treatment, at the Swedish school of textiles, to simulate age/wear exhibited greater fiber loss than new textiles (Åström, 2016). Researchers report that fibers can be numerous in sewage treatment plant effluents, sewage sludge, and sludge products depending on sewage treatment processes (Mintenig et al., 2017). If sludge is returned to the land as fertilizer, this can become a route for microfibers captured in wastewater treatment processes to reach land and eventually waterways via runoff. Hence, due to the various wastewater treatment processes and fates, once effluents leave sewage treatment plants, fibers can be detected in waterways and sediments in the environment (Habib et al., 1998; Zubris and Richards, 2005).

The third abovementioned category of microplastics are neither primary nor secondary; they arise as a consequence of wear during the product lifetime. This includes a wide assortment of microplastics including particles from plastic mulching, abrasion from car tires, fibers released from synthetic textiles, or abrasion of synthetic paints (Duis and Coors, 2016). While these plastics may be defined as secondary source microplastics, they are different from particles that arise due to environmental weathering and, in our view, therefore merit their own category to define their source.

4.3 OCCURRENCE OF MICROPLASTICS IN FRESHWATER SYSTEMS

Historically, microplastics in freshwater systems have received less attention than microplastics in marine systems (Wagner et al., 2014; Eerkes-Medrano et al., 2015). In the marine environment, microplastics have been described in a wide range of habitats: at the sea surface, on shorelines, in the deep sea, in sea ice, and in biota. They are reported from geographic locations around the globe, for example, from the tropics (Costa et al., 2010; Costa and Barletta, 2015) to the poles (Obbard et al., 2014), and widely reported in different depositional environments, for example, from pelagic open ocean habitats (Kooi et al., 2016), to deep-sea benthic habitats (Taylor et al., 2016), to coastlines of all continents (e.g., Browne et al., 2011; Isobe, 2016; Costa et al., 2010). Microplastics are even found within organisms far from the human presence, such as in deep-sea corals (Woodall et al., 2014). These widespread reports of microplastics in marine habitats demonstrate ubiquity and indicate the exponential increase in published research (Figs. 4.1 and 4.3).

In contrast, microplastics in freshwater systems have received less attention with the topic only coming into active discussion in the last 5 years. Spread over three decades from 1974 to 2012, the first studies in North American rivers (Hays and Cormons, 1974; Moore et al., 2011), a lake (Zbyszewski and Corcoran, 2011), and a European lake

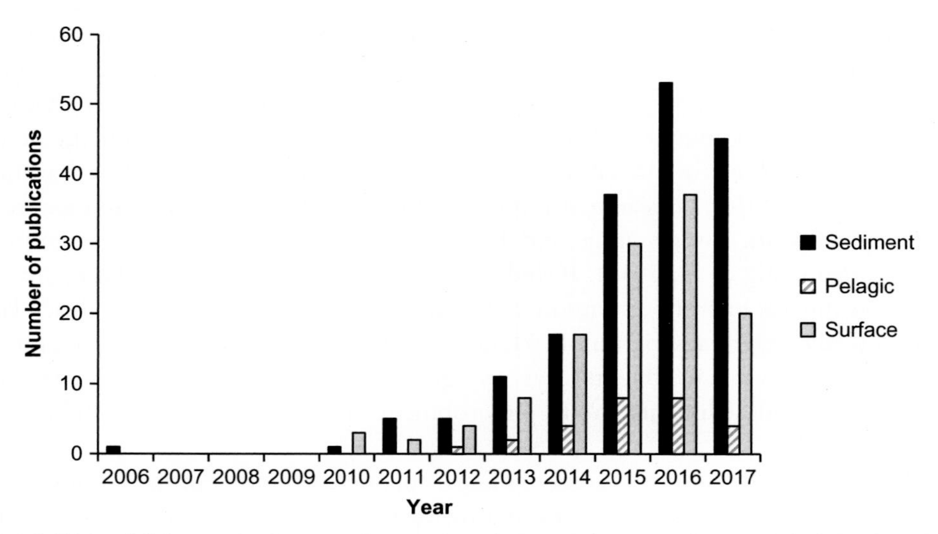

Fig. 4.3 Web of Science citation search on microplastic studies in sediments (*dark bars*), pelagic (*diagonal hatch*), and surface (*light shading*) marine systems. Search terms used for sediment studies were (marine) and (microplastic) and (sediment) not (pelagic) not (freshwater). Search terms used for pelagic studies were (marine) and (microplastic) and (pelagic) not (sediment) not (freshwater). Search terms used for surface studies were (marine) and (microplastic) and (surface) not (sediment) not (freshwater) not (terrestrial). Search was for all years covered by the database, and search was conducted on 5 August 2017.

(Faure et al., 2012) were published. The objective of some of the early studies was to consider sources of microplastics to oceans (and to coastal seabirds) and to determine whether freshwaters acted as upstream sources (Hays and Cormons, 1974; Moore et al., 2011; Faure et al., 2012). Hays and Cormons (1974), having previously discovered plastics in gull and tern pellets from Great Gull Island, New York, the United States, surveyed for possible sources of plastics from manufacturing plants in four eastern coastal states. By sampling river sediments near plastic factory sites on the Connecticut and Chicopee Rivers and by sampling sediments downstream at the mouth of the Connecticut river, they concluded that similarities between samples suggested plastic spherules from factory effluents could be carried to the river mouth and coastal waters (Hays and Cormons, 1974). Moore et al. (2011) quantified the plastic debris load of two major rivers draining from the Los Angeles basin watershed (California, the United States) and confirmed primary and secondary microplastics among the river debris. In Lake Geneva, between Switzerland and France, Faure et al. (2012) recorded the presence of primary and secondary microplastics and, along with Moore et al. (2011), emphasized that microplastics entered the environment upstream of oceans. Research on microplastics in freshwater systems is gaining momentum, and microplastics are now recorded in freshwater systems of several continents around the globe.

4.3.1 Microplastics in Rivers

The majority of the studies in rivers have been conducted in Europe and North America and to a lesser extent in Asia or South America. In Europe, microplastics have been found in the Danube River between Vienna and Slovakia (Lechner et al., 2014); in the Rhine River traversing Switzerland, France, Germany, and the Netherlands (Mani et al., 2015); in the German Elbe, Mosel, Neckar, and Rhine Rivers (Wagner et al., 2014); in the Seine and Marne rivers in France (Dris et al., 2015); and in various Swiss rivers (Faure et al., 2015). In North America, the studies refer to the Connecticut and Chicopee Rivers, the United States (Hays and Cormons, 1974); rivers draining from the Los Angeles basin watershed, California, the United States (Moore et al., 2011); the St. Lawrence River, Canada (Castañeda et al., 2014); the North Shore Channel in Chicago, the United States (McCormick et al., 2014); four tributaries of Lake Ontario (Ballent et al., 2016); 29 tributaries at six states bordering the Great Lakes (Baldwin et al., 2016); and Detroit and Niagara Rivers (Cable et al., 2017). In Asia, they have been recorded in the Yangtze and Hanjiang Rivers (Wang et al., 2017). Studies documenting microplastics in fish have also placed microplastics in various French streams (Sanchez et al., 2014); at the River Thames in England (McGoran et al., 2017); at various rivers in Texas, the United States (Phillips and Bonner, 2015; Peters and Bratton, 2016); and the Pajeú river in Brazil (Silva-Cavalcanti et al., 2017).

4.3.2 Microplastics in Lakes

Microplastics have been reported from lakes in Africa, Asia, Europe, and North America. The range of studies encompasses lakes of different characteristics such as the lake surface area, watershed populations, elevation, remoteness, and lake uses. The focus in North America has mainly been on the Great Lakes, which are characterized by large lake areas and watershed populations, and several of the Great Lakes have industrial activity (e.g., Zbyszewski and Corcoran, 2011; Eriksen et al., 2013; Zbyszewski et al., 2014; Corcoran et al., 2015; Ballent et al., 2016; Anderson et al., 2017; Cable et al., 2017). The lakes in Europe are comparatively small, and the associated human population sizes vary from the highly populated Swiss lakes, Constance and Geneva (Faure et al., 2012, 2015) to the less densely populated lakes, such as the Swiss Lake Brienz and Italian Lakes Bolsena, Chiusi, and Garda (Imhof et al., 2013; Faure et al., 2015; Fischer et al., 2016). In Asia, microplastics have been reported from the remote Lake Hovsgol of Mongolia (Free et al., 2014) and the remote lakes of the Tibetan Plateau, China (Zhang et al., 2016). They have also been reported from urban lakes of Wuhan, China (Wang et al., 2017) and at the third largest lake, Taihu Lake, in one of the most developed areas of China (Su et al., 2016). In Africa, the presence of microplastics in Lake Victoria was indicted by their presence in fish caught in that lake (Biginagwa et al., 2016).

Estuaries are a transition zone between freshwater and marine environments. As a result, these transition zones experience characteristics of both habitats. The presence of microplastics in these environments may be influenced by factors related to freshwaters (e.g., river currents and upstream population densities) and marine habitats (e.g., tides, wave action, and marine currents). There have been a number of studies on estuaries, for example, in the United Kingdom (Tamar estuary; Sadri and Thompson, 2014), Italy (Lagoon of Venice; Vianello et al., 2013), and South Africa (multiple estuaries; Naidoo et al., 2015). Some of these studies focus on multiple estuaries (e.g., Naidoo et al., 2015) and are therefore able to compare patterns of microplastic abundance in relation to factors such as the size of population centers close to the estuary or the level of water exchange with the ocean.

4.4 TYPES OF MICROPLASTICS OCCURRING IN FRESHWATER SYSTEMS

While the oceans are potentially a "mixing pot" for plastics from diverse sources, some *types* of microplastics found in lakes and rivers can be reflective of localized uses and activities of humans nearby or upstream. Associations between the types and concentrations of microplastics found and activities of the watershed populations offers an opportunity to study microplastic origins in freshwaters, where the dispersal distance, on a temporal or spatial scale, between the source and the point of sampling, may be shorter than that in the oceans. In riverine systems, the unidirectional flow also facilitates linking abundance to source, which inevitably lies upstream. Linking abundance to source has been possible in marine studies, when dispersal distances are short, for example, raw pellets on Costa del Sol attributed to plastic factories on the coast (Shiber, 1982), high concentrations of plastic pellets on New Zealand beaches of industrialized areas (Gregory, 1977), and the relatively large presence of microbeads at coastal areas in Japan attributed to regular loading of new microplastics from land areas (Isobe, 2016). In freshwaters, examples where microplastic types can give insights to watershed activities are the industrial microplastics in the Danube and Rhine Rivers of Europe (Lechner et al., 2014; Lechner and Ramler, 2015; Mani et al., 2015); resin pellets at Lake Huron, North America (Zbyszewski and Corcoran, 2011); microbeads in Lake Erie, North America (Eriksen et al., 2013); fragments suggested to arise from postconsumer products in Lake Hovsgol, Asia (Free et al., 2014); and high microplastic concentrations linked to high nearshore population density at the Great Lakes, North America (Cable et al., 2017).

Elaborating on the above examples, in the Danube River, 79% of plastics captured over a 2-year survey using drift samples (2010 and 2012) comprised industrial microplastics such as pellets (cylindrical in shape), flakes (3D, not films/foils), and spherules (Lechner et al., 2014). These were attributed to plastic manufacturing activities along the river's watershed (Lechner and Ramler, 2015). In an 820 km stretch of the Rhine River surveyed for floating microplastics, spherical microplastics (spherules) measuring

300–100 μm in size were cumulatively the most abundant particle type, representing 60% of total sampled microplastics (Mani et al., 2015). Their abundance was heterogeneous and peaked at the two sampling locations ($n = 11$) that accounted for >66% of all microplastics measured in the study. The second most common particle type in the Rhine study was fragments. The authors of the study emphasized the high numbers of plastic manufacturers and plastic processing sites in close proximity to the two sample sites with high microplastic concentrations dominated by spherules (Mani et al., 2015).

At Lake Huron, the resin pellets (<5 mm) dominating 94% of the plastic samples from Sarnia beach were attributed to the plastic industry activity of that city (Zbyszewski and Corcoran, 2011). Zbyszewski et al. (2014) noted that pellets may be transported to the lake via spillage during truck or rail transport or by falling into factory drains. In surface water samples from a different study of the Great Lakes, pellets and fragments were reported as the most abundant microplastics. Lakes Superior, Huron, and Erie were surveyed, but a breakdown of abundance for each particle type at each lake was not provided (Eriksen et al., 2013). Microplastic spheres <1 mm size, at a sampling site with high microplastic abundance from Lake Erie, had similar composition, size, shape, and color to microbeads from personal care consumer products. Eriksen et al. (2013) suggested residents of the watershed population for this site likely used personal care products with microbeads, and these traveled into the lake via escape from wastewater treatment routes, sewage sludge used as fertilizer in agriculture and public lands, or combined sewage overflow during rain events. A more recent study of microplastics in surface waters of the Great Lakes reported that secondary microplastics (fragments) were the most common type of microplastic in samples (Cable et al., 2017). Authors of this study linked the high microplastic load estimates, which at 2 million particles km^{-2} were the highest reported levels for the Great Lakes, with the high nearshore population densities of the lakes (Cable et al., 2017).

At more remote locations where population numbers and industrialization are reduced, studies report dominance of microplastic fragments, fibers, and films. Surface water samples at a remote lake in the mountains of northern Mongolia and sediment samples from lakes of the Tibetan Plateau were dominated by fragments, films, lines/fibers, and foams (Free et al., 2014; Zhang et al., 2016). Weathering of these microplastics indicated that they likely originated from breakdown of larger products (Zhang et al., 2016) such as the macrodebris (bottles, bags, packaging, and fishing gear) recorded on lake shorelines (Free et al., 2014). Microbeads and virgin plastic pellets were not observed in these Asian lakes, suggesting that such contributions from industry and personal care products were largely absent (Free et al., 2014; Zhang et al., 2016). Indeed, these remote lakes have limited human activity and an absence of industry or agriculture in their watersheds. Despite this, they had microplastic levels comparable with those reported for urbanized lakes, and the authors speculated this may be due to the lack of modern waste management systems and lack of wastewater and sewage systems. In a study of the relatively remote Lake Garda, in Italy, microplastics recorded from beach sediment samples included polystyrene from foam, polyethylene foils, thin polyethylene fragments, rigid polyethylene fragments, polypropylene foils, thin fragments, and rigid fragments of unknown sources (Imhof et al., 2013). Only a small number of raw pellets were reported. The authors of this study speculated that microplastics reflected the lakeside activities and uses, such as recreational tourism (water sports, tourist boats, and picnics) and fishing, and speculated that few microplastics were from land-based sources (Imhof et al., 2013).

Rates of environmental fragmentation and degradation are unknown in marine systems (Law and Thompson, 2014) or freshwater systems. There may be greater physical forces that affect microplastics, such as storms and wave action in marine systems, but plastics in freshwater systems still experience physical and chemical degradation (Andrady, 2011). Free et al. (2014) investigating microplastics in Lake Hovsgol suggested that particles may experience relatively high levels of weathering due to high levels of UV light penetration and reduced biofouling in the oligotrophic waters. At lakes in the high Tibetan Plateau, reduced abundance of the smallest size ranges of microplastics was suggested to result from increased UV radiation and high diurnal temperature ranges that led to faster degradation of the smallest size classes (Zhang et al., 2016).

In contrast to the Danube and Rhine Rivers, where researchers related high concentrations of primary microplastics to industrial activities near the studies' sampling locations (Lechner et al., 2014; Lechner and Ramler, 2015; Mani et al., 2015), at other rivers and lakes where surface waters have also been sampled (e.g., tributaries of the Great Lakes, Canada; the Yangtze and Hanjiang Rivers, China; and the Seine River, France), samples have been dominated by fibers (Dris et al., 2015; Baldwin et al., 2016; Ballent et al., 2016; Fischer et al., 2016; Su et al., 2016; Anderson et al., 2017; Wang et al., 2017). Fibers have also been reported as dominant particle types in sediment samples (e.g., Ballent et al., 2016; Su et al., 2016). Researchers discuss potential origins for the fibers based on the chemical composition of fibers; for example, polyethylene terephthalate (PET, aka polyester) is often used in the production of textiles like clothes, blankets, fleece, and furniture, and polypropylene (PP) is used in furniture, geotextiles, carpets, curtains, textiles, sporting equipment, rucksacks, cuddly toys, buildings, agriculture, fishing nets, and ropes (Åström, 2016; Dris et al., 2017; Wang et al., 2017). In some studies, fibers have been suggested to come from domestic sources such as washing machine effluents (Su et al., 2016), fishing gear such as nets and ropes (Wang et al., 2017), sewage plant effluents and surface runoff (Fischer et al., 2016; Wang et al., 2017), and atmospheric fallout on the catchment (Dris et al., 2015). It has also been suggested that fibers are released into the air (Dris et al., 2015, 2017) and eventually accumulate in water bodies through atmospheric deposition (Dris et al., 2015). Many studies reporting high abundances of fibers have watersheds that are urbanized, experience agriculture, and/or have large populations (Dris et al., 2015; Mani et al., 2015; Baldwin et al., 2016; Su et al., 2016;

Wang et al., 2017). In some studies, researchers also emphasize the lack of association between high fiber concentrations in microplastic samples and watershed attributes. For example, at tributaries of the Great Lakes ($n=29$ tributaries), the dominant fibers/lines were not correlated with any watershed attributes, but there were significant correlations between urban land use of a watershed; watershed population density; and concentrations of fragments, foams, pellets/beads, and films in samples (Baldwin et al., 2016). Ballent et al. (2016) stated that this lack of correlation between the fiber presence and hydrology or watershed attributes reflects the need for further study and understanding of the sources of fibers (Baldwin et al., 2016). Cable et al. (2017), in their study of microplastic concentrations and distributions of the Great Lakes, excluded fibers from total particle counts, as fibers could not be quantified with equal confidence across size fractions; for example, fiber count data were influenced by size class of the particles, oxidative treatment of the sample, and effort of sample sorter.

Hidalgo-Ruz et al. (2012) reviewing methods for identification and quantification of microplastics in the marine environment noted that, for seawater samples, the reported microplastic abundances and types are likely related to the mesh size of the net used for collection; an 80 μm net could retain up to 100,000 times higher concentrations of fibers than a 450 μm net. Researchers investigating microplastics in the Seine River obtained higher microplastic concentrations from tows taken with a plankton net (80 μm mesh) than with a manta trawl (330 μm mesh) (Dris et al., 2015). In the latter study, retention of different types of plastics was also affected by sampling device; the 80 μm plankton net captured more fibers, while the 330 μm manta trawl captured a greater diversity of microplastic shapes and types. Wang et al. (2017), who also used a small mesh size (50 μm) in their collection of surface water samples (20L) from 19 lakes and 2 rivers in Wuhan, China, also found fibers dominated their samples. Dris et al. (2015) suggested that the lack of fragments or beads from 80 μm plankton net samples in the Seine may have been partially due to a short exposure time and limited volume sampled. Nevertheless, high fiber abundances have also been reported by studies sampling surface waters with 330 μm mesh nets (Baldwin et al., 2016; Fischer et al., 2016; Su et al., 2016). Baldwin et al. (2016) speculated that using wet peroxide oxidation (WPO), rather than salt water flotation, to isolate plastics in samples may have also led to high fiber abundances. Wang et al. (2017) sampled with a smaller mesh size (50 μm) and used the WPO method and also found high fiber abundance in samples; the WPO method is cited as more effective in capturing dense particles (Baldwin et al., 2016). However, Cable et al. (2017), who sampled surface waters of the Great Lakes with a 100 μm manta trawl and used WPO to process samples, found fragments rather than fiber-dominated samples.

In marine systems, there may be patterns in the types of plastics sampled from surface waters versus sediments, for example, fibers reported with the significant presence in sediments (Thompson et al., 2004; Woodall et al., 2014) and fragments reported in surface waters (Eriksen et al., 2014; Kooi et al., 2016). Whether this pattern also shows up in freshwater systems and whether it is a reflection of true abundances or a reflection of different sampling approaches are uncertain. Surface waters are often sampled with a net, while sediment sampling approaches tend to use bulk sampling. Recent studies of buoyant plastics in the Atlantic Ocean may reflect the influence of mesh size on the types of particles that dominate samples. Kanhai et al. (2017) found fibers dominated North Atlantic seawater samples collected on a 7345-nautical-mile transect via a pump and sieved at a minimum size fraction of 250 μm. In contrast, Kooi et al. (2016), using a minimum sieve size fraction of 500 μm, found that fragments dominated net tows of the North Atlantic accumulation zone. There may also be differences that arise from interacting factors such as particle properties, like density, and water-body hydrology. Cable et al. (2017) discussed the possible reasons behind another study finding that fibers dominated surface samples from 29 Great Lake tributaries (Baldwin et al., 2016), acknowledging that these tributaries provide source waters to the Great Lakes, while their own study found fragments dominated surface waters of the Great Lakes. They suggested that fibers are made of denser polymers and therefore remain suspended in turbulent environments like rivers but sink in the stable waters of a lake (Cable et al., 2017). This suggestion could be supported by previous findings from Ballent et al. (2016) who looked at microplastics in sediments of tributaries and nearshore environments of Lake Ontario. They found a higher proportion of fibers to fragments in nearshore sediments than in tributary sediments.

The above discussions reflect the variability across studies in the types of particles that are reported to be dominant in a water body. In some studies, there are suggestive associations between the types of microplastics that are dominant and human activity near the water body (e.g., Lechner et al., 2014; Mani et al., 2015), but in other cases, there are no clear associations (e.g., in the case of fibers; Baldwin et al., 2016). Speculative reasons for dominance of certain particle types in samples and possible explanations for contrasting results when the same water body is sampled include the use of different sampling devices (Dris et al., 2015), the use of different sample processing (Baldwin et al., 2016), and interactions between particle properties such as density and water-body hydrology (Cable et al., 2017). Taken together, these studies reflect the reality that there is still a lot of uncertainty and more research is needed to understand the link between dominance of certain microplastic particle types in a water body and their likely sources. Resolving this issue may require combined approaches, as suggested by Dris et al. (2015), for detecting a wider range of particle types and sizes.

Given the challenge in establishing the origin of particles once they enter the environment, some freshwater studies have suggested the use of surface degradation patterns such as cracks, fractures, pits, and adhering items to understand a particle's history (Zbyszewski and Corcoran, 2011; Imhof et al., 2013; Zbyszewski et al., 2014; Zhang et al., 2016). It has been suggested that surface patterns, observed with scanning electron microscopy, reveal whether particles experienced mechanical degradation (e.g., wave action and sand friction) and oxidative weathering (e.g., photooxidation from UV-B exposure) or if particles encounter depositional environments (e.g., sandy beaches vs muddy organic-rich shorelines) that marked their surface (Zbyszewski et al., 2014; Zhang et al., 2016). Zbyszewski and Corcoran (2011) studying microplastics from Lake Huron suggested that a high degree of oxidation in polypropylene fragments could indicate a long residence time in the environment and the high mechanical abrasion and stress fractures of polyethylene pellets could suggest abrasion by sand of fragments already exposed to oxidation processes. Degradation patterns may be important to consider since shape and texture, along with size and density contribute to the way particles interact with factors affecting their presence in the environment (Ballent et al., 2012). The use of surface features may be indicative of the sources and history of a fragment, but there is a lack of comparative studies to confirm this, and many of the reported comments need to be considered as speculative at present.

4.5 CONDITIONS INFLUENCING THE QUANTITY OF MICROPLASTICS IN THE ENVIRONMENT

A variety of factors can influence the *quantity* of microplastics present in the environment. These include human population density proximal to the water body, proximity to urban centers, proximity to major river inflows, precipitation events, movement of surface water currents, water residence time, size of the water body, the type of waste management used, amount of sewage overflow, and atmospheric depositions (Moore et al., 2011; Zbyszewski and Corcoran, 2011; Eriksen et al., 2013; Free et al., 2014; Faure et al., 2015; Dris et al., 2016; Cable et al., 2017). Many of these factors are linked to population density at or near the freshwater body. In a 2013 study of the Great Lakes of North America, the highly populated Lake Erie had microplastic particle counts (466,305 particles km^{-2}) that were ~70 and ~38 times greater than those of the less populated Lakes Huron (6541 particles km^{-2}) and Superior (12,645 particles km^{-2}) (Eriksen et al., 2013). The overall trend, of high microplastic concentrations at Lake Erie, holds in 2017 empirical and modeling studies where plastic concentrations from Lake Erie were 4- and 80-fold higher than Lakes Huron and Superior, respectively (Cable et al., 2017; Hoffman and Hittinger, 2017). Within Lake Erie, sites near cities had more microplastics, including more pellets compared with rural shorelines (Zbyszewski et al., 2014). In the most recent study of the Great Lakes, the highest concentrations of microplastics (2 million particles km^{-2}) came from urban sites at Lake Erie or sites at the Detroit River plume that feeds into it (Cable et al., 2017). At Lake Huron, North America, and at Lake Hovsgol, Mongolia, the microplastic counts decreased along a south-to-north gradient, with highest abundances at the southern shores where the lakes experienced greater industrial activity and tourist, respectively (Zbyszewski and Corcoran, 2011; Free et al., 2014). In Lake Winnipeg, temporal variation in microplastic densities has been suggested to result from interannual changes in inflow from the Grand Rapids River (Anderson et al., 2017). This river supports the largest populations in the watershed (Anderson et al., 2017). In Swiss lakes, at Lake Garda, Italy, and at Lake Ontario, Canada, high microplastic concentrations on beaches have been found close to major river inflows (Imhof et al., 2013; Faure et al., 2015; Ballent et al., 2016).

Authors of freshwater studies have observed an association between higher microplastic concentrations and rainfall events (e.g., Moore et al., 2011; Corcoran et al., 2015; Faure et al., 2015). Greater microplastic concentrations recorded in surface waters of Swiss lakes and in sediments of Lake Ontario, Canada, following rainfall events and following rain events with greater water levels have been attributed to increased flow rates of rivers and greater loads of microplastics in the rivers (Corcoran et al., 2015; Faure et al., 2015). In the Swiss study, urbanized rivers, such as the Vauchère urban stream that is fed by street runoff, had particularly high microplastic abundances following rain events. Surface samples collected from the Vauchère urban stream in rainy periods contained microplastic abundances that were 150 times greater than abundances in dry periods; the less urban Venoge River had microplastic abundances that were nine times greater in wet versus dry periods (Faure et al., 2015). During wet periods, microplastics may enter these rivers from urban runoff and sewer overflows (Faure et al., 2015). In a study measuring microplastic load in the Los Angeles River, San Gabriel River, and Coyote Creek of California, the United States, during wet and dry periods, the greatest microplastic abundance occurred after rainfall. Another observation from this study was a change in the relative difference between the plastic item size fractions recorded from the river waters. Microplastics (particles <5 mm in size) became 3–16 times more abundant than larger particles (>5 mm) following rain events (Moore et al., 2011). Moore et al. (2011) concluded that rain events contribute to storm-water loading of plastic debris in rivers and suggested that a river's flood stage would also be linked to microplastic abundances (Moore et al., 2011).

Another factor suggested to affect microplastic concentrations in freshwater bodies is atmospheric fallout of plastic particles (Dris et al., 2016). A yearlong study measuring atmospheric fallout from a dense urban environment and a less dense suburban environment reports high annual variability of fibers from atmospheric fallout and a significantly higher fiber abundance in the urban environment. There was also an association between rainfall and the flux of fibers in atmospheric fallout; fiber abundances were high during rainy weather. In this study, the measured fibers included both natural and synthetic; half of the fibers were natural; the remaining were man-made including 29% that contained plastic polymers. Dris et al. (2016) suggested that the source of synthetic fibers could be from clothes, houses, degradation of macroplastics, landfills, or waste incineration. They estimated that 3–10 t of fibers could be deposited annually from the atmosphere at the scale of the Parisian agglomeration. They suggested that atmospheric fibers may be deposited on cities, agrosystems, and freshwater environments (Dris et al., 2016), and others suggest that fibers in the atmosphere could be transported by wind to more remote freshwater environments (Free et al., 2014).

In some lake studies, microplastic abundances are high despite the lakes having smaller populations and less industrialization or being remote. Examples include high pelagic microplastic abundance in the Swiss Lakes Geneva and Maggiore (Faure et al., 2012, 2015) compared with Lake Huron despite Lake Huron's population being more than three times the size of the Swiss lakes (Eriksen et al., 2013) or high microplastic levels, comparable with those reported for urbanized lakes, in the very remote lakes in Mongolia and Tibet (Free et al., 2014; Zhang et al., 2016). Authors speculating on factors, other than population density, that likely affect microplastic abundance in lakes suggest that the conditions that may concentrate particles are long water residence time, large catchment areas relative to a small lake size, riverine inputs, lakes that form closed drainage basins, and the presence or absence and type of waste management systems and wastewater and sewage management systems used by watershed populations (Free et al., 2014; Faure et al., 2015; Zhang et al., 2016).

Industrial and domestic wastewater is considered to be an important source of microplastics to marine and freshwater systems. Wastewater treatment plants (WWTPs), when present and depending on the type of treatment, can remove significant amounts of microplastics from wastewater (up to 97% to >98% removal), but due to the large volumes of wastewater processed by WWTPs, authors still report large quantities of microplastics released into effluents daily (e.g., 65 million microplastics) (Murphy et al., 2016; Mintenig et al., 2017). The quantities of microplastics that escape WWTP processing vary in size and type of microplastic. In a study of effluents from 12 WWTPs in Germany, microplastic abundances released within >500 μm size fraction ranged from 0 to 50 particles m^{-3}, while in the size fraction of particles <500 μm in size, particle release ranged from 10 to 9000 particles m^{-3}; polyethylene was the most common polymer of particles in both size fractions. Fibers, mostly of polyester, were abundant with 90–1000 particles m^{-3} (Mintenig et al., 2017). In the Seine River, France, raw wastewater samples taken from the Seine Center WWTP had microplastic concentrations of 260,000–320,000 particles m^{-3}, and these consisted of fibers (Dris et al., 2015). Following primary treatment of lamellar settling at the Seine Center WWTP, concentrations reduced to 50,000–120,000 particles m^{-3}, and the size of fibers in samples decreased indicating that the treatment process removed particles >1 mm in size. After undergoing further treatment, concentrations reduced to 14,000–50,000 particles m^{-3} with the largest fibers (>1 mm) eliminated entirely (Dris et al., 2015). In the latter study, Dris et al. (2015) only found one spherical particle in raw wastewater and stated that their study was not able to quantify how effective the WWTP was in removing this type of microplastic (Dris et al., 2015). In addition, even for particles captured in sewage sludge, there is the potential for reentry into the environment if the sludge is returned to the land or dumped at sea (Browne et al., 2011).

Microbeads have been found in wastewater effluent in the Great Lakes (Nalbone, 2015) and in wastewater effluents from 17 WWTPs across the United States (Mason et al., 2016). Microbead abundances are comparatively low in WWTP effluents relative to other microplastics (fibers and fragments), but total discharge can be considerable when daily discharge rates of WWTP are estimated; for example, it is estimated that US municipal treatment facilities release 3–23 billion microbeads day^{-1} (Mason et al., 2016). Distance from point source also affects particle abundance and can also affect sizes of microplastics. Habib et al. (1998) suggested that with increasing distance from WWTPs fiber size decreases (Habib et al., 1998). This may be because lighter particles travel further away. Another possibility is if larger fibers become denser due to clays adsorbing to their greater surface area (Corcoran, *pers. comm.*). In sediment samples from Lake Ontario, many microplastics had claylike particles adhered to their surfaces, especially if they had textured, irregularly shaped, or graded surfaces (Ballent et al., 2016).

4.6 FACTORS AFFECTING DISPERSAL OF MICROPLASTICS

Physical forces influence the *dispersal* of microplastics, and knowledge of wind and water movement is key for estimating transport of microplastics at a range of spatial scales. These forces vary according to spatial scale. For example, in the marine realm, forces observed and modeled to act on large scales are wind-driven surface currents and geostrophic circulation (Law et al., 2010). At smaller scales, experiments and modeling indicate that turbulence and wind-driven mixing can influence the movement of particles through the water column (Ballent et al., 2012; Kukulka et al., 2012); models indicate that turbulence can resuspend particles from the benthos (Ballent et al., 2012, 2013). On beaches and within sediments, field observations suggest that high-energy forces such as storms can influence the accumulation of microplastics and influence the three-dimensional deposition of microplastics within sediments (Shiber, 1982; Turra et al., 2014). Such external forces influencing the movement of particles interact with other environmental properties (e.g., seawater density and seabed). Physical forces that influence microplastic dispersal and transport in freshwaters are likely similar to those of marine waters, though in rivers and streams, there will also be the influence from forces and interactions associated with unidirectional flow.

Currents, convergence zones, and wave action are some of the forces that are likely to influence microplastic dispersal, transport, and their eventual concentrations in lakes. Wind-created surface circulations could result in patterns of high concentrations, the absence of microplastics, and concentration gradients that relate to distance from suspected microplastic sources (e.g., industrial sectors and higher population centers) (Zbyszewski and Corcoran, 2011; Imhof et al., 2013; Free et al., 2014; Cable et al., 2017; Hoffman and Hittinger, 2017). In other lake studies, microplastic accumulations have been speculated to occur by internal currents, wave action, converging currents, and/or gyres (Eriksen et al., 2013; Ballent, 2016; Anderson et al., 2017). Conversely, high winds that cause vertical mixing may lead to reduced microplastics in surface samples (Faure et al., 2015). Similarly, in sediments, the combined interaction of wave action and particle density may lead to fewer particles when nonbuoyant particles that are less dense than sand get suspended by high wave activity that carries particles away (Ballent, 2016). Authors have suggested that the abundance of microplastics in lakeshore sediments can be linked to the presence of large organic debris (e.g., weeds), which can trap microplastics and move them when the debris is transported by physical forces (Corcoran et al., 2015).

One of the main physical factors associated with microplastic transport in rivers is water flow (e.g., Lechner et al., 2014; Dris et al., 2015). Studies of microplastics in rivers suggest that sinks and retentions of particles can be due to a variety of factors including turbulence, still waters, and drift toward riverbanks (Mani et al., 2015; Klein et al., 2015). Authors suggest that microplastics may behave like fine sediments, becoming concentrated in low-energy environments and being absent in high-energy environments (Castañeda et al., 2014). In the case of sediments, grain size sorting and deposition patterns are affected by vertically varying water velocity and by structures of the riverbed that interact with the overlying water flow (Rice and Church, 2010; Bouchez et al., 2011; Ma et al., 2017). These physical forces can lead to concentration and size gradients of suspended particulate matter in the water column, and water velocity over the bottom can affect the efficiency of turbulent eddies that resuspend bed material. This in turn can affect size and concentration gradients of particulate matter in the water column (Bouchez et al., 2011). Particle size and density also determine whether particles are suspended or enter the riverbed and whether or not they are affected by local physical forces (Ma et al., 2017). Ballent et al. (2016) suggested that the high proportion of fibers to fragments of nearshore sediment samples in Lake Ontario in contrast to tributary sediments was due to fibers, as lighter particles, being transported in suspension in tributaries, versus fragments, as denser particles, having settled sooner or transported as bedload. River features and physical structures that create higher elevation, such as weirs, attract small particles due to reduced flow (Rice and Church, 2010; Nizzetto et al., 2016). These particles can then settle out. This has been suggested as the reason for changes in floating microplastic concentrations along the Rhine River. Mani et al. (2015) suggested that reduced surface microplastic concentrations were caused by a low slope in the river that led to sedimentation and by weirs that served as sinks for the lightweight particles. In estuaries, especially those with partially enclosed lagoons, low flow and stratification may lead to particle retention and deposition (e.g., Durban Bay estuary; Naidoo et al., 2015). This is supported by a model study of microplastic movement in the River Thames (Nizzetto et al., 2016). In marine settings, however, a large-scale study of beach sediments from six continents found no significant association between microplastic abundance and mean size of natural particulates (Browne et al., 2011). Once particles end up on the river bottom, the amount of time they spend in the substrate and the depth at which they may be buried can affect whether particles are mobilized again. There can be an active zone, within a certain depth from the surface of the substrate, from which particles are more likely to be entrained again (Hassan et al., 2013). Interactions between environmental factors (physical and nonphysical) and properties of the plastics themselves can influence patterns in microplastic distribution that might appear counterintuitive; therefore, an understanding of such factors is useful for making sense of patterns. For example, in a study of sediment microplastics in the Rhine, correlations between particle abundance and sites with higher population density, industrial activity, or the presence of sewage treatment plants were obscured by hydrodynamic effects (e.g., channel currents, channel geometry, stagnant water zones, and flood events) on microplastic distributions (Klein et al., 2015).

In both rivers and lakes, properties of microplastics themselves, such as the size, the shape, and the type of plastic, are all likely to affect the buoyancy and vertical distributions of microplastics (Kooi et al., 2016; Cable et al., 2017). These properties when interacting with physical factors in the environment (such as different currents in the water column) can in turn affect the residence time, transport, and dispersal of particles between environmental compartments. Authors have combined information from the marine literature, on the physical properties (size, density, and shape) of microplastics, with mathematical principles of geometry and dynamics to substantiate the point of how microplastic density affects whether particles have greater horizontal transport (low-density microplastics) or vertical transport (high-density microplastics) (Chubarenko et al., 2016). In freshwater studies, authors have suggested that changes to sedimentation of particles could also occur if particle density is altered; for example, increased sedimentation could occur through adsorption of minerals (clay or quartz) to particle surface, through fouling of microplastics by organisms, or if tidal exchange at a river mouth brought incoming particles from high-salinity to lower-salinity waters (Corcoran et al., 2015; Mani et al., 2015; Kowalski et al., 2016). Authors also suggest that particle density affects residence time of lakes. In Lake Erie, plastic transport models, floating particles flushed from the lakes approximately five times faster than neutrally buoyant particles (Cable et al., 2017). Particle shape and texture can also influence transport. Roundness increases settling velocity, and particles with surface characteristics that are conducive to biofouling, such as polyethylene fibers, spend less time in surface waters (e.g., 6–8 months) than spherical particles (e.g., 10–15 years) (Chubarenko et al., 2016; Kowalski et al., 2016). A common conclusion from marine studies of microplastics is in the importance of incorporating physical and geophysical forces to quantify microplastic movements in a water body and in the importance of field observations, laboratory experiments, and models for gaining this knowledge (Kukulka et al., 2012; Chubarenko et al., 2016; Kooi et al., 2016; Kowalski et al., 2016).

4.7 DETECTING AND QUANTIFYING MICROPLASTICS IN THE ENVIRONMENT

With the increase in microplastics research, there has been continued investigation into methods for observational and experimental studies in both field and laboratory settings.

The topic of methods for microplastic detection, quantification, and monitoring is large enough to merit its own chapter and will not be covered in detail here. Neither will the topic of methods for studies of microplastic interactions with biota. There are now a number of useful studies and reviews investigating and contrasting methods on these topics, and these may be referred to for further coverage. Examples of such papers include a review of methods for identification and quantification of microplastics in the environment (Hidalgo-Ruz et al., 2012); a review of techniques for extracting microplastics from sediments (Van Cauwenberghe et al., 2015a); a review of impacts of microplastics on marine organisms (Wright et al., 2013a); and a review of methods for extraction, identification, and quantification of microplastic particles ingested by biota Lusher et al. (2017). There are also useful summary documents prepared by expert task groups, such as the guidance report produced by the MSFD Technical Subgroup on Marine Litter (Galgani et al., 2013), which has a chapter on microliter, or the NOAA methods manual for measuring microplastics in the marine environment (Masura et al., 2015). These documents can provide valuable information on the range of methodological considerations in monitoring (e.g., beach litter, floating litter, and litter in biota). The majority of these reviews and recommendation documents are based on a marine knowledge base and cover methods/recommendations for that research direction. International recommendation reports on methods aimed at the issue of microplastics in freshwaters are not yet available, though freshwater studies such as Great Lake study by Cable et al. (2017) are using the NOAA Marine Debris Technical Memorandum (Masura et al., 2015). For freshwaters, the lack of coordination across methods challenges efforts for comparable assessment of litter (González et al., 2016).

The need for standardization. Within the past decade, there's been an exponential growth in peer-reviewed publications of microplastics, but there are challenges in inter-study comparison due to the lack of standardization among studies (Van Cauwenberghe et al., 2015a). The concept of standardization needs to apply across all aspects of methodologies from the sampling techniques to the sample processing techniques to the analyses of results. In most freshwater studies, the consistent use of a manta trawl to sample surface microplastics (even though some studies use timed tows and others use distance tows) allows for an easier comparison between studies. At the same time, there are limitations in the approach such as the following: The mesh size may miss certain sizes of microplastics (Klein et al., 2015; Anderson et al., 2017), mesh size is not necessarily the same across studies (e.g., Eriksen et al., 2013 vs Cable et al., 2017), organic matter and sediments in freshwater environments can lead to clogging of nets (González et al., 2016), and prevailing weather conditions can lead to variability in results (e.g., Mani et al., 2015). The approach for sampling sediment microplastics has been more variable across studies (e.g., variation in the sampling devices and the quantities of sediment sampled). This is something that investigators have been cognizant of; for example, Faure et al. (2012), in one of the first studies of lake sediment microplastics, identified that bias in sample collections could be due to the size and number of samples collected and suggested that the collection could be improved by collecting sediments from a greater surface area and taking more systematic sampling. A useful example of how to direct efforts toward standardization of measuring approaches is in the report by the Technical Subgroup on Marine Litter (Galgani et al., 2013) that provides concise summaries of recommended methodologies at each stage of measuring marine littler.

What metrics of microplastic presence should be measured and reported. In cases of detecting and measuring the microplastic presence in the environment, until approaches become standardized in what is measured (e.g., rates and number density) and in the units used to report their findings, it is essential that methods are clearly reported. It may also be useful for multiple measurement metrics to be taken when time and funding allows. For example, Faure et al. (2015) reported both rates (particles per hour and milligrams per hour) and number densities (particles per meter cubed and milligrams per meter cubed) of microplastics in rivers. They also extrapolated densities as the number of particles per unit area (particles km^{-2}) and mass of all particles in sample per unit area (e.g., mg m^{-2}) since these are the units commonly used in the microplastic literature (Faure et al., 2015). Reporting per unit volume gives a concentration, whereas reporting in 2D is easier in relation to some sampling approaches but has limitations in the value of the data. Faure et al. (2015) approach to reporting multiple measurement metrics has two benefits: (1) Since a consensus has not been established on what aspects of the microplastic presence should be measured to determine environmental status and impact and since there is as yet no standardized approach to reporting the microplastic presence in freshwater studies, Faure et al. (2015) approach to reporting rates and number densities allows wider comparison across studies and makes the information available in more than one way until standardized monitoring and environmental monitoring practices become established. (2) The second benefit to the approach by Faure et al. (2015) was that measuring the presence in different ways (counts and mass) gave different insights into the dominance of microplastics in the environment. For example, while pellet counts had a small presence in Swiss lakes, their presence was significant when their mass was considered (Faure et al., 2015).

It should be noted that the study of microplastics is still in its infancy, and while there is a desire to monitor abundance in order to detect trends, environmental monitoring is typically linked to materials or substances of concern. As yet the types of microplastics that might present harm are not clear and since differing methodologies will bias the type of microplastic detected, there is some justification in not adopting standard methods at the present time. The key consideration in any study must be to ensure it is linked to a question and that the methodology is appropriate to that. For example, it may be appropriate to monitor the abundance of microbeads following legislation, and sampling methods and locations can therefore be optimized for this specific question. At the present time, it would be difficult to advocate a monitoring approach that could directly link to impacts.

Definition of the microplastic size range. Microplastics are commonly defined as particles measuring <5 mm, but this is not consistently applied, and there has been discussion about the definition of the size range (Costa et al., 2010; Van Cauwenberghe et al., 2013; Gouin et al., 2015) and whether alternate classifications should be applied. Of key importance is that studies report size range of particles measured and quantified and the instruments used to sample microplastics (e.g., sampling with a corer, pump, sieve, or net). At present, no method will reliably quantify all size ranges, so the best that can currently be achieved is a consistent approach to give an index of microplastic burden. An upcoming challenge will be in practices for measuring and reporting the smallest size ranges. For example, particles in the nanosize range are yet to be reported but are likely to exist in the environment.

Factors influencing quantities recorded and the importance of metadata to accompany monitoring. For monitoring and management purposes, it is important to consider the journey of microplastics. This includes where microplastics enter the environment (site/location) and whether particles remain and/or are transported elsewhere. It is useful to collect information on factors that may influence quantities recorded and metadata that can give insights on the particle's journey. For example, information such as time of year or season reveals whether physical forces (winds, storms, rainfall, surface currents) and/or human activities (tourist activity, sediment dredging, and beach cleaning) may be dominant factors influencing the microplastic presence and journey in the environment (e.g., Shiber, 1982; Mani et al., 2015).

4.8 INTERACTIONS BETWEEN BIOTA AND MICROPLASTICS

There are various ways in which organisms may interact with microplastics. The first interaction is biofouling, which involves colonization of the particle surface by a biofilm (Reisser et al., 2014). Colonization may lead to transport of biota and/or invasive species (Carpenter and Smith, 1972; Reisser et al., 2014). Another early interaction is ingestion (e.g., oysters, mussels, De Witte et al., 2014; Van Cauwenberghe and Janssen, 2014). Unlike macroplastic debris, microplastics do not cause known effects from entanglement (Gregory, 2009). Other possible interactions could be stress responses such as inflammation or oxidative stress at tissue and cellular levels (Wright et al., 2013b; Jeong et al., 2017). Microplastics may also be a medium to expose biota to environmental contaminants adsorbed to their surfaces (Bakir et al., 2014). Since there is considerably more research on interactions between organisms and microplastics in marine environments compared with freshwater settings, much can be learned from marine studies about the types of biota that interact with microplastics, how the interactions take place, and the impacts of these interactions.

4.8.1 Microplastic Uptake by Invertebrates and Interactions With Bacteria and Algae

Marine organisms with a range of feeding strategies and across a range of trophic levels ingest microplastics. Benthic marine invertebrates shown in laboratory experiments to ingest microplastics are sea cucumbers (Graham and Thompson, 2009), mussels (Browne et al., 2008; Farrell and Nelson, 2013), lobsters (Murray and Cowie, 2011), amphipods, lugworms, and barnacles (Thompson et al., 2004; Browne et al., 2013; Wright et al., 2013b). Even in remote deep-sea habitats, benthic invertebrates have been reported to contain microplastics (Woodall et al., 2014; Taylor et al., 2016). Some benthic invertebrates preferentially ingest microplastics over natural particulates. For example, in controlled laboratory experiments, four species of deposit-feeding and suspension-feeding sea cucumber selectively ingested plastic particles instead of sand or other sediment particles (Graham and Thompson, 2009). Pelagic marine invertebrates shown to ingest microplastics in laboratory experiments include a range of invertebrate holoplankton and meroplankton (Cole et al., 2013; Cole and Galloway, 2015; Setälä et al., 2014). Limited studies show microplastic ingestion by freshwater invertebrates. Most studies focus on single species, but one laboratory study found microplastic ingestion in 30%–100% of exposed invertebrates belonging to five species comprising multiple habitats and feeding guilds (Imhof et al., 2013). These freshwater species included the benthic detritivorous amphipod (*Gammarus pulex*), grazing mud snail (*Potamopyrgus antipodarum*), deposit-feeding clitellate worm (*Lumbriculus variegatus*), planktonic filter-feeding water flea (*Daphnia magna*), and surface film-feeding ostracod (*Notodromas monacha*) (Imhof et al., 2013). While marine invertebrates collected from natural habitats have been shown to contain microplastics (e.g., lobsters, shellfish, and deep-sea fauna; Murray and Cowie, 2011; Woodall et al., 2014; Van Cauwenberghe et al., 2015b), little evidence exists of field-collected freshwater invertebrates containing microplastics in their bodies. Only recently are field studies reporting microplastic ingestion by freshwater invertebrates (e.g., the worm *Tubifex tubifex*; Hurley et al., 2017).

There are fewer studies of plastics at the nanoscale interacting with biota. In experimental studies, Pacific oyster larvae ingest nano- and microplastics without measurable effects on development or feeding capacity (Cole and Galloway, 2015). Rotifers ingesting 50 nm polystyrene particles have reduced growth rate, fecundity, and life span (Jeong et al., 2016). There may be potential for uptake of nanoplastics by bacteria and microalgae. Internalization of nanoparticles has been demonstrated in various bacteria and cyanobacteria (von Moos et al., 2014), and plastics can be manufactured at the nanoscale.

4.8.2 Microplastic Uptake by Fish

Microplastic uptake by fish was documented early in marine studies, for example, Carpenter et al. (1972), who found 8 out of 14 species of fish examined from coastal

New England, the United States, contained microplastics. Percent of fish containing microplastics spanned from 2.1% of winter flounder (*Pseudopleuronectes americanus*) to 33% of white perch (*Roccus americanus*) and silversides (*Menidia menidia*). Even 5 mm-long flounder and grubby larvae contained 0.5 mm-diameter spherules in their guts (Carpenter et al., 1972). Recently, some studies of microplastic ingestion by marine fish have focused on whether microplastic uptake differs based on habitat or consumption habits of the fish. A study of five pelagic and five demersal fish species from the English Channel showed no significant difference in microplastic ingestion based on fish habitat. The percent of fish ingesting plastics ranged from >20% to ~ 50% at average densities of 1.9 microplastics per fish (Lusher et al., 2013). Mizraji et al. (2017) investigating fish with different consumption habits found that in intertidal habitats, omnivorous fish ingested larger quantities of microplastic particles than herbivorous or carnivorous fish. They suggest this may be due to omnivores ingesting a wider range of food items (Mizraji et al., 2017).

In freshwater systems, microplastic ingestion has been reported for fish from a range of water bodies (lakes, rivers, and streams, urbanized and nonurbanized), taxonomic groups, trophic guilds, and feeding habits (e.g., Phillips and Bonner, 2015). Some publications report an association between the number of fish containing microplastics and the level of urbanization of a water body. The first field investigation of microplastic ingestion by freshwater fish reported microplastics in fish sampled from seven French rivers characterized by industry, urban, or agriculture pressures. Fish without microplastics were sampled at low anthropic sites from three of the 11 sampled French rivers (Sanchez et al., 2014). An association between microplastic ingestion by fish and level of urbanization was also reported in studies of fish from Brazil and the United States. The microplastic presence in fish guts was more frequent in fish collected from urbanized areas of the Pajeú river (Silva-Cavalcanti et al., 2017) and of the Brazos River Basin (Peters and Bratton, 2016).

In freshwater drainages of the Gulf of Mexico, where authors studied a large diversity of fish species (44 species and 12 families) spanning across habitat guilds, authors found an interaction between urbanization of streams, habitat guild of fish, and the microplastic presence in fish guts (Phillips and Bonner, 2015). In urbanized streams, more fish of the benthic habitat guild (19%) contained microplastics in guts than those of the pelagic habitat guild (7.7%). In nonurbanized streams, both habitat guilds (< 6%) contained similar amounts of microplastics. Associations may also exist between habitat guild, trophic guild (whether fish are herbivores/omnivores, invertivores, or carnivores), and level of urbanization of the water body. In benthic habitats of urbanized streams, there was a greater difference in the numbers of microplastics between trophic guilds (13%–21.2%) than for nonurbanized streams (5.9%–7.5%). In the pelagic habitat guild of both urbanized and nonurbanized streams, a greater number of microplastics were reported in the herbivore-omnivore trophic guild (21% and 8%) than in the invertivore-carnivore guild (6.3% and 2.4%) (Phillips and Bonner, 2015). There is also a reported difference in the microplastic presence in fish of benthic versus pelagic habitats in the River Thames in London; 75.0% of European flounder (a benthic feeder) contained microplastics in their gut, while only 20.0% of European smelt (a pelagic feeder) contained microplastics in their gut (McGoran et al., 2017). Other field-caught freshwater fish reported to contain microplastics include common dace and bleak from Lake Geneva (Faure et al., 2015) and Nile perch and Nile tilapia from Lake Victoria, Tanzania (Biginagwa et al., 2016).

An important development to note from one of the most recent marine studies on microplastic ingestion by fish is the use of strict quality assurance criteria to control contamination when carrying out studies of microplastic ingestion (Hermsen et al., 2017). After studying 400 fish from four North Sea species and finding 0.25% of individuals (95% CI = 0.09%–1.1%) contained microplastics in their gut, Hermsen et al. (2017) suggest an association between reported levels of microplastic ingestion and quality control. They draw reference to other studies that apply strict quality assurance criteria (e.g., taking airborne contamination into account) and report low incidence of plastic ingestion, and they suggest that contamination may result in reported high levels of microplastic fibers ingestion by fish. Fibers have been a major component of the microplastics that have been reported in freshwater fish (Faure et al., 2015; McGoran et al., 2017; Peters and Bratton, 2016; Silva-Cavalcanti et al., 2017).

4.8.3 Microplastic Uptake by Birds and Mammals

At the higher trophic levels of marine environments, seabirds and pinnipeds have been reported to ingest microplastics. Microplastics have been dissected from bird stomachs (Van Franeker et al., 2011; Tanaka et al., 2013) and regurgitated pellets (Hays and Cormons, 1974), and researchers speculate that these microplastics may enter the seabirds by direct ingestion or via fish prey (Hays and Cormons, 1974; Tanaka et al., 2013). Trophic transfer, by ingesting fish prey, is suggested as the most likely route of microplastic ingestion by fur seals and sea lions from sub-Antarctic islands (McMahon et al., 1999; Eriksson and Burton, 2003) as plastic fragments were only present in sea lion scats when fish otoliths were also present (McMahon et al., 1999). It has been suggested that transfer through the food web could lead to a magnification factor of 22–160 times in fur seals (Eriksson and Burton, 2003), but this would depend upon retention by the seals. Pelagic feeding baleen whales and demersal blackmouth catsharks have also been reported to ingest microplastics (Besseling et al., 2015; Alomar and Deudero, 2017). It seems likely that freshwater vertebrates, such as birds or mammals (e.g., castorids, mustelids, and murids), also ingest microplastics. On the coasts of Lake Geneva, gull feces contain plastic fragments (Faure et al., 2012), and microplastics have been found in digestive tracts from the gray heron (*Ardea cinerea*), mute swans (*Cygnus olor*), and mallards (*Anas platyrhynchos*) (Faure et al., 2015). Most of the particles found in waterfowl from Lake Geneva were fragments, but foams, films, microbeads, and fibers were also ingested at mean particle counts of 4.3 ± 2.6 particles per bird. Most particles were polished and found in the gizzard, suggesting they may have been retained inside the birds over time (Faure et al., 2015).

4.9 IMPACTS OF MICROPLASTICS ON BIOTA

4.9.1 Differential Retention, Translocation and Nutritional Effects

One factor that may influence the impacts of microplastics on an organism is retention. Characteristics of microplastics (size, type, and composition) may affect whether particles are retained and the length of time of retention; not all biota retain microplastics, and some retain them differentially. For example, in the marine literature, laboratory experiments with the sea scallop (*Placopecten magellanicus*) showed retention of larger (20 μm) and lighter (1.05 g mL^{-1}) particles longer than smaller (5 μm) and denser (2.5 g mL^{-1}) particles (Brillant and MacDonald, 2000). In field-collected *Nephrops* lobsters, 83% of individuals contained plastics, and 63% had filamentous balls in their stomachs (Murray and Cowie, 2011). Among pelagic fauna, the copepod (*Paracyclopina nana*) and rotifer (*Brachionus koreanus*) retain smaller (0.05 μm) microbeads for a longer period than larger microbeads (0.5 and 6 μm) (Jeong et al., 2016, 2017).

Once present in the digestive system, microplastics have been shown to translocate across cell membranes. In the marine mussel, *Mytilus edulis*, ingested particles (>0–80 μm) that reach the gastrointestinal tract become endocytosed by the cells of the digestive system (von Moos et al., 2012), and from the digestive system, particles 3 and 9.6 μm have been shown to translocate into the circulatory system (Browne et al., 2008). In the marine copepod, *P. nana* ingested 0.05 μm beads translocate across cell membranes of the digestive system (Jeong et al., 2017). Accumulation of microplastics in digestive tracts can affect nutritional gain and energy reserves by reducing feeding or causing false satiation (Cole et al., 2013; Wright et al., 2013b). There may also be carryover effects, such as those for rotifers, *B. koreanus*, exposed to 0.05 μm microbeads, where authors suggested that insufficient nutrition led to inhibited growth, reduced fecundity, and shortened life spans (Jeong et al., 2016).

The initial freshwater studies on interactions between biota and microplastics span across trophic levels from invertebrates to fish and birds. The water flea, *D. magna*, has been shown to differentially accumulate and retain particles 20 and 1000 nm, which may also cross into cells and translocate into oil storage droplets (Rosenkranz et al., 2009). Zebrafish fed *Artemia* containing microplastic particles (1–5 and 10–20 μm) have been shown to exhibit some microplastic retention between intestinal villi and a small amount of uptake by epithelial cells, but overall accumulation in the gastrointestinal tract was not large, and there was no sign of disease due to exposure (Batel et al., 2016). Retention of microplastics is also reported for waterfowl at Lake Geneva (Faure et al., 2015).

It should be noted that many exposure studies use concentrations of microplastics that are orders of magnitude higher than those reported in the environment. In addition, the effects reported are for microplastics of sizes smaller than those reported from the environment, and therefore, there is a need for microplastic exposure studies that are environmentally realistic (Lenz et al., 2016).

4.9.2 Impacts at Cellular and Tissue Levels

At the tissue and cellular level, ingested microplastics can trigger inflammation responses, such as the phagocytic activity by coelomic cells of the marine lugworm (*Arenicola marina*) (Wright et al., 2013b), the formation of granulocytomas in the digestive tissues of the mussel (*M. edulis*) (von Moos et al., 2012), and the activation of signaling proteins involved in intracellular processes associated with inflammation in the marine rotifer (Jeong et al., 2016). Microplastic ingestion can also lead to reduced stability of cell membranes (e.g., in digestive tissues of *M. edulis*) (von Moos et al., 2012), and inflammatory responses can ultimately affect energy reserves (Browne et al., 2008; Wright et al., 2013b). In the amphidromous Japanese medaka fish, *Oryzias latipes*, ingested microplastics (size, < 0.5 mm) led to sublethal effects to the liver (glycogen depletion, fatty vacuolation, and single-cell necrosis) and early tumor formation (Rochman et al., 2013). In the freshwater, literature Catfish, *Clarias gariepinus*, exposed to LDPE fragments showed biomarker changes that could potentially be explained by the release of ethylene monomers from the LDPE fragments during digestion or from abrasions due to the edges of microplastics (Karami et al., 2016). Transcriptomes in the brain were also affected by LDPE fragment exposure, and there was evidence that microplastics could interact with the effects of contaminant phenanthrene on biomarker response (Karami et al., 2016).

4.9.3 Transfer of Chemicals

Aside from the potential for direct interactions of microplastics with organisms, microplastics can also be a medium for the transfer of chemicals to organisms (Teuten et al., 2007, 2009; Browne et al., 2013). Chemicals may come from the environment and sorb onto the plastic surface, or they may be added to the plastic during the production process (Mato et al., 2001; Talsness et al., 2009). Marine studies have found that hydrophobic contaminants may sorb more readily to plastics than to some natural sediments and be subsequently transferred to organisms (Teuten et al., 2007). Laboratory experiments indicate that chemicals are transferred to marine biota and can cause physiological stress and responses (e.g., inflammation, oxidative stress, and activation of transcription factors), impaired immune functions, and altered mortality (Browne et al., 2013; Oliveira et al., 2013; Rochman et al., 2013; Jeong et al., 2016, 2017). However, a recent review by Koelmans et al. (2016) examined the potential impact of microplastics with sorbed hydrophobic organic chemicals (HOC) and related this to HOC that could be sorbed onto other environmental media (e.g., dissolved organic carbon or detritus). The authors concluded that the amount of HOC sorbed onto plastic is small compared with the amounts sorbed on other media in the ocean and that at present, natural prey with bioaccumulated HOC is likely a greater source of HOC (Koelmans et al., 2016). This is supported by experimental work by Bakir et al. (2016).

4.9.4 Other Impacts

Aside from the direct impacts on organisms and the potential for trophic transfer, there is potential for wider environmental impacts such as indirect effects on biotic communities and ecosystems. Wider environmental impacts span issues such as changes in the behavior or organization of biota (Chen et al., 2011), changes to bioturbation and oxygenation of sediments (Wright et al., 2013b), and impacts on carbon flux (Chen et al., 2011). With regard to the latter, when microorganisms colonize microplastics (Lobelle and Cunliffe, 2011; Zettler et al., 2013), microplastics may act as a transport medium for organic carbon (Chen et al., 2011; Kowalski et al., 2016). Biota colonizing microplastics in marine studies include diatoms, dinoflagellates, bryozoans, coralline algae, serpulid worms, isopods, barnacles, marine insect eggs, and cells presumed to be bacteria, cyanobacteria, and fungi (Carpenter and Smith, 1972; Reisser et al., 2014). Colonization may occur when plastics settle at one part of a water body (e.g., coastline), and resuspension and transport via weather events may transport the plastic and associated biota to different parts of the water body (Kowalski et al., 2016). Transfer of biota between areas might have impacts on population connectivity and the spread of nonnative and invasive species.

There may also be potential for microplastics to increase an organism's exposure to viruses by supporting the survival, transport, and transmission of viruses to biota. In laboratory experiments, viral hemorrhagic septicemia virus (VHSV), a virus that causes disease in salmonid fish, had the potential to survive for 2 weeks after being kept dry at 4°C on plastic surfaces, and at 14°C and at room temperature, the virus survived longer on plastic than on glass. Plastic water bottle pieces that were incubated with the virus and then rinsed, incubated wet for 10 days, and then transferred to cell lines from the fathead minnow were able to transfer the virus to the cell line (Pham, 2009). Pham (2009) suggests that since plastic has the potential to support attachment and survival of the virus, floating plastic debris in aquatic environments has the potential to transport fish viruses. Since transmission of the virus requires close contact between fish and the virus, this could be more likely to occur with microliter (Pham, 2009).

4.10 POTENTIAL INTERACTIONS BETWEEN MICROPLASTICS AND HUMANS

In marine systems, there is evidence of microplastics in commercially important species for humans. Farmed mussels, *M. edulis*, from a German mussel farm, and oysters,

Crassostrea gigas, reared in the Atlantic Ocean contained concentrations of 0.36 ± 0.07 SD and 0.47 ± 0.16 SD (respectively) microplastic particles/g of soft tissue (Van Cauwenberghe and Janssen, 2014). Wild-collected mussels, *M. edulis*, from the French-Belgian-Dutch coast contained 0.2 microplastic particles/g of tissue (Van Cauwenberghe et al., 2015b). In another study, all mussels, *M. edulis*, purchased at Belgian supermarkets and originating from mussel farms on the Eastern Scheldt, the Netherlands, and wild-caught mussels from the Dutch coast contained microplastics. The mean number of microplastic fibers in supermarket mussels was 3.5 per 10 g w.w., and in field-caught mussels, microplastic fibers reached counts of 5.1 per 10 g w.w. (De Witte et al., 2014).

In freshwater habitats, microplastics have been identified in commercially important fish from Lake Victoria, Africa (Biginagwa et al., 2016), and from the Pajeú River, Brazil (Silva-Cavalcanti et al., 2017). However, the microplastics in these fish were found in their gastrointestinal tracts. The potential for humans to ingest microplastics from fish would depend on whether the fish is consumed whole or gutted and/or whether microplastics are translocated to tissues that humans consume. This is in contrast to the bivalves from marine studies where the whole organism is likely to be consumed.

Wright and Kelly (2017) review and discuss the potential risks of microplastics to human health. Avenues for risk include direct interaction between microplastics and human tissues. If ingested, microplastics could have similar effects in tissues (e.g., stress responses and immune responses) as seen for other animals (Wright and Kelly, 2017). As described in studies of other organisms, there's also potential of chemicals, sorbed onto the plastics from the environment or existent in the plastics from the plastic production process, to leach from microplastics to humans. Certain polymers from plastic additives and degradation products have received classifications of carcinogenic, mutagenic, and endocrine disrupting (Koch and Calafat, 2009; Lithner et al., 2011; Galloway, 2015; Wright and Kelly, 2017). The health risk of these chemicals would depend on degradation rates and pathways of the plastics (Galloway, 2015).

The potential for microplastics to transport pathogenic microorganisms could be of concern. In aquatic environments, the physical characteristics of microplastics, such as hydrophobicity, allow them to acquire microbial communities on their surface (Zettler et al., 2013). These microbial communities are unique from the surrounding aquatic environment and may include pathogenic bacteria, for example, the gram-negative *Vibrio*. Authors suggest that humans may come into contact with plastics that have microbes on their surface (Zettler et al., 2013; Kirstein et al., 2016). Whether microplastics could expose humans to viruses is unknown. However, what may be of concern to humans is the potential economic losses if microplastics negatively impact aquaculture or fishery species.

At present, the threat that humans potentially face from consuming biota contaminated with microplastics is likely low. Based on existing evidence for microplastic concentrations in biota and our current practices of food preparation, in our view, there is little evidence for concern relating to the transfer of microplastics to humans via fish and shellfish. However, the widespread occurrence of microplastics in biota, including commercially important species, is an indication of concern and an indication that we need to change our practices to prevent microplastic levels in our food from becoming harmful in the future. Additional areas that humans may want to track, concurrently to the issue of microplastics in freshwaters and their biota, are other potential causes of exposure, such as atmospheric microplastic depositions (Dris et al., 2016) and exposure from internal environments, such as homes and workplaces, which may present higher potential for microplastic exposure (Dris et al., 2017).

The interactions between microplastics and humans have multiple aspects. As Pahl and Wyles (2017) state, "people contribute to the problem, they can help address it, and they may experience negative impacts of microplastics in the environment." In the first part of this chapter, in the sections dealing with "occurrence," we've outlined ways in which humans are associated with the presence of microplastics in the environment, and in this section, we've described some potential impacts of microplastics on humans. Coming from the perspective of the social and behavioral sciences, Pahl and Wyles (2017) remind the microplastic scientific community of the importance of including the human dimension in research of microplastics; the human dimension (behavioral and social science), coupled with our knowledge gained in the natural sciences, will be important in addressing the issue of microplastics in the environment (Pahl and Wyles, 2017).

4.11 CONCLUSIONS AND CONSIDERATIONS FOR FUTURE WORK

The distributions and abundance of microplastics in freshwater systems are likely to increase with the growing input of plastic into the environment. In the oceans where projections exist, the cumulative plastic waste available to enter the marine environment from coastal populations (within 50 km of coast) is predicted to increase by an order of magnitude between the years 2010 and 2025 (Jambeck et al., 2015). The presence of microplastics in freshwaters is also likely to increase in spread and abundance as is predicted for marine systems. Freshwaters may be similar to marine systems wherein studies have shown that various types of biota, from microorganisms to vertebrates, across numerous trophic levels interact with microplastics with some negative impacts documented in laboratory studies. Further work to investigate potential impacts of microplastics on freshwater biota at environmentally realistic concentrations of microplastic is needed to reach any firm conclusions. As marine authors state, there is also a high need of studies on microplastic ingestion, which apply standardized analytic methods that include strict quality assurance criteria (Hermsen et al., 2017).

This chapter highlights the increased research efforts on microplastics in freshwater systems, and the extent to which research on microplastics in aquatic environments are dominated by the larger research base of marine systems. In freshwater systems, there is a need for basic understanding of microplastic sources, microplastic distributions and abundance, drivers of distributions and abundances, and the potential impacts of microplastics on the environment and biota. One can learn from and translate methods and approaches used to address these issues in marine research, but there remain unique characteristics of freshwaters (e.g., aspects of river's and lake's hydrology and geomorphology) that require methods tailored for freshwater systems.

As the issue of plastic waste into the environment continues to grow, there is a need for persistent and coordinated research efforts within and among those nations investing in this topic. The infancy of microplastic research in freshwater systems presents an opportunity to establish harmonized methods that will allow comparability of findings as the field grows and expands.

Early policy initiatives aimed at addressing the issue of microplastics in freshwaters include the microbead bans by countries or states in North America, Europe, and Australia (Council of the European Union, 2014; Environment Protection Amendment Bill, 2016; Schroeck, 2016; Xanthos and Walker, 2017), and the decision by government ministries and regional water boards in the Netherlands to perform risk assessments for nanoplastics and microplastics in freshwaters and coasts (Rochman et al., 2016). An example of a larger-scale effort contributing to management of microplastics in freshwaters is the European Union's Marine Strategy Framework Directive (MSFD). Several products have come out of this directive, of which *Riverine Litter Monitoring—Options and Recommendations* is a thematic report written by the MSFD Technical Group on Marine Litter and intended for use by experts who implement the MSFD in marine regions (González et al., 2016). The authors state that the report should "support EU Member States in the implementation of monitoring programmes and the planning of measures to tackle marine litter." In presenting the current scientific and technical background to litter in river systems, the report moves toward such goals. González et al. (2016) also hope that the information presented in the report allows the formation of datasets to compare litter flows from different rivers into the marine environment. In this context of riverine litter, specific information on microplastics covered in the report is with regard to microplastic sources, the way in which particle shape affects particle movement, how their distributions are affected by riverine morphology and hydrology, and some methods for monitoring microplastics (González et al., 2016). There isn't a European Commission report equivalent to *Riverine Litter Monitoring—Options and Recommendations* (González et al., 2016) focused on microplastics in lakes, and the majority of scientific papers and reports showcased on the European Parliamentary Research Blog (EPRS, 2016) are focused on marine litter and riverine inputs of marine litter. Even within those efforts focused on rivers, methodology for studying wastes can be lacking; for example, the OSPAR waste sorting grid is not adapted for fluvial context, and efforts to improve such methodologies are valuable, for example, the work by Bruge et al. (2017) to develop a Riverine Input protocol for identifying wastes in a fluvial context. In North America, the International Joint Commission of Canada and the United States has identified science needs to develop and adopt harmonized methods for microplastic sampling and quantification in the Great Lakes (IJC, 2016).

Integrated approaches, such as the thematic report on riverine litter, are beneficial in many ways. They allow issues of widespread impact to be addressed, permit limited funding resources to be used more effectively, allow the benefits from growing the knowledge base to be shared among multiple member states, and provide a platform to propose recommendations for monitoring and management. Integrated approaches also enable methodologies and planning measures to be harmonized so that findings from individual research efforts are more widely comparable. In the umbrella of integrated approaches, it is important to keep abreast of the research gaps. The area of microplastics in freshwater systems still has many research gaps, and harmonized methods for both river systems and lakes still need attention.

ACKNOWLEDGMENTS

We acknowledge the very helpful comments of Patricia L. Corcoran and one anonymous reviewer in improving this paper.

REFERENCES

Alomar, C., Deudero, S., 2017. Evidence of microplastic ingestion in the shark *Galeus melastomus* Rafinesque, 1810 in the continental shelf off the western Mediterranean Sea. Environ. Pollut. 223, 223–229.

Anderson, P.J., Warrack, S., Langen, V., Challis, J.K., Hanson, M.L., Rennie, M.D., 2017. Microplastic contamination in Lake Winnipeg, Canada. Environ. Pollut. 225, 223–231.

Andrady, A.L., 2011. Microplastics in the marine environment. Mar. Pollut. Bull. 62 (8), 1596–1605.

Andrady, A.L., Hamid, S.H., Hu, X., Torikai, A., 1998. Effects of increased solar ultraviolet radiation on materials. J. Photochem. Photobiol. B Biol. 46 (1), 96–103.

Andrady, A.L., Hamid, H.S., Torikai, A., 2003. Effects of climate change and UV-B on materials. Photochem. Photobiol. Sci. 2 (1), 68–72.

Åström, L., 2016. Shedding of Synthetic Microfibers From Textiles. University of Gothenburg, Gothenburg, p. 378.

Bakir, A., Rowland, S.J., Thompson, R.C., 2014. Enhanced desorption of persistent organic pollutants from microplastics under simulated physiological conditions. Environ. Pollut. 185, 16–23.

Bakir, A., O'Connor, I.A., Rowland, S.J., Hendriks, A.J., Thompson, R.C., 2016. Relative importance of microplastics as a pathway for the transfer of hydrophobic organic chemicals to marine life. Environ. Pollut. 219, 56–65.

Baldwin, A.K., Corsi, S.R., Mason, S.A., 2016. Plastic debris in 29 Great Lakes tributaries: relations to watershed attributes and hydrology. Environ. Sci. Technol. 50 (19), 10377–10385.

Ballent, A.M., 2016. Anthropogenic Particles in Natural Sediment Sinks: Microplastics Accumulation in Tributary, Beach and Lake Bottom Sediments of Lake Ontario, North America (Master of Science dissertation).The University of Western Ontario

Ballent, A., Purser, A., de Jesus Mendes, P., Pando, S., Thomsen, L., 2012. Physical transport properties of marine microplastic pollution. Biogeosci. Discuss. 9 (12), 18755–18798.

Ballent, A., Pando, S., Purser, A., Juliano, M.F., Thomsen, L., 2013. Modelled transport of benthic marine microplastic-pollution in the Nazaré Canyon. Biogeosciences 10 (12), 7957.

Ballent, A., Corcoran, P.L., Madden, O., Helm, P.A., Longstaffe, F.J., 2016. Sources and sinks of microplastics in Canadian Lake Ontario nearshore, tributary and beach sediments. Mar. Pollut. Bull. 110 (1), 383–395.

Batel, A., Linti, F., Scherer, M., Erdinger, L., Braunbeck, T., 2016. Transfer of benzo [a] pyrene from microplastics to *Artemia* nauplii and further to zebrafish via a trophic food web experiment: CYP1A induction and visual tracking of persistent organic pollutants. Environ. Toxicol. Chem. 35 (7), 1656–1666.

Beach, W.J., 1967. United States Patent 3,645,904 for Skin Cleaner. United States; 3,645,904, Filed July 27, 1967, http://www.google.co.uk/patents/US3645904.

Besseling, E., Foekema, E.M., Van Franeker, J.A., Leopold, M.F., Kühn, S., Rebolledo, E.B., Heße, E., Mielke, L., IJzer, J., Kamminga, P., Koelmans, A.A., 2015. Microplastic in a macro filter feeder: humpback whale *Megaptera novaeangliae*. Mar. Pollut. Bull. 95 (1), 248–252.

Biginagwa, F.J., Mayoma, B.S., Shashoua, Y., Syberg, K., Khan, F.R., 2016. First evidence of microplastics in the African Great Lakes: recovery from Lake Victoria Nile perch and Nile tilapia. J. Great Lakes Res. 42 (1), 146–149.

Bouchez, J., Métivier, F., Lupker, M., Maurice, L., Perez, M., Gaillardet, J., France-Lanord, C., 2011. Prediction of depth-integrated fluxes of suspended sediment in the Amazon River: particle aggregation as a complicating factor. Hydrol. Process. 25 (5), 778–794.

Brillant, M.G.S., MacDonald, B.A., 2000. Postingestive selection in the sea scallop, *Placopecten magellanicus* (Gmelin): the role of particle size and density. J. Exp. Mar. Biol. Ecol. 253 (2), 211–227.

Browne, M.A., Dissanayake, A., Galloway, T.S., Lowe, D.M., Thompson, R.C., 2008. Ingested microscopic plastic translocates to the circulatory system of the mussel, *Mytilus edulis* (L.). Environ. Sci. Technol. 42 (13), 5026–5031.

Browne, M.A., Crump, P., Niven, S.J., Teuten, E., Tonkin, A., Galloway, T., Thompson, R., 2011. Accumulation of microplastic on shorelines worldwide: sources and sinks. Environ. Sci. Technol. 45 (21), 9175–9179.

Browne, M.A., Niven, S.J., Galloway, T.S., Rowland, S.J., Thompson, R.C., 2013. Microplastic moves pollutants and additives to worms, reducing functions linked to health and biodiversity. Curr. Biol. 23 (23), 2388–2392.

Bruge, A., Maison, P., Barreau, C., Dussaussois, J.P., Schaal, A., Peña, L., 2017. Monitoring of the aquatic macro-litter inputs from the ardour river to the marine environment. Technical report riverine input ardour 2014–2016, Surfrider Foundation Europe.

Cable, R.N., Beletsky, D., Beletsky, R., Wigginton, K., Locke, B.W., Duhaime, M.B., 2017. Distribution and modeled transport of plastic pollution in the Great Lakes, the world's largest freshwater resource. Front. Environ. Sci. 5, 45.

Carpenter, E.J., Smith, K.L., 1972. Plastics on the Sargasso Sea surface. Science 175 (4027), 1240–1241.

Carpenter, E.J., Anderson, S.J., Harvey, G.R., Miklas, H.P., Peck, B.B., 1972. Polystyrene spherules in coastal waters. Science 178 (4062), 749–750.

Castañeda, R.A., Avlijas, S., Simard, M.A., Ricciardi, A., 2014. Microplastic pollution in St. Lawrence river sediments. Can. J. Fish. Aquat. Sci. 71 (12), 1767–1771.

Chang, M., 2015. Reducing microplastics from facial exfoliating cleansers in wastewater through treatment versus consumer product decisions. Mar. Pollut. Bull. 101 (1), 330–333.

Chen, C.S., Anaya, J.M., Zhang, S., Spurgin, J., Chuang, C.Y., Xu, C., Miao, A.J., Chen, E.Y., Schwehr, K.A., Jiang, Y., Quigg, A., 2011. Effects of engineered nanoparticles on the assembly of exopolymeric substances from phytoplankton. PLoS One 6(7), e21865.

Chubarenko, I., Bagaev, A., Zobkov, M., Esiukova, E., 2016. On some physical and dynamical properties of microplastic particles in marine environment. Mar. Pollut. Bull. 108 (1), 105–112.

Cole, M., Galloway, T.S., 2015. Ingestion of nanoplastics and microplastics by Pacific oyster larvae. Environ. Sci. Technol. 49 (24), 14625–14632.

Cole, M., Lindeque, P., Fileman, E., Halsband, C., Goodhead, R., Moger, J., Galloway, T.S., 2013. Microplastic ingestion by zooplankton. Environ. Sci. Technol. 47 (12), 6646–6655.

Cooper, D.A., Corcoran, P.L., 2010. Effects of mechanical and chemical processes on the degradation of plastic beach debris on the island of Kauai, Hawaii. Mar. Pollut. Bull. 60 (5), 650–654.

Corcoran, P.L., Norris, T., Ceccanese, T., Walzak, M.J., Helm, P.A., Marvin, C.H., 2015. Hidden plastics of Lake Ontario, Canada and their potential preservation in the sediment record. Environ. Pollut. 204, 17–25.

Costa, M.F., Barletta, M., 2015. Microplastics in coastal and marine environments of the western tropical and sub-tropical Atlantic Ocean. Environ. Sci. Processes Impacts 17 (11), 1868–1879.

Costa, M.F., Do Sul, J.A.I., Silva-Cavalcanti, J.S., Araújo, M.C.B., Spengler, Â., Tourinho, P.S., 2010. On the importance of size of plastic fragments and pellets on the strandline: a snapshot of a Brazilian beach. Environ. Monit. Assess. 168 (1–4), 299–304.

Council of the European Union, 2014. Elimination of micro-plastics in products—an urgent need—information from the Belgian, Dutch, Austrian and Swedish delegations, supported by the Luxembourg delegation (16263/14). Brussels.

De Witte, B., Devriese, L., Bekaert, K., Hoffman, S., Vandermeersch, G., Cooreman, K., Robbens, J., 2014. Quality assessment of the blue mussel (*Mytilus edulis*): comparison between commercial and wild types. Mar. Pollut. Bull. 85 (1), 146–155.

Dris, R., Gasperi, J., Rocher, V., Saad, M., Renault, N., Tassin, B., 2015. Microplastic contamination in an urban area: a case study in greater Paris. Environ. Chem. 12 (5), 592–599.

Dris, R., Gasperi, J., Saad, M., Mirande, C., Tassin, B., 2016. Synthetic fibers in atmospheric fallout: a source of microplastics in the environment? Mar. Pollut. Bull. 104 (1), 290–293.

Dris, R., Gasperi, J., Mirande, C., Mandin, C., Guerrouache, M., Langlois, V., Tassin, B., 2017. A first overview of textile fibers, including microplastics, in indoor and outdoor environments. Environ. Pollut. 221, 453–458.

Duis, K., Coors, A., 2016. Microplastics in the aquatic and terrestrial environment: sources (with a specific focus on personal care products), fate and effects. Environ. Sci. Eur. 28 (1), 2.

Eerkes-Medrano, D., Thompson, R.C., Aldridge, D.C., 2015. Microplastics in freshwater systems: a review of the emerging threats, identification of knowledge gaps and prioritisation of research needs. Water Res. 75, 63–82.

Environment Protection Amendment Bill, 2016. Environment Protection Amendment (Banning Plastic Bags, Packaging and Microbeads) Bill 2016. Parliament of Victoria, Australia. 581PM12B.I-21/6/2016. BILL LC INTRODUCTION 21/6/2016.

EPRS, 2016. Microplastic Pollution. European Parliamentary Research Services Blog. https://epthinktank.eu/2016/06/29/microplastic-pollution/. Accessed 3 August 2017.

Eriksen, M., Mason, S., Wilson, S., Box, C., Zellers, A., Edwards, W., Farley, H., Amato, S., 2013. Microplastic pollution in the surface waters of the Laurentian Great Lakes. Mar. Pollut. Bull. 77 (1), 177–182.

Eriksen, M., Lebreton, L.C., Carson, H.S., Thiel, M., Moore, C.J., Borerro, J.C., Galgani, F., Ryan, P.G., Reisser, J., 2014. Plastic pollution in the world's oceans: more than 5 trillion plastic pieces weighing over 250,000 tons afloat at sea. PLoS One 9 (12), e111913.

Eriksson, C., Burton, H., 2003. Origins and biological accumulation of small plastic particles in fur seals from Macquarie Island. AMBIO J. Hum. Environ. 32 (6), 380–384.

Farrell, P., Nelson, K., 2013. Trophic level transfer of microplastic: *Mytilus edulis* (L.) to *Carcinus maenas* (L.). Environ. Pollut. 177, 1–3.

Faure, F., Corbaz, M., Baecher, H., de Alencastro, L., 2012. Pollution due to plastics and microplastics in Lake Geneva and in the Mediterranean Sea. Arch. Sci. 65 (EPFL-ARTICLE-186320), 157–164.

Faure, F., Demars, C., Wieser, O., Kunz, M., De Alencastro, L.F., 2015. Plastic pollution in Swiss surface waters: nature and concentrations, interaction with pollutants. Environ. Chem. 12 (5), 582–591.

Fendall, L.S., Sewell, M.A., 2009. Contributing to marine pollution by washing your face: microplastics in facial cleansers. Mar. Pollut. Bull. 58 (8), 1225–1228.

Fischer, E.K., Paglialonga, L., Czech, E., Tamminga, M., 2016. Microplastic pollution in lakes and lake shoreline sediments—a case study on Lake Bolsena and Lake Chiusi (central Italy). Environ. Pollut. 213, 648–657.

Free, C.M., Jensen, O.P., Mason, S.A., Eriksen, M., Williamson, N.J., Boldgiv, B., 2014. High-levels of microplastic pollution in a large, remote, mountain lake. Mar. Pollut. Bull. 85 (1), 156–163.

Galgani, F., Hanke, G., Werner, S., Oosterbaan, L., Nilsson, P., Fleet, D., Kinsey, S., Thompson, R.C., van Franeker, J., Vlachogianni, T., Scoullos, M., Veiga, J.M., Palatinus, A., Matiddi, M., Maes, T., Korpinen, S., Budziak, A., Leslie, H., Gago, J., Liebezeit, G., 2013. Guidance on monitoring of marine litter in European Seas. MSFD Technical Subgroup on Marine Litter; JRC technical report; EUR 26113.

Galloway, T.S., 2015. Micro-and nano-plastics and human health. In: Bergmann, M., Gutow, L., Klages, M. (Eds.), MarineAnthropogenic Litter. Springer International Publishing, Cham, pp. 343–366.

González, D., Hanke, G., Tweehuysen, G., Bellert, B., Holzhauer, M., Palatinus, A., Hohenblum, P., Oosterbaan, L., 2016. Riverine litter monitoring—options and recommendations. MSFD GES TG Marine Litter thematic report; JRC technical report; EUR 28307, https://doi.org/10.2788/461233.

Gouin, T., Avalos, J., Brunning, I., Brzuska, K., de Graaf, J., Kaumanns, J., Koning, T., Meyberg, M., Rettinger, K., Schlatter, H., Thomas, J., van Welie, R., Wolf, T., 2015. Use of micro-plastic beads in cosmetic products in Europe and their estimated emissions to the North Sea environment. SOFW J 141 (4), 40–46.

Graham, E.R., Thompson, J.T., 2009. Deposit-and suspension-feeding sea cucumbers (Echinodermata) ingest plastic fragments. J. Exp. Mar. Biol. Ecol. 368 (1), 22–29.

Gregory, M.R., 1977. Plastic pellets on New Zealand beaches. Mar. Pollut. Bull. 8 (4), 82–84.

Gregory, M.R., 1978. Accumulation and distribution of virgin plastic granules on New Zealand beaches. N. Z. J. Mar. Freshw. Res. 12 (4), 399–414.

Gregory, M.R., 1996. Plastic "scrubbers" in hand cleansers: a further (and minor) source for marine pollution identified. Mar. Pollut. Bull. 32 (12), 867–871.

Gregory, M.R., 2009. Environmental implications of plastic debris in marine settings—entanglement, ingestion, smothering, hangers-on, hitch-hiking and alien invasions. Philos. Trans. R. Soc. Lond. B Biol. Sci. 364 (1526), 2013–2025.

Habib, D., Locke, D.C., Cannone, L.J., 1998. Synthetic fibers as indicators of municipal sewage sludge, sludge products, and sewage treatment plant effluents. Water Air Soil Pollut. 103 (1), 1–8.

Hassan, M.A., Voepel, H., Schumer, R., Parker, G., Fraccarollo, L., 2013. Displacement characteristics of coarse fluvial bed sediment. J. Geophys. Res. Earth Surf. 118 (1), 155–165.

Hays, H., Cormons, G., 1974. Plastic particles found in tern pellets, on coastal beaches and at factory sites. Mar. Pollut. Bull. 5 (3), 44–46.

Hermsen, E., Pompe, R., Besseling, E., Koelmans, A.A., 2017. Detection of low numbers of microplastics in North Sea fish using strict quality assurance criteria. Mar. Pollut. Bull. 122 (1–2), 253–258.

Hidalgo-Ruz, V., Gutow, L., Thompson, R.C., Thiel, M., 2012. Microplastics in the marine environment: a review of the methods used for identification and quantification. Environ. Sci. Technol. 46 (6), 3060–3075.

Hoffman, M.J., Hittinger, E., 2017. Inventory and transport of plastic debris in the Laurentian Great Lakes. Mar. Pollut. Bull. 115 (1), 273–281.

Hurley, R.R., Woodward, J.C., Rothwell, J.J., 2017. Ingestion of microplastics by freshwater Tubifex worms. Environ. Sci. Technol. 51 (21),12844–12851.

IJC, 2016. Microplastics in the Great Lakes workshop report. Final report, International Joint Commission, Canada and United States. September 14.

Imhof, H.K., Ivleva, N.P., Schmid, J., Niessner, R., Laforsch, C., 2013. Contamination of beach sediments of a subalpine lake with microplastic particles. Curr. Biol. 23 (19), R867–R868.

Isobe, A., 2016. Percentage of microbeads in pelagic microplastics within Japanese coastal waters. Mar. Pollut. Bull. 110 (1), 432–437.

Jambeck, J.R., Geyer, R., Wilcox, C., Siegler, T.R., Perryman, M., Andrady, A., Narayan, R., Law, K.L., 2015. Plastic waste inputs from land into the ocean. Science 347 (6223), 768–771.

Jeong, C.B., Won, E.J., Kang, H.M., Lee, M.C., Hwang, D.S., Hwang, U.K., Zhou, B., Souissi, S., Lee, S.J., Lee, J.S., 2016. Microplastic size-dependent toxicity, oxidative stress induction, and p-JNK and p-P38 activation in the monogonont rotifer (*Brachionus koreanus*). Environ. Sci. Technol. 50 (16), 8849–8857.

Jeong, C.B., Kang, H.M., Lee, M.C., Kim, D.H., Han, J., Hwang, D.S., Souissi, S., Lee, S.J., Shin, K.H., Park, H.G., Lee, J.S., 2017. Adverse effects of microplastics and oxidative stress-induced MAPK/Nrf2 pathway-mediated defense mechanisms in the marine copepod *Paracyclopina nana*. Sci. Rep. 7, 41323.

Kanhai, L.D.K., Officer, R., Lyashevska, O., Thompson, R.C., O'Connor, I., 2017. Microplastic abundance, distribution and composition along a latitudinal gradient in the Atlantic Ocean. Mar. Pollut. Bull. 115 (1), 307–314.

Karami, A., Romano, N., Galloway, T., Hamzah, H., 2016. Virgin microplastics cause toxicity and modulate the impacts of phenanthrene on biomarker responses in African catfish (*Clarias gariepinus*). Environ. Res. 151, 58–70.

Kirstein, I.V., Kirmizi, S., Wichels, A., Garin-Fernandez, A., Erler, R., Löder, M., Gerdts, G., 2016. Dangerous hitchhikers? Evidence for potentially pathogenic *Vibrio* spp. on microplastic particles. Mar. Environ. Res. 120, 1–8.

Klein, S., Worch, E., Knepper, T.P., 2015. Occurrence and spatial distribution of microplastics in river shore sediments of the Rhine-Main area in Germany. Environ. Sci. Technol. 49 (10), 6070–6076.

Koch, H.M., Calafat, A.M., 2009. Human body burdens of chemicals used in plastic manufacture. Philos. Trans. R. Soc. Lond. B Biol. Sci. 364 (1526), 2063–2078.

Koelmans, A.A., Bakir, A., Burton, G.A., Janssen, C.R., 2016. Microplastic as a vector for chemicals in the aquatic environment: critical review and model-supported reinterpretation of empirical studies. Environ. Sci. Technol. 50 (7), 3315–3326.

Kooi, M., Reisser, J., Slat, B., Ferrari, F.F., Schmid, M.S., Cunsolo, S., Brambini, R., Noble, K., Sirks, L.A., Linders, T.E., Schoeneich-Argent, R.I., Koelmans, A.A., 2016. The effect of particle properties on the depth profile of buoyant plastics in the ocean. Sci. Rep. 6, 33882.

Kowalski, N., Reichardt, A.M., Waniek, J.J., 2016. Sinking rates of microplastics and potential implications of their alteration by physical, biological, and chemical factors. Mar. Pollut. Bull. 109 (1), 310–319.

Kukulka, T., Proskurowski, G., Morét-Ferguson, S., Meyer, D.W., Law, K.L., 2012. The effect of wind mixing on the vertical distribution of buoyant plastic debris. Geophys. Res. Lett. 39(7).

Lambert, S., 2013. Environmental Risk of Polymer and Their Degradation Products (Doctoral dissertation). University of York.

Law, K.L., Thompson, R.C., 2014. Microplastics in the seas. Science 345 (6193), 144–145.

Law, K.L., Morét-Ferguson, S., Maximenko, N.A., Proskurowski, G., Peacock, E.E., Hafner, J., Reddy, C.M., 2010. Plastic accumulation in the North Atlantic subtropical gyre. Science 329 (5996), 1185–1188.

Lechner, A., Ramler, D., 2015. The discharge of certain amounts of industrial microplastic from a production plant into the River Danube is permitted by the Austrian legislation. Environ. Pollut. 200, 159–160.

Lechner, A., Keckeis, H., Lumesberger-Loisl, F., Zens, B., Krusch, R., Tritthart, M., Glas, M., Schludermann, E., 2014. The Danube so colourful: a potpourri of plastic litter outnumbers fish larvae in Europe's second largest river. Environ. Pollut. 188, 177–181.

Lenz, R., Enders, K., Nielsen, T.G., 2016. Microplastic exposure studies should be environmentally realistic. Proc. Natl. Acad. Sci., 201606615.

Lithner, D., Larsson, Å., Dave, G., 2011. Environmental and health hazard ranking and assessment of plastic polymers based on chemical composition. Sci. Total Environ. 409 (18), 3309–3324.

Lobelle, D., Cunliffe, M., 2011. Early microbial biofilm formation on marine plastic debris. Mar. Pollut. Bull. 62 (1), 197–200.

Lusher, A.L., McHugh, M., Thompson, R.C., 2013. Occurrence of microplastics in the gastrointestinal tract of pelagic and demersal fish from the English Channel. Mar. Pollut. Bull. 67 (1), 94–99.

Lusher, A.L., Welden, N.A., Sobral, P., Cole, M., 2017. Sampling, isolating and identifying microplastics ingested by fish and invertebrates. Anal. Methods 9 (9), 1346–1360.

Ma, H., Nittrouer, J.A., Naito, K., Fu, X., Zhang, Y., Moodie, A.J., Wang, Y., Wu, B., Parker, G., 2017. The exceptional sediment load of fine-grained dispersal systems: example of the Yellow River, China. Sci. Adv. 3 (5), e1603114.

Mani, T., Hauk, A., Walter, U., Burkhardt-Holm, P., 2015. Microplastics profile along the Rhine River. Sci. Rep. 5, 17988.

Mason, S.A., Garneau, D., Sutton, R., Chu, Y., Ehmann, K., Barnes, J., Fink, P., Papazissimos, D., Rogers, D.L., 2016. Microplastic pollution is widely detected in US municipal wastewater treatment plant effluent. Environ. Pollut. 218, 1045–1054.

Masura, J., Baker, J., Foster, G., Arthur, C., 2015. Laboratory methods for the analysis of microplastics in the marine environment: recommendations for quantifying synthetic particles in waters and sediments. NOAA Technical Memorandum NOS-OR&R-48.

Mato, Y., Isobe, T., Takada, H., Kanehiro, H., Ohtake, C., Kaminuma, T., 2001. Plastic resin pellets as a transport medium for toxic chemicals in the marine environment. Environ. Sci. Technol. 35 (2), 318–324.

McCormick, A., Hoellein, T.J., Mason, S.A., Schluep, J., Kelly, J.J., 2014. Microplastic is an abundant and distinct microbial habitat in an urban river. Environ. Sci. Technol. 48 (20), 11863–11871.

McGoran, A.R., Clark, P.F., Morritt, D., 2017. Presence of microplastic in the digestive tracts of European flounder, *Platichthys flesus*, and European smelt, *Osmerus eperlanus*, from the river Thames. Environ. Pollut. 220, 744–751.

McMahon, C.R., Holley, D., Robinson, S., 1999. The diet of itinerant male Hooker's sea lions, *Phocarctos hookeri*, at sub-Antarctic Macquarie Island. Wildl. Res. 26 (6), 839–846.

Mintenig, S.M., Int-Veen, I., Löder, M.G., Primpke, S., Gerdts, G., 2017. Identification of microplastic in effluents of waste water treatment plants using focal plane array-based micro-Fourier-transform infrared imaging. Water Res. 108, 365–372.

Mizraji, R., Ahrendt, C., Perez-Venegas, D., Vargas, J., Pulgar, J., Aldana, M., Ojeda, F.P., Duarte, C., Galbán-Malagón, C., 2017. Is the feeding type related with the content of microplastics in intertidal fish gut? Mar. Pollut. Bull. 116 (1), 498–500.

Moore, C.J., Lattin, G.L., Zellers, A.F., 2011. Quantity and type of plastic debris flowing from two urban rivers to coastal waters and beaches of Southern California. J. Integr. Coast. Zone Manag 11 (1), 65–73.

Murphy, F., Ewins, C., Carbonnier, F., Quinn, B., 2016. Wastewater treatment works (WwTW) as a source of microplastics in the aquatic environment. Environ. Sci. Technol. 50 (11), 5800–5808.

Murray, F., Cowie, P.R., 2011. Plastic contamination in the decapod crustacean *Nephrops norvegicus* (Linnaeus, 1758). Mar. Pollut. Bull. 62 (6), 1207–1217.

Naidoo, T., Glassom, D., Smit, A.J., 2015. Plastic pollution in five urban estuaries of KwaZulu-Natal, South Africa. Mar. Pollut. Bull. 101 (1), 473–480.

Nalbone, J., 2015. Discharging microbeads to our waters: an examination of wastewater treatment plants in New York. Environmental Protection Bureau of the New York State Attorney General's Office.

Napper, I.E., Thompson, R.C., 2016. Release of synthetic microplastic plastic fibres from domestic washing machines: effects of fabric type and washing conditions. Mar. Pollut. Bull. 112 (1), 39–45.

Napper, I.E., Bakir, A., Rowland, S.J., Thompson, R.C., 2015. Characterisation, quantity and sorptive properties of microplastics extracted from cosmetics. Mar. Pollut. Bull. 99 (1), 178–185.

Nizzetto, L., Bussi, G., Futter, M.N., Butterfield, D., Whitehead, P.G., 2016. A theoretical assessment of microplastic transport in river catchments and their retention by soils and river sediments. Environ. Sci. Processes Impacts 18 (8), 1050–1059.

Obbard, R.W., Sadri, S., Wong, Y.Q., Khitun, A.A., Baker, I., Thompson, R.C., 2014. Global warming releases microplastic legacy frozen in Arctic Sea ice. Earths Futur. 2 (6), 315–320.

O'Brine, T., Thompson, R.C., 2010. Degradation of plastic carrier bags in the marine environment. Mar. Pollut. Bull. 60 (12), 2279–2283.

Oliveira, M., Ribeiro, A., Hylland, K., Guilhermino, L., 2013. Single and combined effects of microplastics and pyrene on juveniles (0+ group) of the common goby *Pomatoschistus microps* (Teleostei, Gobiidae). Ecol. Indic. 34, 641–647.

Pahl, S., Wyles, K.J., 2017. The human dimension: how social and behavioural research methods can help address microplastics in the environment. Anal. Methods 9 (9), 1404–1411.

Peters, C.A., Bratton, S.P., 2016. Urbanization is a major influence on microplastic ingestion by sunfish in the Brazos River Basin, Central Texas, USA. Environ. Pollut. 210, 380–387.

Pham, P.H., 2009. A New Angle on Plastic Debris in the Aquatic Environment: Investigating Interactions Between Viral Hemorrhagic Septicemia Virus (VHSV) and Inanimate Surfaces (Master's thesis). University of Waterloo.

Phillips, M.B., Bonner, T.H., 2015. Occurrence and amount of microplastic ingested by fishes in watersheds of the Gulf of Mexico. Mar. Pollut. Bull. 100 (1), 264–269.

PlasticsEurope, 2016. Plastics—The Facts 2016. PlasticsEurope, Brussels. http://www.plasticseurope.org/documents/document/20161014113313-plastics_the_facts_2016_final_version.pdf. Accessed 20 January 2017.

Reisser, J., Shaw, J., Hallegraeff, G., Proietti, M., Barnes, D.K., Thums, M., Wilcox, C., Hardesty, B.D., Pattiaratchi, C., 2014. Millimeter-sized marine plastics: a new pelagic habitat for microorganisms and invertebrates. PLoS One 9 (6), e100289.

Rice, S.P., Church, M., 2010. Grain-size sorting within river bars in relation to downstream fining along a wandering channel. Sedimentology 57 (1), 232–251.

Rochman, C.M., Hoh, E., Kurobe, T., Teh, S.J., 2013. Ingested plastic transfers hazardous chemicals to fish and induces hepatic stress. Sci. Rep 3, 3263.

Rochman, C.M., Cook, A.M., Koelmans, A.A., 2016. Plastic debris and policy: using current scientific understanding to invoke positive change. Environ. Toxicol. Chem. 35 (7), 1617–1626.

Rosenkranz, P., Chaudhry, Q., Stone, V., Fernandes, T.F., 2009. A comparison of nanoparticle and fine particle uptake by *Daphnia magna*. Environ. Toxicol. Chem. 28 (10), 2142–2149.

Sadri, S.S., Thompson, R.C., 2014. On the quantity and composition of floating plastic debris entering and leaving the Tamar Estuary, Southwest England. Mar. Pollut. Bull. 81 (1), 55–60.

Sanchez, W., Bender, C., Porcher, J.M., 2014. Wild gudgeons (*Gobio gobio*) from French rivers are contaminated by microplastics: preliminary study and first evidence. Environ. Res. 128, 98–100.

Schroeck, N.J., 2016. Microplastic pollution in the Great Lakes: state, federal, and common law solutions. Univ. Detroit Mercy Law Rev. 93, 273–291.

Setälä, O., Fleming-Lehtinen, V., Lehtiniemi, M., 2014. Ingestion and transfer of microplastics in the planktonic food web. Environ. Pollut. 185, 77–83.

Shiber, J.G., 1979. Plastic pellets on the coast of Lebanon. Mar. Pollut. Bull. 10 (1), 28–30.

Shiber, J.G., 1982. Plastic pellets on Spain's "costa del sol" beaches. Mar. Pollut. Bull. 13 (12), 409–412.

Silva-Cavalcanti, J.S., Silva, J.D.B., de França, E.J., de Araújo, M.C.B., Gusmão, F., 2017. Microplastics ingestion by a common tropical freshwater fishing resource. Environ. Pollut. 221, 218–226.

Su, L., Xue, Y., Li, L., Yang, D., Kolandhasamy, P., Li, D., Shi, H., 2016. Microplastics in taihu lake, China. Environ. Pollut. 216, 711–719.

Talsness, C.E., Andrade, A.J., Kuriyama, S.N., Taylor, J.A., Vom Saal, F.S., 2009. Components of plastic: experimental studies in animals and relevance for human health. Philos. Trans. R. Soc. Lond. B Biol. Sci. 364 (1526), 2079–2096.

Tanaka, K., Takada, H., Yamashita, R., Mizukawa, K., Fukuwaka, M.A., Watanuki, Y., 2013. Accumulation of plastic-derived chemicals in tissues of seabirds ingesting marine plastics. Mar. Pollut. Bull. 69 (1), 219–222.

Taylor, M.L., Gwinnett, C., Robinson, L.F., Woodall, L.C., 2016. Plastic microfibre ingestion by deep-sea organisms. Sci. Rep 6, 33997.

Teuten, E.L., Rowland, S.J., Galloway, T.S., Thompson, R.C., 2007. Potential for plastics to transport hydrophobic contaminants. Environ. Sci. Technol. 41 (22), 7759–7764.

Teuten, E.L., Saquing, J.M., Knappe, D.R., Barlaz, M.A., Jonsson, S., Björn, A., Rowland, S.J., Thompson, R.C., Galloway, T.S., Yamashita, R., Ochi, D., 2009. Transport and release of chemicals from plastics to the environment and to wildlife. Philos. Trans. R. Soc. Lond. B Biol. Sci. 364 (1526), 2027–2045.

Thompson, R.C., Olsen, Y., Mitchell, R.P., Davis, A., Rowland, S.J., John, A.W., McGonigle, D., Russell, A.E., 2004. Lost at sea: where is all the plastic? Science 304 (5672), 838.

Turra, A., Manzano, A.B., Dias, R.J.S., Mahiques, M.M., Barbosa, L., Balthazar-Silva, D., Moreira, F.T., 2014. Three-dimensional distribution of plastic pellets in sandy beaches: shifting paradigms. Sci. Rep 4, 4435.

UNEP, 2015. Plastic in Cosmetics. 38 pp, http://www.unep.org/ourplanet/september-2015/unep-publications/plastic-cosmetics-are-we-polluting-environment-through-our-personal.

Van Cauwenberghe, L., Janssen, C.R., 2014. Microplastics in bivalves cultured for human consumption. Environ. Pollut. 193, 65–70.

Van Cauwenberghe, L., Vanreusel, A., Mees, J., Janssen, C.R., 2013. Microplastic pollution in deep-sea sediments. Environ. Pollut. 182, 495–499.

Van Cauwenberghe, L., Devriese, L., Galgani, F., Robbens, J., Janssen, C.R., 2015a. Microplastics in sediments: a review of techniques, occurrence and effects. Mar. Environ. Res. 111, 5–17.

Van Cauwenberghe, L., Claessens, M., Vandegehuchte, M.B., Janssen, C.R., 2015b. Microplastics are taken up by mussels (*Mytilus edulis*) and lugworms (*Arenicola marina*) living in natural habitats. Environ. Pollut. 199, 10–17.

Van Franeker, J.A., Blaize, C., Danielsen, J., Fairclough, K., Gollan, J., Guse, N., Hansen, P.L., Heubeck, M., Jensen, J.K., Le Guillou, G., Olsen, B., 2011. Monitoring plastic ingestion by the northern fulmar Fulmarus glacialis in the North Sea. Environ. Pollut. 159 (10), 2609–2615.
Vasile, C. (Ed.), 2000. Handbook of Polyolefins, second ed. CRC Press, New York.
Vianello, A., Boldrin, A., Guerriero, P., Moschino, V., Rella, R., Sturaro, A., Da Ros, L., 2013. Microplastic particles in sediments of Lagoon of Venice, Italy: first observations on occurrence, spatial patterns and identification. Estuar. Coast. Shelf Sci. 130, 54–61.
von Moos, N., Burkhardt-Holm, P., Köhler, A., 2012. Uptake and effects of microplastics on cells and tissue of the blue mussel *Mytilus edulis* L. after an experimental exposure. Environ. Sci. Technol. 46 (20), 11327–11335.
von Moos, N., Bowen, P., Slaveykova, V.I., 2014. Bioavailability of inorganic nanoparticles to planktonic bacteria and aquatic microalgae in freshwater. Environ. Sci.: Nano 1 (3), 214–232.
Wagner, M., Scherer, C., Alvarez-Muñoz, D., Brennholt, N., Bourrain, X., Buchinger, S., Fries, E., Grosbois, C., Klasmeier, J., Marti, T., Rodriguez-Mozaz, S., 2014. Microplastics in freshwater ecosystems: what we know and what we need to know. Environ. Sci. Eur. 26 (1), 12.
Wang, W., Ndungu, A.W., Li, Z., Wang, J., 2017. Microplastics pollution in inland freshwaters of China: a case study in urban surface waters of Wuhan, China. Sci. Total Environ. 575, 1369–1374.
Woodall, L.C., Sanchez-Vidal, A., Canals, M., Paterson, G.L., Coppock, R., Sleight, V., Calafat, A., Rogers, A.D., Narayanaswamy, B.E., Thompson, R.C., 2014. The deep sea is a major sink for microplastic debris. R. Soc. Open Sci. 1 (4), 140317.
Wright, S.L., Kelly, F.J., 2017. Plastic and human health: a micro issue? Environ. Sci. Technol. 51 (12), 6634–6647.
Wright, S.L., Thompson, R.C., Galloway, T.S., 2013a. The physical impacts of microplastics on marine organisms: a review. Environ. Pollut. 178, 483–492.
Wright, S.L., Rowe, D., Thompson, R.C., Galloway, T.S., 2013b. Microplastic ingestion decreases energy reserves in marine worms. Curr. Biol. 23 (23), R1031–R1033.
Xanthos, D., Walker, T.R., 2017. International policies to reduce plastic marine pollution from single-use plastics (plastic bags and microbeads): a review. Mar. Pollut. Bull. 118, 17–26.
Zbyszewski, M., Corcoran, P.L., 2011. Distribution and degradation of fresh water plastic particles along the beaches of Lake Huron, Canada. Water Air Soil Pollut. 220 (1–4), 365–372.
Zbyszewski, M., Corcoran, P.L., Hockin, A., 2014. Comparison of the distribution and degradation of plastic debris along shorelines of the Great Lakes, North America. J. Great Lakes Res. 40 (2), 288–299.
Zettler, E.R., Mincer, T.J., Amaral-Zettler, L.A., 2013. Life in the "plastisphere": microbial communities on plastic marine debris. Environ. Sci. Technol. 47 (13), 7137–7146.
Zhang, K., Su, J., Xiong, X., Wu, X., Wu, C., Liu, J., 2016. Microplastic pollution of lakeshore sediments from remote lakes in Tibet plateau, China. Environ. Pollut. 219, 450–455.
Zitko, V., Hanlon, M., 1991. Another source of pollution by plastics: skin cleaners with plastic scrubbers. Mar. Pollut. Bull. 22 (1), 41–42.
Zubris, K.A.V., Richards, B.K., 2005. Synthetic fibers as an indicator of land application of sludge. Environ. Pollut. 138 (2), 201–211.

FURTHER READING

Booth, A.M., Hansen, B.H., Frenzel, M., Johnsen, H., Altin, D., 2016. Uptake and toxicity of methylmethacrylate based nanoplastic particles in aquatic organisms. Environ. Toxicol. Chem. 35 (7), 1641–1649.
Galloway, T.S., Cole, M., Lewis, C., 2017. Interactions of microplastic debris throughout the marine ecosystem. Nat. Ecol. Evol. 1, 0116.

CHAPTER 5

The Occurrence, Fate, and Effects of Microplastics in the Marine Environment

Wai Chin Li
Department of Science and Environmental Studies, The Education University of Hong Kong, Tai Po, Hong Kong, China

5.1 INTRODUCTION

The contamination of microplastics in the marine environment is regarded as a major risk for the health of marine organisms. Numerous studies have shown that many species suffer from plastic ingestion or entanglement (Gregory, 2009; Lusher, 2015; Auta et al., 2017). Because of light weight, high strength/weight ratio, and thermal degradation-resistant and biodegradation-resistant properties of plastic, it has resulted in the use of it in various applications (Table 5.1), for example, packaging, domestic personal cleaning products, or industrial construction materials (Andrady, 2011; Ghosh et al., 2013). According to PlasticsEurope (2016), the global plastic production has increased from 230 t in 2005 to 322 t in 2015. The most commonly used plastic types are polyvinyl chloride (PVC), nylons, and polyethylene terephthalate (PET), which tend to sink, and polyethylene (PE), polypropylene (PP), and polystyrene (PS), which tend to float (Table 5.1) (Avio et al., 2017). Other polymers include polyvinyl alcohol (PA), polyamide, polycarbonate (PC), acrylonitrile butadiene styrene (ABS), and high-impact polystyrene (HIPS) (Avio et al., 2017). Plastic is seldom recovered for recycling (< 5 %), which results in the accumulation in the marine environment (Sutherland et al., 2010). Once plastic enters the marine environment, it is difficult to degrade completely due to its biodegradation-resistant properties. Therefore, large plastic debris degrades into smaller fragments via different mechanisms such as weathering (Arthur et al., 2009; Andrady, 2017), photodegradation (Barnes et al., 2009), and biodegradation (O'Brine and Thompson, 2010) and thus becomes "microplastics" (<5 mm) (Cole et al., 2011). Also, microplastic can directly enter into the marine environment through river, industrial and urban discharge, sewage, and littering by beach goers (Culin and Bielic, 2016).

Microplastics have been studied and reported in the marine environment globally (Nel and Froneman, 2015; Fok and Cheung, 2015; Isobe et al., 2017; Kanhai et al., 2017). Marine organisms, including fish (Lusher et al., 2016), seabirds (Amélineau et al., 2016), sea turtles (Tourinho et al., 2010), invertebrates (Davidson and Dudas, 2016), and marine mammals (Besseling et al., 2015), are directly and indirectly vulnerable

Microplastic Contamination in Aquatic Environments
https://doi.org/10.1016/B978-0-12-813747-5.00005-9

Table 5.1 Types of plastic commonly found in the natural environment (Halden, 2010; Andrady, 2011; Ghosh et al., 2013; Li et al., 2016)

Type	Specific gravity	Use/application	Health effects
Polyester (PES)	1.40	Fibers, textiles	Cause eye and respiratory tract irritation and acute skin rashes (Ecology Center, 1996)
Polyethylene terephthalate (PET)	1.37	Carbonated drinks bottles, peanut butter jars, plastic film, microwavable packaging, tubes, pipes, insulation molding	Potential human carcinogen (Ecology Center, 1996)
Polyethylene (PE)	0.91–0.96	Wide range of inexpensive uses including supermarket bags, plastic bottles	
High-density polyethylene (HDPE)	0.94	Detergent bottles, milk jugs, tubes, pipes, insulation molding	Release of estrogenic chemicals resulting in changes in the structure of human cells (Ecology Center, 1996)
Polyvinyl chloride (PVC)	1.38	Plumbing pipes and guttering, shower curtains, window frames, flooring, films	Lead to cancer, birth defects, genetic changes, chronic bronchitis, ulcers, skin diseases, deafness, vision failure, indigestion, and liver dysfunction (Ecology Center, 1996)
Low-density polyethylene (LDPE)	0.91–0.93	Outdoor furniture, siding, floor tiles, shower curtains, clamshell packaging, films	
Polypropylene (PP)	0.85–0.83	Bottle caps, drinking straws, yogurt containers, appliances, car fenders (bumpers), plastic pressure pipe systems, tanks and jugs	
Polystyrene (PS)	1.05	Packaging foam, food containers, plastic tableware, disposable cups, plates, cutlery, CD, cassette boxes, tanks, jugs, building materials (insulation)	Irritate the eyes, nose, and throat and can cause dizziness and unconsciousness. Migrates into food and stores in body fat. Elevated rates of lymphatic and hematopoietic cancers for workers (Ecology Center, 1996)

Table 5.1 Types of plastic commonly found in the natural environment (Halden, 2010; Andrady, 2011; Ghosh et al., 2013; Li et al., 2016)—cont'd

Type	Specific gravity	Use/application	Health effects
High-impact polystyrene (HIPS)	1.08	Refrigerator liners, food packaging, vending cups, electronics	
Polyamides (PA) (nylons)	1.13–1.35	Fibers, toothbrush bristles, fishing line, under-the-hood car engine moldings, making films for food packaging	Lead to cancer, skin allergies, dizziness, headaches, spine pains, and system dysfunction (Ecology Center, 1996)
Acrylonitrile butadiene styrene (ABS)	1.06–1.08	Electronic equipment cases (e.g., computer monitors, printers, and keyboards), drainage pipe, automotive bumper bars	Airborne ultrafine particle (UFP) concentrations may be generated while printing with ABS, which leads to oxidative stress and inflammatory mediator release and could induce heart disease, lung disease, and other systemic effects (Card et al., 2008)
Polycarbonate (PC)	1.20–1.22	Compact disks, eyeglasses, riot shields, security windows, traffic lights, lenses, construction materials	Bisphenol A could be leached from polycarbonate products, which leads to liver function alternation, changes in insulin resistance, reproductive system, and brain function (Srivastava and Godara, 2012)
Polycarbonate/acrylonitrile butadiene styrene (PC/ABS)		A blend of PC and ABS that creates a stronger plastic. Used in car interior and exterior parts and mobile phone bodies	

to microplastic ingestion. More importantly, microplastics can adsorb hydrophobic contaminants or heavy metals from the surrounding seawater and potentially act as a vector for these contaminants to enter the food web (Reisser et al., 2014). Therefore, it is essential to understand the distribution and the potential hot spots of microplastics, as well as the effects of microplastics on different marine organisms. This article will review the occurrence of microplastics in the marine environment and identify potential hot spots. Moreover, the sources, fate, and effects will be summarized. Finally, recommendations are provided to control the sources of microplastics.

5.1.1 Primary Microplastics

Primary microplastics are regarded as the plastics manufactured to be microscopic (Cole et al., 2011). Industrial and domestic cleaning products generate most of the primary microplastics in the marine environment (Betts, 2008; Moore, 2008). Exfoliants are the most commonly reported primary microplastic types. For example, personal care products such as hand or facial cleaners and toothpaste can contain large amounts of microplastics, which are used as exfoliants (Lassen et al., 2015). According to Beach (1972), the US-patented microplastics used for exfoliants in skin cleaners have amorphous shapes without sharp edges with sizes of 74–420 μm and are materials such as polyolefin particles. The polyolefin particles used are usually composed of PS, PE, and PP plastics. The size, shape, color, and composition of the plastics differ and depend on the type of product (Zitko and Hanlon, 1991; Sundt et al., 2014; Hintersteiner et al., 2015; Lassen et al., 2015; Napper et al., 2015). For example, skin cleaners contain spherical threads and irregular microplastics comprising PE and PS with blue or white colors (Hintersteiner et al., 2015). Additional research has also reported that a single cosmetic product contains PE, PS granules (<5 mm), and PS spheres (<2 mm) (Fendall and Sewell, 2009). Recent research investigated the personal care products in the markets of European countries such as Norway and Switzerland and estimated that approximately 6% contain microplastics (Gouin et al., 2015). The products usually contain 0.05%–12% microplastics, with sizes ranging from 450 to 800 μm (Gouin et al., 2015). Of the microplastic particles, PE was found to be the most common microplastic type (93%) used in personal care cleaning products in the abovementioned countries in 2012.

In addition, industrial activities also generate microplastic particles, particularly in oil and gas industry as drilling fluids and abrasives (Gregory, 1996). For example, microplastics are used for air-blasting media to remove paint from metal surfaces and to clean engines (Sundt et al., 2014). According to Eriksen et al. (2013a, 2013b), acrylic, PS, melamine, polyester (PES), and poly allyl diglycol carbonate microplastics are all found in industrial abrasives. Other industrial activities such as plastic product fabrication use plastic resin pellets or flakes and plastic powder or fluff, which generate primary microplastics (Duis and Coors, 2016). These microplastics were accidently discharged into the marine environment during transportation or the production process (Moore, 2008). Previous research has reported that there was a high concentration of plastic resin pellets in the environment from the 1970s to the 1990s (Ryan et al., 2009; Karapanagioti, 2012), especially in areas near plastic processing and production plants. For example, it was reported that plastic resin pellets were observed on the beaches near plastic production sites that reach to 100,000/m (Mato et al., 2001).

Microplastics are also used in medical applications including dental polishing and are used as carriers to transfer active medicine agents (Lassen et al., 2015, Sundt et al., 2014). These medical residuals enter the marine environment via wastewater. Additionally, washing of domestic clothes can be another important source of microplastic

(Napper and Thompson, 2016). Browne et al. (2011) investigated microplastic contamination along shorelines at 18 different regions in six different continents. The study reported that fibers from washing clothes could not be effectively removed by sewage treatment plants, and thus, they enter the marine environment eventually. Polyester (78%) and acrylic (22%) particles were found on all six continents with microplastic contamination. The observed PES and acrylic were observed to be from washer discharge, rather than plastic cleaning tool fragmentation since the proportions of the PES fibers observed in the waste discharge and marine sediment were similar to those used for cloth production (Oerlikon, 2009). Also, the presence of fiber in the marine environment could be attributed to the degradation of cigarette butts (Wright et al., 2015) and fragmentation of maritime equipment (e.g., ropes and nets) (Cole, 2016).

5.1.2 Secondary Microplastics

Secondary microplastics are fragmented from large plastics into small size of debris (Ryan et al., 2009). Therefore, secondary marine microplastics can be attributed to the large plastic debris from land-based and ocean-based sources. It is estimated that about 75%–90% and 10%–25% of the plastic in the marine environment comes from land-based and ocean-based sources, respectively (Andrady, 2011; Mehlhart and Blepp, 2012; Nelms et al., 2017). For example, windblown plastic might be transported from refuse facilities to the marine environment (Barnes et al., 2009; Mehlhart and Blepp, 2012; Lambert et al., 2014). Additionally, some plastic debris may enter the ocean from uncovered landfills during natural hazards such as tsunamis, hurricanes, and strong seas, especially in developing countries without landfill management (Barnes et al., 2009; Lambert et al., 2014; Jambeck et al., 2015). The longevity of many plastics is still uncertain, and many can remain in the environment for months to centuries (Zheng et al., 2005; Barnes et al., 2009). After entering the marine environment, large plastic debris can fragment into small pieces via physical, biological, or chemical weathering processes resulting in reducing the structural integrity (Browne et al., 2007).

Weathering process is regarded as the most essential factor to induce plastic fragmentation (Arthur et al., 2009). One essential mechanism is photodegradation induced by sunlight. Oxidation of the polymer matrix is the result of ultraviolet radiation from sunlight and leads to the breakage of chemical bonds (Barnes et al., 2009). The lower temperature in the marine environment slows the rate of degradation of plastic, compared with beach or land conditions (Browne et al., 2007; Andrady, 2011; Andrady, 2017). Besides, surface foulants on floating microplastic hinder oxidation reactions induced by UV radiation (Weinstein et al., 2016) and reduce weathering degradation (Andrady, 2017). The degradation rate of plastic in the deep water and sediment is also minimized by the same reason (Muthukumar et al., 2011). Also, plastic debris is susceptible to fragmentation from a combination of mechanical forces including abrasion, wave action, and turbulence (Barnes et al., 2009). Additionally, some research has found that the improper disposal of biodegradable plastics might lead to secondary microplastic accumulation in the marine environment because traditional synthetic polymers are involved in some biodegradable plastics, which are not biodegradable (Thompson et al., 2004; O'Brine and Thompson, 2010). Also, biodegradable plastics require anaerobic conditions that are not present within seawater (Zettler et al., 2013).

5.2 SOURCES OF MICROPLASTICS IN THE MARINE ENVIRONMENT

5.2.1 Land-Based Sources of Microplastics

It is estimated that 75%–90% of the plastic debris in the marine environment can be attributed to land-based sources (Duis and Coors, 2016). For example, activities such as building of ships and ship recycling may introduce plastic debris along with river, industrial and urban discharge, sewage, and littering by beach goers (Culin and Bielic, 2016). The amount of land-based plastics is estimated to significantly increase by 2025 (Jambeck et al., 2015). One of the largest land-based sources is from sewage and storm water, especially in regions near industrial activities (Culin and Bielic, 2016). Those industrial and domestic microplastic residues are transported via sewage to the marine environment or are carried by natural hazards such as hurricanes and flooding. Another land-based source of microplastics is medicine, including ingestible and inhalable medicines. In those medicines, microplastics are used as delivering drugs for humans and animals (Wen et al., 2003; Kockisch et al., 2003; Corbanie et al., 2006). Similar to personal care products, microplastics in medicines are also transported to the marine environment via sewage (Cole et al., 2011). Moreover, a recent study suggested that atmospheric fallout might be another source of synthetic fibers in the marine environment, where 29% of these fibers are microplastics (Dris et al., 2016). Atmospheric fallout rates ranging from 2 to 355 particles/m^2/day were reported at the sampling sites. It is assumed that these fibers in the atmosphere come from several sources, including clothes and houses, degradation of macroplastics, and landfills or waste incineration (Dris et al., 2016). Because of their light weights, microplastics can be transported by the wind to the marine environment (Free et al., 2014).

5.2.2 Ocean-Based Source of Microplastics

The remaining 10%–25% of marine plastic is generated from ocean-based sources. Human activities such as shipping, fishing, recreation, and offshore industries contribute a large amount of plastic debris to the marine environment (Ramirez-Llodra et al., 2013). Research conducted by Good et al. (2010) reported that approximately 640,000 t of fishing gear is discarded in the ocean every year. The study indicated that floating plastics, including microplastics, accumulate in shipping routes, near to fishing areas, and in oceanic convergence zones (Cózar et al., 2014). Discarded and lost fishing items (called ghost gear) can entangle marine organisms. Some microplastics are irregularly shaped, which imply that they are fragments of large plastic debris, such as fibers (Ribic et al., 2010).

5.3 OCCURRENCE

5.3.1 Water Bodies

Microplastics have been reported and found globally in multiple seas, including the Arctic Ocean and Antarctic Ocean (see Table 5.2). Once microplastics enter the sea, they persist and accumulate in water bodies and are transported around the world via winds and surface currents (Lusher et al., 2015). According to Cózar et al. (2014), it has been estimated that approximately 7000–35,000 t of plastic, including microplastics, are floating and persistent in the open ocean. A similar study conducted by Eriksen et al. (2014) indicated that over 250,000 t and more than 5 trillion pieces of plastic have accumulated in the ocean, including in the Atlantic (Law et al. 2010), North Pacific (Carson et al. 2013), South Pacific (Eriksen et al., 2013a, b), and Indian Ocean gyres (Isobe et al., 2014, 2015), and the amount of plastic debris continuously increases. The occurrence of microplastics in the oceans differs, with a high abundance of microplastics found near the regions with high levels of industrial activities or high population densities and in remote areas far from human habitation (Derraik, 2002; Claessens et al., 2011; Vianello et al., 2013; Jambeck et al., 2015; Duis and Coors, 2016).

Ocean gyres and oceanic convergence zones are vulnerable to microplastic accumulation because microplastics can be transported to the center of the region via the rotational patterns of the currents (Karl, 1999). Gyres are found in the oceans globally and lead to the accumulation of microplastics at the global scale, which has been reported since the 1970s (Kukulka et al., 2012). The North Pacific Ocean has contributed significantly to the amount of global plastic debris, including microplastics, which are estimated to make up 33%–35% of the plastics in the area because of the size of the gyre. The abundance of microplastics in the North Pacific Ocean can be attributed to the high level of human activities on East Asia coast (Cózar et al., 2014). The North Pacific Central Gyre is known as the "Great Pacific garbage patch" (Kaiser, 2010), which is one of the largest accumulation zones of microplastics. It has been reported that the abundance of microplastics in the North Pacific Central Gyre has increased by more than two times over the last four decades (Goldstein et al., 2012). According to Eriksen et al. (2013a, b), it has been reported that a high number of microplastics were observed in the South Pacific subtropical gyre. The amount of microplastics in the South Pacific subtropical gyre could be attributed to the transportation of microplastics from the shores of Indonesia and Ecuador via the boundary currents (Eriksen et al., 2013a, b). Another study by Carson et al. (2013) investigated the abundance of plastics in the Eastern North Pacific

Table 5.2 Occurrence of microplastics found in marine environment

Location	Regions	Water column	Amount	Reference
Atlantic Ocean	Atlantic Ocean	Subsurface water	1.15 particles/m^3	Kanhai et al. (2017)
	Offshore, Ireland	Surface water	2.46 particles/m^3	Lusher et al. (2014)
	English Channel, the United Kingdom	Surface water	0.27 particles/m^3	Cole et al. (2014)
	St. Peter and St. Paul Archipelago, Brazil	Surface water	0.01 particles/m^3	Ivar do Sul et al. (2013)
	North Atlantic Gyre	Surface water	0.0041 particles/m^3	Law et al. (2010)
	Goiana estuary, Brazil	Estuary	0.26 particles/m^3	Lima et al. (2014)
	Caribbean Sea	Surface water	0.00028 particles/m^3	Law et al. (2010)
	North Atlantic subtropical gyre	Surface water	13–501 particles/m^3	Enders et al. (2015)
	North Atlantic gyre	Surface water	1.70 particles/m^3	Reisser et al. (2015)
	Portuguese coast	Surface water	0.02–0.04 particles/m^3	Frias et al. (2014)
Mediterranean and European seas	Norderney, Germany	Surface water	150–2400 particles/m^3	Norén (2007)
	Skagerrak, Sweden	Surface water	102,000 particles/m^3	Noren and Naustvoll (2010)
	Bay of Calvi, Corsica, France	Surface water	0.062 particles/m^3	Collignon et al. (2014)
	Gulf of Oristano, Sardinia, Italy	Surface water	0.15 particles/m^3	de Lucia et al. (2014)
	North Sea, Finland	Surface water	0–0.74 particles/m^3	Magnusson (2014)
Pacific Ocean	North Pacific Central Gyre	Surface water	0.017 particles/m^3	Carson et al. (2013)
	Bering Sea	Surface water	0.004–0.19 particles/m^3	Doyle et al. (2011)
	North Pacific subtropical gyre	Surface water	0.02–0.45 particles/m^3	Goldstein and Goodwin (2013)
	Eastern Pacific Ocean	Surface water	>2500 pieces km^{-2}	Law et al. (2014a, 2014b)
	Yangtze estuary system, East China Sea	Surface water	4137 particles/m^3	Zhao et al. (2014)
	Northeastern Pacific Ocean	Subsurface water	8–9200 particles/m^3	Desforges et al. (2014)

	North Pacific	Surface water	0.12 particles/m^3	Goldstein et al. (2012)
	Kuroshio current system	Surface water	0.034 particles/m^3	Yamashita and Tanimura (2007)
	Jinhae Bay	Surface water	182 ± 68 particles/L	Song et al. (2015)
	Geoje Island, South Korea	Surface water	16,000 particles/m^3	Song et al. (2014)
	South Pacific subtropical gyre	Surface water	0.0054 particles/m^3	Eriksen et al. (2013a, b)
	Australian coast	Surface water	0.00085 particles/m^3	Reisser et al. (2013)
Arctic polar waters	Arctic polar waters	Surface water	0.34 particles/m^3	Lusher et al. (2015)
Southern Ocean	Antarctica	Surface water	0.031 particles/m^3	Isobe et al. (2017)
Indian Ocean	East Asian Sea	Surface water	3.70 particles/m^3	Isobe et al. (2015)
	Seto Inland Sea	Surface water	0.39 particles/m^3	Isobe et al. (2014)

Gyre. It has been reported that the concentration of plastic in the Eastern North Pacific Gyre has been estimated to be three times higher than that in the South Pacific subtropical gyre (Carson et al., 2013; Eriksen et al., 2013a, b). Both natural and anthropogenic factors lead to the formation of an accumulation zone (Galgani et al., 2000; Lusher, 2015). Naturally, an accumulation zone could be formed by surface current convergence induced by winds. For example, the North Atlantic and South Pacific gyres are located in subtropical regions and lead to the presence of large-scale convergence zones (Lusher, 2015). It has been reported that there are an estimated 580,000 particles km^{-2} of microplastics in the North Atlantic and Caribbean Sea (Law et al., 2010). Sedimentation factors also are important in the formation of an accumulation zone (Jegou and Salomon, 1991). Sedimentation occurs in an area with low turbulence. Apart from the natural factors, anthropogenic activities, for example, shipping or fishing, will also increase the abundance and occurrence of microplastics in the marine environment (Cózar et al., 2014).

There was not any direct research of microplastics in polar region, for example, in Arctic or Antarctica before 2014. According to Zarfl and Matthies (2010), it is estimated that approximately 62,000–105,000 t of plastic will enter the Arctic Ocean per year, which can vary because of spatial heterogeneity, temporal variability, and different sampling methods (Zarfl and Matthies, 2010). Obbard et al. (2014) reported that 38–234 particles/m^3 of microplastics were collected and found from isolated locations in the Arctic Ocean, which is about 100 times higher than other study reported in the Pacific gyre (Goldstein et al., 2012). Two recent studies reported that an estimated 97.2% and 95% of the collected samples were microplastic fragments in East Greenland (Amélineau et al., 2016) and Svalbard (Lusher et al., 2015), respectively. In research by Lusher et al. (2015), the majority of microplastics were classified as fiber. The abundance of fiber microplastics can be attributed to the fragmentation of large plastic debris from shipping or fishing activities, which is possibly transported over a long distance from remote sites via ocean currents because dense urban development is not present near the Arctic (David et al., 2009).

The occurrence of microplastics is less likely in the southern hemisphere oceanic regions than in the oceans of the northern hemisphere because the Antarctic Ocean is the least populated area in the world, which leads to less improper plastic management (Avio et al., 2017). It is assumed that microplastics have not yet been transported to the oceans of the southern hemisphere. However, Maximenko et al. (2012) modeled that the southern hemisphere possibly suffered from plastic contamination by floating plastic debris. There is limited comprehensive research on the occurrence of microplastics in the southern hemisphere. The occurrence of pelagic microplastics in the Southern Ocean is still uncertain. A previous study has estimated that the current plastic load in the Southern Ocean surface waters is only approximately 10,000–40,000 t and most of that is composed of microplastics. This result provides insight into the amount of plastic in semiclosed seas such as the Southern Ocean (Cózar et al., 2014). In recent research by Isobe et al. (2017), a field survey was conducted in 2016 to collect microplastics from the Southern Ocean. It was reported that 44 pieces of microplastics were collected with five net tows. The total plastics observed at two stations near Antarctica were estimated to be in the order of 100,000 pieces km^{-2} (Isobe et al., 2017). This study suggested that there is a significant concentration of microplastics in the Southern Ocean, indicating the widespread nature of marine microplastics.

Microplastics persist and accumulate through the water column not only in surface waters but also in subsurface waters and on the seafloor (Lattin et al., 2004; Cózar et al., 2014). Similar to the occurrence of macroplastic debris, oceanographic factors highly influence the occurrence of microplastics in open seas, for example, oceanic convergence zones for plastic debris created by areas of upwelling. A study by Van Cauwenberghe et al. (2013) investigated the occurrence of microplastics in deep-sea sediments in the Atlantic Ocean and the Mediterranean Sea. On average, the abundance of microplastics was recorded as 1 piece per 25 cm^2. According to Fischer et al. (2015), 2020 particles/m^2 were found in the benthic sediments of the Kuril-Kamchatka Trench area, which is sparsely populated. Most of the microplastics in the study were fibers that are usually composed of high-density polymers (Browne et al., 2011; Claessens et al., 2011). This type of polymer usually accumulates close to the point where it entered the environment (Fischer et al., 2015). Therefore, it is assumed that these particles are transported by ocean currents (Engler, 2012). Another similar study also reported the presence of microfibers in benthic sediments in the Atlantic Ocean, the Mediterranean Sea, and the Indian Ocean (Woodall et al., 2014). It is estimated that there are 4 billion fibers km^{-2} present in the Indian Ocean seamount sediments.

5.3.2 Potential Hotspots

The East Asian Sea is regarded as one of the largest hot spots for microplastic pollution. According to the research by Jambeck et al. (2015), it is estimated that the amount of plastic debris from East Asian countries into the surrounding ocean is the highest in the world. When the plastic debris enters the ocean, it gradually degrades into microplastics due to photodegradation and mechanical weathering on beaches (Andrady, 2011). This degradation explains the high concentration of microplastics in East Asian seas. For example, the total amount of plastic in the seas around Japan is approximately 1.7 million pieces km^{-2}, which is 16 times higher than in the North Pacific and 27 times higher than in the world oceans, when compared with the research by Eriksen et al. (2014). Additionally, the surface density of plastic in East Asian seas (2422 g km^{-2}) is approximately seven times higher than in the North Pacific (337 g km^{-2}) (Eriksen et al., 2014; Isobe et al., 2015).

5.3.3 Beach and Beach Sediment

Many studies have reported that the primary reason for microplastic contamination on beaches and in beach sediment can be attributed to densely populated areas (Browne et al., 2011; Naidoo, 2015). The occurrence of microplastics on beaches was first reported in the 1970s (Gregory, 1977, 1978; Shiber, 1979). Industrial resin pellets (2–5 mm) were one of the major concerns in many early studies (Gregory, 1977, 1983; Shiber, 1979, 1982) because of their high abundances on beaches (up to 100,000 pellets m^{-1}) (Gregory, 1978). The major point sources were identified as large ports and local plastic industries. Over the last 30 years, plastic resin pellets have been recovered from several Pacific beaches in both Japan and Russia (Kusui and Noda, 2003). However, numerous beaches all over the world are still contaminated by microplastics, as illustrated in Table 5.3 (Martins and Sobral, 2011; Nel and Froneman, 2015; Kunz et al., 2016; Nel et al., 2017).

A positive relationship between the amount of microplastics and the high-density areas was observed by Browne et al. (2011). In addition, a study by Naidoo (2015) also suggested that the abundance of microplastics observed in the water around the city of Durban, KwaZulu-Natal, might be ascribed to the high human population density of the cities. A similar study by Romeo et al. (2015) also recorded higher amounts of microplastics in areas with higher levels of human activities when compared with different control sites. In Singapore, high microplastic abundances were associated with higher levels of human activity (Nor and Obbard, 2014). Recent research reported that an extremely high abundance of microplastics was observed in Fan Lau Tung Wan, Hong Kong, with 92% of the plastics classified as expanded polystyrene (Fok and Cheung, 2015). This finding could be attributed to the high population density and specific living style of Hong Kong people, including the use of insulated boxes for food or transportation of food (Fok and Cheung, 2015).

In contrast, many studies have suggested that there is no direct correlation between human population density and the abundance of microplastics on beaches or in beach sediments (Reisser et al., 2013; Laglbauer et al., 2014; Klein et al., 2015; Alomar et al., 2016). For example, according to Klein et al. (2015), there is no significant correlation between human population density and the amount of microplastics at a sample site close to a sewage discharge area. Additionally, research in the Mediterranean investigated the microplastic abundance near a densely populated coastal area and in a marine protected area (MPA) and did not observe any marked differences (Alomar et al., 2016). Moreover, Reisser et al. (2013) indicated that the amounts of microplastics in cities and remote locations in Australia are similar. Laglbauer et al. (2014) also did not observe significant spatial variation in microplastic accumulation between touristic and nontouristic beaches. This finding may indicate that the presence of microplastics on beaches and in beach sediment is not associated with the land-based sources or human population

Table 5.3 Occurrence of microplastics found in beach sediment

Continent	Location	Occurrence	Reference
Asia	Hong Kong	Average abundance of 5595 items/m^2 and maximum 258,408 items/m^2	Fok and Cheung (2015)
	India	10–180 items/m^2	Jayasiri et al. (2013)
	South Korea	8205 items/m^2	Lee et al. (2013a, b)
	Singapore	36.8 items/kg dry	Nor and Obbard (2014)
	South Korea	56–285,673 items/m^2	Kim et al. (2015)
Africa	Southern African coast	From 86.67 ± 48.68 to 754.7 ± 393 particles/m^2	Nel et al. (2017)
	Southeastern coastline of South Africa	From 688.9 ± 348.2 to 3308 ± 1449 particles/m^2	Nel and Froneman (2015)
	Canary Islands	<1 to >100 g/L	Baztan et al. (2014)
America	Brazil	200 items/0.01 m^2	Costa et al. (2010)
	Brazil	Average density of 82.1 items/m^2	Santos et al. (2009)
	Chile	<1–805 items/m^2	Hidalgo-Ruz and Thiel (2013)
	Nova Scotia	20–80 fibers/10 g	Mathalon and Hill (2014)
	Hawaii	0.12%–3.3% plastic by weight	Carson et al. (2011)
	Brazil	60 items/m^2	Ivar do Sul et al. (2009)
Europe	North Sea beach	0.2–0.8 fibers/50 mL	Browne et al. (2011)
	Island of Norderney	1–2 particles/kg dry sediment	Dekiff et al. (2014)
	German Baltic coast	0–7 particles/kg and 2–11 fibers/kg	Stolte et al. (2015)
	Malta Island	>1000 particles/m^2	Turner and Holmes (2011)
	East Frisian Islands, Germany	Maximum of 50,000 particles/kg	Liebezeit and Dubaish (2012)
	Greece	10–602 items/m^2	Kaberi et al. (2013)
	Portugal	133.3 items/m^2	Martins and Sobral (2011)
	Italy	672–2175 items/kg dry	Vianello et al. (2013)

density nearby; however, they are transported from the open ocean through hydrodynamic processes and ocean currents (Claessens et al., 2011; Nel and Froneman, 2015).

Apart from human activities, natural factors such as seasonal variation could also affect the abundance and distribution of microplastics on beaches. Lee et al. (2013a) investigated the seasonal influence on the abundance of microplastics and found that the occurrence of microplastics in beach sediment varies by season. A similar study by Cheung et al. (2016) found that the amount of microplastics was remarkably higher in the wet season than in the dry season in Hong Kong. This result suggests that the Pearl River Estuary to the west of Hong Kong plays a major role in the amount and distribution of microplastics in Hong Kong. A study conducted by Veerasingam et al. (2016) investigated the distribution and abundance of microplastics before and after flooding in surface sediments along the Chennai coast. The study observed that the abundance of microplastics was estimated to be three times higher after flooding than before flooding, which suggests that the fresh microplastics are washed through the rivers from land during the flood (Veerasingam et al., 2016). Moreover, winds and surface currents can affect the transportation and deposition of microplastics from the sea to beaches. More microplastics were reported on upwind beaches compared with downwind beaches, even though the windward beaches did not have plastic industries or fishing activity nearby (Ivar do Sul et al., 2009). This suggests that the surface currents were the driving forces for the transportation and deposition of microplastics to the windward beaches (Ivar do Sul et al., 2009). This result indicated that the winds and currents have different effects on microplastic and macroplastic distribution (Carson et al., 2013; Nel et al., 2017).

5.4 FATE OF MICROPLASTICS IN THE MARINE ENVIRONMENT

Once microplastics enter the marine environment, they can persist in the ocean for long periods because most plastic polymers are highly "corrosion-resistant" (Tamara, 2015; Duis and Coors, 2016). The occurrence of microplastics in different oceans is summarized in Table 5.2. In the marine environment, the fate of microplastics can be attributed to the plastic density and the degree of natural attenuation (Andrady, 2011; Li et al., 2016).

Buoyant plastics are often submerged but float within the top 1 m of water column. The plastic's density and the degree of biofouling will determine if it is buoyant or submerged in seawater. Plastics with lower densities than seawater (approximately 1.03 g cm^{-3}) are buoyant, while those with higher densities are submerged (Andrady, 2011). Floating microplastics are more easily transported over long distances to isolated regions, depending on the direction of winds, the ocean currents, and the geographic location of the coastline. Floating plastics will gather in the accumulation zones or plastic hot spots, for example, in oceanic gyres and the East Asian Sea. Furthermore, microplastics, even the lower-density plastics such as PE and PS, which are usually buoyant on the sea surface, can be transported to the seafloor. The mechanisms of microplastic transport still remain uncertain (Gregory, 2009). Early research reported that microplastics can be transported to the sea bottom as marine snow, which is produced as a biologically enhanced aggregation of small particles (Alldredge and Silver, 1988). These microaggregates are generally made up of different particles or organisms such as phytoplankton, organic or clay particles, and zooplankton fecal pellets. It is estimated that the sinking rate of marine snow ranges from 1 to 368 m/day (Alldredge and Silver, 1988). Another similar study suggested that biofouling could be the reason for the lower density of plastic found in marine bottom sediments (Lobelle and Cunlife, 2011). A previous study found that biofilm can develop in a short period (approximately 1 week). Therefore, the plastic coated with biofilm would finally sink because of the increase in the density of the plastic (Lobelle and Cunlife 2011). Cole (2016) also found that microplastics have been shown to be incorporated into zooplankton fecal pellets and can alter sinking rates, a possible pathway to the benthos.

Natural environment conditions such as sunlight, temperature, and oxygen play essential roles in the degradation of plastic. The degradation rate of plastic mostly depends on its location (Weinstein et al., 2016). Floating microplastics will be more affected by photodegradation because direct sunlight causes light-induced breakdown and induces the chemical transformation process (Cole et al., 2011; Andrady, 2017). The photodegradation rate is enhanced at higher temperature and humidity and increased level of UV radiation as well as mechanical weathering (Ho et al., 1999; Gregory and Andrady, 2003). However, seawater hinders the photodegradation process and decreases the degradation of microplastics (Gregory and Andrady, 2003; Andrady, 2011). The photodegradation process is decreased due to the low temperature and low oxygen concentration in the marine environment. It is estimated that the complete degradation of microplastics could take over 50 years (Müller et al., 2001). Besides, the additives such as UV stabilizers used in microplastic can hinder the weathering process by UV radiation (Andrady, 2017).

The rate of biodegradation depends on the molecular weight of the plastic. Those plastics with lower molecular weights are more easily biodegraded by microorganisms (Zheng et al., 2005). In contrast, the high molecular plastics such as PE and PS plastics are resistant to biodegradation (Guo et al., 2012). Shah et al. (2008) reported that nonoxidized PE could only be biodegraded by specific microbial strains, while biodegradable plastics are susceptible to biological degradation such as bacteria and fungi activities (Gregory and Andrady, 2003). Also, it is suggested that unstabilized PE is expected to degrade slower than unstabilized PP under comparable exposure because unstabilized PP can form more stable tertiary radicals (Gewert et al., 2015). Additionally, it has been reported that the carbonyl index of plastic determines the degree of polymer oxidation. The carbonyl index is regarded as the aging index of a polymer. Carbonyl indexes between 0.13 and 0.74 indicate that plastic is in an advanced stage of oxidation. For example, the morphology and chemical structure of plastic changes with age (Ter Halle et al., 2016).

5.5 EFFECTS ON ORGANISMS

5.5.1 Physical Effects

Recently, a study reviewed that over 690 species including seabirds, turtles, and fish have been reported to ingest microplastics and macroplastics (Provencher et al., 2017). Due to the ubiquity and small sizes, both benthic and pelagic marine organisms are vulnerable to microplastic ingestion (Tables 5.4 and 5.5) (Rummel et al., 2016). The physical effects of macroplastic ingestion have been widely reported, including internal or external lacerations and damage and digestive tract blockages, which lead to satiation, starvation, and physical deterioration (Lazar and Gracan, 2011; van Franeker et al., 2011; Yamashita et al., 2011; Wright et al., 2013; Brate et al., 2016; Rummel et al., 2016). The blockages of the digestive tract lead to dietary dilution that subsequently leads to the other effects, for example, decreased reproduction, drowning, diminished predator avoidance, impairment of feeding ability, transfer of contaminants from the seawater, and death (Gregory, 2009). Such adverse effects could also apply to marine organisms with smaller size, such as invertebrates, including zooplankton, molluscs, arthropods, shrimp (Welden and Cowie, 2016), and marine worms (Besseling et al., 2013) if they ingest microplastics and exhibit the effects seen by larger organisms ingesting macroplastics.

5.5.1.1 Fish

Microplastic ingestion by fish has been widely reported in the Mediterranean Sea (Romeo et al., 2015; Nadal et al., 2016; Alomar and Deudero, 2017; Guven et al., 2017), Atlantic Ocean (Lusher et al., 2016), North and Baltic Seas (Lenz et al., 2016; Rummel et al., 2016), and along the European Coast (Neves et al., 2015; Brate et al., 2016). The lack of nutrition, starvation, and decreased in fish populations are the potential results of long period of time of the ingested microplastics in the guts of fishes (Boerger et al., 2010; Cole et al., 2011).

The varied occurrence of microplastics in different species of fish could be attributed to their feeding behaviors (Rummel et al., 2016). Fish with unselective filter feeding behaviors, such as mackerels, may ingest more microplastics than other marine fish species. Mackerels are more susceptible due to their feeding habitat, which is mainly located in the pelagic zone and at the sea surface, where floating or neutrally buoyant microplastic particles may be available for ingestion (Lusher et al., 2013). Mackerels are also visual feeders and so select prey based on color or shape that could be confused with microplastics (Nøttestad et al., 2015). Carlos De Sá et al. (2015) reported that young common goby (*Pomatoschistus microps*) ingested microplastics after confusing microplastics with natural prey. The ingestion resulted in a significant reduction in predatory performance and efficiency, and thus, ontogenic developmental conditions may influence the prey selection capability. The mortality rate of young fish that have ingested microplastics is significantly higher when compared with the control in laboratory conditions (Mazurais et al., 2015).

Table 5.4 Occurrence of microplastic ingestion found in fish

Incidence	Locations	Ingested material	Mean number of particles per individual (± s.d.)	Reference
16.8%	Western Mediterranean Sea	Filament (86.36%), fragment (12.12%), and film (1.51%)	0.34 ± 0.07	Alomar and Deudero (2017)
34%	Turkey, Mediterranean	Fibers (70%) and hard plastic (20.8%), nylon (2.7%), rubber (0.8%), and miscellaneous plastic (5.5%)	2.36 ± n.d.	Guven et al. (2017)
18.2%	Italy, Mediterranean	–	1.32 ± n.d.	Romeo et al. (2015)
17.5%	Spain, Atlantic, and Mediterranean	Fibers (71%), spheres (24%), films (3.2%), and fragments (1.6%)	1.56 ± 0.5	Bellas et al. (2016)
68%	Balearic Islands, Mediterranean	–	3.75 ± 0.25	Nadal et al. (2016)
28%	Adriatic Sea	Fragments (57%), line (23%), film (11%), and pellet (9%)	1.39 ± n.d.	Avio et al. (2015)
19.8%	Coast of Portugal	Fibers (65.8%) and fragments (34.2%)	1.4 ± 0.66	Neves et al. (2015)
2.9%	Norwegian coast	–	1.77 ± n.d.	Brate et al. (2016)
11%	North Atlantic	Fibers (93%) and fragments (7%)	1.2 ± 0.54	Lusher et al. (2016)
77%	Tokyo Bay	Fragments (86.0%), beads (7.3%), filaments (5.3%), and foams (1.3%)	3.06 ± n.d.	Tanaka and Takada (2016)
	South Africa urban harbor	Fragments (51.2%), polystyrene (34.6%), films (7.3%), monofilament line (5.0%), and twine (1.5%)	5.1 ± n.d.	Naidoo et al. (2016)

Continued

Table 5.4 Occurrence of microplastic ingestion found in fish—cont'd

Incidence	Locations	Ingested material	Mean number of particles per individual (± s.d.)	Reference
0.3%	Australia and Southern Ocean	–	2 ± n.d.	Cannon et al. (2016)
10.4%	Gulf of Mexico	Filament (3.8%), fragment (2.6%), and film plastics (2.6%)	–	Phillips and Bonner (2015)
23%	North and Baltic Sea	Fiber (83%)	–	Lenz et al. (2016)
5.5%	North and Baltic Sea	PE (40%), PA (22%), PP (13%)	1.44 ± n.d.	Rummel et al. (2016)

Table 5.5 Occurrence of microplastic ingestion found in sea turtle, marine mammals, and marine invertebrates in the natural marine environment

Species	N	Locations	Mean number of particles per individual (± s.d.)	Reference
Marine mammal				
Harbor seal	100	The Netherlands	8 items	Bravo Rebolledo et al. (2013)
Fur seal	145	Macquarie Island, Australia	1–4/scat	Eriksson and Burton (2003)
Humpback whale	–	The Netherlands	17	Besseling et al. (2015)
Beaked whale	3	North and west coast of Ireland	29	Lusher et al. (2015)
Turtle				
Green turtle	24	Rio Grande do Sul, Brazil	11	Tourinho et al. (2010)
Marine invertebrates				
Blue mussels	45	Belgium, The Netherlands	3.7/10 g mussel	De Witte et al. (2014)
Blue mussels	36	North Sea, Germany	$0.36 \pm 0.07\ g^{-1}$	Van Cauwenberghe and Janssen (2014)
Blue mussels	–	Along the French, Belgian, and Dutch North Sea coast	0.2 ± 0.3 particles/g	Van Cauwenberghe et al. (2015)
Blue mussels	10	East coast of Canada	106–126 particles/mussel	Mathalon and Hill (2014)
Manila clams	–	British Columbia	0.9 ± 0.9 particles/g	Davidson and Dudas (2016)
Gooseneck barnacle	385	North Pacific	1–30	Goldstein and Goodwin (2013)
Norway lobster	120	Clyde, the United Kingdom	–	Murray and Cowie (2011)
Brown shrimp	110	Belgium	11.5 fibers per 10 g shrimp	Devriese et al. (2014)
Lugworm	–	Along the French, Belgian, and Dutch North Sea coast	1.2 ± 2.8 particles/g	Van Cauwenberghe et al. (2015)
Copepods	–	Northeast Pacific Ocean	1 particle every 34 copepods	Desforges et al. (2015)
Euphausiids	–	Northeast Pacific Ocean	1 particle every 17 euphausiids	Desforges et al. (2015)

Additionally, the color of the microplastics plays a role in microplastic ingestion by fish. Several studies have suggested that epipelagic and mesopelagic fishes were likely to ingest microplastics during normal feeding activities because the ingested microplastics were the same color as prey items (Boerger et al., 2010; Lusher et al., 2013). For example, plastics with dark colors such as blue or black were most commonly ingested by *Galeus melastomus* (Alomar and Deudero, 2017) and demersal fish from the Spanish Atlantic and Mediterranean coasts (Bellas et al., 2016) and from the English Channel (Lusher et al., 2013).

5.5.1.2 Invertebrates

Microplastics have been reported and found in zooplankton such as arrow worms, copepods, salps, and fish larvae (Cole et al., 2013) and cause various physical effects. Lee et al. (2013b) reported that the sizes and abundance of microplastic ingestion led to fecundity reduction of marine copepod *Tigriopus japonicus* and also affect the survival and development of their second generation. Besides, the PS beads ingested and accumulated in ovigerous females could be transferred to their offspring (Lee et al., 2013b). Additionally, reproductive effects induced by microplastic ingestion appear to occur in copepods and other types of zooplankton, which might lead to population-level consequences (Besseling et al., 2014; Lee et al., 2013b). Another similar study by Cole et al. (2013) also showed that microplastics adversely affect the health of another type of marine copepod *Centropages typicus* by decreasing feeding (Cole et al., 2013). It has been reported that suspension and filter feeders ingested more microplastics because of the specific feeding modes. These kinds of feeding modes are used to concentrate food from large volumes of water (Moore, 2008; Kaposi et al., 2014). For example, in a study conducted by Desforges et al. (2015), the calanoid copepod *Neocalanus cristatus* and the euphausiid *Euphausia pacifica*, which are suspension filter feeders, drew food particles to their feeding basket using the movement of their external appendages to produce a feeding current; therefore, nonmotile prey, including microplastics, were ingested.

Microplastics such as microfibers have been found in benthic sediments (Fischer et al., 2015). Therefore, benthic invertebrates such as lugworms (*Arenicola marina*) (Besseling et al., 2013), amphipods (*Orchestia gammarellus*) (Cole et al., 2013), blue mussels (*Mytilus edulis*) (Van Cauwenberghe and Janssen, 2014), and sea cucumbers (Taylor et al., 2016) have also been reported to ingest microplastics. Physical effects such as inflammatory responses and oxidative stress, which are manifested by the formation of granulocytomas, lysosomal membrane destabilization, increased phagocytic activity, and epithelial cell apoptosis, have appeared in blue mussels exposed to microplastics (Koehler et al. 2008; Von Moos et al., 2012). Moreover, reduction in feeding activity and energy reserves and weight loss were detected in lugworms that ingested microplastics (Besseling et al., 2013; Wright et al., 2013).

Some species such as blue mussels and Manila clams are popular for human consumption, which might pose a risk to human health. Several studies have reported that cultured blue mussels (Mathalon and Hill, 2014) and Manila clams (Davidson and Dudas, 2016) contain higher concentrations of microplastics than wild ones. A study conducted by Mathalon and Hill (2014) reported that the concentration of microplastics found in farmed mussels (178 particles/mussel) is 28%–40% higher than that of wild mussels (106–126 particles/mussel). A recent study also revealed similar findings in other commercial bivalves, Manila clams. It was reported that the amount of microplastics found in cultured Manila clams (0.9 ± 0.9 particles/g) was about two times higher than in wild Manila clams (1.7 ± 1.2 particles/g). It was assumed that cultured clams would have a high abundance of microplastics because the culture region is close to plastic farm infrastructures, such as antipredator netting and oyster culture ropes (Davidson and Dudas, 2016).

5.5.1.3 Sea Birds

Seabirds are affected from plastic ingestion worldwide (Fig. 5.1). A study by Wilcox et al. (2015) predicted the occurrence of plastic in the gastrointestinal tracts of 99 % of all seabird species by 2050, and 95% of the individuals within these species will have ingested plastic (>5 mm) by the same year. The ingestion of microplastics seldom causes the immediate death of seabirds, and the effects are not as severe compared with macroplastic ingestion (Lusher, 2015). The occurrence of microplastics in seabirds (occurrence, abundance in gut content, and spatial distribution) has been determined from foraging practices (van Franeker et al., 2011), feeding techniques, and diet (Teuten et al., 2009). Gulls, shearwaters, and fulmars ingested more floating microplastic because of their surface-foraging behavior, while zooplanktivorous species, such as the little auk (*Alle alle*), are more vulnerable to microplastic ingestion. The little auk is not able to differentiate microplastics from zooplankton leading to the direct ingestion of microplastics (Avery-Gomm et al., 2013). In addition, several studies have indicated that microplastic beads were found in zooplankton, including copepods, which result in the indirect ingestion of microplastics via prey (Cole et al., 2013; Desforges et al., 2015; Lin, 2016). Moreover, it has been found that young seabirds ingested more microplastics than adult seabirds (Kuhn and van Franeker, 2012; Provencher et al., 2014; Acampora et al., 2014). For example, Kuhn and van Franeker (2012) observed that young northern fulmars (*Fulmarus glacialis*) contained more plastic in their intestines than mature birds. A similar study by Carey (2011) reported that the nonbreeding short-tailed shearwater suffered from microplastic ingestion at a level higher than its breeding equivalent. Seabirds such as glaucous-winged gulls (*Larus glaucescens*) are able to eliminate microplastics from their gastrointestinal tracts via regurgitation (Lindborg et al., 2012). However, this suggests that the regurgitated microplastics may be transferred from the adult to the young birds (Carey, 2011; van Franeker et al., 2011; Kuhn and van Franeker, 2012).

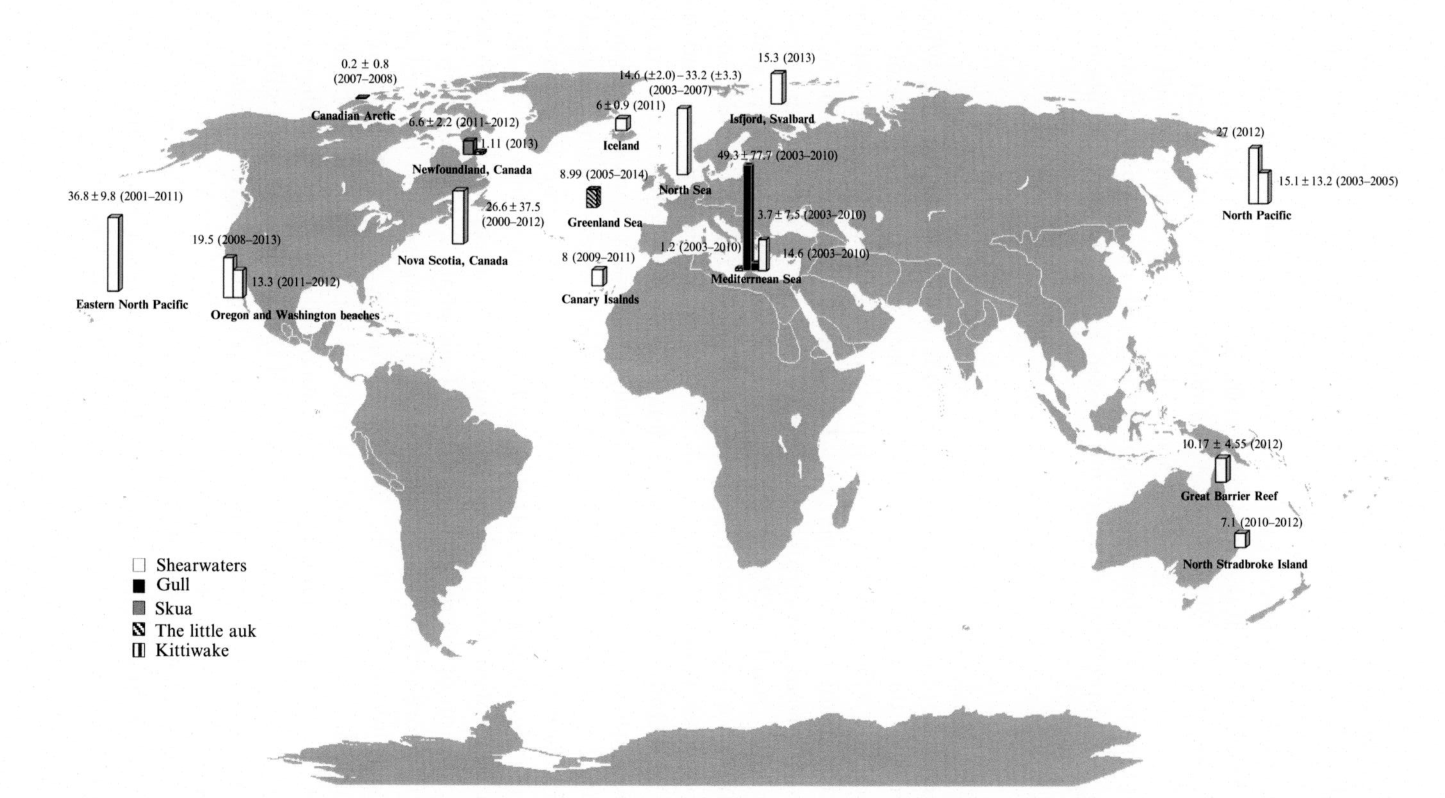

Fig. 5.1 Occurrence (mean pieces) of microplastic ingestion found in seabirds in the year 2000–14 (Provencher et al., 2010; van Franeker et al., 2011; Yamashita et al., 2011; Avery-Gomm et al., 2012; Kuhn and van Franeker, 2012; Rodríguez et al., 2012; Codina-García et al., 2013; Verlis et al., 2013; Acampora et al., 2014; Bond et al., 2013; Tanaka et al., 2013; Bond et al., 2014; Trevail et al., 2015; Fife et al., 2015; Amélineau et al., 2016; Terepocki et al., 2017).

5.5.1.4 Sea Turtles

Schuyler et al. (2014) reviewed the likelihood of a green turtle ingesting debris and found approximately a twofold increase from the estimated 30% likelihood in 1985 to a nearly 50 % likelihood in 2012. Many studies have reported macroplastic ingestion by sea turtles (Carman et al., 2014; Da Silva Mendes et al., 2015; Santos et al., 2015). However, direct research on the occurrence of microplastics in sea turtles is limited. A previous study reported that virgin plastic pellets with sizes between 2 and 5 mm were found in the stomachs of green turtles (Tourinho et al., 2010). It is possible that other species of sea turtle also suffer from microplastic ingestion directly or indirectly, depending on their feeding behavior (Schuyler et al. 2014; Van Houtan et al., 2016; Nelms et al., 2016). Microplastic ingestion by sea turtles could be attributed to a case of mistaken identity (Nelms et al., 2016). Turtles are primarily visual foragers and might misidentify the macroplastics as prey and consume them (Gregory, 2009; Hoarau et al., 2014). It has been reported that white and transparent plastics represent the vast majority of ingested plastics (Tourinho et al., 2010; Schuyler et al., 2012; Camedda et al., 2014; Hoarau et al., 2014). In addition to the visual factor, the formation of microbial biofilm on microplastics and the associated invertebrate grazers could also attract sea turtles to ingest these items because of sensory cues such as smell and taste (Reisser et al., 2014).

5.5.1.5 Trophic Transfer

Zooplankton is regarded as an important energy source in the world's oceans and is heavily preyed upon by fish and various marine organisms (Fig. 5.2). However, the impacts of microplastics transferred from zooplankton to other organisms are still uncertain. A study by Setälä et al. (2014) showed that microplastics were potentially transferred via planktonic organisms from one trophic level (mesozooplankton) to a higher level (macrozooplankton). Moreover, according to Murray and Cowie (2011), microplastics can be transferred from contaminated prey to predator by feeding fish seeded with polypropylene fibers to lobsters, suggesting that these omnivorous feeders can be exposed to microplastics via passive ingestion from sediment or via a trophic pathway. Another laboratory study investigated the trophic transfer of microplastics such as polystyrene spheres from mussels to crabs (Farrell and Nelson, 2013). This study documented that the microspheres were transferred from mussels and accumulated in the tissue samples from the stomach, hepatopancreas, ovary, and gills of crabs. However, the retention efficiency of 0.5 μm fluorescent polystyrene microspheres in mussels and crabs was found to be 0.04% and 0.28%, respectively, indicating that the ingested microspheres decreased after 21 days possibly due to egestion of the microspheres (Setälä et al., 2014). In the wild, trophic transfer of microplastics may be more severe because the prey or predator could be exposed to a range of types and sizes of microplastics and exposed to various concentrations over their lifetime (Farrell and Nelson, 2013).

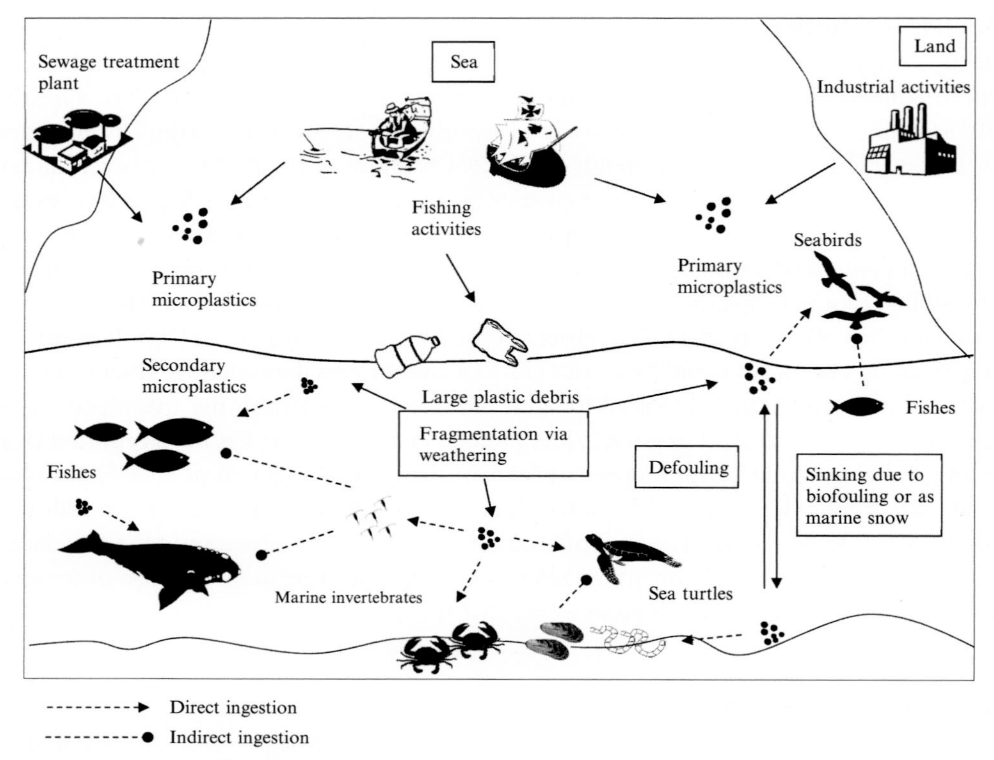

Fig. 5.2 Major sources and potential movement pathways of microplastics and its biological interactions in marine environment (Wright et al., 2013; Ivar do Sul and Costa, 2014; Li et al., 2016).

5.5.2 Chemical Effects

Apart from the physical effects induced by microplastic ingestion, it can, more importantly, be a vector for the absorption of hydrophobic organic pollutants. Multiple studies have reported that pollutants such as hexachlorinated hexanes, polycyclic aromatic hydrocarbons (PAHs), polychlorinated biphenyls (PCBs), and polybrominated diphenyl ethers (PBDEs) as well as organochlorine pesticides such as DDT can adsorb onto and are enriched in microplastics in seas and oceans (Ogata et al., 2009; Hirai et al., 2011). These organic contaminants can persist and accumulate in the environment and can be transported to remote locations, which cause toxic effects on marine organisms (Zarfl and Matthies, 2010). Microplastics are vulnerable to absorbing and carrying these contaminants due to their hydrophobic properties (Ivar do Sul and Costa, 2014). It has been found that the concentration of hydrophobic organic pollutants absorbed to microplastics could be up to million times higher than that in the surrounding seawater because of their hydrophobic properties and large surface-area-to-volume ratios

(Mato et al., 2001; Hong et al., 2017). Multiple studies have reported that hydrophobic organic pollutants were detected in microplastics found all over the world, for example, along US coasts (300–600 ng/g PCBs) (Ogata et al., 2009), in the Portuguese coast (0.02–15.56 ng/g PCBs and 0.16–4.5 ng/g DDT) (Frias et al., 2010), and in the central Pacific gyre (12–868 ng/g PAHs) (Hirai et al., 2011).

There are many factors that influence the sorption or desorption of hydroponic organic contaminants to microplastics, including the type, color, and size of the polymer and the degree of weathering (Frias et al., 2010). For example, Rochman et al. (2013) suggested that PS plastic is a source and sink for PAHs. This study reported that the concentrations of parent PAHs (PPAHs) on PS are estimated to be 8–200 times higher than PET, HDPE, PVC, LDPE, and PP. The concentrations of petrogenic PAHs are often higher in low molecular plastic such as naphthalene and alkyl naphthalenes and phenanthrenes and alkyl phenanthrenes, while the concentrations of pyrogenic PAHs are higher in plastics with higher molecular weight. Generally, PAHs in the microplastics were of lower molecular weight, suggesting petrogenic sources of PAHs in the plastics. It has been estimated that up to 9300 ng/g PAHs have been found in microplastic fragments at Kugenuma Beach. A study by Rochman et al. (2013) also suggested that the sorption rates and concentrations of PCBs and PAHs varied significantly among plastic types. For example, the HDPE, LDPE, and PP absorbed more concentration of contaminants such as PAHs and PCBs when compared with PET and PVC, because HDPE, LDPE, and PP reach equilibrium slower. Bakir et al. (2014) investigated the desorption rate of persistent organic pollutants (POPs) from microplastics under simulated physiological conditions. The study reported that desorption rate of POP was enhanced under simulated gut conditions (38°C and pH 4), which could be up to 30 times higher than in seawater conditions (18°C at seawater pH). The desorption rates were even more enhanced under gastric conditions more representative of warm-blooded organisms with pH 4, 38°C ranging from 2.1 for DDT desorption from PVC to 31.3 for desorption of DDT from PE (Gauthier-Clerc et al., 2002; Thouzeau et al., 2004).

Exposure to a high concentration of hydrophobic organic pollutants may result in disruption of the endocrine system, teratogenicity, liver and kidney problems, pathological and oxidative stress, and inflammation of the liver (Muirhead et al., 2006; Yogui and Sericano, 2009; Rochman et al., 2013). The hydrophobic organic pollutants sorbed onto microplastics can be metabolized in marine microorganisms. Chua et al. (2014) reported that *Allorchestes compresa* assimilated PBDEs into their tissues from microplastics and ingested approximately 45 contaminated particles. A similar result reported the assimilation of PBDEs by fish into their tissues (Wardrop et al., 2016).

Moreover, exposure to polychlorinated dibenzo-p-dioxins (PCDDS) and PCBs resulted in endocrine disruption of fish in arctic (Yogui and Sericano, 2009). Additionally, alterations in steroid hormones, thyroid hormones, and prolactin in the glaucous gull were reported because of exposure to POPs (Verboven et al., 2008).

Plastic additives also play an important role in causing chemical effects on marine organisms. Plastic additives, such as phthalates, bisphenol A (BPA), alkylphenols, and PBDEs, usually added into plastics during the manufacturing processes to enhance their properties or improve heat resistance, reduce oxidative damage and microbial degradation (Browne et al., 2007; Thompson et al., 2009; Avio et al., 2017). It has been reported that some types of plastic contain higher levels of additives than others (Lithner et al., 2011). Polyvinylchloride (PVC) is considered to be the polymer with the highest use of additives. For instance, heat stabilizers are added to maintain the stability of the polymer during production, and plasticizers such as phthalates are used to soften plastics, which can be up to 50 % of the polymer weight in some PVC products (Zweifel, 2001; Oehlmann et al., 2009; Talsness et al., 2009). Sporadic high concentrations of BPA were detected in plastic fragments from remote coasts of Costa Rica (730 ng/g) and open-ocean fragments in the central Pacific gyre (283 ng/g) (Hirai et al., 2011). Oehlmann et al. (2009) reviewed the biological impacts of plasticizers on marine organisms such as molluscs, crustaceans, fish, and amphibians. For example, phthalate esters (PAEs) and PBDE can have various noxious, toxic effects such as endocrine disruption on marine organisms even at very low concentrations (Oehlmann et al., 2009; Tanaka et al., 2013; Mariana et al., 2016; Baini et al., 2017). Plasticizers can interact with hormone synthesis and alter reproduction or other physiological and metabolic functions by causing oxidative stress and immunotoxicity in organisms (Mathieu-Denoncourt et al., 2015). PBDE, including BDE-183 and BDE-209, was detected in the tissues of short-tailed shearwaters possibly resulting in endocrine disruption (Tanaka et al., 2013). Moreover, exposure to phthalates, such as diethyl phthalate (DEP) and benzyl butyl phthalate (BBP), may change the behavior of fish (Barse et al., 2007; Oehlmann et al., 2009), for example, 5 mg of DEP L^{-1} found to alter the diets and feeding behavior of the European carp (*Cyprinus carpio*) (Barse et al., 2007).

In addition, microplastics could be also a vector for heavy-metal contamination. Recently, many studies reported that heavy metals could also be adsorbed to the microplastics suspended in the marine environment (Holmes et al., 2012; Holmes et al., 2014; Rochman et al., 2014; Brennecke et al., 2016). Brennecke et al. (2016) investigated the role of microplastics as a vector for heavy metals by examining the adsorption of heavy metals such as copper (Cu) and zinc (Zn) from an antifouling paint to virgin PS beads and aged PVC fragments in the marine environment. This study indicated that Cu and Zn were adsorbed and accumulated on the PS and PVC and the concentrations of Cu and Zn were around 800 times higher than in the surrounding seawater (Brennecke et al., 2016). The primary source of heavy metals is leaching from antifouling paints and mining waste (Tornero and Hanke, 2016). The surface properties and porosity of microplastics play essential roles in the degree of metal adsorption. It has been reported that there is a relationship between the absorption rate of metal and the surface area and polarity of plastic (Holmes et al., 2014; Rochman et al., 2014). For example, a study by

Brennecke et al. (2016) showed that aged PVC fragments have higher adsorption rates than virgin PS beads possibly due to a higher surface area and polarity of the PVC fragments. In addition, natural attenuation such as weathering, photodegradation, or biodegradation leads to the exhibition of different porosity characteristics in plastics, which might promote the adsorption of metals from the seawaters (Mato et al., 2001; Morét-Ferguson et al., 2010; Holmes et al., 2012). The concentration of metal absorbed to microplastics is much higher than that in seawater, which is toxic and poses a risk to marine organisms. It has been suggested that heavy metals are bioavailable and toxic elements from the heavy metal could be extracted in the acidic digestive tract and accumulate in fatty tissues such as blubber (Holmes, 2013; El-Moselhy et al., 2014). Long-term exposure to the ingestion of heavy-metal polluted plastic would lead to chronic diseases induced by biomagnification (Brennecke et al., 2016).

5.6 RECOMMENDATIONS

As mentioned above, the majority of plastic in the marine environment comes from land-based sources; therefore, sewage management, especially in the developing countries, is one of the essential ways to reduce microplastic entering the marine environment. In some European countries, almost all municipal wastewater is collected and subject to some form of tertiary treatment. On the contrary, it is estimated that 90% of all wastewater generated in developing countries is discharged without primary treatment (Corcoran et al. 2010). According to Carr et al. (2016), the primary treatment zones can remove the majority of microplastics in sewage during the skimming and settling treatment processes. Also, microbes could be applied to sewage wastewater treatment plants to eliminate the abundance of microplastics from wastewater discharge. Microbes could also be utilized to biodegrade microplastic polymers (Auta et al., 2017). It has been reported that microbes can utilize polymers as sources of carbon and energy (Caruso, 2015). Various bacteria species have been found to degrade plastic polymers. For instance, PE could be degraded by *Staphylococcus* sp., *Pseudomonas* sp., and *Bacillus* sp., isolated from soil (Singh et al., 2016). A similar study by Asmita et al. (2015) found that microorganisms such as *Pseudomonas aeruginosa*, *Bacillus subtilis*, *Staphylococcus aureus*, *Streptococcus pyogenes*, and *Aspergillus niger* can potentially degrade plastics such as PET and PS. The use of microbes to degrade plastic is regarded as environmentally safe.

Also, the reinforcement of legislation to control the plastic contamination from ships is important. Recently, garbage pollution, including plastic pollution from ships and mineral resource platforms at sea, has been controlled by Annex V in international marine environmental conventions "International Convention for the Prevention of Pollution from Ships," which was ratified by 147 nations (IMO, 2015). Annex V prohibits the discharge of plastics from ships, including plastic package, plastic residues, and fishing items. However, Annex V can hardly control the garbage discharge from ships under 400 GT

(Culin and Bielic, 2016) because the vast majority of the global fishing vessels are not required to cassette operations of garbage discharge (Chen and Liu, 2013). This could explain why fishing vessels are often the greatest sources of plastic debris in the sea (Chen and Liu, 2013). For example, according to Topping et al. (1997), 75.2% of observed fishing vessels operating along Canada's east coast discharged debris into the sea after the introduction of Annex V. Moreover, it is difficult to detect violations by ships because the linkage between the discarded plastic debris and a particular ship is difficult to define, which might result in the continuous dumping of plastic debris from ships (Topçu et al., 2013; Tubau et al., 2015).

In addition, education for plastic waste management is one of the most important tools to eliminate microplastic pollution in the marine environment. Therefore, awareness should be raised by implementing a series of campaigns for different stakeholders, especially those in the marine business. For example, according to the Seafarers' Training, Certification, and Watchkeeping Code, maritime officers are required to learn about the prevention of pollution to the marine environment. Additionally, the International Maritime Organization (IMO) developed a "marine environmental awareness" course to provide information on sustainability of marine environment and the importance of regulations and procedures of shipping and raise humans' environmental awareness (IMO, 2015). These programs can raise awareness and promote positive behavior and attitude changes among marine crews (Chen and Liu, 2013).

5.7 CONCLUSION

Microplastics are ubiquitous in the marine environment and can be attributed to land-based and ocean-based sources. Industrial and household products are one of the most important land-based sources of primary microplastics entering the marine environment. Ocean-based sources usually include large plastic discarded from shipping vessels or fishing activities in which secondary microplastics are generated from large plastic fragmentation. Microplastics are commonly found in the water columns, surface water, deep-sea sediments, beaches, and beach sediments all over the world. One potential hot spot has been identified in East Asian seas where the microplastic density is approximately seven times higher than in the North Pacific. This hot spot can be attributed to natural factors such as hydrodynamic processes and ocean currents and human factors including high levels of human and industrial activities. The accumulation of microplastics in the marine environment will pose a risk to marine organisms including marine invertebrates, fish, seabirds, sea turtles, and marine mammals such as whales via ingestion. Moreover, ingested microplastics may absorb a high concentration of hydrophobic organic pollutants and heavy metals prior to ingestion, which could be assimilated in the tissues of the organisms. Thus, well-developed sewage management and legislation reinforcement is required to control the plastic entering to the marine environment. Additionally, education for plastic waste management should be promoted to raise the awareness from different stakeholders. In addition, remediation approaches such as the use of microbes to degrade plastic polymers could be applied to sewage wastewater treatment plants.

ACKNOWLEDGMENT

Financial support from the Department of Science and Environmental Studies of The Education University of Hong Kong is gratefully acknowledged.

REFERENCES

Acampora, H., Schuyler, Q.A., Townsend, K.A., Hardesty, B.D., 2014. Comparing plastic ingestion in juvenile and adult stranded short-tailed shearwaters (*Puffinus tenuirostris*) in eastern Australia. Mar. Pollut. Bull. 78 (1-2), 63–68.

Alldredge, A.L., Silver, M.W., 1988. Characteristics, dynamics and significance of marine snow. Prog. Oceanogr. 20, 41–82.

Alomar, C., Deudero, S., 2017. Evidence of microplastic ingestion in the shark *Galeus melastomus* Rafinesque, 1810 in the continental shelf off the western Mediterranean Sea. Environ. Pollut. 223, 223–229.

Alomar, C., Estarellas, F., Deudero, S., 2016. Microplastics in the Mediterranean Sea: deposition in coastal shallow sediments, spatial variation and preferential grain size. Mar. Environ. Res. 115, 1–10.

Amélineau, F., Bonnet, D., Heitz, O., Mortreux, V., Harding, A.M.A., Karnovsky, N., Walkusz, W., Fort, J., Grémillet, D., 2016. Microplastic pollution in the Greenland Sea: background levels and selective contamination of planktivorous diving seabirds. Environ. Pollut. 219, 1131–1139.

Andrady, A.L., 2011. Microplastics in the marine environment. Mar. Pollut. Bull. 62, 1596–1605.

Andrady, A.L., 2017. The plastic in microplastics: a review. Mar. Pollut. Bull. 119 (1), 12–22.

Arthur, C., Baker, J., Bamford, H., 2009. Proceedings of the International Research Workshop on the Occurrence, Effects and Fate of Micro-plastic Marine Debris, September 9–11, 2008. NOAA Technical Memorandum NOS-OR&R-30.

Asmita, K., Shubhamsingh, T., Tejashree, S., 2015. Isolation of plastic degrading micro-organisms from soil samples collected at various locations in Mumbai, India. Curr. World Environ. 4 (3), 77–85.

Auta, H.S., Emenike, C.U., Fauziah, S.H., 2017. Distribution and importance of microplastics in the marine environment: a review of the sources, fate, effects, and potential solutions. Environ. Int. 102, 65–176.

Avery-Gomm, S., O'Hara, P.D., Kleine, L., Bowes, V., Wilson, L.K., Barry, K.L., 2012. Northern Fulmars as biological monitors of trends of plastic pollution in the eastern North Pacific. Mar. Pollut. Bull. 64, 1776–1781.

Avery-Gomm, S., Provencher, J.F., Morgan, K.H., Bertram, D.F., 2013. Plastic ingestion in marine-associated bird species from the eastern North Pacific. Mar. Pollut. Bull. 72, 257–259. https://doi.org/10.1016/j.marpolbul.2013.04.021.

Avio, C.G., Gorbi, S., Regoli, F., 2015. Experimental development of a new protocol for extraction and characterization of microplastics in fish tissues: first observations in commercial species from Adriatic Sea. Mar. Environ. Res. 111, 18–26.

Avio, C.G., Gorbi, S., Regoli, F., 2017. Plastics and microplastics in the oceans: from emerging pollutants to emerged threat. Mar. Environ. Res. 128, 2–11.

Baini, M., Martellini, T., Cincinelli, A., Campani, T., Minutoli, R., Panti, C., Finoia, M., Fossi, M., 2017. First detection of seven phthalate esters (PAEs) as plastic tracers in superficial neustonic/planktonic samples and cetacean blubber. Anal. Methods 9 (9), 1512–1520.

Bakir, A., Rowland, S.J., Thompson, R.C., 2014. Enhanced desorption of persistent organic pollutants from microplastics under simulated physiological conditions. Environ. Pollut. 185, 16–23.

Barnes, D.K., Galgani, F., Thompson, R.C., Barlaz, M., 2009. Accumulation and fragmentation of plastic debris in global environments. Philos. Trans. R. Soc. Lond. B 364, 1985–1998.

Barse, A.V., Chakrabarti, T., Ghosh, T.K., Pal, A.K., Jadhao, S.B., 2007. Endocrine disruption and metabolic changes following exposure of *Cyprinus carpio* to diethyl phthalate. Pestic. Biochem. Physiol. 88, 36–42.

Baztan, J., Carrasco, A., Chouinard, O., Cleaud, M., Gabaldon, J.E., Huck, T., Jaffrès, L., Jorgensen, B., Miguelez, A., Paillard, C., Vanderlinden, J.-P., 2014. Protected areas in the Atlantic facing the hazards of micro-plastic pollution: first diagnosis of three islands in the Canary Current. Mar. Pollut. Bull. 80, 302e311. https://doi.org/10.1016/j.marpolbul.2013.12.052.

Beach, W.J., 1972. United States Patent 3,645,904 for skin cleaner. Patented February 29, 1972. .

Bellas, J., Martnez-Armental, J., Martnez-Camara, A., Besada, V., Martnez-Gomez, C., 2016. Ingestion of microplastics by demersal fish from the Spanish Atlantic and Mediterranean coasts. Mar. Pollut. Bull. 109, 55–60.

Besseling, E., Foekema, E.M., Van Franeker, J.A., Leopold, M.F., Kühn, S., Bravo Rebolledo, E.L., Heße Mielke, L., Ijzer, J., Kamminga, P., Koelmans, A.A., 2015. Microplastic in a macro filter feeder: Humpback whale Megaptera novaeangliae. Mar. Pollut. Bull. 95, 248–252. https://doi.org/10.1016/j.marpolbul.2015.04.007.

Besseling, E., Wang, B., Lurling, M., Koelmans, A.A., 2014. Nanoplastic affects growth of *S. obliquus* and reproduction of *D. magna*. Environ. Sci. Technol. 48, 12336–12343.

Besseling, E., Wegner, A., Foekema, E.M., van den Heuvel-Greve, M., Koelmans, A.A., 2013. Effects of microplastic on fitness and PCB bioaccumulation by the lugworm *Arenicola marina* (L.). Environ. Sci. Technol. 47, 593–600.

Betts, K., 2008. Why small plastic particles may pose a big problem in the oceans. Environ. Sci. Technol. 42, 8995.

Boerger, C.M., Lattin, G.L., Moore, S.L., Moore, C.J., 2010. Plastic ingestion by planktivorous fishes in the north Pacific central gyre. Mar. Pollut. Bull. 60, 2275–2278.

Bond, A.L., Provencher, J.F., Daoust, P.Y., Lucas, Z.N., 2014. Plastic ingestion by fulmars and shearwaters at Sable Island, Nova Scotia, Canada. Mar. Pollut. Bull. 87 (1), 68–75.

Bond, A.L., Provencher, J.F., Elliot, R.D., Ryan, P.C., Rowe, S., Jones, I.L., Robertson, G.J., Wilhelm, S.I., 2013. Ingestion of plastic marine debris by common and thick-billed Murres in the northwestern Atlantic from 1985 to 2012. Mar. Pollut. Bull. 77 (1), 192–195.

Brate, I.L.N., Eidsvoll, D.P., Steindal, C.C., Thomas, K.V., 2016. Plastic ingestion by Atlantic cod (*Gadus morhua*) from the Norwegian coast. Mar. Pollut. Bull. 112, 105–110.

Bravo Rebolledo, E.L., van Franeker, J.A., Jansen, O.E., Brasseur, S.M., 2013. Plastic ingestion by harbour seals (*Phoca vitulina*) in The Netherlands. Mar. Pollut. Bull. 67 (1), 200–202.

Brennecke, D., Duarte, B., Paiva, F., Cacador, I., Clode, J.C., 2016. Microplastics as vector for heavy metal contamination from the marine environment. Estuar. Coast. Shelf Sci. 178, 189–195.

Browne, M.A., Crump, P., Nivens, S.J., Teuten, E., Tonkin, A., Galloway, T., Thompson, R., 2011. Accumulation of microplastics on shorelines worldwide: sources and sinks. Environ. Sci. Technol. 45 (21), 9175–9179.

Browne, M.A., Galloway, T., Thompson, R., 2007. Microplastic—an emerging contaminant of potential concern? Integr. Environ. Assess. Manag. 3, 559–561.

Camedda, A., Marra, S., Matiddi, M., Massaro, G., Coppa, S., Perilli, A., Ruiu, A., et al., 2014. Interaction between loggerhead sea turtles (*Caretta caretta*) and marine litter in Sardinia (Western Mediterranean Sea). Mar. Environ. Res. 100, 25–32.

Cannon, S.M.E., Lavers, J.L., Figueiredo, B., 2016. Plastic ingestion by fish in the Southern Hemisphere: a baseline study and review of methods. Mar. Pollut. Bull. 107, 286–291.

Card, J.W., Zeldin, D.C., Bonner, J.C., Nestmann, E.R., 2008. Pulmonary applications and toxicity of engineered nanoparticles. Lung Cell. Mol. Physiol. 295 (3), 400–411.

Carey, M.J., 2011. Intergenerational transfer of plastic debris by Short-tailed Shearwaters (*Ardenna tenuirostris*). Emu 111, 229–234.

Carlos De Sá, L., Luís, L., Guilhermino, L., 2015. Effects of microplastics on juveniles of the common goby (*Pomatoschistus microps*): confusion with prey, reduction of the predatory performance and efficiency, and possible influence of developmental conditions. Environ. Pollut. 196, 359–362.

Carman, V.G., Acha, E.M., Maxwell, S.M., Albareda, D., Campagna, C., Mianzan, H., 2014. Young green turtles, *Chelonia mydas*, exposed to plastic in a frontal area of the SW Atlantic. Mar. Pollut. Bull. 78 (1–2), 56–62.

Carr, S.A., Liu, J., Tesoro, A.G., 2016. Transport and fate of microplastic particles in wastewater treatment plants. Water Res. 91, 174–182.

Carson, H.S., Colbert, S.L., Kaylor, M.J., McDermid, K.J., 2011. Small plastic debris changes water movement and heat transfer through beach sediments. Mar. Pollut. Bull. 62, 1708e1713. https://doi.org/10.1016/j.marpolbul.2011.05.032.

Carson, H.S., Nerheim, M.S., Carroll, K.A., Eriksen, M., 2013. The plastic-associated microorganisms of the North Pacific Gyre. Mar. Pollut. Bull. 75 (1), 126–132.

Caruso, G., 2015. Plastic degrading microorganisms as a tool for bioremediation of plastic contamination in aquatic environments. Pollut. Effects Control 3, e112. https://doi.org/10.4172/2375-4397.1000e112.

Cauwenberghe, L.V., Vanreusel, A., Mees, J., Janssen, C.R., 2013. Microplastic pollution in deep-sea sediments. Environ. Pollut. 182, 495–499.

Chen, C.-L., Liu, T.-K., 2013. Fill the gap: developing management strategies to control garbage pollution from fishing vessels. Mar. Policy 40, 34–40.

Cheung, P.K., Cheung, T.O., Fok, L., 2016. Seasonal variation in the abundance of marine plastic debris in the estuary of a subtropical macro-scale drainage basin in South China. Sci. Total Environ. 562, 658–665.

Chua, E.M., Shimeta, J., Nugeoda, D., Morrison, P.D., Clarke, B.O., 2014. Assimilation of polybrominated diphenyl ethers from microplastics by the marine amphipod, Allorchestes compressa. Environ. Sci. Technol. 48, 8127–8134.

Claessens, M., De Meester, S., Landuyt, L.V., De Clerck, K., Janssen, C.R., 2011. Occurrence and distribution of microplastics in marine sediments along the Belgian coast. Mar. Pollut. Bull. 62, 2199–2204.

Codina-García, M., Militão, T., Moreno, J., Gonzalez-Solis, J., 2013. Plastic debris in Mediterranean seabirds. Mar. Pollut. Bull. 77, 220–226.

Cole, M., 2016. A novel method for preparing microplastic fibers. Sci. Rep. 6, 34519.

Cole, M., Lindeque, P., Fileman, E., Halsband, C., Goodhead, R., Moger, J., 2013. Microplastic ingestion by zooplankton. Environ. Sci. Technol. 47, 6646–6655.

Cole, M., Lindeque, P., Halsband, C., Galloway, T.S., 2011. Microplastics as contaminants in the marine environment: a review. Mar. Pollut. Bull. 62 (12), 2588–2597.

Cole, M., Webb, H., Lindeque, P.K., Fileman, E.S., Halsband, C., Galloway, T.S., 2014. Isolation of microplastics in biota-rich seawater samples and marine organisms. Sci. Rep. 4 (4528).

Collignon, A., Hecq, J.H., Galgani, F., Collard, F., Goffart, A., 2014. Annual variation in neustonic micro- and meso-plastic particles and zooplankton in the Bay of Calvi (Mediterranean-Corsica). Mar. Pollut. Bull. 79 (1–2), 293–298.

Corbanie, E.A., Matthijs, M.G.R., van Eck, J.H.H., Remon, J.P., Landman, W.J.M., Vervaet, C., 2006. Deposition of differently sized airborne microspheres in the respiratory tract of chickens. Avian Pathol. 35, 475–485.

Corcoran, E., Nellemann, C., Baker, E., Bos, R., Osborn, D., Savelli, H., 2010. Sick Water? The central role of waste-water management in sustainable development. A rapid response assessment. Birkeland Trykkeri AS, Norway.

Costa, M.F., Ivar do Sul, J.A., Silva-Cavalcanti, J.S., Araúja, M.C.B., Spengler, A., Tourinho, P.S., 2010. On the importance of size of plastic fragments and pellets on the strandline: a snapshot of a Brazilian beach. Environ. Monit. Assess. 168, 299–304. https://doi.org/10.1007/s10661-009-1113-4.

Cózar, A., Echevarria, F., Gonzalez-Gordillo, J.I., Irigoien, X., Ubeda, B., Hernandez-Leon, S., 2014. Plastic debris in the open ocean. Proc. Natl. Acad. Sci. U. S. A. 111, 10239–10244.

Culin, J., Bielic, T., 2016. Plastic Pollution from Ships. Pomorski Zbornik 51 (1), 57–66.

Da Silva Mendes, S., de Carvalho, R.H., de Faria, A.F., de Sousa, B.M., 2015. Marine debris ingestion by *Chelonia mydas* (Testudines: Cheloniidae) on the Brazilian coast. Mar. Pollut. Bull. 92 (1–2), 8–10.

David, K.A.B., Francois, G., Richard, C.T., Morton, B., 2009. Accumulation and fragmentation of plastic debris in global environments. Biol. Sci. 364 (1526), 1985–1999.

Davidson, K., Dudas, S.E., 2016. Microplastic ingestion by wild and cultured Manila clams (*Venerupis philippinarum*) from Baynes Sound, British Columbia. Arch. Environ. Contam. Toxicol. 71 (2), 147–156.

de Lucia, G., Caliani, I., Marra, S., Camedda, A., Coppa, S., Alcaro, L., 2014. Amount and distribution of neustonic micro-plastic off the Western Sardinian coast (Central-Western Mediterranean Sea). Mar. Environ. Res. 100, 10–16.

De Witte, B., Devriese, L., Bekaert, K., Hoffman, S., Vandermeersch, G., Cooreman, K., 2014. Quality assessment of the blue mussel (*Mytilus edulis*): comparison between commercial and wild types. Mar. Pollut. Bull. 85 (1), 146–155.
Dekiff, J.H., Remy, D., Klasmeier, J., Fries, E., 2014. Occurrence and spatial distribution of microplastics in sediments from Norderney. Environ. Pollut. 186, 248–256.
Derraik, J.G.B., 2002. The pollution of the marine environment by plastic debris: a review. Mar. Pollut. Bull. 44, 842–852.
Desforges, J.W., Galbraith, M., Dangerfield, N., Ross, P.S., 2014. Widespread distribution of microplastics in subsurface seawater in the NE Pacific Ocean. Mar. Pollut. Bull. 79, 94–99.
Desforges, J.W., Galbraith, M., Ross, P.S., 2015. Ingestion of microplastics by zooplankton in the Northeast Pacific ocean. Arch. Environ. Contam. Toxicol. 69 (3), 320–330.
Devriese, L., Vandendriessche, S., Theetaert, H., Vandermeersch, G., Hostens, K., Robbens, J., 2014. In: Occurrence of synthetic fibres in brown shrimp on the Belgian part of the North Sea. Platform presentation.International Workshop on Fate and Impact of Microplastics in Marine Ecosystems (MICRO2014). Plouzane (France), 13–15 January 2014.
Doyle, M.J., Watson, W., Bowlin, N.M., Sheavly, S.B., 2011. Plastic particles in coastal pelagic ecosystems of the Northeast Pacific Ocean. Mar. Environ. Res. 71 (1), 41–52.
Dris, R., Gasperi, J., Saad, M., Mirande, C., Tassin, B., 2016. Synthetic fibers in atmospheric fallout: a source of microplastics in the environment? Mar. Pollut. Bull. 104 (1-2), 290–293.
Duis, K., Coors, A., 2016. Microplastics in the aquatic and terrestrial environment: sources (with a specific focus on personal care products), fate and effects. Environ. Sci. Eur. 28, 2.
Ecology Center, 1996. Plastic Task Force Report. Ecology Center, Berkeley, CA.
El-Moselhy, K.M., Othman, A.I., Abd El-Azem, H., El-Metwally, M.E.A., 2014. Bioaccumulation of heavy metals in some tissues of fish in the Red Sea, Egypt. Egypt. J. Basic Appl. Sci. 1 (2), 97–105.
Enders, K., Lenz, R., Stedmon, C., Nielsen, T., 2015. Abundance, size and polymer composition of marine microplastics $\geq$10 μm in the Atlantic Ocean and their modelled vertical distribution. Mar. Pollut. Bull. 100 (1), 70–81.
Engler, R.E., 2012. The complex interaction between marine debris and toxic chemicals in the ocean. Environ. Sci. Technol. 46, 12302–12315.
Eriksen, M., Lebreton, L.C.M., Carson, H.S., Thiel, M., Moore, C.J., Borerro, J.C., 2014. Plastic pollution in the world's oceans: more than 5 trillion plastic pieces weighing over 250,000 tons afloat at sea. PLoS One 9, e111913.
Eriksen, M., Mason, S., Wilson, S., Box, C., Zellers, A., Edwards, W., 2013b. Microplastic pollution in the surface waters of the Laurentian Great Lakes. Mar. Pollut. Bull. 77, 177–182.
Eriksen, M., Maximenko, N., Thiel, M., Cummins, A., Lattin, G., Wilson, S., Hafner, J., Zellers, A., Rifman, S., 2013a. Plastic pollution in the South Pacific subtropical gyre. Mar. Pollut. Bull. 68, 71–76.
Eriksson, C., Burton, H., 2003. Origins and biological accumulation of small plastic particles in fur seals from Macquarie Island. AMBIO: J. Hum. Environ. 32, 380–384.
Farrell, P., Nelson, K., 2013. Trophic level transfer of microplastic: *Mytilus edulis* (L.) to *Carcinus maenas* (L.). Environ. Pollut. 177, 1–3.
Fendall, L.S., Sewell, M.A., 2009. Contributing to marine pollution by washing your face: microplastics in facial cleansers. Mar. Pollut. Bull. 58, 1225–1228.
Fife, D.T., Robertson, G.J., Shutler, D., Braune, B.M., Mallorym, M.L., 2015. Trace elements and ingested plastic debris in wintering dovekies (*Alle alle*). Mar. Pollut. Bull. 91 (1), 368–371.
Fischer, V., Elsner, N.O., Brenke, N., Schwabe, E., Brandt, A., 2015. Plastic pollution of the Kuril-Kamchatka Trench area (NW pacific). Deep-Sea Res. II Top. Stud. Oceanogr. 111, 399–405.
Fok, L., Cheung, P.K., 2015. Hong Kong at the Pearl River Estuary: a hotspot of microplastic pollution. Mar. Pollut. Bull. 99 (1–2), 112–118.
Free, C.M., Jensen, O.P., Mason, S.A., Eriksen, M., Williamson, N.J., Boldgiv, B., 2014. High-levels of microplastic pollution in a large, remote, mountain lake. Mar. Pollut. Bull. 85, 156–163. https://doi.org/10.1016/j.marpolbul.2014.06.001.
Frias, J.P.G.L., Otero, V., Sobral, P., 2014. Evidence of microplastics in samples of zooplankton from Portuguese coastal waters. Mar. Environ. Res. 95, 89–95.

Frias, J.P.G.L., Sobral, P., Ferreira, A.M., 2010. Organic pollutants in microplastics from two beaches of the Portuguese coast. Mar. Pollut. Bull. 60, 1988–1992.

Galgani, F., Leaute, J.P., Moguedet, P., Souplet, A., Verin, Y., Carpentier, A., Goraguer, H., Latrouite, D., Andral, B., Cadiou, Y., Mahe, C., Poulard, J.C., Nerisson, P., 2000. Litter on the sea floor along European Coasts. Mar. Pollut. Bull 40 (6), 516–527.

Gauthier-Clerc, M., Le Maho, Y., Clerquin, Y., Bost, C.-A., Handrich, Y., 2002. Seabird reproduction in an unpredictable environment: how king penguins provide their young chicks with food. Mar. Ecol. Prog. Ser. 237, 291–300.

Gewert, B., Plassmann, M.M., MacLeod, M., 2015. Pathways for degradation of plastic polymers floating in the marine environment. Environ. Sci. Process. Impacts 17 (9), 1513–1521.

Ghosh, S.K., Pal, S., Ray, S., 2013. Study of microbes having potentiality for biodegradation of plastics. Environ. Sci. Pollut. Res. 20, 4339–4355.

Goldstein, M.C., Goodwin, D.S., 2013. Gooseneck barnacles (*Lepas* spp.) ingest microplastic debris in the North Pacific Subtropical Gyre. PeerJ 1, 184.

Goldstein, M.C., Rosenberg, M., Cheng, L., 2012. Increased oceanic microplastic debris enhances oviposition in an endemic pelagic insect. Biol. Lett. 8 (5), 817–820.

Good, T.P., June, J.A., Etnier, M.A., Broadhurst, G., 2010. Derelict fishing nets in Puget Sound and the Northwest Straits: patterns and threats to marine fauna. Mar. Pollut. Bull. 60, 39–50.

Gouin, T., Avalos, J., Brunning, I., Brzuska, K., de Graaf, J., Kaumanns, J., Konong, T., Meyberg, M., Rettinger, K., Schlatter, H., Thomas, J., van Welie, R., Wolf, T., 2015. Use of micro-plastic beads in cosmetic products in europe and their estimated emissions to the north sea environment. SOFW J. 141, 1–33.

Gregory, M.R., 1977. Plastic pellets on New Zealand beaches. Mar. Pollut. Bull. 8, 82–84. https://doi.org/10.1016/0025-326X(77)90193-X.

Gregory, M.R., 1978. Accumulation and distribution of virgin plastic granules on New Zealand beaches. N. Z. J. Mar. Freshw. Res. 12, 399–414. https://doi.org/10.1080/00288330.1978.9515768.

Gregory, M.R., 1983. Virgin plastic granules on some beaches of Eastern Canada and Bermuda. Mar. Environ. Res. 10, 73–92. https://doi.org/10.1016/0141-1136(83) 90011-9.

Gregory, M.R., 1996. Plastic "scrubbers" in hand cleansers: a further (and minor) source for marine pollution identified. Mar. Pollut. Bull. 32, 867–871.

Gregory, M.R., 2009. Environmental implications of plastic debris in marine settings entanglement, ingestion, smothering, hangers-on, hitch-hiking and alien invasions. Philos. Trans. R. Soc. Lond. B: Biol. Sci. 364 (1526), 2013–2025.

Gregory, M.R., Andrady, A.L., 2003. Plastic in the marine environment. In: Andrady, A.L. (Ed.), Plastics and the Environment. John Wiley, New York, pp. 379–401.

Guo, W., Tao, J., Yang, C., Song, C., Geng, W., Li, Q., Wang, Y., Kong, M., Wang, S., 2012. Introduction of environmentally degradable parameters to evaluate the biodegradability of biodegradable polymers. PLoS One 7.

Guven, O., Gokdag, K., Jovanovic, B., KıdeySs, A.E., 2017. Microplastic litter composition of the Turkish territorial waters of the Mediterranean Sea, and its occurrence in the gastrointestinal tract of fish. Environ. Pollut. 223, 286–294. https://doi.org/10.1016/j.envpol.2017.01.025.

Halden, R.U., 2010. Plastics and health risks. Annu. Rev. Public Health 31, 179–194.

Hidalgo-Ruz, V., Thiel, M., 2013. Distribution and abundance of small plastic debris on beaches in the SE Pacific (Chile): a study supported by a citizen science project. Mar. Environ. Res. 87–88, 12–18.

Hintersteiner, I., Himmelsbach, M., Buchberger, W.W., 2015. Characterization and quantitation of polyolefin microplastics in personal-care products using high-temperature gel-permeation chromatography. Anal. Bioanal. Chem. 407, 1253–1259.

Hirai, H., Takada, H., Ogata, Y., Yamashita, R., Mizukawa, K., Saha, M., 2011. Organic micropollutants in marine plastics debris from the open ocean and remote and urban beaches. Mar. Pollut. Bull. 62, 1683–1692.

Ho, K., Pometto, A., Hinz, P., 1999. Effects of temperature and relative humidity on polylactic acid plastic degradation. J. Environ. Polym. Degrad. 7 (2), 83–92.

Hoarau, L., Ainley, L., Jean, C., Ciccione, S., 2014. Ingestion and defecation of marine debris by loggerhead sea turtles, Caretta caretta, from by-catches in the South-West Indian Ocean. Mar. Pollut. Bull. 84, 90–96.

Holmes, L.A., 2013. Interactions of Trace Metals with Plastic Production Pellets in the Marine Environment. PhD Thesis, University of Plymouth.
Holmes, L.A., Turner, A., Thompson, R.C., 2012. Adsorption of trace metals to plastic resin pellets in the marine environment. Environ. Pollut. 160, 42–48.
Holmes, L.A., Turner, A., Thompson, R.C., 2014. Interactions between trace metals and plastic production pellets under estuarine conditions. Mar. Chem. 167, 25–32.
Hong, S.H., Shim, W.J., Hong, L., 2017. Methods of analysing chemicals associated with microplastics: a review. Anal. Methods 9 (9), 1361–1368.
IMO, 2015. List of IMO Model Courses. Available from: http://www.imo.org/en/OurWork/HumanElement/TrainingCertification/Documents/list%20of%20IMO%20Model%20Courses.pdf.
Isobe, A., Kubo, K., Tamura, Y., Kako, S., Nakashima, E., Fujii, N., 2014. Selective transport of microplastics and mesoplastics by drifting in coastal waters. Mar. Pollut. Bull. 89, 324–330.
Isobe, A., Uchida, K., Tokai, T., Iwasaki, S., 2015. East Asian seas: a hot spot of pelagic microplastics. Mar. Pollut. Bull. 101, 618–623.
Isobe, A., Uchiyama-Matsumoto, K., Uchida, K., Tokai, T., 2017. Microplastics in the Southern Ocean. Mar. Pollut. Bull. 114 (1), 623–626.
Ivar do Sul, J.A., Costa, M.F., 2014. The present and future of microplastic pollution in the marine environment. Environ. Pollut. 185, 352–364.
Ivar do Sul, J.A., Costa, M.F., Barletta, M., Cysneiros, F.J.A., 2013. Pelagic microplastics around an archipelago of the Equatorial Atlantic. Mar. Pollut. Bull. 75 (1), 305–309.
Ivar do Sul, J.A., Spengler, A., Costa, M.F., 2009. Here, there and everywhere. Small plastic fragments and pellets on beaches of Fernando de Noronha (Equatorial Western Atlantic). Mar. Pollut. Bull. 58, 1229–1244.
Jambeck, J.R., Geyer, R., Wilcox, C., Siegler, T.R., Perryman, M., Andrady, A., Narayan, R., Law, K.L., 2015. Plastic waste inputs from land into the ocean. Science 347, 768–771.
Jayasiri, H.B., Purushothaman, C.S., Vennila, A., 2013. Quantitative analysis of plastic debris on recreational beaches in Mumbai, India. Mar. Pollut. Bull. 77, 107–112. https://doi.org/10.1016/j.marpolbul.2013.10.024.
Jegou, A.M., Salomon, J.C., 1991. Couplage imagerie thermique satellitaire-modeles numeriques. application a la manche. Oceanol. Acta 91 (11), 55–61.
Kaberi, H., Tsangaris, C., Zeri, C., Mousdisd, G., Papadopoulos, A., Streftaris, N., 2013. Microplastics along the shoreline of a Greek island (Kea isl., Aegean Sea): types and densities in relation to beach orientation, characteristics and proximity to sources.4th International Conference on Environmental Management, Engineering, Planning and Economics (CEMEPE) and SECOTOX Conference, Mykonos Island, Greece, pp. 197–202.
Kaiser, J., 2010. The dirt on ocean garbage patches. Science 328 (5985), 1506.
Kanhai, L.S.L., Officer, R., Lyashevska, O., Thompson, R., O'Connor, I., 2017. Microplastic abundance, distribution and composition along a latitudinal gradient in the Atlantic Ocean. Mar. Pollut. Bull, 307–314.
Kaposi, K.L., Mos, B., Kelaher, B.P., Dworjanyn, S.A., 2014. Ingestion of microplastic has limited impact on a marine larva. Environ. Sci. Technol. 48 (3), 1638–1645.
Karapanagioti, H.K., 2012. Floating plastics, plastic pellets, and organic micropollutants in the Mediterranean Sea. In: Stambler, N. (Ed.), Life in the Mediterranean Sea: A Look at Habitat Changes. Nova Science Publishers, New York, pp. 557–593.
Karl, D.M., 1999. A sea of change: biogeochemical variability in the North Pacific Subtropical Gyre. Ecosystems 2, 181–214.
Kim, I.-S., Chae, D.-H., Kim, S.-K., Choi, S., Woo, S.-B., 2015. Factors influencing the spatial variation of microplastics on high-tidal coastal beaches in Korea. Arch. Environ. Contam. Toxicol. https://doi.org/10.1007/s00244-015-0155-6.
Klein, S., Worch, E., Knepper, T.P., 2015. Occurrence and spatial distribution of microplastics in river shore sediments of the Rhine-main area in Germany. Environ. Sci. Technol. 49, 6070–6076.
Kockisch, S., Rees, G.D., Young, S.A., Tsibouklis, J., Smart, J.D., 2003. Polymeric microspheres for drug delivery to the oral cavity: an in vitro evaluation of mucoadhesive potential. J. Pharm. Sci. 92, 1614–1623.

Koehler, A., Marx, U., Broeg, K., Bahns, S., Bressling, J., 2008. Effects of nanoparticles in *Mytilus edulis* gills and hepatopancreas—a new threat to marine life? Mar. Environ. Res. 66, 12–14.
Kuhn, S., van Franeker, J.A., 2012. Plastic ingestion by the northern fulmar (*Fulmarus glacialis*) in Iceland. Mar. Pollut. Bull. 64 (6), 1252–1254.
Kukulka, T., Proskurowski, G., Morét-Ferguson, S., Meyer, D.W., Law, K.L., 2012. The effect of wind mixing on the vertical distribution of buoyant plastic debris. Geophys. Res. Lett. 39 (7), 1–6.
Kunz, A., Walther, B.A., Lowemark, L., Lee, Y.C., 2016. Distribution and quantity of microplastic on sandy beaches along the northern coast of Taiwan. Mar. Pollut. Bull. 111 (1–2), 126–135.
Kusui, T., Noda, M., 2003. International survey on the distribution of stranded and buried litter on beaches along the sea of Japan. Mar. Pollut. Bull. 47, 175–179.
Laglbauer, B.J.L., Franco-Santos, R.M., Andreu-Cazenave, M., Brunelli, L., Papadatou, M., Palatinus, A., Grego, M., Deprez, T., 2014. Macrodebris and microplastics from beaches in Slovenia. Mar. Pollut. Bull. 89, 356–366.
Lambert, S., Sinclair, C.J., Boxall, A.B., 2014. Occurrence, degradation and effect of polymer-based materials in the environment. Rev. Environ. Contam. Toxicol. 227, 1–53.
Lassen, C., Foss Hansen, S., Magnusson, K., Norén, F., Bloch Hartmann, N.I., Rehne Jensen, P., Gissel Nielsen, T., Brinch, A., 2015. Microplastics—occurrence, effects and sources of releases to the environment in Denmark. Environmental Project No. 1793, Environment Protection Agency, Ministry of Environment and Food of Denmark, Copenhagen.
Lattin, G.L., Moore, C.J., Zellers, A.F., Moore, S.L. Weisberg. S.B., 2004. A comparison of neustonic plastic and zooplankton at different depths near the southern California shore Mar. Pollut. Bull. 49, 291–294.
Law, K., Morét-Ferguson, S.E., Goodwin, D., Zettler, E., Deforce, E., Kukulka, T., Proskurowski, G., 2014a. Distribution of surface plastic debris in the eastern pacific ocean from an 11-year data set. Environ. Sci. Technol. 48, 4732–4738.
Law, K.L., Morét-Ferguson, S., Goodwin, D.S., Zettler, E.R., DeForce, E., Kukulka, T., 2014b. Distribution of surface plastic debris in the eastern Pacific Ocean from an 11-year dataset. Environ. Sci. Technol. 48 (9), 44732–44738.
Law, K.L., Moret-Ferguson, S., Maximenko, N.A., Proskurowski, G., Peacock, E.E., Hafner, J., 2010. Plastic accumulation in the North Atlantic subtropical gyre. Science 329, 1185–1188.
Lazar, B., Gracan, R., 2011. Ingestion of marine debris by loggerhead sea turtles, Caretta caretta, in the Adriatic sea. Mar. Pollut. Bull. 62 (1), 43–47.
Lee, J., Hong, S., Song, Y.K., Jang, Y.C., Jiang, M., Heo, N.W., Han, G.M., Kang, D., Shim, W.J., 2013a. Relationships among the abundances of plastic debris in different size, classes on beaches in South Korea. Mar. Pollut. Bull. 77, 349–354.
Lee, K.-W., Shim, W.J., Kwon, O.Y., Kang, J.-H., 2013b. Size-dependent effects of micro polystyrene particles in the marine copepod *Tigriopus japonicus*. Environ. Sci. Technol. 47 (19), 11278–11283.
Lenz, R., Enders, K., Beer, S., Sørensen, T.K., Stedmon, C.A., 2016. Analysis of Microplastic in the Stomachs of Herring and Cod From the North Sea and Baltic Sea. DTU Aqua National Institute of Aquatic Resources. https://doi.org/10.13140/RG.2.1.4246.6168.
Li, W.C., Tse, H.F., Fok, L., 2016. Plastic waste in the marine environment: a review of sources, occurrence and effects. Sci. Total Environ. 566–567, 333–349.
Liebezeit, G., Dubaish, F., 2012. Microplastics in beaches of the East Frisian Islands Spiekeroog and Kachelotplate. Bull. Environ. Contam. Toxicol. 89 (1), 213–217.
Lima, A.R.A., Costa, M.F., Barletta, M., 2014. Distribution patterns of microplastics within the plankton of a tropical estuary. Environ. Res. 132, 146–155.
Lin, V., 2016. Research highlights: impacts of microplastics on plankton. Environ. Sci. Process. Impacts 18, 160.
Lindborg, V.A., Ledbetter, J.F., Walat, J.M., Moffett, C., 2012. Plastic consumption and diet of Glaucous-winged gulls (*Larus glaucescens*). Mar. Pollut. Bull. 64 (11), 2351–2356.
Lithner, D., Larsson, Å., Dave, G., 2011. Environmental and health hazard ranking and assessment of plastic polymers based on chemical composition. Sci. Total Environ. 409, 3309–3324.
Lobelle, D., Cunlife, M., 2011. Early microbial biofilm formation on marine plastic debris. Mar. Pollut. Bull. 62 (1), 197–200.

Lusher, A., 2015. Microplastics in the marine environment: distribution, interactions and effects. In: Bergmann, M., Gutow, L., Klages, M. (Eds.), Marine Anthropogenic Litter. Springer, Berlin, pp. 245–308.
Lusher, A.L., Burke, A., O'Connor, I., Officer, R., 2014. Microplastic pollution in the Northeast Atlantic Ocean: validated and opportunistic sampling. Mar. Pollut. Bull. 88 (1–2), 325–333.
Lusher, A.L., McHugh, M., Thompson, R.C., 2013. Occurrence of microplastics in the gastrointestinal tract of pelagic and demersal fish from the English Channel. Mar. Pollut. Bull. 67, 94–99. https://doi.org/10.1016/j.marpolbul.2012.11.028.
Lusher, A.L., O'Donnell, C., Officer, R., O'Connor, I., 2016. Microplastic interactions with North Atlantic mesopelagic fish. ICES J. Mar. Sci. 73, 1214–1225.
Lusher, A.L., Tirelli, V., Connor, I., Officer, R., 2015. Microplastics in Arctic polar waters: the first reported values of particles in surface and sub-surface samples. Sci. Rep. 5, 14947. https://doi.org/10.1038/srep14947.
Magnusson, K., 2014. Microlitter and Other Microscopic Anthropogenic Particles in the Sea Area Off Auma and Turku, Finland. IVL Swedish Environmental Research Institute. U4645.
Mariana, M., Feiteiro, J., Verde, I., Cairrao, E., 2016. The effects of phthalates in the cardiovascular and reproductive systems: a review. Environ. Int. 2016 (94), 758–776.
Martins, J., Sobral, P., 2011. Plastic marine debris on the Portuguese coastline: a matter of size? Mar. Pollut. Bull. 62 (12), 2649–2653.
Mathalon, A., Hill, P., 2014. Microplastic fibers in the intertidal ecosystem surrounding Halifax Harbor, Nova Scotia. Mar. Pollut. Bull. 81, 69–79.
Mathieu-Denoncourt, J., Wallace, S.J., de Solla, S.R., Langlois, V.S., 2015. Plasticizer endocrine disruption: highlighting developmental and reproductive effects in mammals and non-mammalian aquatic species. Gen. Comp. Endocrinol. 219, 74–88.
Mato, Y., Isobe, T., Takada, H., Kanehiro, H., Ohtake, C., Kaminuma, T., 2001. Plastic resin pellets as a transport medium for toxic chemicals in the marine environment. Environ. Sci. Technol. 35, 318–324.
Maximenko, N.A., Hafner, J., Niiler, P., 2012. Pathways of marine debris from trajectories of Lagrangian drifters. Mar. Pollut. Bull. 65, 51–62.
Mazurais, D., Ernande, B., Quazuguel, P., Severe, A., Huelvan, C., Madec, L., Mouchel, O., Soudant, P., Robbens, J., Huvet, A., Zambonino-Infante, J., 2015. Evaluation of the impact of polyethylene microbeads ingestion in European sea bass (*Dicentrarchus labrax*) larvae. Mar. Environ. Res. 112, 78–85.
Mehlhart, G., Blepp, M., 2012. Study on land-sourced litter (LSL) in the marine environment: review of sources and literature in the context of the initiative of the Declaration of the Global Plastics Associations for Solutions on Marine Litter. Oko-Institut e.V, Darmstadt/Freiburg.
Moore, C.J., 2008. Synthetic polymers in the marine environment: a rapidly increasing, long-term threat. Environ. Res. 108, 131–139.
Morét-Ferguson, S., Law, K.L., Proskurowski, G., Murphy, E.K., Peacock, E.E., Reddy, C.M., 2010. The size, mass, and composition of plastic debris in the western North Atlantic Ocean. Mar. Pollut. Bull. 60 (10), 1873–1878.
Muirhead, E.K., Skillman, A.D., Hook, S.E., Schulz, I.R., 2006. Oral exposure of PBDE-47 in Fish: toxicokinetics and reproductive effects in Japanese medaka (*Oryzias latipes*) and fathead minnows (*Pimephales promelas*). Environ. Sci. Technol. 40 (2), 523–528.
Müller, R.-J., Kleeberg, I., Deckwer, W.-D., 2001. Biodegradation of polyesters containing aromatic constituents. J. Biotechnol. 86, 87–95.
Murray, F., Cowie, P.R., 2011. Plastic contamination in the Decapod Crustacean Nephrops norvegicus (Linnaeus 1758). Mar. Pollut. Bull. 62, 1207–1217.
Muthukumar, T., Aravinthan, A., Lakshmi, K., Venkatesan, R., Vedaprakash, L., Doble, M., 2011. Fouling and stability of polymers and composites in marine environment. Int. Biodeterior. Biodegrad 65 (2), 276–284.
Nadal, M.A., Alomar, C., Deudero, S., 2016. High levels of microplastic ingestion by the semipelagic fish bogue *Boops boops* (L.) around the Balearic Islands. Environ. Pollut. 214, 517–523.
Naidoo, T., 2015. Plastic Pollution in Five Urban Estuaries of KwaZulu-Natal, South Africa (MSc Thesis). University of KwaZulu-Natal, pp. 1–69.
Naidoo, T., Smit, A.J., Glassom, D., 2016. Plastic ingestion by estuarine mullet *Mugil cephalus* (Mugilidae) in an urban harbour, KwaZulu-Natal, South Africa. Afr. J. Mar. Sci. 38, 145–149.

Napper, I.E., Thompson, R.C., 2016. Release of synthetic microplastic plastic fibres from domestic washing machines: effects of fabric type and washing conditions. Mar. Pollut. Bull. 112 (1–2), 39–45.
Napper, I.E., Bakir, A., Rowland, S.J., Thompson, R.C., 2015. Characterisation, quantity and sorptive properties of microplastics extracted from cosmetics. Mar. Pollut. Bull. https://doi.org/10.1016/j.marpolbul.2015.07.029.
Nel, H., Hean, J., Noundou, X., Froneman, P., 2017. Do microplastic loads reflect the population demographics along the southern African coastline? Mar. Pollut. Bull. 115 (1-2), 115–119.
Nel, H.A., Froneman, P.W., 2015. A quantitative analysis of microplastic pollution along the south-eastern coastline of South Africa. Mar. Pollut. Bull. 101, 274–279.
Nelms, S., Duncan, E., Broderick, A., Galloway, T., Godfrey, M., Hamann, M., Lindeque, P., Godley, B., 2016. Plastic and marine turtles: a review and call for research. ICES J. Mar. Sci. 73 (2), 165–181.
Nelms, S.E., Coombes, C., Foster, L.C., Galloway, T.S., Godley, B.J., Lindeque, P.K., Witt, M.J., 2017. Marine anthropogenic litter on British beaches: A 10-year nationwide assessment using citizen science data. Sci Total Environ 579, 1399–1409.
Neves, D., Sobral, P., Ferreira, J.L., Pereira, T., 2015. Ingestion of microplastics by commercial fish off the Portuguese coast. Mar. Pollut. Bull. 101, 119–126.
Nor, N.H.M., Obbard, J.P., 2014. Microplastics in Singapore's coastal mangrove system. Mar. Pollut. Bull. 79, 278–283. https://doi.org/10.1016/j.marpolbul.2013.11.025.
Norén, F., 2007. Small Plastic Particles in Coastal Swedish Waters. KIMO Sweden, Lysekil, p. 11.
Noren, F., Naustvoll, F., 2010. Survey of Microscopic Anthropogenic Particles in Skagerrak. Report Commissioned by Klimaog Forurensningsdirektoratet (Oslo, Norway). .
Nøttestad, L., Diaz, J., Penã, H., Søiland, H., Huse, G., Fernö, A., 2015. Feeding strategy of mackerel in the Norwegian Sea relative to currents, temperature, and prey. ICES J. Mar. Sci. 73 (4), 1127–1137.
Obbard, R.W., Sadri, S., Wong, Y.Q., Khitun, A.A., Baker, I., Thompson, R.C., 2014. Global warming releases microplastic legacy frozen in Arctic Sea ice. Earth's Future 2 (6), 315–320.
O'Brine, T., Thompson, R.C., 2010. Degradation of plastic carrier bags in the marine environment. Mar. Pollut. Bull. 60, 2279–2283.
Oehlmann, J.R., Schulte-Oehlmann, U., Kloas, W., Jagnytsch, O., Lutz, I., Kusk, K.O., Wollenberger, L., Santos, E.M., Paull, G.C., Van Look, K.J.W., Tyler, C.R., 2009. A critical analysis of the biological impacts of plasticizers on wildlife. Philos. Trans. R. Soc. B 364, 2047–2062.
Oerlikon, 2009. The Fiber Year 2008/09: A world Survey on Textile and Nonwovens Industry, Oerlikon, Switzerland.
Ogata, Y., Takada, H., Mizukawa, K., Hirai, H., Iwasa, S., Endo, S., Mato, Y., Saha, M., Okuda, K., Nakashima, A., Murakami, M., Zurcher, N., Booyatumanondo, R., Zakaria, M.P., Dung, L.Q., Gordon, M., Miguez, C., Suzuki, S., Moore, C., Karapanagioti, H.K., Weerts, S., McClurg, T., Burres, E., Smith, W., Van Velkenburg, M., Lang, J.S., Lang, R.C., Laursen, D., Danner, B., Stewardson, N., Thompson, R.C., 2009. International Pellet Watch: Global monitoring of persistent organic pollutants (POPs) in coastal waters. 1. Initial phase data on PCBs, DDTs, and HCHs. Mar. Pollut. Bull. 58, 1437–1446.
Phillips, M.B., Bonner, T.H., 2015. Occurrence and amount of microplastic ingested by fishes in watersheds of the Gulf of Mexico. Mar. Pollut. Bull. 100, 264–269.
PlasticsEurope, 2016. Plastics—The Facts 2015/2016: An Analysis of European Plastics Production. Demand Waste Data. 2015.
Provencher, J.F., Bond, A., Avery-Gomm, S., Borrelle, S., Bravo, R., Hammer, S., Khn, S., Lavers, J., Mallory, M., Trevail, A., Van Franeker, J., 2017. Quantifying ingested debris in marine megafauna: a review and recommendations for standardization. Anal. Methods 9 (9), 1454–1469.
Provencher, J.F., Bond, A.L., Mallory, M.L., 2014. Marine birds and plastic debris in Canada: a national synthesis, and a way forward. Environ. Rev. https://doi.org/10.1139/er-2014-0039.
Provencher, J.F., Gaston, A.J., Mallory, M.L., O'hara, P.D., 2010. Gilchrist H.G. Ingested plastic in a diving seabird, the thick-billed murre (*Uria lomvia*), in the eastern Canadian Arctic. Mar. Pollut. Bull 60 (9), 1406–1411.
Ramirez-Llodra, E., Company, J.B., Sard, F., De Mol, B., Coll, M., Sardà, F., 2013. Effects of natural and anthropogenic processes in the distribution of marine litter in the deep Mediterranean Sea. Prog. Oceanogr. 118, 273–287.

Reisser, J., Shaw, J., Hallegraeff, G., Proietti, M., Barnes, D.K.A., Thums, M., et al., 2014. Millimeter-sized marine plastics: a new pelagic habitat for microorganisms and invertebrates. PLoS One 9, e100289.

Reisser, J., Shaw, J., Wilcox, C., Hardesty, B.D., Proiett, M., Thums, M., Pattiaratchi, C., 2013. Marine plastic pollution in waters around Australia: characteristics, concentrations, and pathways. PLoS One 8 (11), e80466.

Reisser, J., Slat, B., Noble, K., du Plessis, K., Epp, M., Proietti, M., de Sonneville, J., Becker, T., Pattiaratchi, C., 2015. The vertical distribution of buoyant plastics at sea: an observational study in the North Atlantic Gyre. Biogeosciences 12, 1249–1256.

Ribic, C.A., Sheavly, S.B., Rugg, D.J., Erdmann, E.S., 2010. Trends and drivers of marine debris on the Atlantic coast of the United States 1997–2007. Mar. Pollut. Bull. 60, 1231–1234.

Rochman, C.M., Hoh, E., Hentschel, B.T., Kaye, S., 2013. Long-term field measurement of sorption of organic contaminants to five types of plastic pellets: implications for plastic marine debris. Environ. Sci. Technol. 47, 1646–1654.

Rochman, C., Hentschel, B., Teh, S., 2014. Long-term sorption of metals is similar among plastic types: implications for plastic debris in aquatic environments. PLoS ONE 9 (1), 85433.

Rodríguez, A., Rodríguez, B., Nazaret Carrasco, M., 2012. High prevalence of parental delivery of plastic debris in Cory's shearwaters (*Calonectris diomedea*). Mar. Pollut. Bull. 64 (10), 2219–2223.

Romeo, T., Pietro, B., Peda, C., Consoli, P., Andaloro, F., Fossi, M.C., 2015. First evidence of presence of plastic debris in stomach of large pelagic fish in the Mediterranean Sea. Mar. Pollut. Bull. 95, 358–361.

Rummel, C.D., Löder, M.G.L., Fricke, N.F., Lang, T., Griebeler, E.M., Janke, M., Gerdts, G., 2016. Plastic ingestion by pelagic and demersal fish from the North Sea and Baltic Sea. Mar. Pollut. Bull. 102 (1), 134–141.

Ryan, P.G., Moore, C.J., van Franeker, J.A., Moloney, C.L., 2009. Monitoring the abundance of plastic debris in the marine environment. Philos. Trans. R. Soc. Lond. B 364, 1999–2012.

Santos, I.R., Friedrich, A.C., Ivar do Sul, J.A., 2009. Marine debris contamination along undeveloped tropical beaches from northeast Brazil. Environ. Monit. Assess. 148, 455–462.

Santos, R.G., Andrades, R., Boldrini, M.A., Martins, A., S., 2015. Debris ingestion by juvenile marine turtles: an underestimated problem. Mar. Pollut. Bull. 93 (1–2), 37–43.

Schuyler, Q., Hardesty, B.D., Wilcox, C., Kathy, T., 2012. To eat or not to eat? Debris selectivity by marine turtles. PLoS One 7 (7), e40884.

Schuyler, Q., Hatderty, B.D., Wilcox, C., Townsend, K., 2014. Global analysis of anthropogenic debris ingestion by sea turtles. Conserv. Biol. 28 (1), 129.

Setälä, O., Fleming-Lehtinen, V., Lehtiniemi, M., 2014. Ingestion and transfer of microplastics in the planktonic food web. Environ. Pollut. 185, 77–83.

Shah, A.A., Fariha, H., Abdul, H., Safia, A., 2008. Biological degradation of plastics: a comprehensive review. Biotechnol. Adv. 26, 2467–2650.

Shiber, J.G., 1979. Plastic pellets on the coast of Lebanon. Mar. Pollut. Bull. 10, 28–30. https://doi.org/10.1016/0025-326X(79)90321-7.

Shiber, J.G., 1982. Plastic pellets on Spain's 'Costa del Sol' beaches. Mar. Pollut. Bull. 13, 409–412. https://doi.org/10.1016/0025-326X(82)90014-5.

Singh, G., Singh, A.K., Bhatt, K., 2016. Biodegradation of polyethylene by bacteria isolated from soil. Int. J. Res. Dev. Pharm. Life Sci. 5 (2), 2056–2062.

Song, Y.K., Hong, S.H., Jang, M., Shim, W.J., 2015. Occurrence and distribution of microplastics in the sea surface microlayer in Jinhae Bay, South Korea. Environ. Contam. Toxicol. 69 (3), 121.

Song, Y.K., Hong, S.H., Kang, J.H., Kwon, O.Y., Jang, M., Han, G.M., 2014. Large accumulation of micro-sized synthetic polymer particles in the sea surface microlayer. Environ. Sci. Technol. 48 (16), 9014–9021.

Srivastava, R.K., Godara, S., 2012. Use of polycarbonate plastic products and human health. Int. J. Basic Clin. Pharmacol. 2 (1), 12–17.

Stolte, A., Forster, S., Gerdts, G., Schubert, H., 2015. Microplastic concentrations in beach sediments along the German Baltic coast. Mar. Pollut. Bull. 99 (1-2), 216–229.

Sundt, P., Schulze, P.-E., Syversen, F., 2014. Sources of microplastic-pollution to the marine environment. Report no M-321/2015, Mepex Consult, Asker.

Sutherland, W.J., Clout, M., Côté, I.M., Daszak, P., Depledge, M.H., Fellman, L., Fleishman, E., Garthwaite, R., Gibbons, D.W., De Lurio, J., Impey, A.J., Lickorish, F., Lindenmayer, D., Madgwick, J., Margerison, C., Maynard, T., Peck, L.S., Pretty, J., Prior, S., Redford, K.H., Scharlemann, J.P.W., Spalding, M., Watkinson, A.R., 2010. A horizon scan of global conservation issues for 2010. Trends Ecol. Evol. 25, 1–7.

Talsness, C.E., Andrade, A.J.M., Kuriyama, S.N., Taylor, J.A., vom Saal, F.S., 2009. Components of plastic: experimental studies in animals and relevance for human health. Philos. Trans. R. Soc. B 364, 2079–2096.

Tamara, S.G., 2015. Micro- and nano-plastics and human health. In: Bergmann, M., Gutow, L., Klages, M. (Eds.), Marine Anthropogenic Litter. Springer, Berlin, p. 343.

Tanaka, K., Takada, H., 2016. Microplastic fragments and microbeads in digestive tracts of planktivorous fish from urban coastal waters. Sci. Rep. 6, 34351. https://doi.org/10.1038/srep34351.

Tanaka, K., Takada, H., Yamashita, R., Mizukawa, K., Fukuwaka, M.A., Watanuki, Y., 2013. Accumulation of plastic-derived chemicals in tissues of seabirds ingesting marine plastics. Mar. Pollut. Bull. 69, 219–222.

Taylor, M.L., Gwinnett, C., Robinson, L.F., Woodall, L.C., 2016. Plastic microfibre ingestion by deep-sea organisms. Sci. Rep. 6.

Ter Halle, A., Ladirat, L., Gendre, X., Goudouneche, D., Pusineri, C., Routaboul, C., Tenailleau, C., Duployer, B., Perez, E., 2016. Understanding the fragmentation pattern of marine plastic debris. Environ. Sci. Technol. 50 (11), 5668–5675.

Terepocki, A., Brush, A., Kleine, L., Shugart, G., Hodum, P., 2017. Size and dynamics of microplastic in gastrointestinal tracts of Northern Fulmars (*Fulmarus glacialis*) and Sooty Shearwaters (*Ardenna grisea*). Mar. Pollut. Bull. 116 (1-2), 143–150.

Teuten, E.L., Saquing, J.M., Knappe, D.R.U., Barlaz, M.A., Jonsson, S., BjÃrn, A., Rowland, S.J., Thompson, R.C., Galloway, T.S., Yamashita, R., Ochi, D., Watanuki, Y., Moore, C., Viet, P.H., Tana, T.S., Prudente, M., Boonyatumanond, R., Zakaria, M.P., Akkhavong, K., Ogata, Y., Hirai, H., Iwasa, S., Mizukawa, K., Hagino, Y., Imamura, A., Saha, M., Takada, H., 2009. Transport and release of chemicals from plastics to the environment and to wildlife. Philos. Trans. R. Soc. B: Biol. Sci. 364, 2027–2045.

Thompson, R.C., Moore, C.J., vom Saal, F.S., Swan, S.H., 2009. Plastics, the environment and human health: current consensus and future trends. Philos. Trans. R. Soc. B 364, 2153–2216.

Thompson, R.C., Olsen, Y., Mitchell, R.P., Davis, A., Rowland, S.J., John, A.W.G., McGonigle, D., Russell, A.E., 2004. Lost at sea: where is all the plastic? Science 304 (5672), 838.

Thouzeau, C., Peters, G., Le Bohec, C., Le Maho, Y., 2004. Adjustments of gastric pH, motility and temperature during long-term preservation of stomach contents in free-ranging incubating king penguins. J. Exp. Biol. 207, 2715–2724.

Topçu, E.N., Tonay, A.M., Dede, A., Öztürk, A.A., Öztürk, B., 2013. Origin and abundance of marine litter along sandy beaches of the Turkish Western Black Sea Coast. Mar. Environ. Res. 85, 21–28.

Topping, P., Morantz, D., Lang, G., 1997. Waste disposal practices of fishing vessels: Canada's east coast, 1990–1991. In: Coe, J.M., Roger, D.B. (Eds.), Marine Debris: Sources, Impacts, and Solutions. Springer, NY, pp. 253–262.

Tornero, V., Hanke, G., 2016. Chemical contaminants entering the marine environment from sea-based sources: a review with a focus on European seas. Mar. Pollut. Bull 112 (1–2), 17–38.

Tourinho, P.S., Ivar do Sul, J.A., Fillmann, G., 2010. Is marine debris ingestion still a problem for the coastal marine biota of southern Brazil? Mar. Pollut. Bull. 60 (3), 396–401.

Trevail, A.M., Gabrielsen, G.W., Kuhn, S., van Franeker, J.A., 2015. Elevated levels of ingested plastic in a high Arctic seabird, the northern fulmar (*Fulmarus glacialis*). Polar Biol. 38, 975–981.

Tubau, X., Canals, M., Lastras, G., Rayo, X., Rivera, J., Amblas, D., 2015. Marine litter on the floor of deep submarine canyons of the Northwestern Mediterranean Sea: the role of hydrodynamic processes. Prog. Oceanogr. 134, 379–403. https://doi.org/10.1016/j.pocean.2015.03.013.

Turner, A., Holmes, L., 2011. Occurrence, distribution and characteristics of beached plastic production pellets on the island of Malta (central Mediterranean). Mar. Pollut. Bull. 62, 377–381. https://doi.org/10.1016/j.marpolbul.2010.09.027.

Van Cauwenberghe, L., Claessens, M., Vandegehuchte, M.B., Janssen, C.R., 2015. Microplastics are taken up by mussels (*Mytilus edulis*) and lugworms (*Arenicola marina*) living in natural habitats. Environ. Pollut. 199, 10–17.

Van Cauwenberghe, L., Janssen, C.R., 2014. Microplastics in bivalves cultured for human consumption. Environ. Pollut. 193, 65–70.

van Franeker, J.A., Blaize, C., Danielsen, J., Fairclough, K., Gollan, J., Guse, N., Hansen, P.L., Heubeck, M., Jensen, J.K., Le Guillou, G., Olsen, B., Olsen, K.O., Pedersen, J., Stienen, E.W., Turner, D.M., 2011. Monitoring plastic ingestion by the northern fulmar *Fulmarus glacialis* in the North Sea. Environ. Pollut. 159 (10), 2609–2615.

Van Houtan, K., Francke, D., Alessi, S., Jones, T., Martin, S., Kurpita, L., King, C., Baird, R., 2016. The developmental biogeography of hawksbill sea turtles in the North Pacific. Ecol. Evolut. 6 (8), 2378–2389.

Veerasingam, S., Mugilarasanb, M., Venkatachalapathyc, R., Vethamonya, P., 2016. Influence of 2015 flood on the distribution and occurrence of microplastic pellets along the Chennai coast, India. Mar. Pollut. Bull. 109 (1), 196–204.

Verboven, N., Verreault, J., Letcher, R.J., Gabrielsen, G.W., Evans, E., 2008. Maternally derived testosterone and 17b-estradiol in the eggs of Arctic-breeding glaucous gulls in relation to persistent organic pollutants. Comp. Biochem. Physiol. C: Toxicol. Pharmacol. 148, 143–151.

Verlis, K.M.M., Campbell, M.L.L., Wilson, S.P.P., 2013. Ingestion of marine debris plastic by the wedge-tailed shearwater, Ardenna pacifica, in the Great Barrier Reef, Australia. Mar. Pollut. Bull. 72, 244–249.

Vianello, A., Boldrin, A., Guerriero, P., Moschino, V., Rella, R., Sturaro, A., Da Ros, L., 2013. Microplastic particles in sediments of Lagoon of Venice, Italy: first observations on occurrence, spatial patterns and identification. Estuar. Coast. Shelf Sci. 130, 54–61.

Von Moos, N., Burkhardt-Holm, P., Kohle, A., 2012. Uptake and effects of microplastics on cells and tissue of the blue mussel *Mytilus edulis* L. after an experimental exposure. Environ. Sci. Technol. 46, 11327–11335.

Wardrop, P., Shimeta, J., Nugegoda, D., Morrison, P., Miranda, A., Tang, M., Clarke, B., 2016. Chemical pollutants sorbed to ingested microbeads from personal care products accumulate in fish. Environ. Sci. Technol. https://doi.org/10.1021/acs.est.5b06280.

Weinstein, J., Crocker, B., Gray, A., 2016. From macroplastic to microplastic: degradation of high-density polyethylene, polypropylene, and polystyrene in a salt marsh habitat. Environ. Toxicol. Chem. 35 (7), 1632–1640.

Welden, N., Cowie, P.R., 2016. Environment and gut morphology influence microplastic retention in langoustine, Nephrops norvegicus. Environ. Pollut. 214, 859–865. https://doi.org/10.1016/j.envpol.2016.03.067.

Wen, J., Kim, G.J.A., Leong, K.W., 2003. Poly(D, lactide-co-ethyl ethylene phosphate)s as new drug carriers. J. Control. Release 92, 39–48.

Wilcox, C., Heathcote, G., Goldberg, J., Gunn, R., Peel, D., Hardesty, D., 2015. Understanding the sources and effects of abandoned, lost, and discarded fishing gear on marine turtles in northern Australia. Conserv. Biol. 29 (1), 198–206.

Woodall, L.C., Sanchez-Vidal, A., Canals, M., Paterson, G.L.J., Coppock, R., Sleight, V., Calafat, A., Rogers, A.D., Narayanaswamy, B.E., Thompson, R.C., 2014. The deep sea is a major sink for microplastic debris. R. Soc. Open Sci. 1 (4), 140317.

Wright, S.L., Rowe, D., Reid, M.J., Thomas, K.V., Galloway, T.S., 2015. Bioaccumulation and biological effects of cigarette litter in marine worms. Sci. Rep. 5.

Wright, S.L., Thompson, R.C., Galloway, T.S., 2013. The physical impacts of microplastics on marine organisms: a review. Environ. Pollut. 178, 483–492.

Yamashita, R., Takada, H., Fukuwaka, M., Watanuki, Y., 2011. Physical and chemical effects of ingested plastic debris on short-tailed shearwaters, *Puffinus tenuirostris*, in the North Pacific ocean. Mar. Pollut. Bull. 62 (12), 2845–2849.

Yamashita, R., Tanimura, A., 2007. Floating plastic in the Kuroshio Current area, western North Pacific Ocean. Mar. Pollut. Bull. 54 (4), 485–488.

Yogui, G.T., Sericano, J.L., 2009. Polybrominated diphenyl ether flame retardants in the U.S. marine environment: a review. Environ. Int. 35, 655–666.
Zarfl, C., Matthies, M., 2010. Are marine plastic particles transport vectors for organic pollutants to the Arctic? Mar. Pollut. Bull. 60 (10), 1810–1814.
Zettler, E.R., Mincer, T.J., Amaral-Zettler, L.A., 2013. Life in the "plastisphere": microbial communities on plastic marine debris. Environ Sci Technol 47 (13), 7137–7146.
Zhao, S., Zhu, L., Wang, T., Li, D., 2014. Suspended microplastics in the surface water of the Yangtze Estuary System, China: first observations on occurrence, distribution. Mar. Pollut. Bull. 86 (1-2), 562–568.
Zheng, Y., Yanful, E., Bassi, A., 2005. A review of plastic waste biodegradation. Crit. Rev. Biotechnol. 25 (4), 243–250.
Zitko, V., Hanlon, M., 1991. Another source of pollution by plastics: skin cleaners with plastic scrubbers. Mar. Pollut. Bull. 22, 41–42.
Zweifel, H., 2001. Plastics Additives Handbook, fifth ed. Carl Hanser Verlag, Munich.

FURTHER READING

Bhattacharya, P., Turner, J.P., Ke, P.-C., 2010. Physical adsorption of charged plastic nanoparticles affects algal photosynthesis. J. Phys. Chem. C 114 (39), 16556–16561.
Cousin, H.R., Auman, H.J., Alderman, R., Virtue, P., 2015. The frequency of ingested plastic debris and its effects on body condition of Short-tailed Shearwater (*Puffinus tenuirostris*) pre-fledging chicks in Tasmania, Australia. Meu 115, 6–11.
Hong, S., Lee, J., Kang, D., Choi, H.-W., Ko, S.-H., 2014. Quantities, composition, and sources of beach debris in Korea from the results of nationwide monitoring. Mar. Pollut. Bull. 84, 27–34.
Kako, S., Isobe, A., Magome, S., 2010. Sequential monitoring of beach litter using webcams. Mar. Pollut. Bull. 60, 775–779.
Kuo, F.-J., Huang, H.-W., 2014. Strategy for mitigation of marine debris: analysis of sources and composition of marine debris in northern Taiwan. Mar. Pollut. Bull. 83, 70–78.
Nakashima, E., Isobe, A., Magome, S., Kako, S., Deki, N., 2011. Using aerial photography and in-situ measurements to estimate the quantity of macro-litter on beaches. Mar. Pollut. Bull. 62, 762–769.
Neulicht, R., Shular, J., 1997. Emission factor documentation for AP-42. Section 13.2.6. Abrasive blasting final report. EPA, Washington, USA, pp. 1–38. http://www.epa.gov/ttnchie1/ap42/ch13/bgdocs/b13s02-6.pdf.
Santos, R.G., Andrades, R., Fardim, L.M., Martins, A.S., 2016. Marine debris ingestion and Thayer's law—the importance of plastic color. Environ. Pollut. 214, 585–588.
Thayer, A.H., 1896. The law which underlies protective coloration. Auk 13, 124–129.
Vlietstra, L.S., Parga, J.A., 2002. Long-term changes in the type, but not amount, of ingested plastic particles in short-tailed shearwaters in the southeastern Bering Sea. Mar. Pollut. Bull. 44, 945–955.
Zhou, P., Huang, C., Fang, H., Cai, W., Li, D., Li, X., Yu, H., 2011. The abundance, composition and sources of marine debris in coastal seawaters or beaches around the northern South China Sea (China). Mar. Pollut. Bull. 62, 1998–2007.

CHAPTER 6

Behavior of Microplastics in Coastal Zones

Irina Chubarenko*, Elena Esiukova*, Andrei Bagaev†, Igor Isachenko*, Natalia Demchenko*, Mikhail Zobkov‡, Irina Efimova*, Margarita Bagaeva*, Lilia Khatmullina*

*Shirshov Institute of Oceanology of Russian Academy of Sciences, Moscow, Russia
†Marine Hydrophysical Institute of Russian Academy of Sciences, Sevastopol, Russia
‡Northern Water Problems Institute of the Karelian Research Centre of the Russian Academy of Sciences, Petrozavodsk, Russia

6.1 INTRODUCTION

The presence of plastic debris in costal zones is reported widely around the world (Claessens et al., 2011; Cole et al., 2011; Eriksen et al., 2014; Jambeck et al., 2015; Li et al., 2016). Larger and heavier litter pieces are typically found in higher concentrations near urbanized cites, densely populated coasts, or popular resort areas (Barnes et al., 2009; Li et al., 2016; UN Environment Annual Report, 2016), that is, closer to their sources of origin. As for smaller microplastic (MP) particles (<5 mm), their presence in the deepest ocean sediments (Thompson et al., 2004; Van Cauwenberghe et al., 2013b; Woodall et al., 2014; Bergmann et al., 2017), in Arctic ice (Obbard et al., 2014), on distant oceanic islands (Ivar do Sul et al., 2013), and in highly protected beaches of national parks (Cooper, 2012) indicates a rapidly advancing plastic pandemic that has embraced all of Earth's environments today. Monitoring of oceanic and marine beaches worldwide (e.g., Gregory, 1978; Ogata et al., 2009; Claessens et al., 2011; Hidalgo-Ruz and Thiel, 2013; Van Cauwenberghe et al., 2013a; Nor and Obbard, 2014; Esiukova, 2017) points out that, in contrast to larger plastic litter, MP concentrations are largely untied with local sources of contamination and particular regional human presence/activity: they are equally present everywhere, showing out their highest mobility even at oceanic scales. Obviously, this extraordinary mobility is a combined result of the properties of particles and external environmental conditions.

Coastal zones, hosting both the main sources of plastics and the main "end users" of the marine environment (humans and biota) at the same time, exhibit very complicated physical forcing. Thus, even the understanding of some general features of the MP behavior in the coastal zone is both highly desirable and highly challenging, especially taking into account our present-day very limited knowledge on physical characteristics of the MP particles per se. Indeed, the majority of studies on MPs are focused on their abundance, chemical composition, and biological threat, while the physical properties are

Microplastic Contamination in Aquatic Environments
https://doi.org/10.1016/B978-0-12-813747-5.00006-0

analyzed relatively rarely (e.g., Chubarenko et al., 2016; Fazey and Ryan, 2016; Chubarenko and Stepanova, 2017; Zhang, 2017).

Only a few experimental studies have been published up to now exploring one or another aspect of the motion of the very (micro) plastic particles (Ballent et al., 2012, 2013; Kowalski et al., 2016; Khatmullina and Isachenko, 2017). At the same time, "motion of particles in fluids" is a classical problem with lots of developed applications (e.g., Shields, 1936; Gibbs, 1974; van Rijn, 1993; Le Roux, 2005; Blasco et al., 2015). This review is intended to integrate the achievements of several branches of aquatic science, with the goal to clarify the place of MPs and its characteristic properties in the general field of particle motions in hydraulics, hydrodynamics, hydrobiology, sediment transport, and coastal physical oceanography. With this idea in mind, we address physical and dynamic properties of marine (i.e., found in marine environments) MP particles: (i) their density range, size classes, and typical shapes; (ii) variations of these properties with time due to biofouling and mechanical fragmentation in the sea swash zone; (iii) the present-day (very limited) experimental basis on MP motion; (iv) classical hydrodynamics and applied works toward the sinking velocity and resuspension threshold for particles with MP characteristics; (v) oceanographic applications on particle motions under surface waves, in roll strictures, during convective mixing; and, finally, as a natural case study, we (vi) analyze observations on the Baltic amber migrations in the sea coastal zone, since its properties fall in the range of those common for most typical MPs.

When discussing the motion of MPs here, we shall refer to the *coastal zone* as to a region where the motion of MP particles in water differs somehow from that in the open sea. Such a definition is quite typical for problems in physical oceanography, because the balance between (generally the same) driving forces near the (coastal) boundary differs from that far from it. For example, currents in the ocean interior (far from any boundaries) are in the geostrophic balance, that is, pressure gradient balances the Coriolis force, other forces being negligibly small. Meanwhile, closer to the water surface, wind-induced currents appear, which is considered to be important only within the Ekman boundary layer (see, e.g., Gill, 1982). The same way, near the coastal boundary, the balance between driving forces becomes different from that in open-sea waters—either due to wind setup near the coast or surface wave deformation or bottom friction, etc. Every one of these factors is important within its own characteristic distance from the coast. In application to the analysis of motion of MP particles, this highlights that the "coastal zone for MPs" is not the same (i) under different external environmental conditions (e.g., stormy or calm) and (ii) for different prevailing transport mechanisms (e.g., for developed surface waves, in river-dominated areas, during episodes of strong vertical convection, and in the case of wind-induced currents). In Fig. 6.1, the offshore limit of the coastal zone is related to the field of surface waves: deep-water waves are considered to begin deformation due to bottom friction when water depth becomes smaller than about a half of the wave length (e.g., State and Evolution of the Baltic Sea, 2008). Below, it will be

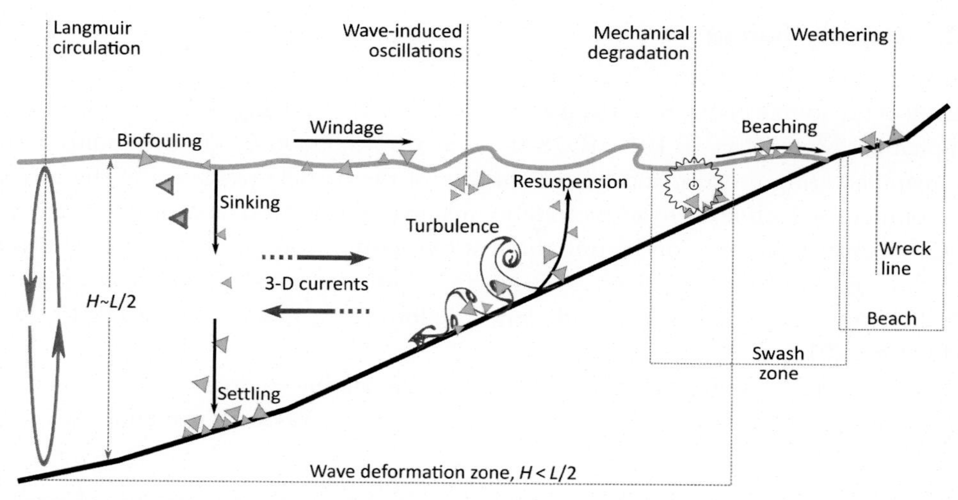

Fig. 6.1 Sketch of the coastal zone, with physical processes relevant to the problem of MP behavior.

shown that deformation of waves in the sea coastal zone plays a significant role in transport of not only bottom sediments and algae but also plastic objects and particles. Discussion of differences and similarities in physical behavior of MPs and natural particles is the main motivation of this review.

6.2 PHYSICAL PROPERTIES OF MARINE MP PARTICLES

The term "microplastics" was coined in 2004 to describe tiny fragments of plastic (~50 μm) in the water column and in sediments (Thompson et al., 2004; Hidalgo-Ruz et al., 2012). Nowadays, it is used in a rather general context, as a collective term to describe a heterogeneous class of synthetic particles ranging in size from a few microns to several millimeters, of various specific densities, chemical composition, shapes, origin, and other characteristics, found in any of the marine environments or biota (e.g., Hidalgo-Ruz et al., 2012; Filella, 2015; Thompson, 2015). This heterogeneity makes it difficult to describe the physical behavior and fate of all the types of MPs at once, requiring certain grouping of particles by their physically important properties. For example, floating MPs are shown to be concentrated in central parts of oceanic gyres (Lebreton et al., 2012; Maximenko et al., 2012; Hardesty et al., 2017). In a highly energetic sea coastal zone, classification by only sinking/floating behavior is obviously not sufficient to predict the particles' fate. As the first step for such a complicated environment, the ranges of the main physical properties are to be realized: they drive further particle dynamics and define the principal time/space scales of the MP "life cycle." In this section, we consider the density range, the typical size classes, and the variability of the MP particle shapes.

6.2.1 Density Range

The density of MP particles found in marine environments is defined by (i) the material density of the source plastic material (i.e., type of the plastic, see Fig. 6.2 and Table S6.1 in Supplement Material in 10.1016/B978-0-12-813747-5.00006-0), (ii) the additives and the manufacturing process, and (iii) the history of the particle residence in the marine environment, which leads to its biofouling, material degradation, and further fragmentation to smaller pieces. Considering the first two points, the specific density of plastic particles can vary in range from $<0.01\,\mathrm{g\,cm^{-3}}$ (for expanded polystyrene foam) to $2.1–2.3\,\mathrm{g\,cm^{-3}}$ for polytetrafluoroethylene (Teflon) (e.g., Driedger et al., 2015; Duis and Coors, 2016).

However, sinking/floating of a MP particle in the marine environment is defined not only by the specific density of the very material the particle was made from but also by the

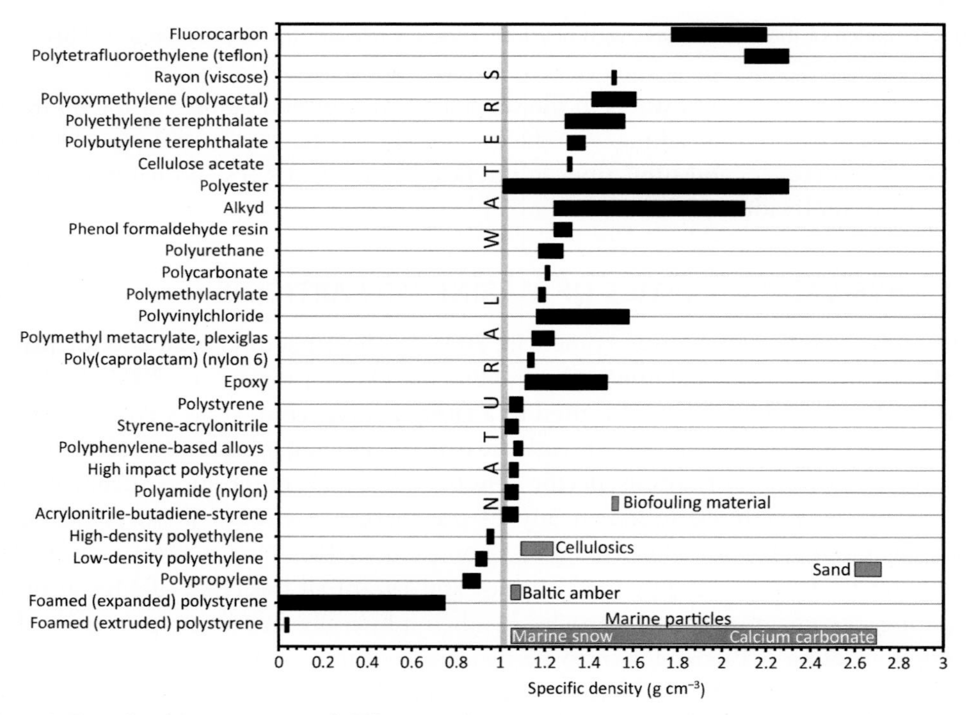

Fig. 6.2 Specific density ranges of different polymer types commonly encountered in the marine environment (see Table S6.1 in Supplement Material for the values in 10.1016/B978-0-12-813747-5.00006-0). For physical understanding, the ranges of density of the following natural materials are shown in *gray* shading: natural waters (from fresh to oceanic ones with salinity of 34 psu); natural marine particles, from marine snow to calcium carbonate (Maggi and Tang, 2015); Baltic amber (Chubarenko and Stepanova, 2017); cellulosics (US EPA, 1992); sand material density (Chubarenko et al., 2016); and density of the biofouling material (dry density of microbial biomass (Bratback and Dundas, 1984)).

integral particle buoyancy. The latter is the difference between the gravity force acting on the entire particle and the Archimedean buoyancy force depending on the particle's volume and the density of the surrounding waters. Thus, physically important is the mean, integral particle density, which may significantly differ from that of a primary polymer. Examples of composition of particles floating at the sea surface and those from the bottom sediments are illuminating in this regard. At the surface of an open ocean, microparticles made of buoyant polymers are typically most abundant, for example, the high- and low-density polyethylene (0.89–0.97 g cm^{-3}) and polypropylene (0.83–0.85 g cm^{-3}). Still, particles denser than water are also reported—those made of polystyrene (1.04–1.1 g cm^{-3}), polyvinyl chloride (1.16–1.58 g cm^{-3}), and polyethylene terephthalate (1.29–1.45 g cm^{-3}) (Hidalgo-Ruz et al., 2012). At the same time, the MPs from low-density materials have been identified in bottom sediments: for example, polyethylene (0.89–0.98 g cm^{-3}), poly(ethylene-propylene) (0.92–1.0 g cm^{-3}), and poly(ethylene-vinyl acetate) (0.93–0.94 g cm^{-3}) were reported in sediments of the Southern Baltic Sea (Graca et al., 2017). Holmström (1975) reported that plastic sheets from low-density polyethylene were found by Swedish fishermen in great abundance at the bottom of Skagerrak, being encrusted with calcareous bryozoan and brown algae. The presence of many plastic bags on the bottom of the Gotland Deep in the Baltic Sea was confirmed by oceanographers who did trawling at the bottom in order to find equipment lost at sea (Chubarenko et al., 2016). As an illustration for the effect of biofouling, the example in Fig. 6.3 shows two similar small floats from foamed plastic that were cut from fishing net found at sea: One is floating, while the other one has sunk to the bottom, being heavily

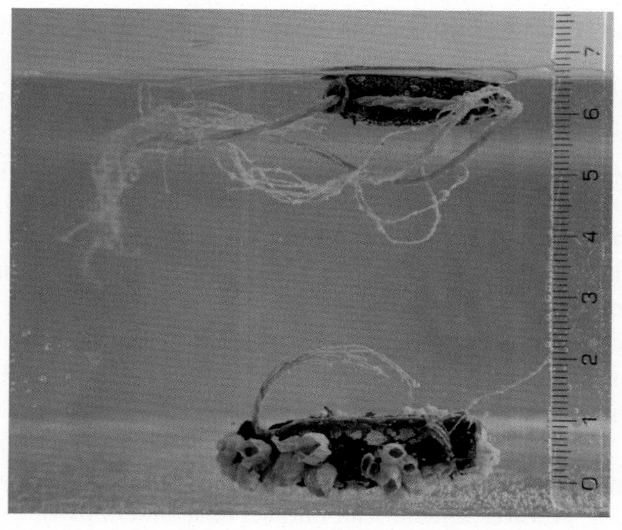

Fig. 6.3 Foamed plastic floats cut off the fishing net found at sea illustrate why buoyant plastics may sink: here, the float at the bottom is colonized by *Balanus* barnacles. *Photo by A. Bagaev.*

occupied by a colony of *Balanus* barnacles. Effects of biofouling are discussed deeper below (Section 6.2.4). Along with the biofouling, the aggregation of MPs with organic matter or surrounding sediments leads to an increase in the integral particle density, while attaching of air bubbles and weathering of the polymer material could lead to its decrease.

It is worth noting that the reported densities of marine MPs (i.e., of the particles obtained from natural samples) may differ from those listed in Fig. 6.2 because of methodical reasons as well, since sampling and analyzing procedures are still inconsistent: they could be weighed in wet or in dry state, before or after complete or incomplete removal of the organic matter, etc. (Andrady, 1998; International Organization for Standardization, 1999; Morét-Ferguson et al., 2010; Hidalgo-Ruz et al., 2012; Zobkov and Esiukova, 2017a,b).

Given severe lack of observations on real behavior of MPs under sea conditions, it is profitable at the present-day level of understanding to relate MPs to other marine particles with similar (in certain aspects) characteristics, like marine snow, Baltic amber, or sand particles. The densities of such materials are shown in Fig. 6.2 for convenience. Similar to MPs, the density of natural marine particles depends on their composition. The composition can be divided into mineral and organic fractions (Maggi and Tang, 2015; Monroy et al., 2017). The mineral part consists of either terrigenic or biogenic minerals; the latter typically is calcium carbonate with the density of 2700 $kg\,m^{-3}$ (Francois et al., 2002) produced by coccoliths and significantly less dense (1950 $kg\,m^{-3}$) biogenic silica produced by diatoms (Balch et al., 2010). The density of particulate organic matter also ranges widely depending on its origin. For instance, the density of cytoplasm spans from 1030 to 1100 $kg\,m^{-3}$, while the one of fecal pellets ranges between 1230 and 1174 $kg\,m^{-3}$ (Komar et al., 1981; Monroy et al., 2017). Thus, in general, the density range of marine organic matter can be assumed to be from 1050 to 1500 $kg\,m^{-3}$, while for all the natural marine particles, the density ranges approximately between 1050 and 2700 $kg\,m^{-3}$ (Maggi, 2013; Monroy et al., 2017). So some investigations on, say, marine snow, can also be representative for MPs.

The density range of the Baltic amber is close to that of widespread plastics like polyamide, polystyrene, or acrylic (Chubarenko and Stepanova, 2017). This makes it possible to use the observations of the Baltic citizens who have been monitoring the migrations of amber stones in the sea coastal zone and their massive washing ashore for ages. This topic is explored in more detail in Section 6.3.3.

Practical conclusion toward physical description here can be formulated as follows. The density of marine MP particles can significantly deviate from the tabular values for the original plastic material due to many factors experienced by the particle on its long-lasting way from the factory to the sea bottom. Many of these factors are related to external environmental conditions and vary in time and space (e.g., weathering, degradation, biofouling, and attachment of air bubbles or sediment particles), making the life history of the particle also significant. Thus, such a fundamental property as the specific density of the material, the particle was made of seems to play the role of the initial reference value only. It is the process of the integral particle density alteration with time that is important for the problems of general MP transport and fate. This way, for the purposes of, for example, numerical modeling of transport of such a heterogeneous community in the ocean or in the coastal zone, the use of the density range and distribution, and the time rate of the density change in the particular environment might be practically convenient. This suggests the development and use of probabilistic (statistical) models along with presently used deterministic ones (e.g., the Lagrangian particles approach (Maximenko et al., 2012; Bagaev et al., 2017b; Hardesty et al., 2017)). Probabilistic models are used, for example, in sediment transport modeling: they take into account grain size statistics (mean, sorting, and skewness) to predict sediment transport of different sediment fractions (e.g., Li et al., 2004; McLaren et al., 2007).

6.2.2 Size Classes

Physical description of the behavior of a particle in water depends substantially on its size. With MPs, however, the definition of size still raises enough confusion. In most studies, the definition by Arthur et al. (2009) is accepted: MPs are defined as "fragments and primary-sourced plastics that are smaller than 5 mm"; see Fig. 6.4. Still, there is ongoing debate on the minimum size limit (Andrady, 2011; Cole et al., 2011; Sundt et al., 2014; Kershaw, 2015; Ryan, 2015; Rocha-Santos and Duarte, 2015; Thompson, 2015; Van Cauwenberghe et al., 2015; Rodríguez-Seijo and Pereira, 2017). This uncertainty is objectively based on the choice of the leading factor, that is, what aspect is taken as the most significant: sampling and processing methods (Hidalgo-Ruz et al., 2012; Filella, 2015; Van Cauwenberghe et al., 2015; Law, 2017; Rodríguez-Seijo and Pereira, 2017), correlation with microobjects in other science branches (Nerland et al., 2014), abundance in environment (McDermid and McMullen, 2004; Costa et al., 2010; Ivar do Sul et al., 2009; Eriksen et al., 2014; Phillips and Bonner, 2015), or ingestion by marine biota (e.g., Perkin et al., 2009; Cole et al., 2013; Foekema et al., 2013; Lusher et al., 2013; Li et al., 2015; Neves et al., 2015; Tanaka and Takada, 2016).

In order to make the method-based classification more flexible, Andrady (2011) suggested adding the term "mesoplastics" to differentiate between small plastics visible to the naked eye and those only discernible with use of the complimentary technique (microscopy). The European MSFD technical subgroup on marine litter (Galgani et al., 2013) proposed the classification that differentiates between small MPs (<1 mm) and large MPs (1–5 mm). Numerically, if sediment samples are sieved, the minimum sizes of collected MP range from 0.5 to 2 mm, depending on the mesh size of the sieve employed (Fujieda and Sasaki, 2005; Costa et al., 2010), while the minimum size of MPs collected from seawater samples is determined by the mesh size of the net—typically

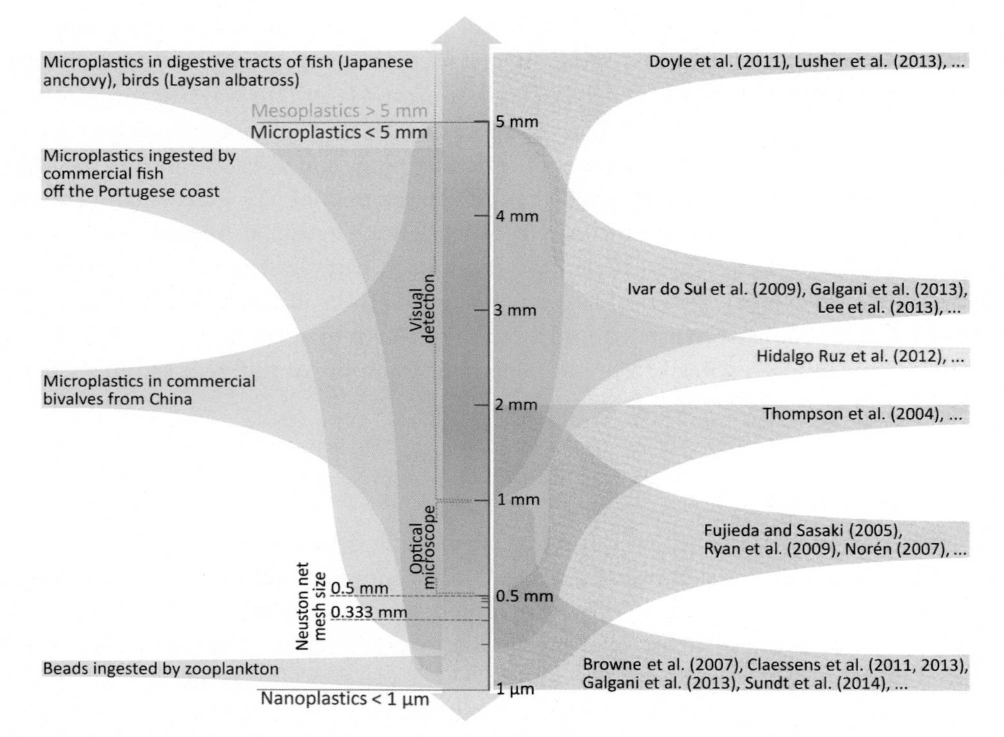

Fig. 6.4 Size ranges for MPs vary between the authors depending mainly on the applied sampling/processing methods and studied biota. Left: examples of living organisms ingesting MPs are given from the studies by Cole et al. (2013), Li et al. (2015), Neves et al. (2015), Plastic ingestion by birds (2016), Tanaka and Takada (2016). Right: the authors and the size ranges used (see Table S6.2 in Supplement Material for broader information and more references in 10.1016/B978-0-12-813747-5.00006-0).

from 53 to 333 µm (Thompson et al., 2004; Norén, 2007; Hidalgo-Ruz et al., 2012). Since it is common practice to use neuston nets with the mesh size of 333 µm (0.33 mm) to collect samples (Arthur et al., 2009; Andrady, 2011; Rocha-Santos and Duarte, 2015; Kershaw, 2015), this value can also be taken as the lower limit for the size of MP particles. Following the sampling and processing methods, Hidalgo-Ruz et al. (2012) proposed to distinguish two main size categories of MPs, (i) < 500 µm and (ii) from 500 µm to 5 mm.

As for numerical abundance of MPs of various sizes, it was shown that in sediment samples, the largest proportion of plastic fragments was obtained in the size classes of 1–5 mm (McDermid and McMullen, 2004; Ivar do Sul et al., 2009; Costa et al., 2010). Phillips and Bonner (2015), with reference also to Eriksen et al. (2014), reported that "MPs ranging in linear length from 0.33 to 4.75 mm comprise up to 92% of the available plastics in marine environments," although the lower size limit was defined by the mesh size of the net used. Such evaluations, however, are objectively dependent on the sampling and processing methods.

Finally, different living creatures are exposed to MPs of different sizes. Among birds, 59% of examined seabird species have ingested plastics (Wilcox et al., 2015). The sizes of plastic items ingested by Laysan albatross (*Phoebastria immutabilis*) chicks have been reported to vary between 0.5 and 51.5 mm and up to 11.3 cm (Auman et al., 1997; Gilbert et al., 2016). The first records of plastic ingestion in fish date back to the 1970s, when fish were observed to consume opaque polystyrene plastic spherules present in the coastal waters of southern New England, the United States (Carpenter et al., 1972). Since then, at least 92 fish species have been reported to ingest plastic (Li et al., 2015; Lusher, 2015). Neves et al. (2015) examined 263 individuals from 26 species of commercial fish off the Portuguese coast and found that the size of the ingested MP particles varied between 0.217 and 4.81 mm (mean 2.11 ± 1.67 mm). In a study by Foekema et al. (2013), plastic sizes in fish stomachs ranged from <5 to 14.3 mm, with the most common size class being 1–2 mm (Lusher et al., 2013). Perkin et al. (2009) found that the gape size among the most common fish family (*Cyprinidae*) is <6% of their total length, which constrains consumption of MPs >6 mm. The maximum length of the MP pieces in the digestive tracts of 64 Japanese anchovy (*Engraulis japonicus*) sampled in Tokyo Bay ranged from 150 to 6830 μm (average 783 ± 1020 μm) (Tanaka and Takada, 2016). Ory et al. (2017) report that 80% of *Decapterus muroadsi* fish captured along the coast of Rapa Nui (Easter Island, southeastern Pacific Ocean) had ingested one to five particles, ranging in length from 0.2 to 5.0 mm (median size 1.3 ± 0.1 mm). The most common plastic types found inside different commercial bivalve species in China were reported to vary in size between 5 μm and 5 mm, but the MPs <250 μm were the most common (Li et al., 2015). Cole et al. (2013) found that zooplankton ingests MP beads 1.7–30.6 μm in size.

Thus, there is no agreement up to now what the size of MP particles is; often, it is linked to the sampling/processing methods or goals of the investigation. For hydrodynamic evaluations below, we assume the particle size in the range from 0.5 to 5 mm. The lower limit still varies in different applications further in this chapter and is based on physical reasoning. Typically, we consider properties of particles at the scale much above the molecular one and their motion in the transitional flow regime, that is, at the Reynolds numbers much above the Stokes case $Re \sim 1$.

6.2.3 Typical Shapes

The shape of a particle is of fundamental significance for every kind of its motion in water, so it is not surprising that much attention is paid to this question in sedimentology, and many goal-specific approaches are developed. Still, shape remains one of the most difficult parameters to characterize and quantify for all but the simplest of shapes, let alone those of MPs. Having in mind that physical description of MP particles in the marine environment is today at its infancy, we review here information from both classical works (mainly—hydrodynamics applied to natural sedimentology) and recent observations and experience with shapes of marine MPs.

Despite of a large body of literature on the subject in sedimentology, there still remains a widespread confusion regarding the meaning and relative value of different measures of a particle shape (Blott and Pye, 2008; Rodriguez et al., 2013). Quantification of form requires the measurement of the length, breadth, and thickness of a particle (Wentworth, 1923a,b; Zingg, 1935; Krumbein, 1941; Sneed and Folk, 1958). In Blott and Pye (2008) it is noted that most workers have agreed that the three dimensions should be perpendicular (orthogonal) to each other, although they need not intersect at a common point (Krumbein, 1941; Flemming, 1965).

Wadell (1933, 1935), Krumbein (1941), Powers (1953), and others formulated sphericity-roundness classification schemes in which drawings of exemplar shapes (based on real sand grains) represented "type" class members (MacLeod, 2002). Also Zingg (1935) developed a more versatile shape classification scheme that uses the particle dimensions along the three principal axes. It is important to note that the same value of sphericity may be applied to differently shaped particles; thus, sphericity and geometric form are different measures of a particle shape. Krumbein's sphericity index (Krumbein, 1941) and Cailleux's flatness index (Cailleux, 1945) were also used for the description of the particle shapes. Kowalski et al. (2016) noted that for irregularly shaped MPs, an equivalent spherical diameter is useful, being calculated on the basis of a best-fit ellipse using the measured major, minor, and intermediate axes as defined by Kumar et al. (2010).

Camenen (2007) found that the deviation of the shape of a particle from a sphere can be generally quantified by a shape factor. Many shape factors have been proposed in literature (e.g., Graf, 1971). The most commonly used one is the Corey (1949) shape factor csf, which is given by $csf = D_s/(D_l D_i)^{0.5}$, where D_l, D_i, and D_s are the longest, intermediate, and the shortest axes of the particle assumed to be an ellipsoid, respectively. The smaller the csf is, the flatter the particle is. Another shape factor is the particle roundness P (Powers, 1953), which is usually considered by analyzing naturally rounded and crushed particles and is a measure of curvature variations along the grain surface. For spheres, $csf = 1$, and $P = 6$, while, for example, for a typical flake, $csf = 0.6$, and $P = 1$ (Camenen, 2007).

Importantly, since different shape factors describe different aspects of the particle shape, they are often used in some combination, for example, sphericity and roundness (e.g., Wadell, 1933, 1935; Krumbein, 1941; Powers, 1953), Corey shape factor and Powers roundness (Dietrich, 1982; Camenen, 2007), and Krumbein's sphericity index and Cailleux's flatness index (Kowalski et al., 2016).

Recently, Blott and Pye (2008) and Rodriguez et al. (2013) have reexamined the basic concepts of the particle shape and suggested a number of new and modified methods that are widely applicable to a range of sedimentologic problems. In particular,

Table 6.1 Developed ways of description of the particle shape

	Visual characteristics	Measurements
General form of a particle	– Spherically isotropic, orthotropic, axisymmetric (Clift et al., 1978) – Spheres, oblate and prolate spheroids, cylinders, and disks (e.g., Lerman et al., 1974) – Cylinders, disks, flat, ovoid, spheruloids (e.g., Hidalgo-Ruz et al., 2012) – 3-D, 2-D, 1-D (Chubarenko et al., 2016)	– Three diameters: for example, Krumbein's sphericity index (Krumbein, 1941), Cailleux's flatness index (Cailleux, 1945), Corey shape factor (Corey, 1949); the way of measurement varies among the authors – Two values out of four: volume, surface area, projected area, projected perimeter (Clift et al., 1978): for example, the volumetric shape factor (Heywood, 1962), sphericity, and circularity (Wadell, 1932)
Irregularity	– Powers roundness: six classes, from very angular ($P=1$), angular ($P=2$)...to well rounded ($P=6$), using comparison with standardized set of photos (Powers, 1953) – Irregular, elongated, degraded, rough, and broken edges (Hidalgo-Ruz et al., 2012)	– Measurement of corners: for example, Wadell roundness (Wadell, 1932) and many others, measuring inscribed and circumscribed circles in angles of the enlarged particle' projection (Wentworth, 1919; Krumbein, 1940; Sneed and Folk, 1958; Sukumaran and Ashmawy, 2001)

Rodriguez et al. (2013) suggested that the particle shape could be classified in three categories: (i) sphericity, the overall particle shape and similitude with a sphere; (ii) roundness, the description of the particle's corners; and (iii) roughness, the surface texture of the particle. A short summary of the developed ways of description of a particle shape is presented in Table 6.1.

From the MP side, worldwide observations indicate that they may appear in the environment in a great diversity of shapes. Moreover, changes of the particle shape with residence time in the environment become also significant. In general, the marine MP shapes are shown to depend on (1) the original form of primary plastics, (2) degradation and erosion processes of the particle surface (biological breakdown, photodegrading, physical forcing, etc.), and (3) residence time in the environment, with sharp edges denouncing a recent introduction and smooth edges associated with a large residence time (e.g., Hidalgo-Ruz et al., 2012; Rodríguez-Seijo and Pereira, 2017). The main shapes of MPs on the ocean surface are fragments, films, and pellets (e.g., Cózar et al., 2017), while fibers are typical of water column (e.g., Bagaev et al., 2017a,b) and deep-water sediments (e.g., Van Cauwenberghe et al., 2013b), and many other shapes are found in marine environments; see Fig. 6.5.

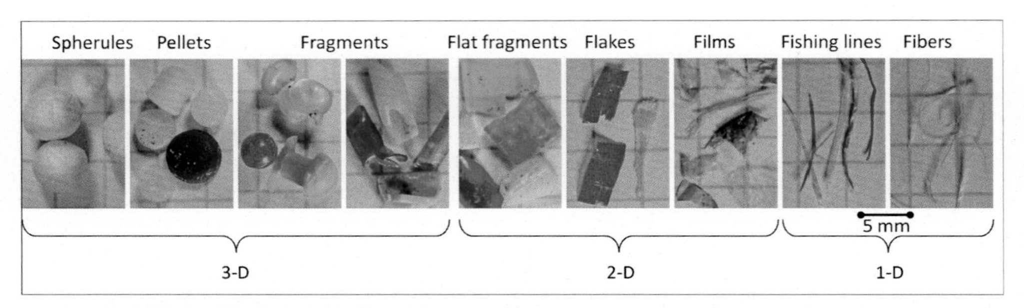

Fig. 6.5 Collection of MP shapes. Typical shapes of marine MPs. With all the observed variability, from the general point of view, particles having all three dimensions of the same order ($a \sim b \sim c$) can be classified as three-dimensional; flat fragments, flakes, and films have one dimension much smaller than the other two ($a \sim b \gg c$) and are thus generally two-dimensional; fishing line cuts, threads, fibers, etc., having $a \gg b \sim c$, are one-dimensional. Such differentiation is suggested by differences in biofouling rate (see Section 6.2.4) and principally different sinking behavior (Section 6.3.1).

Tanaka and Takada (2016) gave a description of the main shapes of MPs as follows: (i) fragments: particles produced by fragmentation of larger materials; (ii) beads: particles manufactured as microsized products, either spherical or an aggregate of spheres; (iii) pellets: granules manufactured as a raw material of larger plastic products, generally in the size range of 2–5 mm with the shape of a cylinder or a disk; (iv) foams: foams made from polymer; (v) films: soft fragments of thin polymers derived from plastic bags or wrapping paper and so on; (vi) sheets: hard fragments of thin polymers; (vii) filaments: threadlike polymers produced by fragmentation of ropes or lines used in fishing, >50 μm; and (viii) fibers: threadlike polymers derived from textiles, including clothing and furnishings, <50 μm. The 50 μm threshold for fibers was chosen by Tanaka and Takada (2016) because typical textile fibers have a diameter of 10–20 μm (up to 50 μm) (Sinclair, 2015), and the diameter of monofilaments used in fishing ropes or lines is larger than several hundred micrometer (Chattopadhyay, 2010).

In the review by Hidalgo-Ruz et al. (2012), the following categories are listed used by researchers to visually describe the shapes of MPs found in the marine environment: (i) for pellets: cylindrical, disk-shaped, flat, ovoid, and spheruloid; (ii) for fragments: rounded, subrounded, subangular, and angular; and (iii) general: irregular, elongated, degraded, rough, and with broken edges.

An important question is which shapes are more abundant in different marine environments. In particular, at the water surface and on the beaches, *fragments* from plastic products were often numerically dominant, followed by plastic pellets (e.g., Ryan, 1988; Moore et al., 2001; Browne et al., 2010; Cózar et al., 2017) and styrofoam as the second most abundant material (Yamashita and Tanimura, 2007). Most fragments found in subtidal and estuarine sediments were fibers (Thompson et al., 2004; Browne et al., 2010, 2011). It is suggested (Hidalgo-Ruz et al., 2012) that the shape of plastic fragments depends on the fragmentation process and residence time in the environment: sharp edges might indicate either recent introduction into the sea or recent breakup of larger pieces, while smooth edges are often associated with older fragments that have been continuously polished by other particles or sediment (Carpenter and Smith, 1972; Doyle et al., 2011; Chubarenko et al., 2016). Indeed, Gilfillan et al. (2009) noted that larger particles had more elongated shapes and/or irregular surfaces, while progressively smaller particles were consistently more circular. In deep-water bottom sediments, Woodall et al. (2014) reported that MPs were all fibrous in shape, commonly 2–3 mm in length and <0.1 mm in diameter.

Plastic films are reported in abundance on the beaches, at the water and bottom surfaces in their macrosize form: shopping and package bags or various greenhouse, construction, insulation, or package films. Among MPs, they are almost never reported, most probable reason being that they break up mainly into threads and filaments (Chubarenko et al., 2016), and sink when breaking up.

The surface texture (or roughness) is also considered as a characteristic of a particle shape (Rodriguez et al., 2013). Field evidence proves that MP particles found in the marine environment have quite rough, mechanically damaged, crackled surfaces (Shaw and Day, 1994; Corcoran et al., 2009). This suggests that for estimations of the particle's surface areas, the fractal description might be more suitable, and fractal dimensionality of the surface might be used to evaluate the residence time (age) that the particle had spent in the sea. It seems that quasi-spherical particles (pellets and pieces) may require such a description; films are shown to have fractal ends after mechanical ruptures (Eran et al., 2002); plastic fibers, however, are reported to be smooth (Hidalgo-Ruz et al., 2012). The idea to link the fractal dimensionality of the surface with the particle's residence time at sea looks quite promising but requires many more laboratory and field investigations. At the same time, an increase in the surface area due to its fractal structure seems to have limited influence on the MP particle biofouling rate: on the one hand, the corrupted surface begins fouling faster, but, on the other hand, too small fractures will soon be covered by relatively large algae/bacteria cells (Chubarenko et al., 2016).

6.2.4 Variations of Properties With Time: Biofouling, Mechanical Degradation

Plastics longevity, being both a great merit and a main threat, does not mean that physical properties of marine MP particles are known and fixed. "Marine MP particles" may consist of both plastic and natural materials whose qualitative properties and proportions may vary with residence time in the environment. This way, plastic material density and surface texture can be changed by photodegradation, thermal degradation, chemical degradation, or biological degradation (Andrady, 1994, 1998, 2015a,b; Gregory and Andrady, 2003; Shah et al., 2008; Cooper, 2012; Muthukumar and Veerappapillai, 2015; Wang et al., 2016; Law, 2017). The average particle density (considering the plastic + biofilm as a unit) varies due to biofouling or aggregation with other marine particles (Ryan, 2015; Chubarenko et al., 2016; Duis and Coors, 2016; Fazey and Ryan, 2016; Wang et al., 2016). Sizes and shapes of particles change due to fragmentation caused by exposure to UV radiation, oxygen and seawater, or mechanical forcing (Andrady, 2005, 2011, 2015a; Duis and Coors, 2016; Wang et al., 2016). Thus, principal variations of properties with time along with the "life history" of every particular particle become of importance for understanding of its behavior under environmental conditions. Addressing here physical questions of MP migrations in the coastal zone, we shall restrict our analysis to relatively short (for plastics and oceans) time scales and will briefly consider only two aspects of variation of physical properties of MP particles: rate of biofouling of particles of different sizes and shapes and celerity of mechanical degradation of different kinds of plastics in the sea swash zone.

6.2.4.1 Biofouling of Particles of Different Sizes and Shapes

Observations show that MP particles with low specific density have also been found in sediments (Thompson et al., 2004; Browne et al., 2011; Claessens et al., 2011, cited after Hidalgo-Ruz et al., 2012). The CHN elemental analysis revealed relatively high contents of nitrogen (N) on MPs, which suggests abundant epibiont overgrowth because N is not a component of synthetic polymers (Morét-Ferguson et al., 2010). Indeed, an overgrowth by heavier micro- and macroorganisms leads to an increase in an average particle density and eventually causes its sinking, for example, the light-foamed float occupied by *Balanus* barnacles in Fig. 6.3 (Holmström, 1975; Harrison et al., 2011; Hidalgo-Ruz et al., 2012; Chubarenko et al., 2016). Studies by Cózar et al. (2014) and Eriksen et al. (2014) report a conspicuous paucity of floating MPs at the sea surface, suggesting this very reason for its explanation (see also Woodall et al., 2014; Ryan, 2015; Cózar et al., 2017). Biofouling should remove small fragments faster than large ones as a result of their higher surface area to volume ratios (Cózar et al., 2014; Ryan, 2015), because while the buoyancy of an item is a function of its volume, its susceptibility to fouling is dependent largely on its surface area (Ryan, 2015). Fazey and Ryan (2016) showed experimentally that small plastic items lost buoyancy much faster than larger ones and that there was a direct relationship between an item's volume (buoyancy) and the time to attain a 50% probability of sinking. More complicated behavior is suggested in the model study by Kooi et al. (2017): an MP particle can either float or sink to the ocean floor or oscillate vertically, depending on the size and density of the particle. Such behavior is driven by the biofilm growth and respiration, which depend on environmental conditions.

Chubarenko et al. (2016) examined the process of biofouling of MPs of different shapes analytically: the dependence on the particle size and shape of the (i) surface area and (ii) the time when the average (plastic + biofilm) particle density reaches that of water are considered while applying simple geometric rules to the particles of different sizes and shapes. It is shown for a floating spherical particle of the radius R_0 and the material density ρ_0 ($\rho_0 < \rho_w$) that, when it is fouled by the film of the density ρ_f ($\rho_f > \rho_0$) and the thickness

$h_f(t)$, the average density $\overline{\rho}(t)$ of the fouled particle (of the radius $R(t) = R_0 + h_f(t)$) is inversely proportional to the initial radius of the sphere and directly proportional to the thickness of the biofilm (growing with time):

$$\overline{\rho}(t) \approx \rho_0 + \rho_f \times \frac{3h_f(t)}{R_0}, \quad h_f(t) \ll R_0 \tag{6.1}$$

This expression shows explicitly that the mean particle density increases with time (parameterized by the thickness of the biofilm) from ρ_0 to some $\overline{\rho}(t)$. When, in time interval τ_{sink}, the mean density of the particle becomes equal to the surrounding water density, ρ_w, the particle begins sinking. The larger the initial radius is, the thicker film d_f is required in order to increase the mean particle density $\overline{\rho}(t)$ up to the water density, that is, the smaller the particle is, the faster (under the same external conditions) it becomes biofouled and begins to sink; see Fig. 6.6. An illustrative example: if the spherical particle of the mass m_0 is broken into smaller spherical pieces of radii $R_0/2$ and $R_0/3$, there will be 8 and 27 smaller spheres, respectively; their total surface area will be $2S_0$ and

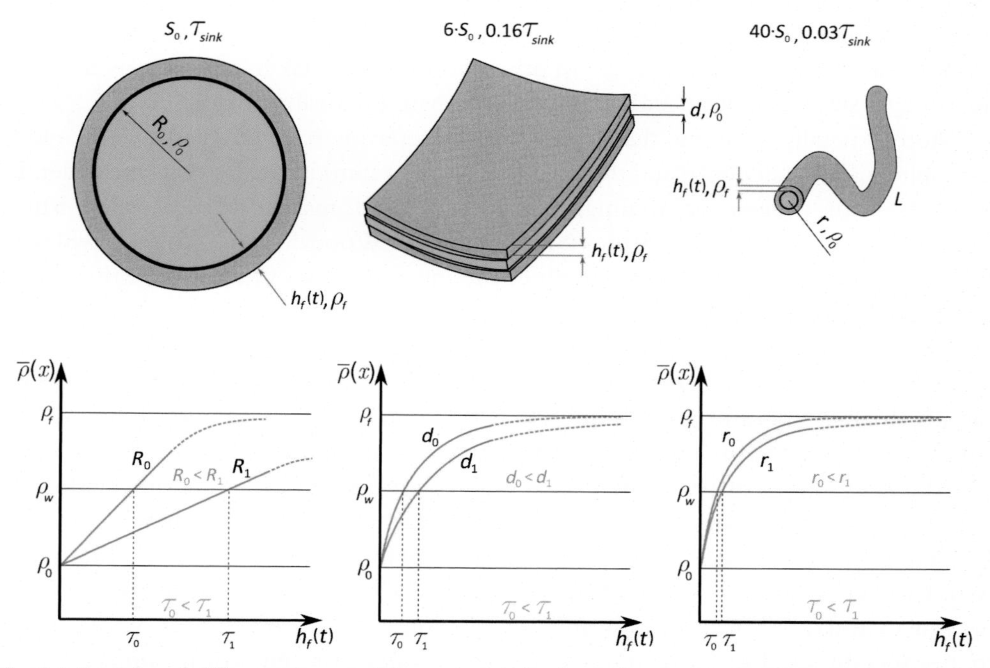

Fig. 6.6 Particles of different shapes with a material density ρ_0 ($\rho_0 < \rho_w$) coated by growing with time surface film of a density ρ_f ($\rho_f > \rho_w$). The graphs show the resulting increase in the particle mean density for spheres/films/fibers of different characteristic scales. Above the particles, surface areas and times of biofouling up to the freshwater density for particles of the same mass m_0 but different shapes are shown. Evaluations are performed for polyethylene (density 0.9 g cm^{-3}) particles of typical spatial scales: diameter of the sphere 5 mm, thickness of the film 30 μm, and diameter of the fiber 10 μm. *Modified from Chubarenko, I., Bagaev, A., Zobkov, M., Esiukova, E., 2016. On some physical and dynamical properties of microplastic particles in marine environment. Mar. Pollut. Bull. 108, 105–112. https://doi.org/10.1016/j.marpolbul.2016.04.048.*

$3S_0$, and time till the beginning of sinking—$\tau_{sink}/6$ and $\tau_{sink}/9$, correspondingly (Chubarenko et al., 2016). Smaller particles also respond faster to the loss of mass, thus oscillating in water column with higher frequency (Kooi et al., 2017). In the coastal zone, attaching/detaching of sediment particles and air bubbles is obviously the case, especially in the swash zone. This highlights a more complicated behavior of smaller MPs in comparison with larger items.

For the given mass m_0, the spherical shape has the minimum surface area among all other possible shapes in three-dimensional (3-D) space. For marine MPs, most common alternative shapes are films and fibers. Comparison of surface areas and corresponding times of biofouling till water density for particles of these shapes, having the same mass m_0, is presented in Fig. 6.6 (see also Chubarenko et al., 2016). Calculations depend on numerical values, so we take as an example the LDPE spherical pellet of a diameter of 5 mm (density $0.9\,\mathrm{g\ cm^{-3}}$), a film of a typical thickness of 30 μm, and a fiber of a diameter of 10 μm, which all get fouled by the microbial biomass $\rho_f = 1.5\,\mathrm{g\ cm^{-3}}$ (Bratback and Dundas, 1984). Then, the ratio of the surface areas is $S_{sphere}/S_{film}/S_{fiber} \sim 1{:}6{:}40$ (Fig. 6.6). Designating as τ_{sink}^{pellet} the time of biofouling up to freshwater density of the spherical pellet, we obtain biofouling times for the film and the fiber (see Chubarenko et al., 2016 for derivation of formulas) as $0.16\tau_{sink}^{pellet}$ and $0.03\tau_{sink}^{pellet}$. In other words, for the taken typical case, films and fibers begin sinking about 6 and 30 times faster than spherical particles.

More generally speaking, the smaller the characteristic scale of a particle (the radius for a sphere (R) or a fiber (r) and the thickness (h) for a film) is, the faster its mean density increases with progressive fouling. Since for typical marine MPs $R \sim 0.5$–5 mm, $h \sim 15$–30 μm, and $r \sim 10$–100 μm (Sinclair, 2015; Woodall et al., 2014), typical is $R > h \sim r$; thus, the time rate of growth of the mean density of the fibers is the largest, while that of spheres is the smallest.

Holmström (1975) argued that plastic bags found on the bottom of the Kattegat area of the Baltic Sea should have spent about 3–4 months in the euphotic zone before biofouling thickness became large enough to force them to sink. For the same (Baltic Sea) conditions, the polyethylene fibers of the same density and the radius r similar to the film thickness h (say, 30 μm) should then spend about the same 3–4 months within the upper layers of the Baltic Sea while spherical particles and plastic fragments (with $R \sim 1$ mm as the scale) up to 6–10 years before sinking due to biofouling (Chubarenko et al., 2016, their formulas (5) and (7)). These comparisons are based on just geometric considerations and assume constant biofilm growth without link to any specific kind of biota. Time scale here comes from empirical estimates made by the observer in the relatively cold Baltic Sea. In comparison with 17–66 days to begin sinking obtained for polyethylene MPs in the direct exposure experiments performed in the False Bay, South Africa, by Fazey and Ryan (2016), several months and even years spent in surface layers look like an overestimate. However, coming solar radiation in the Kattegat area (57° N) and the False Bay (34° S) is obviously different, as well as the composition of species, water temperature, and amount of nutrients, defining the intensity of biofilm growth. Thus, biofouling time estimates are largely variable due to natural variability of external conditions and provide only rough scales, while principal geometric considerations provide only comparative characteristics for MPs of different shapes.

6.2.4.2 Mechanical Degradation in the Sea Swash Zone

Conventionally, oxygen and sunlight are considered as the most important factors of abiotic degradation of plastic objects (Andrady, 1998; Gewert et al., 2015): light-induced oxidation is shown to be orders of magnitude faster compared with other types of degradation processes in the terrestrial environment (Wang et al., 2016). In water, however, plastics are effectively protected from solar UV radiation and high temperatures, so the rate of degradation becomes especially slow (Gregory and Andrady, 2003; Barnes and Milner, 2005; Corcoran et al., 2009; Ryan et al., 2009). It seems that in the marine environment, the mechanical degradation in the shore swash zone—under energetic action of breaking waves and moving sediments—becomes of primary importance for generation of MPs from larger objects. A number of investigations (Debrot et al., 1999; Convey et al., 2002; Eriksson and Burton, 2003; Thiel et al., 2013) have pointed out the role of rocky shores exposed to surface waves: they capture larger debris items and serve as a "grinding mill," exporting the fragments back to the sea. Rocky shores and cobble beaches are also shown to capture MPs not at the beach surface, but at some depth under the cobbles and small rocks (McWilliams et al., 2017). Still, however, very little published information exists concerning mechanical degradation of plastics in the coastal zone (Cooper, 2012).

We performed a set of laboratory experiments with the idea to model the process of mechanical degradation of most common plastics in the sea swash zone. Typically found MPs on the beaches and in the sea/ocean coastal zone tend to reflect the common production trends. For example, a study from Italy (Vianello et al., 2013) found the most predominant MPs to be polyethylene (PE, 48%) and polypropylene (PP, 34%); Frias et al. (2014) found PE, PP, and polyacrylates dominating along the Portuguese coast; on the beaches of Hawaii, PE (85%) and PP (14%) are dominating (Carson et al., 2011); in Claessens et al. (2011), the analyzed MPs in coastal and offshore sediments consisted of PP, PS, and PP; and Esiukova (2017) found that foamed PS is the most common type of MPs on beaches of the Southeastern Baltic Sea. Based on these observations, we chose PE, PP, and PS for our degradation experiments. Since about half of the produced plastic is used today in low-value products designed for disposable single use (PlasticsEurope, 2013), single-use PP cups (material density $0.86\,g\,cm^{-3}$) and PS plates ($1.05\,g\,cm^{-3}$) and LDPE garbage bags ($0.92\,g\,cm^{-3}$) were used as the source of macroplastics for our experiments. In addition, building heat insulator styrofoam was also included ($0.01\,g\,cm^{-3}$), because foamed PS spherules are mentioned in many studies as abundant among the beached MPs (e.g., Claessens et al., 2011; Duis and Coors, 2016; Esiukova, 2017).

In order to model wave breaking in the water edge of the beach, the laboratory mixer with an inclined axis of rotation (35 degrees; frequency of rotation 30 rev/min) was

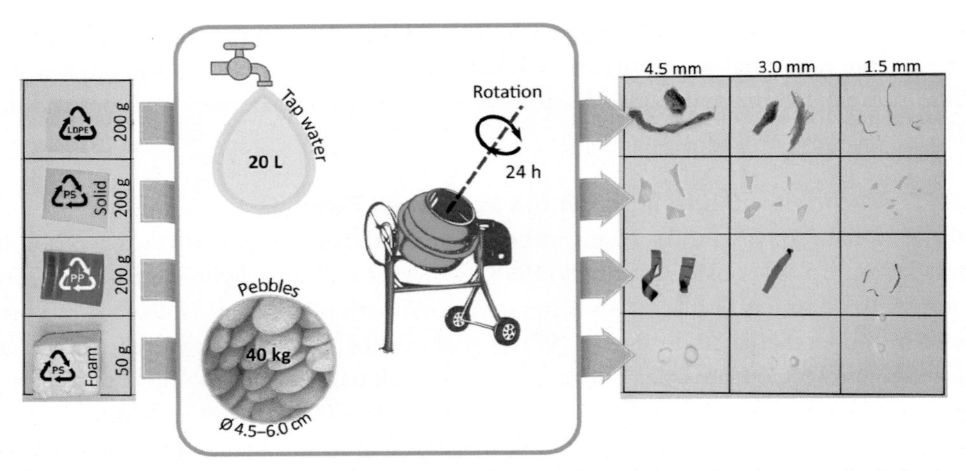

Fig. 6.7 Scheme of laboratory experiment on mechanical degradation of typical beached plastics in the sea swash zone. Plastic samples (2 cm × 2 cm, left-hand side photos) were placed for 24 h in the rotating mixer with an inclined axis of rotation, filled with natural marine pebbles and water. Every 3 h of mixing, plastic material was washed through a set of sieves, and typical examples of particles at particular sieves are shown on the right.

loaded with 40 kg of natural 4–6.4 cm pebbles and 20 L of water (Fig. 6.7). The inclination of the axis was carefully adjusted to the given amount of the mixing material in order to reproduce the wave run-up and pebbles rolling over the surface at one (left) side of the mixer, with the "wave height" of about 20 cm (calculated as doubled maximum water depth above the pebbles, i.e., when they are in their lowermost point during the revolution of the mixer). This way, pebbles and plastic samples at every location inside the mixer experienced flooding and drying at every mixing cycle; characteristic "wave breaking" at the plunging side of the mixture and rolling surface pebbles at its rising side was intended to mimic real wave swash of quite moderate intensity.

The selected plastic objects were cut into 2 cm × 2 cm^2 samples (total weight 200 g of every material) and placed into the mixer for 24 h (eight 3 h-long runs). The aim of these tests was both qualitative and quantitative. Firstly, we wanted to evaluate and compare the rates of destruction and shapes of resulting particles for every type of material under the given exposure. Secondly, quantitative distribution of mass and number of particles over the MP size ranges was obtained. Such stepwise distribution was helpful in the understanding of the MP generation process.

As a result of "wave breaking" and rolling of pebbles in the mixer, plastic samples broke into smaller pieces. Every 3 h of rotation, the plastic material was removed from the mixer, washed through a set of 10 sieves (from 5 to 0.5 mm), dried, weighted, and put back into the mixer for the next 3 h. The fractions passing through a 5 mm sieve were considered as MPs while those passing through the 0.5 mm sieve as nanoplastics. Increase in mass of the generated MPs of all four types during the experiments is presented in Fig. 6.8A. Solid PS (single-use plates) appeared to be the most brittle: about 50% of

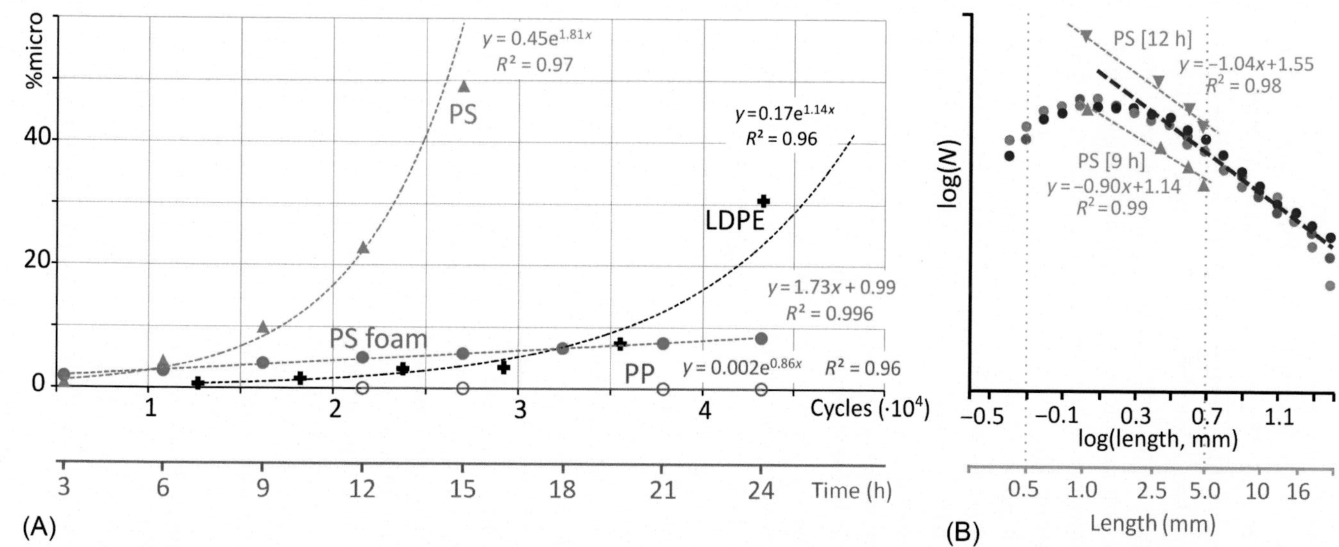

Fig. 6.8 Results of degradation of PP, PS, LDPE, and PS foam samples in the rotating mixer, mimicking the sea swash zone with coarse bottom sediments: (A) exponential increase of mass of MPs (size range from 0.5 to 5 mm) with time of the experiment, expressed both in hours and in cycles of rotation ("wave periods"); (B) the log-log graph of the dependence of the MP particle number on their size; the distribution curve for plastics floating at the ocean surface from Cózar et al. (2017) demonstrates the same linear trend in its right-hand part.

the initial mass was transferred to the MP class in 15 h already (about 2.7×10^4 cycles). Oppositely, PP samples (single-use cups) were most resistant: only 0.1% of initial mass became MPs even after 24 h. Characteristic of LDPE films was that the transfer from large samples to MP class proceeded mainly not via braking in parts, but through four to eight times folding and/or stretching; see typical MP shapes in Fig. 6.7. Foamed PS, being composed from bubbles, first broke into individual spherules of 4–6 mm in diameter, and only then these spherules began breaking into smaller pieces very slowly. Interestingly, a significant portion of all the samples, including those from relatively light (floating) LDPE and PP, were found in abundance below the sediment surface, stuffed in-between pebbles down to the mixer bottom. This is exactly the phenomenon observed at rocky shores of Fogo Island, Newfoundland and Labrador (McWilliams et al., 2017).

Important outcome of these experiments is that, for the used PP, PS, and LDPE samples, the total mass of all MPs increases with time exponentially (R^2 above 0.96, see Fig. 6.8A). Foamed PS showed a specific mechanic behavior: First, it disintegrated to composing spherules demonstrating practically linear increase of MP mass with time ($R^2 \sim 0.99$ for linear trend); linear dependence, however, can also be considered as an approximation of very slowly growing exponent within the given time span. Within 24 h of the experiment, into the class of MPs was transferred about 31% and 82% of the initial mass of LDPE and solid PS, correspondingly. The other two materials showed much smaller degradation rate, so that in the class of MPs came about 8% of mass of the foamed PS and <0.1% of PP. The formulas for the corresponding trends (see Fig. 6.8A) indicate that for disintegration to MPs in the described experiment of 50% of mass of the LDPE and PS samples, 28 and 15 h, correspondingly, are required. Disintegration of 10% of foamed PS samples requires about 29 h, while the extrapolation of fragmentation of PP on the base of 24 h-long experiment is hardly reliable.

Straightforward application of these results for estimation of time rate of mechanical disintegration of plastics in the real coastal zone is quite difficult: both wave periods and amplitudes (energy) are much different. Still, one may suggest considering environmental conditions with moderate wave energy (able to roll only *some* pebbles at the beach surface) and some typical for the given coast surface wave period. For the Baltic Sea, for example, the wave period could be about 5–6 s (Leppäranta and Myrberg, 2009) or about three times as large as the period of rotation in the above experiments. This way, as a rough estimate of intensity of mechanical fragmentation in the swash zone with coarse bottom sediments under rather moderate wind-wave conditions, this gives about 2–4 days for disintegration into MPs of 50% of the mass of PS and LDPE and 10% of the mass of foamed PS. Disintegration of PP samples should last much longer. In order to simplify possible application of the results to another wave periods, the time axis on Fig. 6.8A shows also the number of cycles as a sort of dimensionless time.

The distribution of the number of fragments by size in Fig. 6.8B suggests an interesting conclusion. The number of particles at every sieve was estimated by weighting a part of particles from it on analytic scales and then calculating the number of particles in this subset and obtaining a mass of an individual particle. It appeared that, as long as there are still enough macroparticles to be disintegrated, the number of microparticles depends on the size linearly (in log-log presentation). Mathematically, the log-log linear dependency means the relation between the number of particles, N, and their size, L, of the form $N \sim 1/L^{\alpha}$, where $\alpha > 0$ is a coefficient of proportionality of the log-log dependence. Surprisingly, the curve of the size distribution for the plastic wreckage floating at the ocean surface (Cózar et al., 2017) also presents such a linear part (in log-log scale) in the size range from 2.5 to 20 mm. Fragmentation in open-sea waters is suggested to be driven by (i) material weathering and (ii) mechanical destruction due to motions in surface waves (e.g., Kershaw, 2015). To the authors' knowledge, no experimental data on plastic fragmentation under open-sea conditions are available yet. At the same time, currents and waves in the coastal zone are mechanically much more powerful than those in the open sea, and thus, supposedly, more effective. Experience with fragmentation due to material weathering on land allows a suggestion that large items tend to crumble into small pieces almost without intermediate size distribution. Degradation experiments with plastic pellets immersed in calm water and exposed to visible and ultraviolet light in weathering chamber (Lambert and Wagner, 2016) also produced small particles (up to 60 μm), and particle concentration increased with decreasing particle diameter. Thus, there are certain arguments to suggest that the open-ocean distribution (Cózar et al., 2017) can be formed by plastics mechanically broken at ocean shores. Deviation from linear trend (in log-log presentation) in the range of smaller sizes on the curve by Cózar et al. (2017) is supposed to be caused by faster biofouling of smaller particles and their removal from the surface due to sinking (Fazey and Ryan, 2016).

6.2.5 Summary for Physical Properties

From the main physical properties of marine MP particles, the density and the size per se cannot be considered as the definitive characteristics. The reasons are (i) "natural" wide range of densities and sizes of MPs entering the sea and (ii) significant variations of density and size of every particular particle with residence time in marine environment due to biofouling, aggregation with sediment or organic particles, mechanical destruction in the coastal zone, degradation due to oxidation, UV radiation, weathering, and many other external factors. It seems reasonable to consider for physical applications only the range of variations of MP sizes (say, from 0.5 to 5 mm) and the density classes (say, definitely floating in natural waters, definitely sinking, and able to change the behavior within certain time periods). Then, the distribution curves within these limits/classes can be used to refine the description. "Life history" of a particle is important for MPs, which speaks in favor of the Lagrangian tracers approach in numerical modeling.

In contrast, the particle shape becomes of primary importance for the behavior and fate of a MP particle under real-sea conditions: particle biofouling is significantly different from particles of different shapes, so floating fibers and films sink faster than fragments and pellets. However, again, due to considerable variability of shapes of MPs sampled from natural marine environments, it seems reasonable to distinguish only between the principal shapes, for example, 3-D (for pellets, beads, spherules, fragments, etc.), 2-D (for films, flakes, flat fragments, etc.), and 1-D (for fibers, fishing line cuts, long filaments, etc.). The distribution of all the marine MPs between these principal shapes seems to provide the possibility to describe the biofouling rate, degradation manner, and certain dynamic features of a community of MP particles in the ocean or in the given area.

Biofouling of a particle leads to an increase of its average density. Time of fouling of a particle up to the water density is linearly dependent on the characteristic length scale of a particle (radius of a sphere, thickness of a film, or radius of a fiber): the smaller the scale of the particle is, the faster it is fouled up to the water density. For marine MP particles of typical dimensions but different shapes, fibers and films begin sinking due to overgrowth much faster than 3-D particles.

Due to mechanical destruction of typical plastics in the swash zone with coarse sediments, the mass of the generated MPs increases exponentially with the number of wave breaks. The distribution of the number of MP particles, N, over their size, L, follows the dependency $N \sim 1/L^{\alpha}$, where $\alpha > 0$ is a coefficient of proportionality of the log-log dependence. This leads to the suggestion that the linear portion of the Cózar's curve (Cózar et al., 2017) of $N(L)$ distribution for MPs floating on the ocean surface may be formed by plastics broken at coasts.

6.3 DYNAMICAL PROPERTIES OF MP PARTICLES

Dynamics of MP particles, with their intrinsically wide diversity of physical properties, in an environment as changeable as is the coastal zone, must be infinitely variable. On top of that, no direct in situ observations of MP particle motion at time/space scales of the coastal zone are available up to now. At this level, fundamental works of classical hydrodynamics and sedimentology provide valuable general guidelines, while field experience of physical oceanography assists in applications. In this section, we make an attempt to characterize the settling velocity in still water and the resuspension threshold in unidirectional steady current for the particles with the described above physical properties. Our second aim is to grasp the main features of MP behavior in some flow regimes typical of the coastal zone: oscillating motion under deforming surface waves, mixing in Langmuir circulation (LC), and during vertical convection. Finally, as the case study for MP migrations in the coastal zone, the behavior of the Baltic amber during and after stormy events is discussed. The main outcome of this summary is the general understanding of the time/space scales for MP migrations in the coastal zone and order-of-magnitude evaluations of the corresponding physical parameters.

6.3.1 Sinking Behavior and Terminal Settling Velocity

Settling (or terminal settling or free fall) velocity is one of the fundamental dynamic properties of particles, which characterizes their sinking behavior in motionless fluid. It depends on both particle properties (density, size, and shape) and fluid properties (density and viscosity). In the marine environment, the settling velocity of MP particles falling through the water column or suspended from the sea bed is generally assumed to be the same as the terminal velocity in still water (Ballent et al., 2013; Chubarenko et al., 2016). This is not exactly correct, especially in the surf zone or in an oscillating flow. Still, the value of the settling velocity is a very useful parameter characterizing motion of a particle in water.

When a particle exhibits a steady fall, the meaning of the settling velocity is straightforward. However, MP particles have various shapes and may display very complicated motions—either before they reach their quasi-steady falling regime or due to high Reynolds numbers, *Re* (Fig. 6.9). The initial orientations are stable only for very low *Re* numbers ($0.1 < Re < 5.5$, i.e., in the Stokes flow), while at higher *Re* numbers ($5.5 < Re < 200$, Hazzab et al., 2008) the stable orientation of particles in movement corresponds to the maximum drag. Generally in falling, a particle does not follow a rectilinear trajectory, and the orientation of the particle during its falling depends on its shape (e.g., Hazzab et al., 2008). Oscillatory movements of a particle around its average position and sometimes some rotations are observed; more plate/thin particles (e.g., disks) are most stroked by these oscillations (Hazzab et al., 2008). Disk-shaped particles fall generally broadside on and (at increasing *Re* numbers) experience oscillations about a diameter and glide from side to side, following a zigzag trajectory composed of concave-up arcs (Willmarth et al., 1964; Stringham et al., 1969; Allen, 1985). The sinking behavior of cylinders resembles that of disks (Jayaweera and Mason, 1965; Allen, 1985): falling generally with the long axis maintained horizontally, before reaching the steady-state fall or with increasing *Re* the cylinder may pitch about a horizontal axis normal to the long dimension, glide from side to side, oscillate or spin about the vertical axis; stubby cylinders may tumble during fall. Tiny MP fibers seem to fall in the Stokes regime, which is characterized by keeping the initial orientation of a particle (Bagaev et al., 2017b) that eventually leads to terminal sinking velocities several times different from the very same fiber in different orientations with respect to the flow. Applying to MPs the idea of Allen (1985), who investigated stones dropping from melting icebergs, the smaller particles (i.e., tending to fall more steadily and with the broadside on) should provide more favorable conditions for biofouling and accumulation of other suspended matter on their surface.

Classically, the terminal settling velocity, w_s, of a small solid particle falling in a viscous fluid is defined as a motion with constant speed attained when the gravitational force and the hydrodynamic drag force are balanced (Allen, 1900; Wadell, 1934; Clift et al., 1978;

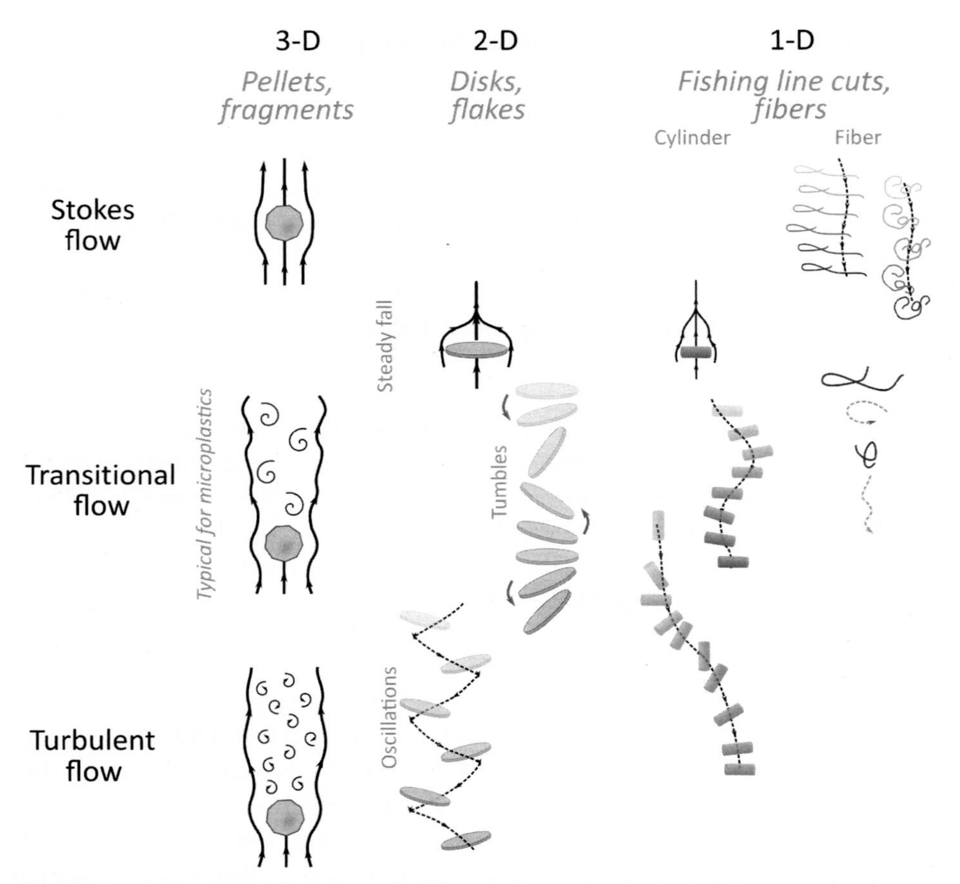

Fig. 6.9 Sinking behavior of particles with different shapes. In the Stokes flow regime, the particle keeps its original orientation while sinking. In transitional flow regime (typical for MP particles), oscillations, rotation, and tumbles are observed. For disks and cylinders, the terminal steady fall is in the orientation characterized by maximum drag force, that is, with flat side/long axis horizontally. *Based on Allen, J.R.L., 1985. Sink or swim. In: Physical Sedimentology. Allen, Unwin, London, pp. 39–54 (Chapter 3); Bagaev, A., Mizyuk, A., Khatmullina, L., Isachenko, I., Chubarenko, I., 2017. Anthropogenic fibres in the Baltic Sea water column: field data, laboratory and numerical testing of their motion. Sci. Total Environ. 599, 560–571; Khatmullina, L., Isachenko, I., 2017. Settling velocity of microplastic particles of regular shapes. Mar. Pollut. Bull. 114, 871–880.*

Hallermeier, 1981; Allen, 1985). On the assumption that no other forces are acting on the particle in vertical direction, the following equation applies:

$$\frac{1}{2}C_D\rho S_{cs}w_s^2 = (\rho_s - \rho)gV \tag{6.2}$$

where C_D denotes the dimensionless drag coefficient or coefficient of resistance; V and S_{cs}, the volume of the particle and the area of its cross section perpendicular to the direction of the motion; ρ and ρ_s, the fluid and particle density, respectively; and, g acceleration due to gravity.

The ratio $(\rho_s - \rho)/\rho$, often denoted by $\Delta\rho_0$, is the dimensionless relative density difference between the particle and the fluid, while the ratio V/S is proportional to the particle diameter d; thus, one can write $w_s^2 \sim \Delta\rho_0 dg/C_D$. Hence, in an ideal case, the square of the particle velocity increases linearly with the relative density difference and particle size but decreases with the drag coefficient. The latter, however, appears to be a very complicated parameter, depending in particular on the particle shape and the flow regime around it. The C_D in different flow regimes can be predicted using the so-called particle Reynolds number (Clift et al., 1978):

$$Re_p = \frac{w_s d}{\nu},$$

where d is the characteristic particle diameter, which should properly determine the size of the particle, and ν is the kinematic viscosity of fluid.

In the case when a small sphere is slowly sinking in a viscous fluid with $Re_p < 1$ (the so-called Stokes flow), the drag depends only on Re_p^{-1}, leading to the $C_D = 24/Re_p$. In the case of high Re_p numbers (the turbulent flow), the drag coefficient becomes insensitive to Re_p and equals $0.455 \pm 13\%$ for spheres (Clift et al., 1978). The intermediate flow regime lies in between; see also Fig. 5.12 in Clift et al. (1978, p. 124) and Fig. 3.6 in Allen (1985, p. 43) for details of the dependence of C_D on Re_p and the corresponding flow patterns. Knowing C_D, the dependence of w_s on the parameters of the particle and the fluid can be obtained using Eq. (6.2).

The drag coefficient also depends on the particle's shape, and consequently, different particles such as spheres, oblate and prolate spheroids, cylinders and disks with the same volume, and weight have different w_s (Wadell, 1934; Lerman et al., 1974). In addition, angular particles experience larger drag in comparison with ideally round particles with a smooth surface (Dietrich, 1982). As practical rule, when considering the settling velocities of differently shaped particles, it is convenient to define the particle diameter in the form of the diameter of a sphere of the same volume as the particle (Wadell, 1932). And then one can subsequently use another quantity to describe the differences in shape and surface properties between the given particle and the equivalent sphere (Blott and Pye, 2008; Rodriguez et al., 2013). The applicability of these methods in context of MP settling is an open question.

Only very few studies up to now have dealt with the settling velocity of MPs. Ballent et al. (2013) measured mean settling velocity and resuspension characteristics of beach pellets of about 5 mm in size. Measurements were conducted in saltwater with density close to that of seawater in the Nazaré Canyon of Portugal, since the research was focused on the numerical simulation of benthic MP transport in that area. Kowalski et al. (2016) conducted sinking experiments with particles of diverse polymer types, density, size, and shape in deionized water and natural seawater. Artificial particles of irregular shapes were used. Khatmullina and Isachenko (2017) also employed artificial MPs of different shapes in their sinking experiments, but the study was focused on regularly shaped particles, namely, quasi-spheres and cylinders. Thus, available information on the MP sinking behavior is very limited and mostly concerns artificial grains produced by authors in the laboratory. That is why the ability of available formulas (Dietrich, 1982; Cheng, 1997; Camenen, 2007; Zhiyao et al., 2008; and many others) for settling velocity of marine MPs is still poorly understood, but some suggestions can be already made according to those studies.

As marine MPs demonstrate a broad range of physical properties (Morét-Ferguson et al., 2010; Hidalgo-Ruz et al., 2012; Chubarenko et al., 2016), it is difficult to predict, without preliminary laboratory tests, the flow regime around the particle. Therefore, the universal formulations (Dietrich, 1982; Camenen, 2007; Zhiyao et al., 2008) are more useful for practical purposes than those developed for the specific case such as the Stokes settling (Lerman et al., 1974). The empirical settling velocity formula for spherical particles (Dietrich, 1982) predicts quite well the settling of certified spheres (Kowalski et al., 2016) and the quasi-spherical MPs with limited shape imperfections (Khatmullina and Isachenko, 2017). Shape dependence is smaller for smaller particles (<0.5–1 mm), so for small rounded MP granules and cylinders, the same formula for spheres is applicable with an error of no more than 10% (Khatmullina and Isachenko, 2017). However, it is not applicable for small but coarse MPs, as was shown by Kowalski et al. (2016). Approximations that explicitly account for the particle shape and roundness (Dietrich, 1982; Camenen, 2007) showed a good agreement with the experimental settling rates of non-spherical MPs (Khatmullina and Isachenko, 2017). They probably can predict the settling rate of coarse MPs. The imperfections of the shape and the peculiarities of the surface texture, so characteristic of real marine MPs, are able to cause deviations in the settling velocity, and these deviations might be one order of magnitude smaller than the values predicted by already known formulas for the given general particle shapes (e.g., spheres and cylinders) (Khatmullina and Isachenko, 2017).

As a possible solution, the dimensionless settling velocity can be calculated using the equations of Dietrich (1982); see Table 6.2. There, $D_* = \frac{(\rho_s - \rho) g D_n^3}{\rho \nu^2}$ is the dimensionless particle diameter, and D_n is the nominal diameter, which is the diameter of a sphere of the same volume as the particle (Wadell, 1932). For ellipsoidal particles, the nominal diameter can be estimated after the measurements of the particle's shortest D_s, longest D_l, and intermediate D_i axis by $D_n = (D_s D_i D_l)^{1/3}$. Baba and Komar (1981) proposed that the D_i itself is a reasonable approximation of sand particles nominal diameter.

The listed equations (a)–(d) are applicable at $0.05 \leq D_* \leq 5 \times 10^9$, $0.2 \leq csf \leq 1.0$, and $2.0 \leq P \leq 6.0$. The settling velocity of a sphere can be calculated by setting $R_2 = 0$ and $R_3 = 1$. Dimensional value of the settling velocity $w_s = \left(W_* \frac{\rho_s - \rho}{\rho} g \nu \right)^{1/3}$, where W_* is the dimensionless settling velocity.

Table 6.2 Formulas for calculation of dimensionless settling velocity for particles of different size, shape, and density by Dietrich (1982)

(a)	$W_* = R_3 10^{R_1+R_2}$	
(b)	$R_1 = \log_{10} W_* = -3.76715 + 1.92944(\log_{10} D_*) - 0.09815(\log_{10} D_*)^{2.0} - 0.00575(\log_{10} D_*)^{3.0} + 0.00056(\log_{10} D_*)^{4.0}$	Size and density
(c)	$R_2 = \log_{10}(1 - (1 - csf)/0.85) - (1 - csf)^{2.3} \tanh(\log_{10} D_* - 4.6) + 0.3(0.5 - csf)(1 - csf)^2(\log_{10} D_* - 4.6)$	Shape
(d)	$R_3 = \left[0.65 - \left(\frac{csf}{2.83}\tanh\left(\log_{10} D_* - 4.6\right)\right)\right]^{(1+(3.5-P)/2.5)}$	Roundness (angularity)

The settling velocity of a particle decreases if its shape deviates from a sphere. Dietrich (1982) estimated the maximum reduction in settling velocity from that of a sphere caused by shape and roundness of the quartz particles of $D_n = 0.4$ mm and $D_n = 2$ mm in about 0.65 and 0.4, correspondingly. For less dense (1.05 g cm^{-3}) MP particles of the same size and $P = 2.0$ and $csf = 0.2$ (very angular and flat/elongate grains), calculations give 0.45–0.44. Thus, as a thumb-rule estimate, we can conclude that imperfections of 3-D-shaped particles (like those of natural sands) can reduce the particle settling velocity twice.

One more factor reducing the settling velocity of a particle is motion of water. Experiments on settling of plastic (Baird et al., 1967) and sapphire/glass (Tunstall and Houghton, 1968) spheres falling in vertically oscillating liquid and the theoretical study (Hwang, 1985) suggest that the settling velocity in the surf zone can be much different (generally smaller) from that measured in the laboratory in still water.

In order to evaluate the range of variation of settling velocities of MP particles, Chubarenko et al. (2016) applied the formula suggested by Zhiyao et al. (2008), deduced on the base of many laboratory experiments and said to take into account "various sizes, shapes, roundnesses, and densities." Taking as contrasting numerical examples the polystyrene particle (1.05 g cm^{-3}) of the diameter 0.5 mm and the polyoxymethylene particle (1.6 g cm^{-3}) of the diameter 5 mm, Chubarenko et al. (2016) obtained the settling velocities of 4 mm s^{-1} and 18 cm s^{-1}, correspondingly.

Observations on the settling of natural biogenic marine particles may also assist in attaining the scale of the settling velocity of MPs in marine environment. According to Maggi (2013), the settling velocities of mineral, biomineral, and biological particles and aggregates in water range from about 1 mm day^{-1} to 1 km day^{-1}. In the field study of McDonnell and Buesseler (2010), average sinking velocities of aggregates of biogenic particulate matter appeared to be the smallest (25 m day^{-1} or 0.3 mm s^{-1}) for particles of about 120–320 mm, while smaller particles sank more quickly (up to 150 m day^{-1} or 1.7 mm s^{-1}). In their review of the measurements of settling velocities of marine particles, McDonnell and Buesseler (2010) reported a huge range from ~5 to 2700 m day^{-1}

($0.06\,\mathrm{mm\,s^{-1}}$ to $3\,\mathrm{cm\,s^{-1}}$) but typically between tens and a few hundred of meters per day (from $0.1\,\mathrm{mm\,s^{-1}}$ to a few $\mathrm{cm\,s^{-1}}$).

Since for marine biogenic particles both the density and the size can be larger than for MPs, the values of the settling velocity obtained via laboratory formulas (Zhiyao et al., 2008; Chubarenko et al., 2016) obviously look as several times overestimates. This overestimation is anticipated: natural turbulence, variable water currents, imperfections of the particle shape, and other factors—they all contribute toward decreasing the terminal settling velocity. As an overall estimate for MP terminal settling velocity, we will come out with the range from eventually zero (for floating particles subject to biofouling) to units of centimeters per second. At the same time, one should keep in mind in applications that, under natural conditions (especially in coastal zone), the regime of terminal (i.e., steady state) settling of a particle may never be reached because of (i) principally complicated sinking behavior of MPs of various shapes and (ii) turbulent environment and never steady currents.

6.3.2 Critical Velocity of Resuspension by a Unidirectional Flow

Initiation of motion of a particle that is currently at rest on the bottom is an important topic of marine sedimentology, providing today several theoretical approaches, huge amounts of laboratory efforts, lots of field data, and a wide range of publications. Here, we strongly limit ourselves to aspects related directly to MP motion, with the idea to just get a feeling of order-of-magnitude velocity thresholds. Only very limited information on direct measurements of a resuspension threshold (initiation of motion) in the water current can be found for MPs; however, other bed-load materials of the "plastic" density range can be used as well.

In a seminal work by Shields (1936) on sediment resuspension and transport by a unidirectional water current, "of the MP range" of density and size are particles from brown coal ($1.27\,\mathrm{g\,cm^{-3}}$, mean diameters 1.77, 1.88, and 2.53 mm) and amber crumbs ($1.06\,\mathrm{g\,cm^{-3}}$, grain size from 0.3 to 3 mm with mean diameter $d = 1.56$ mm). From the MP side, Ballent et al. (2012, 2013) measured in the laboratory dynamic characteristics for three kinds of industrial preproduction high-density plastic pellets collected on sandy beaches of Los Angeles County (California). They were of rounded shape and somewhat larger (3.3–5.1 mm) than the Shields amber particles, with densities from 1.06 to $1.13\,\mathrm{g\,cm^{-3}}$. For more detail on intercomparison of the Shields' and the Ballent' experiments (see Chubarenko and Stepanova, 2017). The experimental points for these studies are presented in Fig. 6.10 in the form of the classical "Shields curve," which expresses in dimensionless form the border between regimes of "motion" and "no motion" of the bed-load material under action of the fluid flow (Chubarenko and Stepanova, 2017).

Conditions of "initiation of motion" of particles in the fluid flow are shown in Fig. 6.10 in the form suggested by Shields (1936): as the dependence of the dimensionless

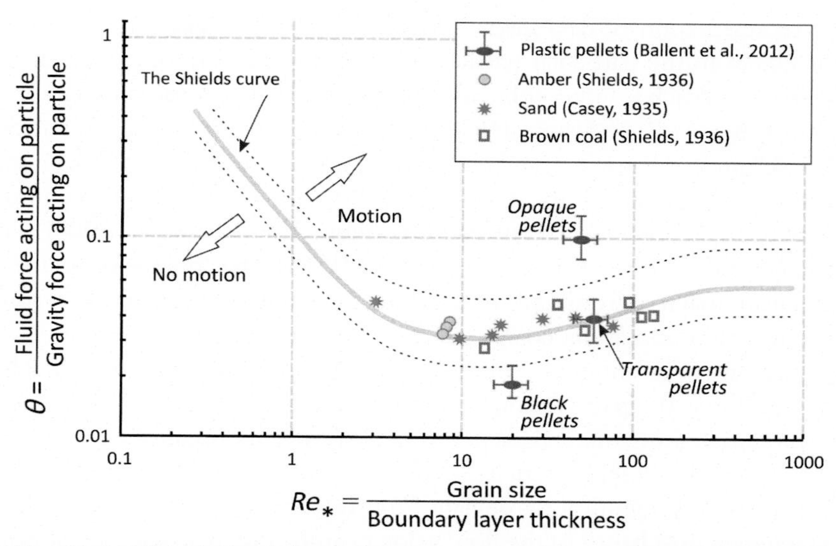

Fig. 6.10 The classical Shields (1936) diagram, showing the dependence of dimensionless shear stress on the particle Reynolds number, is used to describe the initiation of motion of natural bed-load material of different grain size in the fluid flow. Points for amber and lignite from Shields (1936) and for high-density preproduction pellets from Ballent et al. (2012) are shown. Note that Ballent et al. (2012) used a different experimental technique. *Reprinted from Microplastics in sea coastal zone: Lessons learned from the Baltic amber, Chubarenko, I., Stepanova, N., 2017, Environmental Pollution, vol. 224, p. 247, with permission from Elsevier.*

Shields critical shear, $\theta = \frac{\tau_*}{(\rho_p - \rho_w) \cdot g \cdot d}$ on the dimensionless particle Reynolds number, $Re_* = \frac{u_* \cdot d}{\nu}$. Here, $\tau*$ is dimensional critical shear stress, ρ_p particle density, ρ_w water density, g acceleration due to gravity, $u*$ critical velocity, d diameter of the particle, and ν water kinematic viscosity. His experimental curve, obtained using natural sand, barite, and granite particles of various dimensions, is shown in gray. The Shields (1936) data points for amber and brown coal (density 1.27 g cm^{-3}) and three kinds of plastic pellets from Ballent et al. (2012) are plotted. For comparison, points for coarse sand (density 2.65 g cm^{-3}, diameter 0.5–1 mm, data of Casey (1935), as cited by Shields (1936)) are also indicated (Chubarenko and Stepanova, 2017).

As follows from Fig. 6.10, transparent MP pellets from Ballent et al. (2012) with material density of 1.13 g cm^{-3} fit the Shields curve perfectly, while black and opaque pellets (density 1.06 and 1.07 g cm^{-3}, correspondingly) are slightly below/above it. At the same time, points for amber with the same density of 1.06 g cm^{-3} are in agreement with the Shields dependence. This discrepancy may be related to differences in experimental methodology, in particular the experimental definition by the observer of motion/no-motion state. Anyway, generally, amber, brown coal, sands, and MP particles in this selection obey the Shields dependency for the threshold of the initiation of motion obtained for natural loose bed-load materials.

In the representation of data in the Shields (1936) form, evaluation of the critical velocity is linked to the particular particle size and density. So, for the given selection of "MP-like" particles, one obtains the following critical velocities: amber (0.3–3 mm), 4–6 $mm\,s^{-1}$; brown coal (from 1.77 to 2.53 mm), 5–20 $mm\,s^{-1}$; black plastic pellets (3.5–4.7 mm), 4.9 $mm\,s^{-1}$; and transparent and opaque pellets (3.3–5.1 mm), more than 13 $mm\,s^{-1}$ (Chubarenko and Stepanova, 2017). For coarse sand (0.5–1 mm), a critical velocity of 16–22 $mm\,s^{-1}$ is reported in Berenbrock and Tranmer (2008). Thus, the order of magnitude of the critical velocity for MP particles, obtained via the Shields experimental approach, is from units of millimeter to a few centimeters per second, with the upper limit close to natural coarse sand.

One more outcome, important in the MP context, follows from the Shields (1936) work: he found that the particle angularity (sharp edges of amber, rounded corners of angular coal particles, or rounded shapes of sand) is not important for resuspension. At the same time, all the Shields particles were 3-D, and it could be expected that films, flakes, and fibers may significantly deviate from the Shields curve. Indeed, the critical shear characterizes the change of the flow velocity at the distance from the bottom equal to the particle height above it. Films, flakes, and fibers may lie on the (generally rough) bottom both very close to it (with the film thickness or fiber diameter as the vertical scale) and in an inclined orientation, with their parts much above the bottom (film diameter or fiber length as the vertical scale), so that their general response to the flow can be manifold.

A quite serious question for the application of the Shields approach to MP motion arises also from the methodology used: for sediment transport problems, the examined bed load is exposed to the flow moving above it. On the contrary, for the case of MP resuspending from natural bottom sediment, the bed-roughness scale is different from that of a particle. Thus, the particle $Re*$ number, which is the ratio of the particle size to the thickness of the boundary layer, should be based on the dimensions of both MP particle and bottom material. More laboratory experiments on MP resuspension from bottoms covered by silt, sand, cobbles, etc. are very desirable to clarify this question (Chubarenko and Stepanova, 2017).

6.3.3 Oceanographic Applications

Among various mixing mechanisms present in the coastal zone, there are some allowing for analysis of the MP particles behavior. In this section, we make an attempt to apply to MP motion well-known facts about the particle behavior (i) under shallowing surface waves, (ii) within the rolls of the LC, and (iii) under strong vertical convection.

6.3.3.1 Behavior in an Oscillatory Flow

Motion of water parcels and suspended particles under surface waves is investigated quite well. In the coastal zone, wave deformation becomes important, leading to a massive

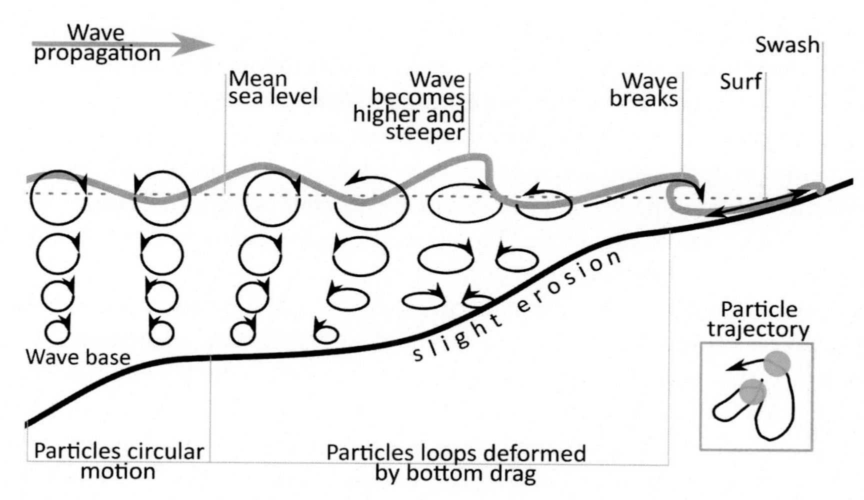

Fig. 6.11 Orbital motions of water particles under a surface wave deforming in the coastal zone. In insert: the path of an individual (neutrally buoyant) particle under real sea waves in the coastal zone.

sediment transport. With the idea to extract some information applicable to MPs motion in sea coastal zone, we shall first concentrate on the principal points only.

Wind-induced surface waves in the open sea are not perfectly sinusoidal, which eventually causes the downwind depth-dependent water mass transport (called the Stokes wave drift). Water parcels below the wave describe circles whose diameters decrease with depth (see Fig. 6.11). When a wave is approaching the coast, these motions begin interacting with the bottom, causing an increase in the wave height, H, and decrease in its length, L. It is commonly accepted that this interaction begins from the depth of about $H \sim L/2$. Onshore from this depth, water parcel trajectories become more and more elongated, until the breaking point is reached, where the wave breaks and water moves further as a chaotic surf until it spreads onto the beach as swash. The strongest sediment erosion due to wave action is observed in the surf zone, but all the area of wave deformation exhibits the sediment transport.

Now, the question is what the behavior of different MP particles in such a flow could be. Neutrally buoyant particles should follow the trajectories of water parcels, so that, integrally, the parcels closer to the surface should drift onshore (with the Stokes drift), while those closer to the bottom should move offshore (with compensating current). This integral motion, however, in the real sea is composed of many different primary movements, because the instantaneous velocity field is formed by not a single wave, but a wave spectrum, and every particular wave experiences deformations slightly different from the other ones, etc. As a result, the path of an individual (neutrally buoyant!) particle is usually sketched like in insert in Fig. 6.11 (see, e.g., Southard, 2006), that is, illuminating just chaos. The diversity of physical properties of MP particles can only contribute to this chaos, because the variety of their densities, sizes, and shapes causes the differences in applied forces, so that every particle will follow its own trajectory. This supports energetic interaction of MPs with sediment particles, algae, etc., explaining both severe mechanical damage of MPs in the coastal zone and its typical entrainment into the root vegetation on the bottom and floating seaweed patches. In fact, no other place in the ocean but the coastal zone provides such intense forcing, so it looks quite logical that larger plastic objects could mainly be fragmented along the coasts (see Fig. 6.8 for distribution of MPs in the open ocean).

Description of general transport and trajectories of sediment particles under surface waves has received much attention; however, it still is a very challenging problem. Particle-based numerical modeling discloses some aspects, which could be of importance also for MP migrations. Finn et al. (2016) analyzed motion of particles of coarse and fine sands under passing surface waves and found strong spatiotemporal particle-size sorting patterns, both near the bed and in suspension. They discovered that in a very complicated field of water velocity under waves, smaller suspended sand grains manage to pass through more areas with upward directed flow velocities in their trajectory, which delays particle settling and enhances the vertical size sorting of grains in suspension (Finn et al., 2016). The vertical sorting, in its turn, can lead to a very different transport capacity for different size fractions. This way, the overall net transport direction for fine sand particles is often found against the wave propagation direction, while medium and coarse sands tend to be transported in line with the wave propagation direction (Camenen and Larson, 2006; Hassan and Ribberink, 2005; Kranenburg et al., 2014; Finn et al., 2016).

Along with separation of particles in suspension, strong sorting was found also within the upper sediment layer. Differentiating between the vortex-ripple (low-flow) conditions and the sheet-flow regime (under large stormy waves (van Rijn, 1993)), Finn et al. (2016) analyzed the composition of the near-bed layer. It appeared that in a low-flow regime this layer remains predominantly coarse throughout most of the wave cycle, thus limiting mobility of finer grains below the coarser ones. Meanwhile, under stormy waves the near-bed layer composition is much more dynamic, and all fractions participate in the pickup events during different phases of the flow (Finn et al., 2016). The latter again means that particles with different properties may be transported in opposite directions under the same wave conditions. The effect of arresting of finer particles below the coarser ones in the low-flow regime of wave-driven motion may be considered as an analog for capturing of MPs under the surface of cobble beaches (McWilliams et al., 2017). This draws attention of field monitoring of MP contamination in coastal waters to sampling in undersurface sediment layers.

Thus, surface waves, their deformation and breaking in the coastal zone, contribute to MP fate in three aspects at least. Firstly, intense mixing of MPs (and larger plastic objects) with bottom sediments causes mechanical degradation and load by mineral particles, while chaotic trajectories of MP particles among rooted or floating algae leads to their filtering out of the general water flow. Secondly, sorting of MPs takes place; criteria of the sorting are not yet known. And thirdly, transport of different MPs in opposite directions is highly probable. Presumably, the offshore (against the wave propagation) direction is prevalent for smaller and lighter MPs, while larger and heavier particles tend to be transported onshore (in line with the wave propagation direction).

6.3.3.2 Suspension by Convective Motions

Surface cooling, strong evaporation, or faster ice formation in the coastal zone makes surface waters denser, which causes both vertical and horizontal exchange above underwater coastal slopes (see, e.g., Ivanov et al., 2004; Chubarenko et al., 2003; Chubarenko, 2010). The denser water makes its routes downward in two principal ways (Fig. 6.12): (i) via vertical convection, creating and maintaining the upper mixed layer, and (ii) via horizontal exchange, generating density currents along the sloping bottom. It appears that the velocity scale for these motions is quite close to the sinking/rising/resuspension velocity of MP particles, so convective motions are able to influence their transport and fate. We shall compare these velocity scales for typical environmental conditions.

Both horizontal and vertical convective mixings are driven by a destabilizing buoyancy flux through the surface, which in the case of, say, surface cooling, can be expressed as $B_0 = g\alpha Q/\rho c_p$, m^2 s^{-3}, where g is acceleration of gravity, α is the coefficient of thermal expansion of water, Q is the heat flux per unit surface area, ρ is density of water, and c_p is water heat capacity. Then, typical vertical velocity of convective motions is

$$w \sim (B_0 H)^{1/3}$$

where H is the depth of unstratified fluid, associated with the bottom depth in the coastal zone. For typical oceanographic values in high-latitude coastal seas where $B = 5 \times 10^{-6}$ - m^2 s^{-3} (Schott et al., 1993; Morawitz et al., 1996) and $H = 10$–100 m, vertical velocity values are approximately 2–4 cm s^{-1}. For typical values at lower latitudes, for example, for

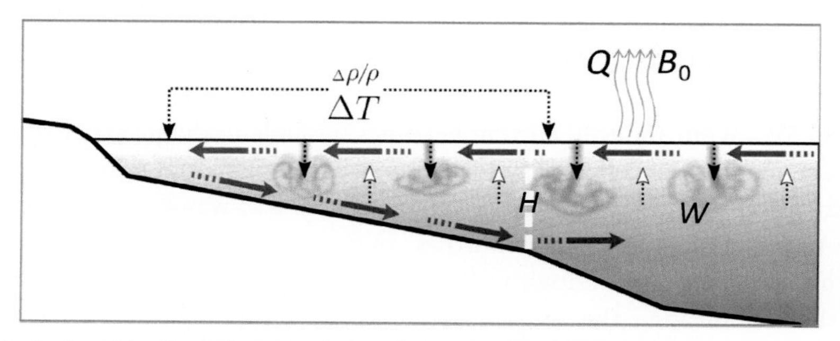

Fig. 6.12 Vertical and horizontal convective exchange in the coastal zone.

the Golfe du Lion region in the North-West Mediterranean, where $B = 10^{-7}\,\mathrm{m^2\,s^{-3}}$ (Jacobs and Ivey, 1998), the vertical velocity value is around 1–2 $\mathrm{cm\,s^{-1}}$. In the case of deep ocean convection in polar oceans, vertical velocities may reach 2–10 $\mathrm{cm\,s^{-1}}$ (Stommel et al. 1971; Schott and Leaman, 1991; Schott et al., 1993; Gawarkiewicz and Chapman, 1995). Comparison of these values with typical sinking velocities of MP particles (up to a few centimeters per second maximum, see Section 6.2.1) indicates that the majority of MP particles can be retained within the upper mixed layer by upward/downward convective motions. In particular, floating MPs can easily be captured into mixing, like in the case of wind-induced vertical mixing reported by Kukulka et al. (2012).

Horizontal exchange of convective nature (Chubarenko and Hutter, 2005; Chubarenko, 2010), arising in coastal zones during cooling episodes due to the horizontal temperature/density difference between shallow and deep waters, is characterized by offshore/onshore currents with the velocity scale, u, of the scale of

$$u \sim (\alpha \Delta T g H)^{1/2} = (\Delta\rho/\rho g H)^{½}$$

where ΔT is the temperature difference between coastal and open waters and $\Delta\rho/\rho$ is the corresponding relative density difference. In this case, horizontal exchange flows have a velocity of about units of centimeter per second (Farrow, 2004; Fer et al., 2002; Monismith et al., 1990). Field measurements in reservoirs and large lakes (e.g., Monismith et al. (1990) in Wellington reservoir, Fer et al. (2002) in Lake Geneva, and Wüest et al. (2005) in Lake Baikal) show velocities of about 2–7 $\mathrm{cm\,s^{-1}}$ and in the coastal ocean (Ivanov et al., 2004) about 2–4 $\mathrm{cm\,s^{-1}}$. Thus, horizontal convective exchange flows, which are shown to be present in the coastal zone in day/night, synoptical, and seasonal timescales (Chubarenko, 2010; Chubarenko, Demchenko, 2010; Chubarenko et al., 2013), are definitely able to contribute to the onshore/offshore transport of MP particles.

6.3.3.3 Behavior of a Particle in Roll Structures

Roll structures are typical of natural water flows, most known examples being fronts and the LC. Behavior of nonneutrally buoyant particles in such flows depends on the size and density and can thus be important for transport of MP particles.

As shown by many authors (e.g., Langmuir, 1938; Craik and Leibovich, 1976; Thorpe, 2005), a wind-induced current becomes unstable in the presence of the Stokes wave drift, generating the coherent system of counterrotating longitudinal rolls with their axes directed downwind (Fig. 6.13). As the consequence, water surface during windy weather is typically ruled by long "windrows" of floating foam and algae, associated with the surface convergence (and downwelling) zones and slick lines (upwelling zones) in between them, indicating the surface water divergence (Langmuir, 1938; Craik and Leibovich, 1976; Thorpe, 2005; Dethleff et al., 2009; Chubarenko et al., 2010).

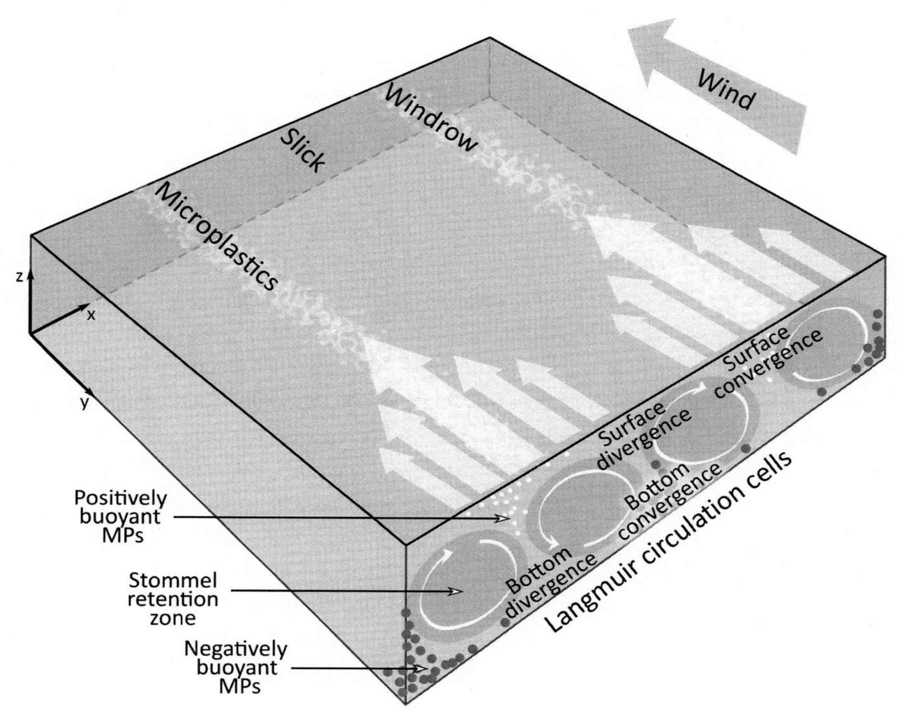

Fig. 6.13 Schematic view of the Langmuir circulation, with floating material collected within windrows at the water surface, slightly positively buoyant particles entrained into the downwelling flow, slightly negatively buoyant particles lifted from the bottom in upwelling zones, and close to neutrally buoyant particles circulating in the Stommel retention zone.

In the lower LC part, vice versa, this circulation implies convergence/upwelling and divergence/downwelling zones (see Fig. 6.13).

Two questions are important here for application to MP behavior: (i) whether the LC is able to carry the MPs and (ii) what are the paths of different particles within the LC rolls.

The settling velocity of MPs (0.5–5 mm) (see Section 6.2.1) is up to a few centimeters per second. The maximum upwelling velocity in LC under moderate winds is reported (Leibovich, 1983) to be 1–1.5 $\mathrm{cm\,s^{-1}}$. Thus, the LC per se can definitely retain the major part, but not all the MPs, in the upper mixed layer. However, in a highly turbulent sea after the storm and in coastal areas of surface wave deformation, the currents are obviously much stronger. Thus, the LC is an important factor in transporting the MPs in both the open ocean and the coastal zone.

As for the MP particle excursions in the LC, the following information can be extracted from previous research. The neutrally buoyant particles suspended in the LC flow follow the water stream lines. It was Stommel (1949) who first deduced that nonneutrally buoyant particles (both sinking and rising) will describe closed trajectories within their "zones of retention" in upwelling and downwelling areas (Fig. 6.13). More recently, quite many studies have addressed the details of particle excursions in LC rolls (e.g., Faller and Woodcock, 1964; Stavn, 1971; Titman and Kilham, 1976; Ledbetter, 1979; Woodcock, 1993; Bees, 1996). In particular, Dethleff et al. (2009), investigating the process of suspension freezing, found that particle trajectories depend of their size and density. Larger ice crystals (near the surface) and sediment particles (near the bottom) describe smaller circuits within the LC cell, while smaller particles, having less buoyancy excess/deficit in water, describe larger circles. As applied to MPs and coastal sea conditions, this means that particles of MPs of different size, shape, and density (having thus different gravity, buoyancy, and drag forces) will describe different orbits, recirculating also through the suspended algae and sands. Close to neutrally buoyant particles should recirculate along closed orbits within the Stommel retention zone, while particles, whose buoyancy deviates from that of surrounding water, should describe different paths. The deviation from water streamlines for such particles depends on the particle acceleration, that is, on the balance of the forces acting on it (Dethleff et al., 2009). This process brings in contact the MP particles, sediments, and biota and definitely contributes to both MP fragmentation and its entangling into the seaweed and algae patches. Such strongly intermixed patches of algae, seaweed, and plastics are typically observed in the coastal sea and on the beach wrack lines (e.g., Esiukova, 2017; Chubarenko and Stepanova, 2017).

6.3.3.4 Natural Case Study: Wash-Outs of the Baltic Amber

Amount and distribution of plastic litter on the beaches are shown to be linked with currents, waves, and hydrometeorological conditions in the coastal zone (e.g., Kataoka et al., 2013, 2015; Hinata et al., 2017). Thus, some information on the behavior of plastic particles in coastal waters can be deduced from time variations and spatial pattern of plastic contamination. Baltic amber (succinite) has its material density (1.05–1.1 g cm^{-3}) close to that of widely used plastics (e.g., polyamide, 1.02–1.05 g cm^{-3}; polystyrene, 1.04–1.1 g cm^{-3}; and acrylic, 1.09–1.20 g cm^{-3}) (Chubarenko and Stepanova, 2017). Deposits of amber in the Southeastern Baltic have openings to the sea bottom at depths of about 4–15 m, and, quite often, amber crumbs and stones are found on regional beaches. Massive amber washouts have been monitored by local citizens for many centuries, making it possible to analyze typical external conditions leading to such events.

Sandy beaches of the Southeastern Baltic are typically clean of beached seaweed and amber, and only after windy weather the patches are washed ashore, bringing also MP particles (see Fig. 6.14). The detailed intercomparison of physical and transport properties of amber crumbs and MPs and the analysis of environmental conditions leading to such washouts are provided in Chubarenko and Stepanova (2017) and here, we only summarize the outcomes.

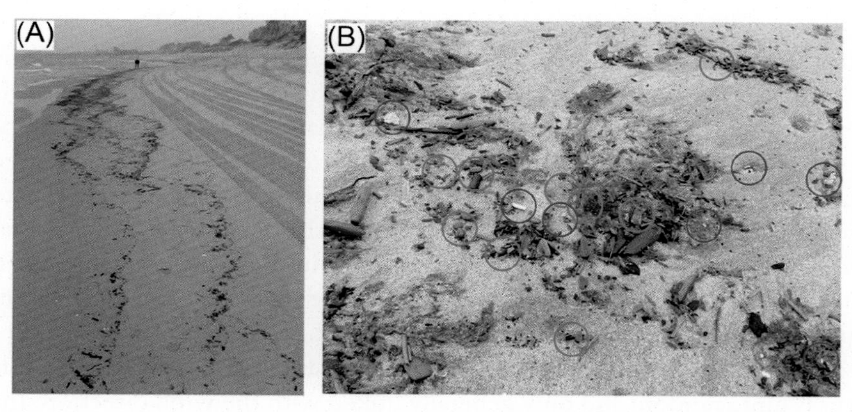

Fig. 6.14 Parches of seaweed, flotsam, and amber appear on the Baltic beaches after windy-weather episodes. (A) View of the beach with washed ashore miscellaneous "marine treasures" (the Vistula Spit, 25 November 2015). (B) Amber crumbs (highlighted in *green ovals*) and MPs (in *red ovals*) are washed ashore together (the Vistula Spit, 10 July 2016, photos by I. Chubarenko).

Within the patch of the beached material, the MPs, larger plastic objects, and also amber crumbs are remarkably mechanically damaged, which indicates both the experienced strong forcings in the sea coastal zone and long residence time of plastics migrations there. The production and expiry dates on some pieces of package of market products suggest 5–7 years of residence in the coastal zone (Wang et al., 2016; Chubarenko and Stepanova, 2017). Most of the beached material is not buoyant: wet wood, amber, PET bottles, and aluminum cans are heavier than local waters (typical salinity 7 psu). Usually, floating pieces, like PE films, foamed PS particles, and parts of PP tableware, are severely loaded by sand particles, mechanically crumpled, and deeply entangled between the seaweed branches. This obviously shows the effect of mixing due to passing and deforming surface waves. Neither prevalent shapes nor types of (floating or sinking) plastics are observed, especially within the patches after stormy winds. Rather, all the particles, fragments, sheets, flakes, fishing lines, nets, etc. are mixed together with algae and wooden sticks. Sometimes, the densely tangled rope/net/algae/plastics garland weights 10–30 kg, and still, it is washed ashore, together with light floating material. Thus, at least at stormy conditions, neither MP shape, density, nor size are decisive for the behavior of a particle—all the material is just conveyed by powerful dynamic processes in the sea coastal zone.

Environmental conditions, leading to events of especially massive washouts of marine flotsam to the Baltic beaches, are of interest here and can be summarized as follows (Chubarenko and Stepanova, 2017). Strong winds, lasting for several days, with the direction perpendicular to the coast line, should raise particularly high waves. Washing ashore of suspended material happens not during the storm, but right after it—several hours after the beginning of wind weakening, when the storm is still vivid and waves are still high, but the sea already begins to calm down. The most characteristic is the patchiness of the picture. Amber and MPs arrive to the shore together with a compact cloud of seaweed, wooden sticks, and other suspended material, which is observed to suddenly rise from the bottom in a separate spot a few kilometers away from the shore. Then, the patch gets distributed throughout the entire water depth, from the bottom to the surface, and drifts toward the shore, where all the material is for some time kept mixed and suspended in the swash zone or gets beached as a patch of some 30–70 m long (see photos and videos in the Supplement Material to Chubarenko and Stepanova (2017)). Sometimes, the main (the largest) patch may be accompanied by 1–2 smaller ones, 400–500 m apart. A stormy seaweed patch in the swash zone may begin drifting offshore already in an hour or two. The location of the event is quite random and only generally linked with the wind direction. Interestingly, local citizens say "the Baltic withdraws its treasures": the shoaled seaweed cloud may stay on the beach up to several days; small (3–4 mm) amber crumbs washed out after moderate-wind events can be found on the beach for about 1 day. Under usual/everyday conditions, no amber is present on/in the beach sands.

For the discussion of the MP behavior in the coastal zone, a new and important feature here is the patchiness of the general picture, which is in fact quite usual for hydrophysical, sedimentologic, biological, and many other natural processes. In the case of the Baltic amber washouts, the patchiness is presumably related (Chubarenko and Stepanova, 2017) to the LCs, arising during the phase of weakening of the stormy winds. The analysis of characteristics of stormy waves and the algae patches washed ashore together with amber indicates that initial suspension of bottom material in the Southeastern Baltic arises at depths down to 15–20 m depth, with the offshore limit related to the half-length of the stormy waves.

6.4 CONCLUSIONS

MP is a recently emerged pollutant with still very poorly known transport properties. We have made an attempt to uncover useful information, bit by bit, from several neighboring disciplines—classical hydrodynamics, marine sedimentology, and physical oceanography. Eventually, we found a natural case study that illuminates the main features of MP migrations in the sea coastal zone—washouts of amber crumbs on the Baltic Sea beaches.

The main physical properties of MP particles are density, shape, and size. With residence in environment, effective integral density and size of particles vary significantly, due to biofouling, weathering, aggregation with sediment or organic particles, mechanical damage, etc. In order to describe general transport of and pollution by MPs, the distribution curves for size and density and their variation in time are definitely a better choice. The particle shape, however, seems to be more important, because it defines

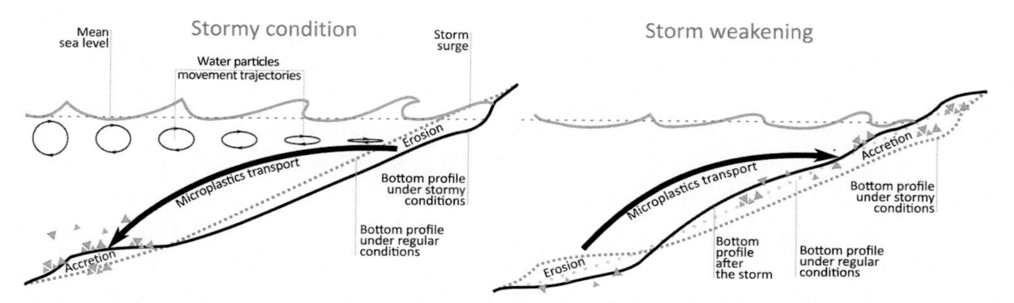

Fig. 6.15 **Migrations of beach plastics in the sea coastal zone, driven by windy/stormy events: offshore transport under the developing storm versus deposition and onshore move when wind weakens.** *Modified from Chubarenko, I., Stepanova, N., 2017. Microplastics in sea coastal zone: lessons learned from the Baltic amber. Environ. Pollut. 224, 243–254.*

the principal manner of the particle sinking and the celerity of its biofouling. From three categories, suggested by Rodriguez et al. (2013) for the description of the particle shape, sphericity, roundness, and the surface texture, the second seems to be the least important. Throughout the analysis, classification of MP shapes into 3-D (particles, pellets, and fragments), 2-D (flat fragments, films, and flakes), and 1-D (filaments, threads, and fibers) was found quite convenient.

The behavior of amber pieces in the sea coastal zone discloses several important aspects of MP migration. It seems that plastic objects in general, and MPs in particular, have a long and complicated "afterlife" within quite a limited depths range between coasts and the deep sea (Chubarenko and Stepanova, 2017). Much like sand particles, plastics should be retained in the area of surface wave transformation, where stronger coastal currents are able to transport and redeposit them. It looks probable that only easily transported fibers and very small particles can be exported from the coastal zone to the deep-sea area, and the asymmetry of sinking behavior of lighter and heavier particles under surface waves supports this "MP pump." Importantly, under strong surface waves, given a wide diversity of MP properties, the particles circulate in wave-induced velocity field along various different trajectories, being effectively mixed with sediments, algae, and other suspended material. This enhances plastic degradation and makes particular MP properties less important in favor of external oceanographic conditions.

On the base of the analyzed oceanographic situations and the case study with the Baltic amber, the following life cycle of plastic objects in the coastal zone can be proposed (Fig. 6.15; modified from Chubarenko and Stepanova, 2017). Thrown on the beach, nonbuoyant plastics are soon captured by waves and carried seaward (and probably along the coast) as far as stormy currents can retain and carry them. Since nonbuoyant macropieces tend to sink faster than smaller ones and wind-wave currents weaken with the distance from the shore, they settle at some (hydrodynamically prescribed) depth, still in the area of deformation of surface waves. Later, with some other storm, they can be further damaged and redeposited within the coastal zone. As amber washouts show, nonbuoyant plastics are able to come back upslope (to the swash zone and the shore line) to be further mechanically destroyed and UV degraded and washed to the sea again and again during the next wind event—until MP pieces will be small and light enough to leave the coastal zone with stormy currents and be deposited in the deeper sea, out of reach of stormy waves. Synthetic fibers, found everywhere in deep-sea sediments, seem to be an example of such easily transported MPs. This way, negatively buoyant plastic objects should repeatedly migrate up- and downslope within the coastal zone, composing the underwater "plastic rim" along the shore and exporting only very small particles toward the deep-sea area. Dynamics of positively buoyant plastics is additionally influenced by wind and stronger surface currents, suggesting much larger spatial scale of their migrations. The observed decrease of amount of floating MP particles smaller than 1 mm (Cózar et al., 2017) seems to be indicative of biofouling and sinking (i.e., removal from water surface). Calculations show that particle biofouling to the integral density equal to that of surrounding water and its subsequent sinking significantly depends (with other conditions being equal) on the particle shape, so that tiny fibers are first to begin sinking, followed by films and flakes, while 3-D fragments are the latest. The terminal (i.e., steady state) settling velocity of the particle sinking may never be reached due to the complicated shape of MPs, turbulent nature and stratification of marine environment.

REFERENCES

Allen, H.S., 1900. The motion of a sphere in a viscous fluid. Philos. Mag. 50, 519–534.

Allen, J.R.L., 1985. Sink or swim. In: Physical Sedimentology, Allen & Unwin, London, 39–54 (Chapter 3).

Andrady, A.L., 1994. Assessment of environmental biodegradation of synthetic polymers: a review. J. Macromol. Sci. R. M. C. 34, 25–75.

Andrady, A.L., 1998. Biodegradation of plastics. In: Pritchard, G. (Ed.), Plastics Additives. Chapman Hall, London.

Andrady, A.L., 2005. Plastics in marine environment. A technical perspective. In: Proceedings of the Plastic Rivers to Sea Conference. Algalita Marine Research Foundation, Long Beach, California.

Andrady, A.L., 2011. Microplastics in the marine environment. Mar. Pollut. Bull. 62, 1596–1605.

Andrady, A.L., 2015a. Persistence of plastic litter in the ocean. In: Bergmann, M., Gutow, L., Klages, M. (Eds.), Marine Anthropogenic Litter. Springer, Berlin, pp. 57–72. https://doi.org/10.1007/978-3-319-16510-3_3.

Andrady, A.L., 2015b. Plastics and Environmental Sustainability. John Wiley & Sons, Hoboken, p. 352.

Arthur C., Baker, J., Bamford, H. 2009. In: *Proceedings of the International Research Workshop on the Occurrence, Effects and Fate of Microplastic Marine Debris*, 9–11 September 2008, NOAA Technical, Memorandum NOS-OR&R30.

Auman, H.J., Ludwig, J.P., Giesy, J.P., Colborn, T., 1997. Plastic ingestion by Laysan Albatross chicks on Sand Island, Midway Atoll, in 1994 and 1995. In: Robinson, G., Gales, R. (Eds.), Albatross Biology and Conservation. Surrey Beatty & Sons, Chipping Norton, pp. 239–244 (Chapter 20).

Baba, J., Komar, P.D., 1981. Measurements and analysis of settling velocities of natural quartz sand grains. J. Sediment. Res. 51, 0631–0640.

Bagaev, A., Khatmullina, L., Chubarenko, I., 2017a. Anthropogenic microlitter in the Baltic Sea water column. Mar. Pollut. Bull. 2017. https://doi.org/10.1016/j.marpolbul.2017.10.049.

Bagaev, A., Mizyuk, A., Khatmullina, L., Isachenko, I., Chubarenko, I., 2017b. Anthropogenic fibres in the Baltic Sea water column: field data, laboratory and numerical testing of their motion. Sci. Total Environ. 599, 560–571.

Baird, M.H.I., Senior, M.G., Thompson, R.J., 1967. Terminal velocities of spherical particles in a vertically oscillating liquid. Chem. Eng. Sci. 22, 551–558.

Balch, W.M., Bowler, B.C., Drapeau, D.T., Poulton, A.J., Holligan, P.M., 2010. Biominerals and the vertical flux of particulate organic carbon from the surface ocean. Geophys. Res. Lett. 37, L22605.

Ballent, A., Purser, A., Mendes, P.D.J., Pando, S., Thomsen, L., 2012. Physical transport properties of marine microplastic pollution. Biogeosci. Discuss. 9, 18755–18798. https://doi.org/10.5194/bgd-9-18755-2012.

Ballent, A., Pando, S., Purser, A., Juliano, M.F., Thomsen, L., 2013. Modelled transport of benthic marine microplastic pollution in the Nazaré Canyon. Biogeosciences 10, 7957–7970.

Barnes, D.K.A., Milner, P., 2005. Drifting plastic and its consequences for sessile organism dispersal in the Atlantic Ocean. Mar. Biol. 146, 815–825.

Barnes, D.K.A., Galgani, F., Thompson, R.C., Barlaz, M., 2009. Accumulation and fragmentation of plastic debris in global environments. Philos. Trans. R. Soc. Lond. B. Biol. Sci. 364, 1985–1998. https://doi.org/10.1098/rstb.2008.0205.

Bees, M.A., 1996. Non-Linear Pattern Generation by Swimming Micro-Organisms (Doctoral dissertation). University of Leeds.

Berenbrock, C., Tranmer, A.W., 2008. Simulation of flow, sediment transport, and sediment mobility of the lower Coeur d'Alene River, Idaho. U.S. Geological Survey Scientific Investigations Report 2008-5093, p. 164.

Bergmann, M., Wirzberger, V., Krumpen, T., Lorenz, C., Primpke, S., Tekman, M.B., Gerdts, G., 2017. High quantities of microplastic in Arctic deep-sea sediments from the HAUSGARTEN observatory. Environ. Sci. Technol. 51 (19), 11000–11010. https://doi.org/10.1021/acs.est.7b03331.

Blasco, J., Corsi, I., Matranga, V., 2015. Particles in the oceans: implication for a safe marine environment. Mar. Environ. Res. 111, 1–4.

Blott, S.J., Pye, K., 2008. Particle shape: a review and new methods of characterization and classification. Sedimentology 55, 31–63.

Bratback, G., Dundas, I., 1984. Bacterial dry-matter content and biomass estimations. Appl. Environ. Microbiol. 48, 755–757.

Browne, M.A., Galloway, T.S., Thompson, R.C., 2010. Spatial patterns of plastic debris along estuarine shorelines. Environ. Sci. Technol. 44, 3404–3409.

Browne, M.A., Galloway, T., Thompson, R., 2007. Microplastic—an emerging contaminant of potential concern? Integr. Environ. Assess. Manag. 3, 559–561.

Browne, M.A., Crump, P., Niven, S.J., Teuten, E.L., Tonkin, A., Galloway, T., Thompson, R.C., 2011. Accumulations of microplastic on shorelines worldwide: sources and sinks. Environ. Sci. Technol. 45, 9175–9179.

Cailleux, A., 1945. Distinction des galets marins et fluviatiles. Bull. Soc. Geol. Fr. 15, 375–404.

Camenen, B., 2007. Simple and general formula for the settling velocity of particles. J. Hydraul. Eng. ASCE 133, 229–233.

Camenen, B., Larson, M., 2006. Phase-lag effects in sheet flow transport. Coast. Eng. 53, 531–542.

Carpenter, E.J., Smith, K., 1972. Plastic on the sargasso sea surface. Science 175, 2–3.

Carpenter, E.J., Anderson, S.J., Harvey, G.R., Miklas, H.P., Peck, B.B., 1972. Polystyrene spherules in coastal waters. Science 178 (4062), 749–750.

Carson, H.S., Colbert, S.L., Kaylor, M.J., McDermid, K.J., 2011. Small plastic debris changes water movement and heat transfer through beach sediments. Mar. Pollut. Bull. 62, 1708–1713.

Casey H., 1935. Berlin, Heft. ber Geschiebebewegung (Concerning Bed-Load Movement). *Mitteilung der Preuss. Versuchsanstalt für Wasserbau und Schiffbau* (VWS) 19.

Chattopadhyay, R., 2010. Introduction: Types of technical textile yarn. In: Alagirusamy, R., Das, A. (Eds.), Technical Textile Yarns—Industrial and Medical Applications. Woodhead Publishing, Cambridge, UK, pp. 3–55. https://doi.org/10.1533/9781845699475.1.3.

Cheng, N.S., 1997. Simplified settling velocity formula for sediment particle. J. Hydraul. Eng. ASCE 123, 149–152.

Chubarenko, I., 2010. Horizontal convective water exchange above sloping bottom: the mechanism of its formation and an analysis of its development. Oceanology 50, 166–174.
Chubarenko, I.P., Demchenko, N.Y., 2010. On contribution of horizontal and intra-layer convection to the formation of the Baltic Sea cold intermediate layer. Ocean Sci. 6 (1), 285–299. http://www.ocean-sci.net/6/285/2010/os-6-285-2010.html.
Chubarenko, I., Hutter, K., 2005. Thermally driven interaction of the littoral and limnetic zones by autumnal cooling process. J. Limnol. 64 (1), 31–42. http://www.jlimnol.it/index.php/jlimnol/article/viewFile/jlimnol.2005.31/258.
Chubarenko, I., Stepanova, N., 2017. Microplastics in sea coastal zone: lessons learned from the Baltic amber. Environ. Pollut. 224, 243–254.
Chubarenko, I., Chubarenko, B., Bäuerle, E., Wang, Y., Hutter, K., 2003. Autumn physical limnological experimental campaign in the Island Mainau littoral zone of Lake Constance. J. Limnol. 62 (1), 115–119. https://doi.org/10.4081/jlimnol.2003.115.
Chubarenko, I., Chubarenko, B., Esiukova, E., Baudler, H., 2010. Mixing by Langmuir circulation in shallow lagoons. Baltica 23, 13–24.
Chubarenko, I., Esiukova, E., Stepanova, N., Chubarenko, B., Baudler, H., 2013. Down-slope cascading modulated by day/night variations of solar heating. J. Limnol. 72 (2), 240–252. https://doi.org/10.4081/jlimnol.2013.e19.
Chubarenko, I., Bagaev, A., Zobkov, M., Esiukova, E., 2016. On some physical and dynamical properties of microplastic particles in marine environment. Mar. Pollut. Bull. 108, 105–112. https://doi.org/10.1016/j.marpolbul.2016.04.048.
Claessens, M., De Meester, S., Van Landuyt, L., De Clerck, K., Janssen, C.R., 2011. Occurrence and distribution of microplastics in marine sediments along the Belgian coast. Mar. Pollut. Bull. 62, 2199–2204.
Claessens, M., Van Cauwenberghe, L., Vandegehuchte, M.B., Janssen, C.R., 2013. New techniques for the detection of microplastics in sediments and field collected organisms. Mar. Pollut. Bull. 70, 227–233.
Clift, R., Grace, J., Weber, M., 1978. Nonspherical rigid particles at higher reynolds number. In: Bubbles, Drops, and Particles. Academic Press, New York, pp. 142–168.
Cole, M., Lindeque, P., Halsband, C., Galloway, T.S., 2011. Microplastics as contaminants in the marine environment: a review. Mar. Pollut. Bull. 62, 2588–2597. https://doi.org/10.1016/j.marpolbul.2011.09.025.
Cole, M., Lindeque, P., Fileman, E., Halsband, C., Goodhead, R., Moger, J., Galloway, T.S., 2013. Microplastic ingestion by zooplankton. Environ. Sci. Technol. 47, 6646–6655.
Convey, P., Barnes, D., Morton, A., 2002. Debris accumulation on oceanic island shores of the Scotia Arc, Antarctica. Polar Biol. 25, 612–617.
Cooper, D.A., 2012. Effects of Chemical and Mechanical Weathering Processes on the Degradation of Plastic Debris on Marine Beaches (Doctoral dissertation). The University of Western Ontario. Retrieved from: http://ir.lib.uwo.ca/cgi/viewcontent.cgi?article=1539&context=etd.
Corcoran, P.L., Biesinger, M.C., Grifi, M., 2009. Plastics and beaches: a degrading relationship. Mar. Pollut. Bull. 58, 80–84. https://doi.org/10.1016/j.marpolbul.2008.08.022.
Corey, A.T., 1949. Influence of Shape on the Fall Velocity of Sand Grains (Master's thesis). A&M College, Colorado.
Costa, M.F., Ivar do Sul, J.A., Silva-Cavalcanti, J.S., Araúja, M.C.B., Spengler, A., Tourinho, P.S., 2010. On the importance of size of plastic fragments and pellets on the strandline: a snapshot of a Brazilian beach. Environ. Monit. Assess. 168, 299–304. https://doi.org/10.1007/s10661-009-1113-4.
Cózar, A., Echevarria, F., González-Gordillo, I.J., Irigoien, X., Úbeda, B., Hernández-León, S., Palma, A.T., Navarro, S., García-de-Lomas, J., Ruiz, A., Fernández-de-Puelles, M.L., Duarte, C.M., 2014. Plastic debris in the open ocean. Proc. Natl. Acad. Sci. 111, 10239–10244.
Cózar, A., Martí, E., Duarte, C.M., García-de-Lomas, J., Van Sebille, E., Ballatore, T.J., Eguíluz, V.M., González-Gordillo, J.I., Pedrotti, M.L., Echevarría, F., Troublè, R., Irigoien, X., 2017. The Arctic Ocean as a dead end for floating plastics in the North Atlantic branch of the thermohaline circulation. Sci. Adv. 3, e1600582. https://doi.org/10.1126/sciadv.1600582.
Craik, A.D.D., Leibovich, S., 1976. A rational model for Langmuir circulations. J. Fluid Mech. 73, 401–426.

Debrot, A.O., Tiel, A.B., Bradshaw, J.E., 1999. Beach debris in Curacao. Mar. Pollut. Bull. 38, 795–801.

Dethleff, D., Kempema, E.W., Koch, R., Chubarenko, I., 2009. On the helical flow of Langmuir circulation: approaching the process of suspension freezing. Cold Reg. Sci. Technol. 56, 50–57.

Dietrich, W.E., 1982. Settling velocity of natural particles. Water Resour. Res. 18, 1615–1626.

Doyle, M.J., Watson, W., Bowlin, N.M., Sheavly, S.B., 2011. Plastic particles in coastal pelagic ecosystems of the Northeast Pacific ocean. Mar. Environ. Res. 71, 41–52.

Driedger, A.G.J., Dürr, H.H., Mitchell, K., Van Cappellen, P., 2015. Plastic debris in the Laurentian Great Lakes: a review. J. Great Lakes Res. 41, 9–19.

Duis, K., Coors, A., 2016. Microplastics in the aquatic and terrestrial environment: sources (with a specific focus on personal care products), fate and effects. Environ. Sci. Eur. 28, 2. https://doi.org/10.1186/s12302-015-0069-y.

Eran, S., Roman, B., Marder, M., Shin, G.S., Swinney, H.L., 2002. Mechanics: buckling cascades in free sheets. Nature 419, 579.

Eriksen, M., Lebreton, L.C.M., Carson, H.S., Thiel, M., Moore, C.J., Borerro, J.C., Galgani, F., Ryan, P.G., Reisser, J., 2014. Plastic pollution in the world's oceans: more than 5 trillion plastic pieces weighing over 250,000 tons afloat at sea. PLoS One 9, e111913. https://doi.org/10.1371/journal.pone.0111913.

Eriksson, C., Burton, H., 2003. Origins and biological accumulation of small plastic particles in fur seals from Macquarie Island. Ambio 32, 380–385.

Esiukova, E., 2017. Plastic pollution on the Baltic beaches of Kaliningrad region, Russia. Mar. Pollut. Bull. 114, 1072–1080. https://doi.org/10.1016/j.marpolbul.2016.10.001.

Faller, A.J., Woodcock, A.H., 1964. The spacing of windrows of *Sargassum* in the ocean. J. Mar. Res. 22 (1), 22–29.

Farrow, D.E., 2004. Periodically forced natural convection over slowly varying topography. J. Fluid Mech. 508, 1–21.

Fazey, F.M., Ryan, P.G., 2016. Biofouling on buoyant marine plastics: an experimental study into the effect of size on surface longevity. Environ. Pollut. 210, 354–360.

Fer, I., Lemmin, U., Thorpe, S.A., 2002. Winter cascading of cold water in Lake Geneva. J. Geophys. Res. Oceans 107 (C6), 13-1–13-16

Filella, M., 2015. Questions of size and numbers in environmental research on microplastics: methodological and conceptual aspects. Environ. Chem. 12, 527–538.

Finn, J.R., Li, M., Apte, S.V., 2016. Particle based modelling and simulation of natural sand dynamics in the wave bottom boundary layer. J. Fluid Mech. 796, 340–385.

Flemming, N.C., 1965. Form and function of sedimentary particles. J. Sediment. Petrol. 35, 381–390.

Foekema, E.M., De Gruijter, C., Mergia, M.T., van Franeker, J.A., Murk, A.J., Koelmans, A.A., 2013. Plastic in north sea fish. Environ. Sci. Technol. 47, 8818–8824.

Francois, R., Honjo, S., Krishfield, R., Manganini, S., 2002. Factors controlling the flux of organic carbon to the bathypelagic zone of the ocean. Glob. Biogeochem. Cycles 16, 1087.

Frias, J.P.G.L., Otero, V., Sobral, P., 2014. Evidence of microplastics in samples of zooplankton from Portuguese coastal waters. Mar. Environ. Res. 95, 89–95.

Fujieda, S., Sasaki, K., 2005. Stranded debris of foamed plastic on the coast of ETA Island and Kurahashi Island in Hiroshima Bay. Nippon Suisan Gakkaishi 71, 755–761.

Galgani, F., Hanke, G., Werner, S., Oosterbaan, L., Nilsson, P., Fleet, D., Kinsey, S., Thompson, R.C., van Franeker, J., Vlachogianni, T., Scoullos, M., Veiga, J.M., Palatinus, A., Matiddi, M., Maes, T., Korpinen, S., Budziak, A., Leslie, H., Gago, J., Liebezeit, G., 2013. Guidance on Monitoring of Marine Litter in European Seas. EUR Scientific and Technical Research Series, Publications Office of the European Union, Luxembourg, p. 128. https://doi.org/10.2788/99475.

Gawarkiewicz, G., Chapman, D.C., 1995. A numerical study of dense water formation and transport on a shallow, sloping continental shelf. J. Geophys. Res. 100 (C3), 4489–4507.

Gewert, B., Plassmann, M.M., MacLeod, M., 2015. Pathways for degradation of plastic polymers floating in the marine environment. Environ. Sci.: Processes Impacts 17, 1513–1521. https://doi.org/10.1039/C5EM00207A.

Gibbs, R.J., 1974. Principles of studying suspended materials in water. In: Gibbs, R.J. (Ed.), Suspended Solids in Water. Plenum, New York, pp. 3–15.

Gilbert, J.M., Reichelt-Brushett, A.J., Bowling, A.C., Christidis, L., 2016. Plastic ingestion in marine and coastal bird species of southeastern Australia. Mar. Ornithol. 44, 21–26.
Gilfillan, L.R., Ohman, M.D., Doyle, M.J., Watson, W., 2009. Occurrence of plastic micro-debris in the southern California current system. CalCOFI Rep. 50, 123–133.
Gill, A.E., 1982. Atmosphere-Ocean Dynamics. International Geophysics Series, vol. 30. Academic Press, New York, London. 662 pp.
Graca, B., Szewc, K., Zakrzewska, D., Dołęga, A., Szczerbowska-Boruchowska, M., 2017. Sources and fate of microplastics in marine and beach sediments of the Southern Baltic Sea—a preliminary study. Environ. Sci. Pollut. R. 24, 7650–7661.
Graf, W., 1971. Hydraulics of Sediment Transport. McGraw-Hill, New York, p. 513.
Gregory, M.R., 1978. Accumulation and distribution of virgin plastic granules on New Zealand beaches. N. Z. J. Mar. Freshw. Res. 12, 399–414. https://doi.org/10.1080/00288330.1978.9515768.
Gregory, M.R., Andrady, A.L., 2003. Plastics in the marine environment. In: Andrady, A.L. (Ed.), Plastics and the Environment. John Wiley & Sons, Hoboken, pp. 379–401.
Hallermeier, R.J., 1981. Terminal settling velocity of commonly occurring sand grains. Sedimentology 28, 859–865.
Hardesty, B.D., Harari, J., Isobe, A., Lebreton, L., Maximenko, N., Potemra, J., van Sebille, E., Vethaak, D., Wilcox, C., 2017. Using numerical model simulations to improve the understanding of microplastic distribution and pathways in the marine environment. Front. Mar. Sci. 4, 30.
Harrison, J.P., Sapp, M., Schratzberger, M., Osborn, A.M., 2011. Interactions between microorganisms and marine microplastics; a call for research. Mar. Technol. Soc. J. 45, 12–20.
Hassan, W.N., Ribberink, J.S., 2005. Transport processes of uniform and mixed sands in oscillatory sheet flow. Coast. Eng. 52, 745–770.
Hazzab, A., Terfous, A., Ghenaim, A., 2008. Measurement and modeling of the settling velocity of isometric particles. Powder Technol. 184, 105–113.
Heywood, H., 1962. In: Uniform and non-uniform motion of particles in fluids.Proc Symp Interaction Fluids and Parts. Inst. Chem. Eng., London, pp. 1–8.
Hidalgo-Ruz, V., Gutow, L., Thompson, R.C., Thiel, M., 2012. Microplastics in the marine environment: a review of the methods used for identification and quantification. Environ. Sci. Technol. 46, 3060–3075.
Hidalgo-Ruz, V., Thiel, M., 2013. Distribution and abundance of small plastic debris on beaches in the SE Pacific (Chile): a study supported by a citizen science project. Mar. Environ. Res. 459, 121–134. https://doi.org/10.1016/j.marenvres.2013.02.015.
Hinata, H., Mori, K., Ohno, K., Miyao, Y., Kataoka, T., 2017. An estimation of the average residence times and onshore-offshore diffusivities of beached microplastics based on the population decay of tagged meso-and macrolitter. Mar. Pollut. Bull. 122, 17–25.
Holmström, A., 1975. Plastic films on the bottom of the Skagerrak. Nature 255, 622–623. https://doi.org/10.1038/255622a0.
Hwang, P.A., 1985. Fall velocity of particles in oscillating flow. J. Hydraul. Eng. 111, 485–502.
International Organization for Standardization, 1999. Plastics—determination of the effects of exposure to damp heat, water spray and salt mist. (ISO/DIN Standard No. *4611*). Retrieved from: http://211.67.52.20:8088/xitong/BZ%5C9064971.pdf.
Ivanov, V.V., Shapiro, G.I., Huthnance, J.M., Aleynik, D.L., Golovin, P.N., 2004. Cascades of dense water around the world ocean. Prog. Oceanogr. 60, 47–98.
Ivar do Sul, J.A., Spengler, A., Costa, M.F., 2009. Here, there and everywhere. Small plastic fragments and pellets on beaches of Fernando de Noronha (Equatorial Western Atlantic). Mar. Pollut. Bull. 58, 1236–1238.
Ivar do Sul, J.A., Costa, M.F., Barletta, M., Cysneiros, F.J.A., 2013. Pelagic microplastics around an archipelago of the Equatorial Atlantic. Mar. Pollut. Bull. 75, 305–309.
Jacobs, P., Ivey, G.N., 1998. The influence of rotation on shelf convection. J. Fluid Mech. 369, 23–48.
Jambeck, J.R., Geyer, R., Wilcox, C., Siegler, T.R., Perryman, M., Andrady, A., Narayan, R., Law, K.L., 2015. Plastic waste inputs from land into the ocean. Science 347, 768–771. https://doi.org/10.1126/science.1260352.
Jayaweera, K.O.L.F., Mason, B.J., 1965. The behaviour of freely falling cylinders and cones in a viscous fluid. J. Fluid Mech. 22, 709–720.

Kataoka, T., Hinata, H., Kato, S., 2013. Analysis of a beach as a time-invariant linear input/output system of marine litter. Mar. Pollut. Bull. 77 (1), 266–273.
Kataoka, T., Hinata, H., Kato, S., 2015. Backwash process of marine macroplastics from a beach by nearshore currents around a submerged breakwater. Mar. Pollut. Bull. 101 (2), 539–548.
Kershaw, P.J. (Ed.), 2015. Sources, fate and effects of microplastics in the marine environment: a global assessment. Rep. Stud. GESAMP 90, 96 p.
Khatmullina, L., Isachenko, I., 2017. Settling velocity of microplastic particles of regular shapes. Mar. Pollut. Bull. 114, 871–880.
Komar, P.D., Morse, A.P., Small, L.F., Fowler, S.W., 1981. An analysis of sinking rates of natural copepod and euphausiid fecal pellets. Limnol. Oceanogr. 26, 172–180. https://doi.org/10.4319/lo.1981.26.1.0172, 1981.
Kooi, M., Van Nes, E.H., Scheffer, M., Koelmans, A.A., 2017. Ups and downs in the ocean: effects of biofouling on the vertical transport of microplastics. Environ. Sci. Technol. 51 (14), 7963–7971. https://doi.org/10.1021/acs.est.6b04702.
Kowalski, N., Reichardt, A.M., Waniek, J.J., 2016. Sinking rates of microplastics and potential implications of their alteration by physical, biological, and chemical factors. Mar. Pollut. Bull. 109, 310–319. https://doi.org/10.1016/j.marpolbul.2016.05.064.
Kranenburg, W.M., Hsu, T.J., Ribberink, J.S., 2014. Two-phase modeling of sheet-flow beneath waves and its dependence on grain size and streaming. Adv. Water Resour. 72, 57–70.
Krumbein, W.C., 1940. Flood gravel of San Gabriel Canyon, California. Geol. Soc. Am. Bull. 51, 639–676.
Krumbein, W.C., 1941. Measurement and geological significance of shape and roundness of sedimentary particles. *J. Sediment. Petrol.* 11, 64–72. https://doi.org/10.1306/D42690F3-2B26-11D7-8648000102C1865D.
Kukulka, T., Proskurowski, G., Morét-Ferguson, S., Meyer, D.W., Law, K.L., 2012. The effect of wind mixing on the vertical distribution of buoyant plastic debris. Geophys. Res. Lett. 39, 1–6.
Kumar, R.G., Strom, K.B., Keyvani, A., 2010. Floc properties and settling velocity of San Jacinto estuary mud under variable shear and salinity conditions. Cont. Shelf Res. 30, 2067–2081.
Lambert, S., Wagner, M., 2016. Formation of microscopic particles during the degradation of different polymers. Chemosphere 161, 510–517. https://doi.org/10.1016/j.chemosphere.2016.07.042.
Langmuir, I., 1938. Surface motion of water induced by wind. Science 87, 119–123.
Law, K.L., 2017. Plastics in the marine environment. Annu. Rev. Mar. Sci. 9, 205–229.
Le Roux, J.P., 2005. Grains in motion: a review. Sediment. Geol. 178, 285–313.
Lebreton, L.C.-M., Greer, S.D., Borrero, J.C., 2012. Numerical modelling of floating debris in the world's oceans. Mar. Pollut. Bull. 64, 653–661.
Ledbetter, M., 1979. Langmuir circulations and plankton patchiness. Ecol. Model. 7, 289–310.
Lee, J., Hong, S., Song, Y.K., Hong, S.H., Jang, Y.C., Jang, M., Heo, N.W., Han, G.M., Lee, M.J., Kang, D., Shim, W.J., 2013. Relationships among the abundances of plastic debris in different size classes on beaches in South Korea. Mar. Pollut. Bull. 77, 349–354. https://doi.org/10.1016/j.marpolbul.2013.08.013.
Leibovich, S., 1983. The form and dynamics of Langmuir circulations. Annu. Rev. Fluid Mech. 15, 391–427.
Leppäranta, M., Myrberg, K., 2009. Physical Oceanography of the Baltic Sea. Springer, Chicester, UK, p. 378.
Lerman, A., Lal, D., Dacey, M.F., 1974. Stokes' settling and chemical reactivity of suspended particles in natural waters. In: Gibbs, R.G. (Ed.), Suspended Solids in Water. Plenum, New York, pp. 17–47.
Li, X., Zhang, J., Lee, J.H.W., 2004. Modelling particle size distribution dynamics in marine waters. Water Res. 38, 1305–1317.
Li, J., Yang, D., Li, L., Jabeen, K., Shi, H., 2015. Microplastics in commercial bivalves from China. Environ. Pollut. 207, 190–195.
Li, W.C., Tse, H.F., Fok, L., 2016. Plastic waste in the marine environment: a review of sources, occurrence and effects. Sci. Total Environ. 566, 333–349.
Lusher, A., 2015. Microplastics in the marine environment: Distribution, interactions and effects. In: Marine Anthropogenic Litter. Springer International Publishing, Cham, pp. 245–307.

Lusher, A.L., McHugh, M., Thompson, R.C., 2013. Occurrence of microplastics in the gastrointestinal tract of pelagic and demersal fish from the English Channel. Mar. Pollut. Bull. 67, 94–99.

MacLeod, N., 2002. Geometric morphometrics and geological form-classification systems. Earth-Sci. Rev. 59, 27–47. https://doi.org/10.1016/S0012-8252(02)00068-5.

Maggi, F., 2013. The settling velocity of mineral, biomineral, and biological particles and aggregates in water. J. Geophys. Res. Oceans 118, 2118–2132. https://doi.org/10.1002/jgrc.20086, 2013.

Maggi, F., Tang, F.H., 2015. Analysis of the effect of organic matter content on the architecture and sinking of sediment aggregates. Mar. Geol. 363, 102–111.

Maximenko, N.A., Hafner, J., Niiler, P., 2012. Pathways of marine debris from trajectories of Lagrangian drifters. Mar. Pollut. Bull. 65 (1–3), 51–62. https://doi.org/10.1016/j.marpolbul.2011.04.016.

McDermid, K.J., McMullen, T.L., 2004. Quantitative analysis of small-plastic debris on beaches in the Hawaiian archipelago. Mar. Pollut. Bull. 48, 790–794.

McDonnell, A.M., Buesseler, K.O., 2010. Variability in the average sinking velocity of marine particles. Limnol. Oceanogr. 55, 2085–2096.

McLaren, P., Hill, S.H., Bowles, D., 2007. Deriving transport pathways in a sediment trend analysis (STA). Sediment. Geol. 202, 489–498.

McWilliams, M., Liboiron, M., Wiersma, Y., 2017. Rocky shoreline protocols miss microplastics in marine debris surveys (Fogo Island, Newfoundland and Labrador). Mar. Pollut. Bull. Available online 13 October 2017, https://doi.org/10.1016/j.marpolbul.2017.10.018.

Monismith, S.G., Imberger, J., Morison, M.L., 1990. Convective motions in the sidearm of a small reservoir. Limnol. Oceanogr. 35, 1676–1702.

Monroy, P., Hernández-García, E., Rossi, V., López, C., 2017. Modeling the dynamical sinking of biogenic particles in oceanic flow. Nonlinear Process. Geophys 24, 293–305. https://doi.org/10.5194/npg-24-293-2017.

Moore, C.J., Moore, S.L., Leecaster, M.K., Weisberg, S.B., 2001. A comparison of plastic and plankton in the North Pacific central gyre. Mar. Pollut. Bull. 42, 1297–1300.

Morawitz, W.M.L., Sutton, P.J., Worcester, P.F., Cornuelle, B.D., Lynch, J.F., Pawlowicz, R., 1996. Three-dimensional observations of a deep convective chimney in the Greenland Sea during winter 1988/89. J. Phys. Oceanogr. 26, 2316–2343.

Morét-Ferguson, S., Law, K.L., Proskurowski, G., Murphy, E.K., Peacock, E.E., Reddy, C.M., 2010. The size, mass, and composition of plastic debris in the western North Atlantic Ocean. Mar. Pollut. Bull. 60, 1873–1878.

Muthukumar, A., Veerappapillai, S., 2015. Biodegradation of plastics—a brief review. Int. J. Pharm. Sci. Rev. Res. 31, 204–209. Retrieved from:http://www.globalresearchonline.net/pharmajournal/vol31iss2.aspx.

Nerland, I.L., Halsband, C., Allan, I., Thomas, K.V., 2014. Microplastics in marine environments: occurrence, distribution and effects. Norwegian Institute for Water Research, p. 71.

Neves, D., Sobral, P., Ferreira, J.L., Pereira, T., 2015. Ingestion of microplastics by commercial fish off the Portuguese coast. Mar. Pollut. Bull. 101, 119–126.

Nor, N.H.M., Obbard, J.P., 2014. Microplastics in Singapore's coastal mangrove ecosystems. Mar. Pollut. Bull. 79, 278–283. https://doi.org/10.1016/j.marpolbul.2013.11.025.

Norén, F., 2007. Small plastic particles in coastal swedish waters. N-Research report commissioned by KIMO, Sweden, p. 11.

Obbard, R.W., Sadri, S., Wong, Y.Q., Khitun, A.A., Baker, I., Thompson, R.C., 2014. Global warming releases microplastic legacy frozen in Arctic Sea ice. Earth's Future 2, 315–320.

Ogata, Y., Takada, H., Mizukawa, K., Hirai, H., Iwasa, S., Endo, S., Mato, Y., Saha, M., Okuda, K., Nakashima, A., Murakami, M., Zurcher, N., Booyatumanondo, R., Zakaria, M.P., Dung, L., Gordon, M., Miguez, C., Suzuki, S., Moore, C., Karapanagioti, H.K., 2009. International pellet watch: global monitoring of persistent organic pollutants (POPs) in coastal waters. Initial phase data on PCBs, DDTs, and HCHs1. Mar. Pollut. Bull. 58, 1437–1446.

Ory, N.C., Sobral, P., Ferreira, J.L., Thiel, M., 2017. Amberstripe scad Decapterus muroadsi (Carangidae) fish ingest blue microplastics resembling their copepod prey along the coast of Rapa Nui (Easter Island) in the South Pacific subtropical gyre. Sci. Total Environ. 586, 430–437.

Perkin, J.S., Williams, C.S., Bonner, T.H., 2009. Aspects of chub shiner Notropis potteri life history with comments on native distribution and conservation status. Am. Midl. Nat. 162, 276–288. https://doi.org/10.1674/0003-0031-162.2.276.

Phillips, M.B., Bonner, T.H., 2015. Occurrence and amount of microplastic ingested by fishes in watersheds of the Gulf of Mexico. Mar. Pollut. Bull. 100, 264–269. https://doi.org/10.1016/j.marpolbul.2015.08.041.

Plastic ingestion by birds, 2016. https://www.blastic.eu/knowledge-bank/impacts/plastic-ingestion/birds/. Accessed 1 April 2017.

PlasticsEurope, 2013. Plastics—The Facts 2013. An Analysis of European Plastics Production, Demand and Waste Data. Plastics Europe: Association of Plastic Manufacturers, Brussels, p. 40.

Powers, M.C., 1953. A new roundness scale for sedimentary particles. J. Sediment. Petrol. 23, 117–119.

Rocha-Santos, T., Duarte, A.C., 2015. A critical overview of the analytical approaches to the occurrence, the fate and the behavior of microplastics in the environment. TrAC Trends Anal. Chem. 65, 47–53.

Rodriguez, J., Edeskär, T., Knutsson, S., 2013. Particle shape quantities and measurement techniques: a review. Electron. J. Geotech. Eng. 18, 169–198.

Rodríguez-Seijo, A., Pereira, R., 2017. Morphological and physical characterization of microplastics. In: Rocha-Santos, T.A.P., Duarte, A.C. (Eds.), In: Characterization and Analysis of Microplastics, Comprehensive Analytical Chemistry, vol. 75. Elsevier, pp. 49–66. https://doi.org/10.1016/bs.coac.2016.10.007.

Ryan, P.G., 1988. Effects of ingested plastic on seabird feeding: evidence from chickens. Mar. Pollut. Bull. 19, 125–128.

Ryan, P.G., 2015. The importance of size and buoyancy for long-distance transport of marine debris. Environ. Res. Lett. 10084019.

Ryan, P.G., Moore, C.J., van Franeker, J.A., Moloney, C.L., 2009. Monitoring the abundance of plastic debris in the marine environment. *Philos. Trans. R. Soc.* B 364, 1999–2012.

Schott, F., Leaman, K.D., 1991. Observation with moored acoustic Doppler current profiles in the convection regime in the Golfe du Lion. J. Phys. Oceanogr. 21, 558–574.

Schott, F., Visbeck, M., Fischer, J., 1993. Observations of vertical currents and convection in the central Greenland Sea during the winter of 1988/89. J. Geophys. Res. 98, 14401–14421.

Shah, A.A., Hasan, F., Hameed, A., Ahmed, S., 2008. Biological degradation of plastic: a comprehensive review. Biotechnol. Adv. 26, 246–265. https://doi.org/10.1016/j.biotechadv.2007.12.005.

Shaw, D.G., Day, R.H., 1994. Colour- and form-dependent loss of plastic micro-debris from the North Pacific Ocean. Mar. Pollut. Bull. 28, 39–43.

Shields A., 1936. Application of Similarity Principles and Turbulence Research to Bed-Load Movement. Publication No 167, 47. (Translated from: Shields A. Anwendung der Aehnlichkeitsmechanilr und der Turbulenzforschung auf die Geschiebe-bewegung. Mitteilungen der Preussischen Versuchsanstalt fur Wasserbau und Schiffbau, Berlin, 1936. Translated from the German by W.P. Ott and J.C. van Uchelen). California Institute of Technology Pasadena.

Sinclair, R., 2015. Textiles and Fashion: Materials, Design and Technology. Woodhead, Cambridge.

Sneed, E.D., Folk, R.L., 1958. Pebbles in the lower Colorado River, Texas, a study in particle morphogenesis. J. Geol. 66, 114–150.

Southard, J., 2006. Oscillatory flow. (Chapter 6). 12.090, In: Introduction to Fluid Motions, Sediment Transport, and Current-Generated Sedimentary Structures. Massachusetts Institute of Technology. MIT OpenCourseWare, https://ocw.mit.edu License: Creative Commons BY-NC-SA. Available from: https://ocw.mit.edu/courses/earth-atmospheric-and-planetary-sciences/12-090-introduction-to-fluid-motions-sediment-transport-and-current-generated-sedimentary-structures-fall-2006/course-textbook/ch6.pdf.

Stavn, R.H., 1971. The horizontal–vertical distribution hypothesis: langmuir circulations and Daphnia distributions. Limnol. Oceanogr. 16, 453–466.

Stommel, H., 1949. Trajectories of small bodies sinking slowly through convective cells. J. Mar. Res. 8, 24–29.

Stommel, H., Voorhis, A., Webb, D., 1971. Submarine clouds in the deep ocean. Am. Sci. 59, 717–723.

Stringham, G.E., Simons, D.B., Guy, H.P., 1969. The behavior of large particles falling in quiescent liquids. US Geol. Surv. Prof. Pap. 562-C, US Geol. Surv., Reston.

Sukumaran, B., Ashmawy, A.K., 2001. Quantitative characterisation of the geometry of discrete particles. Geotechnique 51, 619–627.

Sundt, P., Schulze, P.E., Syversen, F., 2014. Sources of Microplastic-Pollution to the Marine Environment. Norwegian Environment Agency. Available from: http://www.miljodirektoratet.no/Documents/publikasjoner/M321/M321.pdf.

Tanaka, K., Takada, H., 2016. Microplastic fragments and microbeads in digestive tracts of planktivorous fish from urban coastal waters. Sci. Rep 6, 34351. https://doi.org/10.1038/srep34351.

Thiel, M., Hinojosa, I.A., Miranda, L., Pantoja, J.F., Rivadeneira, M.M., Vásquez, N., 2013. Anthropogenic marine debris in the coastal environment: a multi-year comparison between coastal waters and local shores. Mar. Pollut. Bull. 71 (1), 307–316.

Thompson, R.C., 2015. Microplastics in the marine environment: sources, consequences and solutions. In: Bergmann, M., Gutow, L., Klages, M. (Eds.), Marine Anthropogenic Litter. Springer, Berlin, pp. 185–200.

Thompson, R.C., Olsen, Y., Mitchell, R.P., Davis, A., Rowland, S.J., John, A.W.G., McGonigle, D., Russell, A.E., 2004. Lost at sea: where is all the plastic? Science 304, 838.

Thorpe, S., 2005. The Turbulent Ocean. vol. 484. Cambridge University Press, Cambridge.

Titman, D., Kilham, P., 1976. Sinking in freshwater phytoplankton: some ecological implications of cell nutrient status and physical mixing processes. Limnol. Oceanogr. 21, 409–417.

Tunstall, E.B., Houghton, G., 1968. Retardation of falling spheres by hydrodynamic oscillations. Chem. Eng. Sci. 23, 1067–1081.

UN Environment Annual Report, 2016. Marine plastic debris and microplastics—global lessons and research to inspire action and guide policy change. United Nations Environment Programme, Nairobi. Accessed from: http://www.unep.org/annualreport/2016/index.php?page=6&lang=en.

US EPA, 1992. Plastic Pellets in the Aquatic Environment: Sources and Recommendations, Final Report, EPA-842-B-92-010, U.S. Environmental Protection Agency, Office of Water, Washington, DC, http://www.globalgarbage.org/13%20EPA%20Plastic%20Pellets.pdf.

Van Cauwenberghe, L., Claessens, M., Vandegehuchte, M.B., Mees, J., Janssen, C.R., 2013a. Assessment of marine debris on the Belgian continental shelf. Mar. Pollut. Bull. 73, 161–169.

Van Cauwenberghe, L., Vanreusel, A., Mees, J., Janssen, C.R., 2013b. Microplastic pollution in deep-sea sediments. Environ. Pollut. 182, 495–499.

Van Cauwenberghe, L., Devriese, L., Galgani, F., Robbens, J.R., Janssen, C., 2015. Microplastics in sediments: a review of techniques, occurrence and effects. Mar. Environ. Res. 111, 5–17. https://doi.org/10.1016/j.marenvres.2015.06.007.

van Rijn, L.C., 1993. Principles of Sediment Transport in Rivers, Estuaries and Coastal Seas. Aqua Publications, Amsterdam, The Netherlands.

Vianello, A., Boldrin, A., Guerriero, P., Moschino, V., Rella, R., Sturaro, A., Da Ros, L., 2013. Microplastic particles in sediments of Lagoon of Venice, Italy: first observations on occurrence, spatial patterns and identification. Estuar. Coast. Shelf Sci. 130, 54–61.

Wadell, H., 1932. Volume, shape, and roundness of rock particles. J. Geol. 40, 443–451.

Wadell, H., 1933. Sphericity and roundness of rock particles. J. Geol. 41, 310–331.

Wadell, H., 1934. The coefficient of resistance as a function of Reynolds number for solids of various shapes. J. Frankl. Inst. 217, 459–490.

Wadell, H., 1935. Volume, shape and roundness of quartz particles. J. Geol. 43, 250–280.

Wang, J., Tan, Z., Peng, J., Qiu, Q., Li, M., 2016. The behaviors of microplastics in the marine environment. Mar. Environ. Res. 113, 7–17. https://doi.org/10.1016/j.marenvres.2015.10.014.

Wentworth, C.K., 1919. A laboratory and field study of cobble abrasion. J. Geol. 27, 507–521.

Wentworth, C.K., 1923a. The shapes of beach pebbles. US Geol. Surv. Prof. Pap. 131-C, 75–83.

Wentworth, C.K., 1923b. A method of measuring and plotting the shapes of pebbles. In: The Shapes of Pebbles, pp. 91–102. US Geol. Surv. Bull. 730-C.

Wilcox, C., Van Sebille, E., Hardesty, B.D., 2015. Threat of plastic pollution to seabirds is global, pervasive, and increasing. Proc. Natl. Acad. Sci. 112 (38), 11899–11904. https://doi.org/10.1073/pnas.1502108112.

Willmarth, W.W., Hawk, N.E., Harvey, R.L., 1964. Steady and unsteady motions and wakes of freely falling disks. Phys. Fluids 7, 197–208.

Woodall, L.C., Sanchez-Vidal, A., Canals, M., Paterson, G.L., Coppock, R., Sleight, V., Calafat, A., Rogers, A.D., Narayanaswamy, B.E., Thompson, R.C., 2014. The deep sea is a major sink for microplastic debris. Royal Soc. Open Sci. 1, 140317.
Woodcock, A.H., 1993. Winds subsurface pelagic *Sargassum* and Langmuir circulations. J. Exp. Mar. Biol. Ecol. 170 (1), 117–125.
Wüest, A., Ravens, T.M., Granin, N.G., Kocsis, O., Schurter, M., Sturm, M., 2005. Cold intrusions in Lake Baikal: direct observational evidence for deep-water renewal. Limnol. Oceanogr. 50, 184–196.
Yamashita, R., Tanimura, A., 2007. Floating plastic in the Kuroshio Current area, western North Pacific Ocean. Mar. Pollut. Bull. 54, 485–488.
Zhang, H., 2017. Transport of microplastics in coastal seas. Estuar. Coast. Shelf Sci 199, 74–86. https://doi.org/10.1016/j.ecss.2017.09.032.
Zhiyao, S., Tingting, W., Fumin, X., Ruijie, L., 2008. A simple formula for predicting settling velocity of sediment particles. Water Sci. Eng. 1, 37–43.
Zingg, T., 1935. Beitrag zur schotteranalyse. Schweiz. Mineral. Petrogr. Mitt. 15, 39–140.
Zobkov, M., Esiukova, E., 2017a. Microplastics in Baltic bottom sediments: quantification procedures and first results. Mar. Pollut. Bull. 114 (2), 724–732. https://doi.org/10.1016/j.marpolbul.2016.10.060.
Zobkov, M., Esiukova, E., 2017b. Evaluation of the Munich Plastic Sediment Separator efficiency in extraction of microplastics from natural marine bottom sediments. Limnol. Oceanogr. Methods 15, 967–978. https://doi.org/10.1002/lom3.10217.

FURTHER READING

Feistel, R., Naush, G., Wastmund, N.J. (Eds.), 2008. State and Evolution of the Baltic Sea, 1952–2005. A Detailed 50-Year Survey of Meteorology and Climate, Physics, Chemistry, Biology, and Marine Environment. Wiley & Sons, Hoboken, NJ. 712 pp.
US Environmental Protection Agency, 1992. Supplemental Guidance to RAGS: Calculating the Concentration Term. Office of Solid Waste and Emergency Response, US Environmental Protection Agency, Washington, DC, pp. 9285–9287.

CHAPTER 7

Sorption of Toxic Chemicals on Microplastics

Fen Wang, Fei Wang, Eddy Y. Zeng
Jinan University, Guangzhou, China

7.1 INTRODUCTION

In the recent years, plastic pollution in the marine environment has continued to worsen with increasing amounts and use of plastics in industrial and consumer products. As lightweight, strong, versatile, durable, and low-cost materials (Derraik, 2002; Laist, 1987) with fantabulous oxygen/moisture barrier properties and resistance to corrosion and bioinertness, plastics have been widely used in packaging and other applications (Andrady, 2011; Bockhorn et al., 1999; Roy et al., 2011; Teuten et al., 2009). On the other hand, such popularity has also caused severe plastic pollution to the marine environment (Laist, 1987; Pruter, 1987). Plastic is an important part of marine waste, occupying about 60%–80% of the ocean garbage, and in some areas can be as high as 90%–95% (Derraik, 2002; Gregory and Ryan, 1997; Moore, 2008). Plastic residues in the marine environment can be transported over long distances to reach remote regions such as Antarctica by external forces (such as wind, sand blasting, and ocean current) (Andrady, 2011; Cole et al., 2011; Cozar et al., 2014). Marine plastic pollution has become a global environmental problem (Andrady, 2003; Barnes et al., 2009; Woodall et al., 2014).

Microplastics (MPs), that is, plastic particles with diameters <5 mm, are ubiquitously distributed in the marine environment, accounting for 13.2% of the global marine plastic debris mass and 92.4% of the number of global plastic pieces (Eriksen et al., 2014). The minute sizes of MPs make them difficult to identify and easily ingested by various aquatic organisms. In addition, nanoplastics (NPs) have larger specific surface and mobility than MPs, so they may sorb much more chemicals from ambient environment (Ma et al., 2016). The sorption dynamics on MPs may influence the uptake of harmful chemicals (such as polycyclic aromatic hydrocarbons (PAHs), polychlorinated biphenyls (PCBs), dichlorodiphenyltrichloroethanes (DDTs), and hexachlorocyclohexanes (HCHs)) into and transfer in marine food chains (Bakir et al., 2012; Carpenter et al., 1972; Carpenter and Smith, 1972; Mato et al., 2001; Rochman et al., 2013b; Rockstrom et al., 2009; Rothstein, 1973; Zitko and Hanlon, 1991). In addition, MPs may release toxic and harmful chemicals (such as plasticizers), which are used as additive to improve the physical and chemical

Microplastic Contamination in Aquatic Environments
https://doi.org/10.1016/B978-0-12-813747-5.00007-2

properties of plastics (Barnes et al., 2009; Lithner et al., 2011; Talsness et al., 2009; Thompson et al., 2009; Ziccardi et al., 2016). Leachates of plasticizer, such as nonylphenol (NP) and polybrominated diphenyl ethers (PBDEs), are even related to cancer (Browne et al., 2007). Microplastics ingested by turtles, fish, and seabirds can reduce food intake and fitness, block digestive tract and intestinal tract, cause interior damage, impact biodiversity, and even result in death (Carpenter et al., 1972; Derraik, 2002; Rochman et al., 2013b; Rothstein, 1973; Thompson, 2007; Zitko and Hanlon, 1991). The intake process of MPs by organisms can pose endocrine disruption effects on organisms, which in turn can cause delayed ovulation, reproductive failure, and other developmental disorders (Azzarello and Van Vleet, 1987; Barnes et al., 2009; Lithner et al., 2009, 2011). Microplastics are also believed to be harmful to normal marine ecosystem functioning. For example, the accumulation of MPs on the seafloor can prevent gas exchange between surface waters and interstitial water of the sediments and inhibit the oxygen transfer, causing hypoxia in aquatic organisms (Goldberg, 1994; Reddy et al., 2006). In addition, alien or nonnative species attached to floating MPs may enter the marine environment, resulting in biological invasion (Barnes, 2002; Derraik, 2002; Winston, 1982). All these findings imply that MPs, as a transport vector for various chemicals and some alien species, can enter the marine food web and directly or indirectly impact food security and human health (Tanaka et al., 2013; Teuten et al., 2009; Vandermeersch et al., 2015).

7.2 MICROPLASTICS AND CHEMICALS

7.2.1 The Chemicals Adhere on Microplastics

Microplastics are ubiquitous in the global environment, widely found on beaches and estuaries, in the oceans and sediments, on remote islands, in the abysmal seas, and at polar regions and equator (Barnes et al., 2009; Browne et al., 2011; Claessens et al., 2011; Gregory and Ryan, 1997; Mato et al., 2001; Thompson et al., 2004; Van Cauwenberghe et al., 2013; Woodall et al., 2014; Zarfl and Matthies, 2010). The type of chemicals contained in MP samples were mostly trace metals (Ashton et al., 2010; Holmes et al., 2012, 2014) and organic chemicals (Bakir et al., 2014b; Browne et al., 2013; Hirai et al., 2011; Lee et al., 2014; Mato et al., 2001; Ng and Obbard, 2006; Rios et al., 2007; Rochman et al., 2013b, 2014b; Teuten et al., 2007, 2009; Van et al., 2012; Yeo et al., 2015). Specifically, they include toxic metals, for example, Al^{3+}, Fe^{3+}, Cu^{2+}, Mn^{2+}, Ni^{2+}, Zn^{2+}, Ti^{2+}, Cd^{2+}, Cr^{2+}, Pb^{2+}, Co^{2+}, Mo^{2+}, Sn^{2+}, Sb^{2+}, Ag^{2+}, U^{2+}, and Hg^{2+} (Akhbarizadeh et al., 2017; Ashton et al., 2010; Brennecke et al., 2016; Graca et al., 2014; Holmes et al., 2012, 2014; Rochman et al., 2014a; Turner and Holmes, 2015; Wang et al., 2017), and organic chemicals such as PCBs, PAHs, DDTs, and HCHs, while chemical ingredients released from MPs are mainly PBDEs, bisphenol A (BPA) (Rochman et al., 2014c; Teuten et al., 2009), and NP (Hirai et al., 2011; Mato et al., 2001). Table 7.1 presents some basic information of commonly found chemicals sorbed to and released from MPs.

Table 7.1 Basic information of common chemicals adhered to microplastics and released from microplastics

Chemical type	Characteristic	Source	Application	Toxicity
PCBs[a]	High-chemical durability and thermostability, acid and alkali resistance, and high liposolubility	Anthropogenic sources	Industrial purposes (e.g., dielectric fluid in transformers and capacitors, hydraulic oils, paints, sliding agents, and bonding agents)	Carcinogenic: easily accumulate in adipose tissue; cause brain, skin, and visceral diseases; and affect the reproductive, nerve, and immune system
PAHs[b]	Chemical stability and low solubility in water	Natural sources and anthropogenic sources: petroleum pollution, combustion of fossil fuels, thermal decomposition process, refuse incineration, and oil leak	Plastic additives in electric and electronic manufacturing; making release agent and dyestuff; even used in plastic, insecticides, and medicines	Strong carcinogenic effect: cause cancer in humans by contact
DDTs[c]	Chemical stability, resistance to acids, and insolubility in water	Anthropogenic sources	Agricultural insecticide	Disrupt the hormonal secretion of the organism; affect people's liver function and reproductive and nervous systems; even have carcinogenic properties
PBDEs[d]	Lipophilicity and low solubility in water	Anthropogenic sources	Flame retardants in various consumer products (such as electric appliances, furniture, and automobile seats)	Thyroid hormone disruptors and have toxic effects on the organisms and humans

Continued

Table 7.1 Basic information of common chemicals adhered to microplastics and released from microplastics—cont'd

Chemical type	Characteristic	Source	Application	Toxicity
BPA[e]		Anthropogenic sources	Organic chemical materials (polycarbonate, epoxy resin, polysulfone resin, polyphenylene ether resin, unsaturated polyester resin, plasticizer, flame retardant, antioxidant, heat stabilizer, rubber protective agent, pesticide, paint)	Lead to endocrine disorders, threaten the health of the fetus and children, and even cause cancer
NP[f]		Natural sources and anthropogenic sources	Surfactant intermediates, mainly used in the production of surface active agent, antioxidant, textile auxiliaries, oil additives, pesticide emulsifier, resin modifying agent, resin and rubber stabilizer, etc.	Environmental hormones, which can mimic estrogen; have an effect on the sexual development of the organism; and interfere with the endocrine of the organism, which is toxic to the reproductive system

[a]General information about PCBs summarized from Antunes et al. (2013), Frias et al. (2010), Rios Mendoza and Jones (2015), Miller (1985), and Pascall et al. (2005).
[b]General information about PAHs summarized from Blumer (1976), Budzinski et al. (1997), Rios et al. (2007), and Ravindra et al. (2008).
[c]General information about DDTs summarized from Antunes et al. (2013), Endo et al. (2005), Fry and Toone (1981), and Rios et al. (2010).
[d]General information about PBDEs summarized from Choi et al. (2009), Darnerud (2008), de Wit (2002), and Talsness (2008).
[e]General information about BPA summarized from Endo et al. (2005), Teuten et al. (2007), and Oehlmann et al. (2009).
[f]General information about NP summarized from Hirai et al. (2011), Sonnenschein and Soto (1998), and Soto et al. (1991).

Table 7.4 Log distribution coefficient (K_{ow}, K_f, K_d, $K_{polymer\text{-}W}$, and $K_{polymer\text{-}SW}$) of various contaminants on different types of plastic polymers (Ziccardi et al., 2016)

		$\log K_{ow}$[a]	$\log K_f$[b]	$\log K_d$[c]	$\log K_{polymer\text{-}W}$[d]	$\log K_{polymer\text{-}SW}$[e]
PE	PCBs		4.93–10.43		4.19–6.88	
	Pentachlorobenzene	5.17				4.63 (4.49–4.75)
	Hexachlorobenzene	5.31			5.43	5.22 (5.08–5.34)
	Naphthalene				2.81	
	Acenaphthene				3.62	
	Fluorene				3.77	
	Phenanthrene	4.6 (4.5)	4.6 ± 0.12 (4.9)	4.71 ± 4.08	4.22	4.44 (4.33–4.54)
	Anthracene	4.5			4.33	4.77 (4.67–4.87)
	Fluoranthene	5.1			4.93	5.52 (5.41–5.62)
	Pyrene	5			5.1	5.57 (5.45–5.67)
	Chrysene	5.86			5.78	6.39 (6.27–6.50)
	Benzo[*a*]pyrene	6.35			6.75	7.17 (7.03–7.30)
	Benz[*a*]anthracene				5.73	
	Dibenz[*a*,*h*]anthracene	6.75				7.87 (7.72–8.00)
	Benz[*k*]fluoranthene				6.66	
	Benzo[*ghi*]perylene	6.9			7.27	7.61 (7.46–7.75)
	DDT	6.79 (6.36)	5.9	4.99 ± 4.31	5.59	
	DDE				5.77	
	Methoxychlor				4.39	
	Heptachlor				5.22	
	Aldrin				4.71	
	Dieldrin				4.75	
	α-Hexachlorohexane	3.8			2.8	2.41 (2.36–2.46)
	β-Hexachlorohexane	3.81				2.04 (1.99–2.09)
	γ-Hexachlorohexane	3.55			2.2	2.33 (2.28–2.38)
	δ-Hexachlorohexane	4.14				2.08 (2.03–2.12)
	Perfluorooctanoic acid		3	2.70 ± 2.55		
	Di-2-ethylhexyl phthalate	7.5	2.4	4.99 ± 4.41		
	PFOS			1.52		
	FOSA			2.47		

Continued

Table 7.4 Log distribution coefficient (K_{ow}, K_f, K_d, $K_{polymer\text{-}W}$, and $K_{polymer\text{-}SW}$) of various contaminants on different types of plastic polymers (Ziccardi et al., 2016)—cont'd

		log K_{ow}	log K_f	log K_d	log $K_{polymer\text{-}W}$	log $K_{polymer\text{-}SW}$
PP	PCBs			4.98–5.45		
	Pentachlorobenzene	5.17				4.50 (4.39–4.59)
	Hexachlorobenzene	5.31				5.01 (4.89–5.10)
	Phenanthrene	4.6	3.33 ± 0.01	3.34 ± 2.23		4.00 (3.88–4.11)
	Anthracene	4.5				4.29 (4.18–4.38)
	Fluoranthene	5.1				4.79 (4.67–4.90)
	Pyrene	5				4.80 (4.68–4.90)
	Chrysene	5.86				5.51 (5.40–5.61)
	Benzo[*a*]pyrene	6.35				6.10 (5.97–6.20)
	Dibenz[*a,h*]anthracene	6.75				7.00 (6.89–7.10)
	Benzo[*ghi*]perylene	6.9				6.69 (6.56–6.81)
	DDE			5.44		
	α-Hexachlorohexane	3.8				2.69 (2.64–2.75)
	β-Hexachlorohexane	3.81				2.18 (2.08–2.28)
	γ-Hexachlorohexane	3.55				2.58 (2.52–2.64)
	δ-Hexachlorohexane	4.14				2.23 (2.13–2.34)
	Nonylphenol			4.92		
PS	Pentachlorobenzene	5.17				5.10 (4.99–5.20)
	Hexachlorobenzene	5.31				5.28 (5.17–5.38)
	Phenanthrene	4.6				5.39 (5.27–5.49)
	Anthracene	4.5				5.61 (5.51–5.70)
	Pyrene	5				5.84 (5.71–5.96)
	Fluoranthene	5.1				5.91 (5.79–6.01)
	Chrysene	5.86				6.63 (6.52–6.72)
	Benzo[*a*]pyrene	6.35				6.92 (6.80–7.02)
	Dibenz[*a,h*]anthracene	6.75				7.52 (7.41–7.61)
	Benzo[*ghi*]perylene	6.9				7.15 (7.03–7.24)
	α-Hexachlorohexane	3.8				3.19 (3.15–3.23)
	β-Hexachlorohexane	3.81				2.63 (2.59–2.67)

	γ-Hexachlorohexane	3.55				3.01 (2.97–3.05)
	δ-Hexachlorohexane	4.14				2.80 (2.75–2.85)
	FOSA			1.93		
PVC	Phenanthrene	4.6	3.3	3.36 ± 2.81		
	DDT	6.79	5.4	5.02 ± 4.18		
	Perfluorooctanoic acid		2.2	0.85 ± 0.20		
	Di-2-ethylhexyl phthalate	7.5	4.5	4.08 ± 3.64		
	PFOS			2		
	FOSA			2.06		
$PVC_{200–250}$	Phenanthrene	4.6	3.00 ± 0.03	3.22 ± 2.30		
PVC_{130}	Phenanthrene		3.24 ± 0.03	3.23 ± 2.49		
uPVC	Phenanthrene at 10°C		3.998 ± 0.004			
	Phenanthrene at 20°C		4.141 ± 0.003			

[a] K_{ow} = octanol-water partition coefficients, data extracted from some papers (Bakir et al., 2014a,b; Beckingham and Ghosh, 2017; Lee et al., 2014; Mackay et al., 1993; Walker, 2001).
[b] K_f = Freundlich constants, data collected from Bakir et al. (2014a), Teuten et al. (2007), and Velzeboer et al. (2014).
[c] K_d = partition coefficient derived from linear equilibrium sorption model, data extracted from Mato et al. (2001), calculated according to K_d from Wang et al. (2015).
[d] $K_{polymer-W}$ = equilibrium partition coefficient between polymer and water, data extracted from Gouin et al. (2011).
[e] $K_{polymer-SW}$ = equilibrium partition coefficient between polymer and seawater, data extracted from Lee et al. (2014).

7.2.2.2 The Effects of Physical-Chemical Characteristics of Microplastics

Microplastics are known to have different types, colors, sizes, and constituents. The most common types of MPs in the marine environment are PE, PP, PS, and PVC (Endo et al., 2005). Hydrophobic organic chemicals were easily sorbed on MPs due to their hydrophobic nature (Takada, 2006). Polyethylene was also reported to absorb more chemicals than other plastics (Karapanagioti and Klontza, 2008; Mato et al., 2001, 2002; Teuten et al., 2007). For example, the sorption of perfluorooctane sulfonate (PFOS) and perfluorooctanesulfonamide (FOSA) was greater on PE and PVC than on PS (Wang et al., 2015), while the partition coefficient of Phe on PE was an order of magnitude higher than that on PP, followed by PVC (Teuten et al., 2007). Hence, the sorption of chemicals was related to the type of plastics (Lee et al., 2014; Rochman et al., 2013a).

Microplastic samples collected from the environment were generally in different colors, mainly black, white, aging, and colored ones based on the classifications by Endo et al. (2005). Previous studies showed that concentrations of chemicals (such as PCBs and PAHs) were much higher on black and aged plastic pellets than on white and colored ones (Antunes et al., 2013; Frias et al., 2013, 2010; Rios et al., 2007). For example, the highest concentrations of PAH and PCB congeners were detected on black plastic particles, while PCB congeners were the least abundant on white pellets (Frias et al., 2010). Black plastic pellets may have more additives such as polyurethane than white pellets, which would enhance the sorption level (Frias et al., 2013).

The size of plastic particles is also an important factor for chemical sorption. Plastic particles with small sizes possess larger surface areas for sorption compared with large particles, which hence smaller plastics sorb higher concentrations of chemicals than large ones (Lee et al., 2014). For example, NPs ($84.96 \pm 2.91\,m^2\,g^{-1}$) with the size of 50 nm had an order-of-magnitude-larger Brunauer-Emmett-Teller (BET) surface area than 10 μm MPs ($3.71 \pm 0.49\,m^2\,g^{-1}$) (Ma et al., 2016). In fact, the distribution coefficients of NPs ($K_d = 6.54 \times 10^5\,L\,kg^{-1}$) were substantially greater than that of MPs ($K_d = 1.69 \times 10^4\,L\,kg^{-1}$). Therefore, the particle size of plastics was crucial for organic pollutant sorption (Velzeboer et al., 2014).

Table 7.5 shows some physical characteristics and partition coefficients for sorption of Phe on various polymers (Teuten et al., 2007). Clearly, PE, PP, and PVC with the same sizes ranging from 200 to 250 μm possess different BET surface areas and distribution coefficients. The $\log K_d$ of PE ($\log K_d = 4.58 \pm 3.75$) is greater than those of PP ($\log K_d = 3.34 \pm 2.23$) and PVC ($\log K_d = 3.22 \pm 2.30$), obviously attributed to the larger surface area of PE than those of PP and PVC.

The four popular types of plastics (PE, PP, PVC, and PS) can be divided into two categories: rubbery plastics (PE and PP) and glassy plastics (PVC and PS) (Kurtz, 2004; Teuten et al., 2009; Wilkinson and Ryan, 1998). Some studies showed that rubbery plastics had higher affinity toward organic chemicals than glassy plastics (Guo et al., 2012; Rochman et al., 2013a; Wu et al., 2001). George and Thomas (2001) proposed that

Table 7.5 Physical characteristics and partition coefficients for phenanthrene partition on various polymers (Teuten et al., 2007)

Solid phase	Particle size (μm)	BET surface area ($m^2\ g^{-1}$)	$\log K_f$	$\log K_d$[a]	Texture
PE	200–250	4.37	4.60 ± 0.12	4.58 ± 3.75	Granular polymer
PP	200–250	1.56	3.33 ± 0.01	3.34 ± 2.23	Granular polymer
$PVC_{200-250}$[b]	200–250	n.d.[c]	3.00 ± 0.03	3.22 ± 2.30	Granular polymer
PVC_{130}[d]	127[e]	1.76	3.24 ± 0.03	3.23 ± 2.49	Granular polymer

[a]Calculated according to K_d from Teuten et al. (2007).
[b]$PVC_{200-250}$ = PVC with the size range of 200–250 μm.
[c]n.d., not determined.
[d]PVC_{130} = PVC with the size of 130 μm.
[e]Median particle size, determined by low-angle laser light scattering (LALLS).

glassy polymers have a more dense structure with little void space and present higher cohesive forces than rubbery polymers that exhibit greater mobility and flexibility. The structure of glassy polymers poses slower diffusion rates and lower sorption levels of chemicals in glassy polymers than in rubbery plastics (Teuten et al., 2009). In general, rubbery plastics, small particulate plastic particles, and black plastic pellets absorb more contaminants.

7.2.2.3 The Degree of Weathering/Aging Effect

As mentioned in the previous section, aged plastic increases its ability to sorb contaminants (Endo et al., 2005; Ogata et al., 2009; Rios et al., 2007). Plastic aging is a greater state of degradation, which is mainly due to long-term photooxidation, seawater corrosion, friction, or other processes (Brennecke et al., 2016; Eriksson and Burton, 2003). Prolonged soaking in the marine environment would increase the sorption time of chemicals to plastics, further increasing their sorption capacities (Antunes et al., 2013; Frias et al., 2013; Karapanagioti and Klontza, 2008; Mato et al., 2001). Photooxidation and friction substantially enhance plastic decomposition within MPs, thus increasing specific surface area. Pérez et al. (2010) indicated that aging process led to decreased molecular weights of plastic polymers, which indirectly affect the sorbent properties. Holmes et al. (2012) and Holmes et al. (2014) reported that the sorption of metals were visibly higher on beached pellets (aged/weathered plastic pellets) than on virgin pellets, likely due to the effects of aging process and organic matter sorbed on plastic surface. In addition, aged PVC fragments were found to sorb much larger amounts of Cu than virgin PS beads, possibly due to the higher surface area and polarity of PVC (Brennecke et al., 2016). In summary, the effects of weathering/aging on plastics would enhance the sorption capacity of contaminants due to increased specific surface areas of plastics, extended sorption time, and surface coating of organic matter.

7.2.2.4 The Salinity Influence

The $\log K_{PE\text{-}SW}$ and $\log K_{PE\text{-}W}$ of the same pollutant are a little different (Table 7.6). For most PAHs, $\log K_{PE\text{-}SW}$ is slightly greater than $\log K_{PE\text{-}W}$, while $\log K_{PE\text{-}SW}$ of hexachlorobenzene and α-hexachlorohexane are smaller than $\log K_{PE\text{-}W}$. Thus, it remains unclear whether salinity promotes or inhibits sorption processes. Several studies have reported the effects of salinity on sorption between plastics and contaminants (Adams et al., 2007; Bakir et al., 2014b; Velzeboer et al., 2014; Wang et al., 2015). Table 7.6 shows the log K_f of Phe and DDT sorption onto PVC and PE at a series of salinity gradients. The desorption rates in Milli-Q water (0) and in seawater (35) are presented in Table 7.6 (Bakir et al., 2014b). Salinity was 0, 8.8, 17.5, 26.3, and 35 practical salinity scale unit (psu), which is 0%, 25%, 50%, 75%, and 100% of that in seawater. There was no significant change in the $\log K_f$ of PVC-Phe, PE-Phe, PVC-DDT, and PE-DDT with increasing salinity. Their desorption rates in Milli-Q water are slightly inferior to those in seawater, while the desorption rate between plastic and DDT is distinctly less than that between plastic and Phe. Wang et al. (2015) observed that increased salinity enhanced the sorption capacity of PFOS on PE but exerted no evident impact on FOSA sorption. Their results further showed that PFOS cannot be sorbed on PS pellets in pure water, but can be sorbed in water with high salts. As for metals, increased salinity enhances the sorption of increased Cr but diminishes the sorption of Cd, Co, and Ni (Holmes et al., 2014). Apparently, the effects of salinity on sorption processes are variable with different chemicals.

7.2.2.5 The Impacts of pH and Temperature

Some reports about the effects of pH on chemical sorption in MPs have been examined lately. Wang et al. (2015) reported that PFOS sorption on PE and PS particles increased with decreasing pH, but no effect on FOSA was observed. Holmes et al. (2014) showed that the sorption dynamics of different metals on plastic pellets exhibited different patterns as the pH of river water increased.

Table 7.6 $\log K_f$ ($L\,kg^{-1}$) from the sorption of Phe and DDT onto PVC and PE for different salinities and desorption rate (k, $day^{-1} \pm SD$) in Milli-Q water (0) and in seawater (35) (Bakir et al., 2014b)

	Salinity range (PSU)	PVC-Phe[a]	PE-Phe	PVC-DDT	PE-DDT
$\log K_f$ ($L\,kg^{-1}$)	0 (Milli-Q)	3.3	4.7	5.8	5.6
	8.8	3.4	4.8	5.9	5.6
	17.5	3.3	4.9	5.5	5.6
	26.3	3.3	5.1	5.3	5.3
	35	3.3	4.9	5.4	5.9
Desorption rate	0 (Milli-Q)	0.73 ± 0.26	1.15 ± 0.12	0.21 ± 0.01	0.20 ± 0.04
(k, $day^{-1} \pm SD$)	35 (seawater)	0.88 ± 0.56	1.37 ± 0.45	0.26 ± 0.06	0.23 ± 0.08

[a]Phe = phenanthrene.

Table 7.7 The $\log K_{PE\text{-}W}$ of HOCs at different temperatures (Adams et al., 2007)

HOCs	$\log K_{PE\text{-}W}$[a] T = 16.4–18.7°C	$\log K_{PE\text{-}W}$[b] $T \sim 18$°C	$\log K_{PE\text{-}W}$ $T = 23 \pm 1$°C	$\log K_{PE\text{-}W}$ $T = 24 \pm 1$°C	$\log K_{PE\text{-}W}$[c] T = 30°C
Phenanthrene		4.2	4.23 ± 0.02	4.3 ± 0.1	4.16 ± 0.02
2-Methylphenanthrene				4.8 ± 0.2	
Fluoranthene	4.52			4.9 ± 0.1	4.75 ± 0.02
Pyrene	4.62		5.02 ± 0.03	5.0 ± 0.1	4.90 ± 0.01
Benz(*a*)anthracene				5.7 ± 0.1	
D12-benz(*a*)anthracene				5.7 ± 0.1	
Chrysene				5.7 ± 0.1	5.53 ± 0.02
Benzo(*e*)pyrene				6.2 ± 0.1	5.94 ± 0.04
Perylene				6.5 ± 0.2	
PCB-29				5.1 ± 0.1	
PCB-52		4.6	5.4 ± 0.1		5.55 ± 0.01
PCB-69				5.6 ± 0.2	
PCB-97				6.3 ± 0.1	
PCB-118					6.18 ± 0.05
PCB-143				6.8 ± 0.2	

[a]Data extracted and calculated from Müller et al. (2001).
[b]Data extracted from Hucklns et al. (1993).
[c]Data extracted and calculated from Booij et al. (2003).

Table 7.7 shows log $K_{PE\text{-}W}$ values of some chemicals at different temperatures. The $\log K_{PE\text{-}W}$ of Phe increases with increasing temperature from 18°C to 24°C (4.2 at 18°C, 4.23 ± 0.02 at 23°C, and 4.3 ± 0.1 at 24°C) but was slightly lower at 30°C (4.16 ± 0.02) than at 24°C. Pyrene presents a similar trend as Phe with a $\log K_{PE\text{-}W}$ value of 4.62 at 16.4–18.7°C, which is smaller than those at 23°C (5.02 ± 0.03) and 24°C (5.0 ± 0.1). However, it slightly decreased when temperature reached 30°C (4.90 ± 0.01). For most HOCs (e.g., Phe, fluoranthene, pyrene, chrysene, and benzo(*e*)pyrene), $\log K_{PE\text{-}W}$ values at 30°C were less than those at 24°C. PCB-52 is an exception, with $\log K_{PE\text{-}W}$ at 30°C being slightly higher than that at 23°C. The $\log K_{PE\text{-}W}$ of PAH isomers at the same temperature (24°C) follows the sequence of Phe < 2-methyl phenanthrene < fluoranthene < pyrene < benz(*a*)anthracene = d12-benz(*a*)anthracene = chrysene < benzo(*e*)pyrene < perylene. In addition, the $\log K_{PE\text{-}W}$ of PCB congeners at 24°C is in the sequence of PCB-29 < PCB-69 < PCB-97 < PCB-143 (Table 7.7). Clearly, plastic polymers tend to affiliate with high-ring PAHs and high-chlorinated PCBs relative to low-ring PAHs and low-chlorinated PCBs.

7.3 UPTAKE OF MPs IN MARINE ORGANISMS

At present, MPs have been found in a large number of marine organisms (Gall and Thompson, 2015), which mistake and swallow tiny MPs as food, and are transferred through the food chain (Cole et al., 2013; Laist, 1997; Lusher et al., 2013). It was estimated that 693 species in 2015 were affected by marine plastic pollution while 267 species were recorded to have marine plastic debris in their bodies (Gall and Thompson, 2015; Laist, 1997). For example, MPs were identified in the guts, the stomach, and tissues of marine organisms such as seabirds (Blight and Burger, 1997), fishes (Collard et al., 2015; Miranda and de Carvalho-Souza, 2016), turtles (Bjorndal et al., 1994), whales (Fossi et al., 2012; Tarpley and Marwitz, 1993), and crabs (Watts et al., 2016) and even the alimentary tract of gudgeons (*Gobio gobio*) (Sanchez et al., 2014). Table 7.8 shows the spatial occurrence of MPs in several aquatic organisms from different areas. Bivalves contained MPs in the range of 4.3–57.2 items/individual, while the mean number of MPs in *Scapharca subcrenata* was up to 10.5 items g^{-1} from a Chinese fishery market (Li et al., 2015). In addition, the number of plastic fibers was up to 178 items/individual in *Mytilus edulis* from Canada (Mathalon and Hill, 2014), which was much higher than that (0.18 items/individual) in demersal fish from the North Sea and Baltic Sea (Rummel et al., 2016).

The sorption of chemicals on MPs may facilitate transport of harmful chemicals such as PAHs, PCBs, DDTs, HCHs, and metal into the food chain and cause strong ecological effects (Bakir et al., 2012; Carpenter et al., 1972; Koelmans et al., 2016; Mato et al., 2001;

Table 7.8 Microplastic pollution in several aquatic organisms from different areas in the present study

Species and sources	Types of microplastics	Levels of microplastics	References
Fish, the North Pacific		2.1 items/individual	Moore et al. (2001)
Oyster Larvae, Pacific	Particles	10.5 items/ individual	Cole and Galloway (2015)
Planktivorous fishes, North Pacific Central Gyre	Plastic pieces	2.1 ± 5.78 items/ individual	Boerger et al. (2010)
Pelagic and demersal fish, the English Channel	Fibers, fragments, and beads	1.90 ± 0.10 items/ individual	Lusher et al. (2013)
Commercial bivalves, China	Fibers, fragments, and pellets	4.3–57.2 items/ individual	Li et al. (2015)
Mytilus edulis, Canada	Fibers	34–178 items/ individual	Mathalon and Hill (2014)
Commercial fish, Portuguese coast	Fibers and fragments	0.27 ± 0.63 items/ individual	Neves et al. (2015)
Pelagic and demersal fish, the North Sea and Baltic Sea	Particles	0.03 ± 0.18 items/ individual 0.19 ± 0.61 items/ individual	Rummel et al. (2016)
Adult *Crassostrea gigas*, the Atlantic Northeast	Particles	0.47 ± 0.16 items g^{-1}	Van Cauwenberghe and Janssen (2014)

Rochman et al., 2013b). Similarly, desorption of chemicals from MPs in the stomach or gut of marine organisms is also harmful to biotic community. Bakir et al. (2014a) demonstrated that desorption processes can be 30 times faster in intestinal fluids than in seawater. The solubility and bioavailability of chemicals were reported to be higher in *Arenicola marina* digestive solutions than in seawater (Voparil and Mayer, 2000). The desorption of chemicals from MPs and the release of additives in MPs are both potentially harmful to marine organisms.

7.4 CONCLUSIONS

Microplastics have larger specific surface areas than large-sized plastics, so they may sorb more organic chemicals from aquatic system. The sorption of chemicals on MPs may facilitate harmful chemicals into the food chain. Numerous chemicals have been found in MP samples, including trace metals and organic chemicals. Sorption of organic chemicals on MPs is positively correlated with the K_{ow} of chemicals and weathering degree of MPs, that is, sorption of organic chemicals is driven by partitioning. The same chemicals have different partition coefficients on MPs with different materials. The solution properties (such as salinity and pH) and temperature are important parameters for the sorption of chemicals on MPs. Finally, MPs may act as a carrier to transfer chemicals into the body of biota.

REFERENCES

Adams, R.G., Lohmann, R., Fernandez, L.A., Macfarlane, J.K., Gschwend, P.M., 2007. Polyethylene devices: passive samplers for measuring dissolved hydrophobic organic compounds in aquatic environments. Environ. Sci. Technol. 41, 1317–1323.

Akhbarizadeh, R., Moore, F., Keshavarzi, B., Moeinpour, A., 2017. Microplastics and potentially toxic elements in coastal sediments of Iran's main oil terminal (Khark Island). Environ. Pollut. 220, 720–731.

Andrady, A.L., 2003. Plastics and the environment. Paper Film Foil Converter 51, 23–30.

Andrady, A.L., 2011. Microplastics in the marine environment. Mar. Pollut. Bull. 62, 1596–1605.

Antunes, J.C., Frias, J.G.L., Micaelo, A.C., Sobral, P., 2013. Resin pellets from beaches of the Portuguese coast and adsorbed persistent organic pollutants. Estuar. Coast. Shelf Sci. 130, 62–69.

Ashton, K., Holmes, L., Turner, A., 2010. Association of metals with plastic production pellets in the marine environment. Mar. Pollut. Bull. 60, 2050–2055.

Azzarello, M.Y., Van Vleet, E.S., 1987. Marine birds and plastic pollution. Mar. Ecol. Prog. Ser. 37, 295–303.

Bakir, A., Rowland, S.J., Thompson, R.C., 2012. Competitive sorption of persistent organic pollutants onto microplastics in the marine environment. Mar. Pollut. Bull. 64, 2782–2789.

Bakir, A., Rowland, S.J., Thompson, R.C., 2014a. Enhanced desorption of persistent organic pollutants from microplastics under simulated physiological conditions. Environ. Pollut. 185, 16–23.

Bakir, A., Rowland, S.J., Thompson, R.C., 2014b. Transport of persistent organic pollutants by microplastics in estuarine conditions. Estuar. Coast. Shelf Sci. 140, 14–21.

Barnes, D.K.A., 2002. Biodiversity: invasions by marine life on plastic debris. Nature 416, 808–809.

Barnes, D.K., Galgani, F., Thompson, R.C., Barlaz, M., 2009. Accumulation and fragmentation of plastic debris in global environments. Philos. Trans. R. Soc. B 364, 1985–1998.

Beckingham, B., Ghosh, U., 2017. Differential bioavailability of polychlorinated biphenyls associated with environmental particles: microplastic in comparison to wood, coal and biochar. Environ. Pollut. 220, 150–158.

Bjorndal, K.A., Bolten, A.B., Lagueux, C.J., 1994. Ingestion of marine debris by juvenile sea turtles in coastal Florida habitats. Mar. Pollut. Bull. 28, 154–158.

Blight, L.K., Burger, A.E., 1997. Occurrence of plastic particles in seabirds from the eastern North Pacific. Mar. Pollut. Bull. 34, 323–325.

Blumer, M., 1976. Polycyclic aromatic compounds in nature. Sci. Am. 234, 35–45.

Bockhorn, H., Hornung, A., Hornung, U., Schawaller, D., 1999. Kinetic study on the thermal degradation of polypropylene and polyethylene. J. Anal. Appl. Pyrolysis 48, 93–109.

Boerger, C.M., Lattin, G.L., Moore, S.L., Moore, C.J., 2010. Plastic ingestion by planktivorous fishes in the North Pacific Central Gyre. Mar. Pollut. Bull. 60, 2275–2278.

Booij, K., Hofmans, H.E., Fischer, C.V., Van Weerlee, E.M., 2003. Temperature-dependent uptake rates of nonpolar organic compounds by semipermeable membrane devices and low-density polyethylene membranes. Environ. Sci. Technol. 37, 361–366.

Bowmer, T., Kershaw, P., 2010. Proceedings of GESAMP International Workshop on Microplastic Particles as a Vector in Transporting Persistent, Bioaccumulating and Toxic Substances in the Ocean. GESAMP Rep. Stud. 82, 68.

Brennecke, D., Duarte, B., Paiva, F., Caçador, I., Canning-Clode, J., 2016. Microplastics as vector for heavy metal contamination from the marine environment. Estuar. Coast. Shelf Sci. 178, 189–195.

Browne, M.A., Galloway, T., Thompson, R., 2007. Microplastic—an emerging contaminant of potential concern? Integr. Environ. Assess. Manag. 3, 559–566.

Browne, M.A., Crump, P., Niven, S.J., Teuten, E., Tonkin, A., Galloway, T., Thompson, R., 2011. Accumulation of microplastic on shorelines worldwide: sources and sinks. Environ. Sci. Technol. 45, 9175–9179.

Browne, M.A., Niven, S.J., Galloway, T.S., Rowland, S.J., Thompson, R.C., 2013. Microplastic moves pollutants and additives to worms, reducing functions linked to health and biodiversity. Curr. Biol. 23, 2388–2392.

Budzinski, H., Jones, I., Bellocq, J., Pierard, C., Garrigues, P., 1997. Evaluation of sediment contamination by polycyclic aromatic hydrocarbons in the Gironde estuary. Mar. Chem. 58, 85–97.

Carpenter, E.J., Smith, J.K.L., 1972. Plastics on the Sargasso sea surface. Science 175, 1240–1241.

Carpenter, E.J., Anderson, S.J., Harvey, G.R., Miklas, H.P., Peck, B.B., 1972. Polystyrene spherules in coastal waters. Science 178, 749–750.

Choi, K.I., Lee, S.H., Osako, M., 2009. Leaching of brominated flame retardants from TV housing plastics in the presence of dissolved humic matter. Chemosphere 74, 460–466.

Claessens, M., De Meester, S., Van Landuyt, L., De Clerck, K., Janssen, C.R., 2011. Occurrence and distribution of microplastics in marine sediments along the Belgian coast. Mar. Pollut. Bull. 62, 2199–2204.

Cole, M., Galloway, T.S., 2015. Ingestion of nanoplastics and microplastics by Pacific Oyster Larvae. Environ. Sci. Technol. 49, 14625–14632.

Cole, M., Lindeque, P., Halsband, C., Galloway, T.S., 2011. Microplastics as contaminants in the marine environment: a review. Mar. Pollut. Bull. 62, 2588–2597.

Cole, M., Lindeque, P., Fileman, E., Halsband, C., Goodhead, R., Moger, J., Galloway, T.S., 2013. Microplastic ingestion by zooplankton. Environ. Sci. Technol. 47, 6646–6655.

Collard, F., Gilbert, B., Eppe, G., Parmentier, E., Das, K., 2015. Detection of anthropogenic particles in fish stomachs: an isolation method adapted to identification by Raman spectroscopy. Arch. Environ. Contam. Toxicol. 69, 331–339.

Cozar, A., Echevarria, F., Gonzalez-Gordillo, J.I., Irigoien, X., Ubeda, B., Hernandez-Leon, S., Palma, A.T., Navarro, S., Garcia-de-Lomas, J., Ruiz, A., Fernandez-de-Puelles, M.L., Duarte, C.M., 2014. Plastic debris in the open ocean. PNAS 111, 10239–10244.

Darnerud, P.O., 2008. Brominated flame retardants as possible endocrine disrupters. Int. J. Androl. 31, 152–160.

de Lucia, G.A., Caliani, I., Marra, S., Camedda, A., Coppa, S., Alcaro, L., Campani, T., Giannetti, M., Coppola, D., Cicero, A.M., Panti, C., Baini, M., Guerranti, C., Mars, L., Massaro, G., Fossi, M.C., Matiddi, M., 2014. Amount and distribution of neustonic micro-plastic off the western Sardinian coast (Central-Western Mediterranean Sea). Mar. Environ. Res. 100, 10–16.

de Wit, C.A., 2002. An overview of brominated flame retardants in the environment. Chemosphere 46, 583–624.

Derraik, J.G.B., 2002. The pollution of the marine environment by plastic debris: a review. Mar. Pollut. Bull. 44, 842–852.

Endo, S., Takizawa, R., Okuda, K., Takada, H., Chiba, K., Kanehiro, H., Ogi, H., Yamashita, R., Date, T., 2005. Concentration of polychlorinated biphenyls (PCBs) in beached resin pellets: variability among individual particles and regional differences. Mar. Pollut. Bull. 50, 1103–1114.

Eriksen, M., Lebreton, L.C.M., Carson, H.S., Thiel, M., Moore, C.J., Borerro, J.C., Galgani, F., Ryan, P.G., Reisser, J., 2014. Plastic pollution in the world's oceans: more than 5 trillion plastic pieces weighing over 250,000 tons Afloat at Sea. PLoS One 9, e111913.

Eriksson, C., Burton, H., 2003. Origins and biological accumulation of small plastic particles in fur seals from Macquarie Island. Ambio 32, 380–384.

Fossi, M.C., Panti, C., Guerranti, C., Coppola, D., Giannetti, M., Marsili, L., Minutoli, R., 2012. Are baleen whales exposed to the threat of microplastics? A case study of the Mediterranean fin whale (*Balaenoptera physalus*). Mar. Pollut. Bull. 64, 2374–2379.

Frias, J.P.G.L., Sobral, P., Ferreira, A.M., 2010. Organic pollutants in microplastics from two beaches of the Portuguese coast. Mar. Pollut. Bull. 60, 1988–1992.

Frias, J.P.G.L., Antunes, J.C., Sobral, P., 2013. Local marine litter survey—a case study in Alcobaça municipality, Portugal. J. Integr. Coast. Zone Manag. 13, 169–179.

Fries, E., Dekiff, J.H., Willmeyer, J., Nuelle, M.T., Ebert, M., Remy, D., 2013. Identification of polymer types and additives in marine microplastic particles using pyrolysis-GC/MS and scanning electron microscopy. Environ. Sci. Process. Impacts 15, 1949–1956.

Fry, D.M., Toone, C.K., 1981. DDT-induced feminization of gull embryos. Science 213, 922–924.

Gall, S.C., Thompson, R.C., 2015. The impact of debris on marine life. Mar. Pollut. Bull. 92, 170–179.

George, S.C., Thomas, S., 2001. Transport phenomena through polymeric systems. Prog. Polym. Sci. 26, 985–1017.

Goldberg, E.D., 1994. Diamonds and plastics are forever? Mar. Pollut. Bull. 28, 466.

Gouin, T., Roche, N., Lohmann, R., Hodges, G., 2011. A thermodynamic approach for assessing the environmental exposure of chemicals absorbed to microplastic. Environ. Sci. Technol. 45, 1466–1472.

Graca, B., Beldowska, M., Wrzesien, P., Zgrundo, A., 2014. Styrofoam debris as a potential carrier of mercury within ecosystems. Environ. Sci. Pollut. Res. 21, 2263–2271.

Gregory, M.R., Ryan, P.G., 1997. Pelagic plastics and other seaborne persistent synthetic debris: a review of southern hemisphere perspectives. In: Coe, J.M., Rogers, D.B. (Eds.), Marine Debris: Sources, Impacts, and Solutions. Springer, New York, pp. 49–66.

Guo, X.Y., Wang, X., Zhou, X.Z., Kong, X.Z., Tao, S., Xing, B.H., 2012. Sorption of four hydrophobic organic compounds by three chemically distinct polymers: the role of chemical and physical composition. Environ. Sci. Technol. 46, 7252.

Hidalgo-Ruz, V., Gutow, L., Thompson, R.C., Thiel, M., 2012. Microplastics in the marine environment: a review of the methods used for identification and quantification. Environ. Sci. Technol. 46, 3060–3075.

Hirai, H., Takada, H., Ogata, Y., Yamashita, R., Mizukawa, K., Saha, M., Kwan, C., Moore, C., Gray, H., Laursen, D., Zettler, E.R., Farrington, J.W., Reddy, C.M., Peacock, E.E., Ward, M.W., 2011. Organic micropollutants in marine plastics debris from the open ocean and remote and urban beaches. Mar. Pollut. Bull. 62, 1683–1692.

Holmes, L.A., Turner, A., Thompson, R.C., 2012. Adsorption of trace metals to plastic resin pellets in the marine environment. Environ. Pollut. 160, 42–48.

Holmes, L.A., Turner, A., Thompson, R.C., 2014. Interactions between trace metals and plastic production pellets under estuarine conditions. Mar. Chem. 167, 25–32.

Hucklns, J.N., Manuweera, G.K., Petty, J.D., Mackay, D., Lebot, J.A., 1993. Lipid-containing semipermeable membrane devices for monitoring organic contaminants in water. Environ. Sci. Technol. 27, 2489–2496.

Karapanagioti, H.K., Klontza, I., 2008. Testing phenanthrene distribution properties of virgin plastic pellets and plastic eroded pellets found on Lesvos island beaches (Greece). Mar. Environ. Res. 65, 283–290.

Karapanagioti, H.K., Endo, S., Ogata, Y., Takada, H., 2011. Diffuse pollution by persistent organic pollutants as measured in plastic pellets sampled from various beaches in Greece. Mar. Pollut. Bull. 62, 312–317.

Klein, S., Worch, E., Knepper, T.P., 2015. Occurrence and spatial distribution of microplastics in river shore sediments of the Rhine-Main area in Germany. Environ. Sci. Technol. 49, 6070–6076.

Koelmans, A.A., Bakir, A., Burton, G.A., Janssen, C.R., 2016. Microplastic as a vector for chemicals in the aquatic environment: critical review and model-supported reinterpretation of empirical studies. Environ. Sci. Technol. 50, 3315–3326.

Kurtz, S.M., 2004. The UHMWPE handbook: ultra-high molecular weight polyethylene in total joint replacement. J. Bone Joint Surg. 87, 1906.

Laist, D.W., 1987. Overview of the biological effects of lost and discarded plastic debris in the marine environment. Mar. Pollut. Bull. 18, 319–326.

Laist, D.W., 1997. Impacts of marine debris: entanglement of marine life in marine debris including a comprehensive list of species with entanglement and ingestion records. In: Coe, J.M., Rogers, D.B. (Eds.), Marine Debris: Sources, Impacts, and Solutions. Springer, Berlin.

Lee, H., Shim, W.J., Kwon, J.H., 2014. Sorption capacity of plastic debris for hydrophobic organic chemicals. Sci. Total Environ. 470–471, 1545–1552.

Li, J., Yang, D., Li, L., Jabeen, K., Shi, H., 2015. Microplastics in commercial bivalves from China. Environ. Pollut. 207, 190–195.

Lithner, D., Damberg, J., Dave, G., Larsson, K., 2009. Leachates from plastic consumer products-screening for toxicity with *Daphnia magna*. Chemosphere 74, 1195–1200.

Lithner, D., Larsson, A., Dave, G., 2011. Environmental and health hazard ranking and assessment of plastic polymers based on chemical composition. Sci. Total Environ. 409, 3309–3324.

Lusher, A.L., McHugh, M., Thompson, R.C., 2013. Occurrence of microplastics in the gastrointestinal tract of pelagic and demersal fish from the English Channel. Mar. Pollut. Bull. 67, 94–99.

Ma, Y., Huang, A., Cao, S., Sun, F., Wang, L., Guo, H., Ji, R., 2016. Effects of nanoplastics and microplastics on toxicity, bioaccumulation, and environmental fate of phenanthrene in fresh water. Environ. Pollut. 219, 166–173.

Mackay, D., Shiu, W.Y., Ma, K.C., 1993. Volatile Organic Chemicals. Illustrated Handbook of Physical-Chemical Properties and Environmental Fate for Organic Chemicals. Lewis, Boca Raton, p. 916.

Mathalon, A., Hill, P., 2014. Microplastic fibers in the intertidal ecosystem surrounding Halifax Harbor, Nova Scotia. Mar. Pollut. Bull. 81, 69–79.

Mato, Y., Isobe, T., Takada, H., Kanehiro, H., Ohtake, C., Kaminuma, T., 2001. Plastic resin pellets as a transport medium for toxic chemicals in the marine environment. Environ. Sci. Technol. 35, 318–324.

Mato, Y., Takada, H., Zakaria, M.P., Kuriyama, Y., Kanehiro, H., 2002. Toxic chemicals contained in plastic resin pellets in the marine environment-spatial difference in pollutant concentrations and the effects of resin type. Environ. Sci. 15, 415–423.

Miller, R.W., 1985. Congenital PCB poisoning: a reevaluation. Environ. Health Perspect. 60, 211–214.

Miranda, D.A., de Carvalho-Souza, G.F., 2016. Are we eating plastic-ingesting fish? Mar. Pollut. Bull. 103, 109–114.

Mizukawa, K., Takada, H., Ito, M., Geok, Y.B., Hosoda, J., Yamashita, R., Saha, M., Suzuki, S., Miguez, C., Frias, J., Antunes, J.C., Sobral, P., Santos, I., Micaelo, C., Ferreira, A.M., 2013. Monitoring of a wide range of organic micropollutants on the Portuguese coast using plastic resin pellets. Mar. Pollut. Bull. 70, 296–302.

Moore, C.J., 2008. Synthetic polymers in the marine environment: a rapidly increasing, long-term threat. Environ. Res. 108, 131–139.

Moore, C.J., Moore, S.L., Leecaster, M.K., Weisberg, S.B., 2001. A comparison of plastic and plankton in the North Pacific Central Gyre. Mar. Pollut. Bull. 42, 1297–1300.

Müller, J.F., Manomanii, K., Mortimer, M.R., Mclachlan, M.S., 2001. Partitioning of polycyclic aromatic hydrocarbons in the polyethylene/water system. Anal. Bioanal. Chem. 371, 816–822.

Neves, D., Sobral, P., Ferreira, J.L., Pereira, T., 2015. Ingestion of microplastics by commercial fish off the Portuguese coast. Mar. Pollut. Bull. 101, 119–126.

Ng, K.L., Obbard, J.P., 2006. Prevalence of microplastics in Singapore's coastal marine environment. Mar. Pollut. Bull. 52, 761–767.

Oehlmann, J., Schulte-Oehlmann, U., Kloas, W., Jagnytsch, O., Lutz, I., Kusk, K.O., Wollenberger, L., Santos, E.M., Paull, G.C., Van Look, K.J., Tyler, C.R., 2009. A critical analysis of the biological impacts of plasticizers on wildlife. Philos. Trans. R. Soc. B 364, 2047–2062.

Ogata, Y., Takada, H., Mizukawa, K., Hirai, H., Iwasa, S., Endo, S., Mato, Y., Saha, M., Okuda, K., Nakashima, A., Murakami, M., Zurcher, N., Booyatumanondo, R., Zakaria, M.P., Dung, L.Q., Gordon, M., Miguez, C., Suzuki, S., Moore, C., Karapanagioti, H.K., Weerts, S., McClurg, T., Burres, E., Smith, W., Velkenburg, M.V., Lang, J.S., Lang, R.C., Laursen, D., Danner, B., Stewardson, N., Thompson, R.C., 2009. International Pellet Watch: global monitoring of persistent organic pollutants (POPs) in coastal waters. 1. Initial phase data on PCBs, DDTs, and HCHs. Mar. Pollut. Bull. 58, 1437–1446.

Pascall, M.A., Zabik, M.E., Zabik, M.J., Hernandez, R.J., 2005. Uptake of polychlorinated biphenyls (PCBs) from an aqueous medium by polyethylene, polyvinyl chloride, and polystyrene films. J. Agric. Food Chem. 53, 164–169.

Pérez, J.M., Vilas, J.L., Laza, J.M., Arnáiz, S., Mijangos, F., Bilbao, E., Rodríguez, M., León, L.M., 2010. Effect of reprocessing and accelerated ageing on thermal and mechanical polycarbonate properties. J. Mater. Process. Technol. 210, 727–733.

Pruter, A.T., 1987. Sources, quantities and distribution of persistent plastics in the marine environment. Mar. Pollut. Bull. 18, 305–310.

Ravindra, K., Sokhi, R., Vangrieken, R., 2008. Atmospheric polycyclic aromatic hydrocarbons: source attribution, emission factors and regulation. Atmos. Environ. 42, 2895–2921.

Reddy, M.S., Shaik, B., Adimurthy, S., Ramachandraiah, G., 2006. Description of the small plastics fragments in marine sediments along the Alang-Sosiya ship-breaking yard, India. Estuar. Coast. Shelf Sci. 68, 656–660.

Rios Mendoza, L.M., Jones, P.R., 2015. Characterisation of microplastics and toxic chemicals extracted from microplastic samples from the North Pacific Gyre. Environ. Chem. 12, 611–617.

Rios, L.M., Moore, C., Jones, P.R., 2007. Persistent organic pollutants carried by synthetic polymers in the ocean environment. Mar. Pollut. Bull. 54, 1230–1237.

Rios, L.M., Jones, P.R., Moore, C., Narayan, U.V., 2010. Quantitation of persistent organic pollutants adsorbed on plastic debris from the Northern Pacific Gyre's "eastern garbage patch" J. Environ. Monit. 12, 2226–2236.

Rochman, C.M., Hoh, E., Hentschel, B.T., Kaye, S., 2013a. Long-term field measurement of sorption of organic contaminants to five types of plastic pellets: implications for plastic marine debris. Environ. Sci. Technol. 47, 1646–1654.

Rochman, C.M., Hoh, E., Kurobe, T., Teh, S.J., 2013b. Ingested plastic transfers hazardous chemicals to fish and induces hepatic stress. Sci. Rep. 3, 3263–3269.

Rochman, C.M., Hentschel, B.T., Teh, S.J., 2014a. Long-term sorption of metals is similar among plastic types: implications for plastic debris in aquatic environments. PLoS One 9, e85433.

Rochman, C.M., Kurobe, T., Flores, I., Teh, S.J., 2014b. Early warning signs of endocrine disruption in adult fish from the ingestion of polyethylene with and without sorbed chemical pollutants from the marine environment. Sci. Total Environ. 493, 656–661.

Rochman, C.M., Lewison, R.L., Eriksen, M., Allen, H., Cook, A.M., Teh, S.J., 2014c. Polybrominated diphenyl ethers (PBDEs) in fish tissue may be an indicator of plastic contamination in marine habitats. Sci. Total Environ. 476–477, 622–633.

Rockstrom, J., Steffen, W., Noone, K., Persson, A., 2009. A safe operating space for humanity. Nature 461, 472–475.

Rothstein, S.I., 1973. Plastic particle pollution of the surface of the Atlantic Ocean: evidence from a seabird. Condor 75, 344–345.

Roy, P.K., Hakkarainen, M., Varma, I.K., Albertsson, A.C., 2011. Degradable polyethylene: fantasy or reality. Environ. Sci. Technol. 45, 4217–4227.

Rummel, C.D., Loder, M.G., Fricke, N.F., Lang, T., Griebeler, E.M., Janke, M., Gerdts, G., 2016. Plastic ingestion by pelagic and demersal fish from the North Sea and Baltic Sea. Mar. Pollut. Bull. 102, 134–141.

Sanchez, W., Bender, C., Porcher, J.M., 2014. Wild gudgeons (*Gobio gobio*) from French rivers are contaminated by microplastics: preliminary study and first evidence. Environ. Res. 128, 98–100.

Sonnenschein, C., Soto, A.M., 1998. An updated review of environmental estrogen and androgen mimics and antagonists. J. Steroid Biochem. 65, 143–150.

Soto, A.M., Justicia, H., Wray, J.W., Sonnenschein, C., 1991. p-Nonyl-phenol: an estrogenic xenobiotic released from "modified" polystyrene. Environ. Health Perspect. 92, 167–173.

Takada, H., 2006. Call for pellets! International Pellet Watch global monitoring of POPs using beached plastic resin pellets. Mar. Pollut. Bull. 52, 1547–1548.
Takada, H., Mato, Y., Endo, S., Yamashita, R., Zakaria, M.P., 2005. Moore, C., David, S. (Eds.), Pellet Watch: global monitoring of persistent organic pollutants (POPs) using beached plastic resin pellets. The Plastic Debris Rivers to Sea Conference: Focusing on the Land-Based Sources of Marine Debris, Redondo Beach, CA, pp. 1–16.
Talsness, C.E., 2008. Overview of toxicological aspects of polybrominated diphenyl ethers: a flame-retardant additive in several consumer products. Environ. Res. 108, 158–167.
Talsness, C.E., Andrade, A.J., Kuriyama, S.N., Taylor, J.A., vom Saal, F.S., 2009. Components of plastic: experimental studies in animals and relevance for human health. Philos. Trans. R. Soc. B 364, 2079–2096.
Tanaka, K., Takada, H., Yamashita, R., Mizukawa, K., Fukuwaka, M.A., Watanuki, Y., 2013. Accumulation of plastic-derived chemicals in tissues of seabirds ingesting marine plastics. Mar. Pollut. Bull. 69, 219–222.
Tarpley, R.J., Marwitz, S., 1993. Plastic debris ingestion of cetaceans along the Texas coast: two case reports. Aquat. Mamm. 19, 93–98.
Teuten, E.L., Rowland, S.J., Galloway, T.S., Thompson, R.C., 2007. Potential for plastics to transport hydrophobic contaminants. Environ. Sci. Technol. 41, 7759–7764.
Teuten, E.L., Saquing, J.M., Knappe, D.R.U., Barlaz, M.A., Jonsson, S., Bjorn, A., Rowland, S.J., Thompson, R.C., Galloway, T.S., Yamashita, R., Ochi, D., Watanuki, Y., Moore, C., Viet, P.H., Tana, T.S., Prudente, M., Boonyatumanond, R., Zakaria, M.P., Akkhavong, K., Ogata, Y., Hirai, H., Iwasa, S., Mizukawa, K., Hagino, Y., Imamura, A., Saha, M., Takada, H., 2009. Transport and release of chemicals from plastics to the environment and to wildlife. Philos. Trans. R. Soc. B 364, 2027–2045.
Thompson, R.C., 2007. Krause, J.C., von Nordheim, H., Brager, S. (Eds.), Plastic debris in the marine environment: consequences and solutions. Marine Nature Conservation in Europe 2006. Bundesamt für Naturschutz, Bonn, pp. 107–115.
Thompson, R.C., Olsen, Y., Mitchell, R.P., Davis, A., Rowland, S.J., John, A.W.G., McGonigle, D., Russell, A.E., 2004. Lost at sea: where is all the plastic. Science 304, 838.
Thompson, R.C., Swan, S.H., Moore, C.J., vom Saal, F.S., 2009. Our plastic age. Philos. Trans. R. Soc. B 364, 1973–1976.
Turner, A., Holmes, L.A., 2015. Adsorption of trace metals by microplastic pellets in fresh water. Environ. Chem. 12, 600–610.
Van Cauwenberghe, L., Janssen, C.R., 2014. Microplastics in bivalves cultured for human consumption. Environ. Pollut. 193, 65–70.
Van Cauwenberghe, L., Vanreusel, A., Mees, J., Janssen, C.R., 2013. Microplastic pollution in deep-sea sediments. Environ. Pollut. 182, 495–499.
Van, A., Rochman, C.M., Flores, E.M., Hill, K.L., Vargas, E., Vargas, S.A., Hoh, E., 2012. Persistent organic pollutants in plastic marine debris found on beaches in San Diego, California. Chemosphere 86, 258–263.
Vandermeersch, G., Van Cauwenberghe, L., Janssen, C.R., Marques, A., Granby, K., Fait, G., Kotterman, M.J.J., Diogène, J., Bekaert, K., Robbens, J., Devriese, L., 2015. A critical view on microplastic quantification in aquatic organisms. Environ. Res. 143, 46–55.
Velzeboer, I., Kwadijk, C.J.A.F., Koelmans, A.A., 2014. Strong sorption of PCBs to nanoplastics, microplastics, carbon nanotubes, and fullerenes. Environ. Sci. Technol. 48, 4869–4876.
Voparil, I.M., Mayer, L.M., 2000. Dissolution of sedimentary polycyclic aromatic hydrocarbons into the lugworm's (*Arenicola marina*) digestive fluids. Environ. Sci. Technol. 34, 1221–1228.
Walker, C.H., 2001. Organic Pollutants: An Ecotoxicological Perspective. Taylor & Francis, London.
Wang, F., Shih, K.M., Li, X.Y., 2015. The partition behavior of perfluorooctanesulfonate (PFOS) and perfluorooctanesulfonamide (FOSA) on microplastics. Chemosphere 119, 841–847.
Wang, J., Peng, J., Tan, Z., Gao, Y., Zhan, Z., Chen, Q., Cai, L., 2017. Microplastics in the surface sediments from the Beijiang River littoral zone: composition, abundance, surface textures and interaction with heavy metals. Chemosphere 171, 248–258.

Watts, A.J., Urbina, M.A., Goodhead, R., Moger, J., Lewis, C., Galloway, T.S., 2016. Effect of microplastic on the gills of the shore crab *Carcinus maenas*. Environ. Sci. Technol. 50, 5364–5369.
Wilkinson, A.N., Ryan, A.J., 1998. Polymer Processing and Structure Development. Kluwer Academic Publishers, New York, NY.
Winston, J.E., 1982. Drift plastic—an expanding niche for a marine invertebrate? Mar. Pollut. Bull. 13, 348–351.
Woodall, L.C., Sanchez-Vidal, A., Canals, M., Paterson, G.L., Coppock, R., Sleight, V., Calafat, A., Rogers, A.D., Narayanaswamy, B.E., Thompson, R.C., 2014. The deep sea is a major sink for microplastic debris. R. Soc. Open Sci. 1140317.
Wu, B., Taylor, C.M., Knappe, D.R., Nanny, M.A., Barlaz, M.A., 2001. Factors controlling alkylbenzene sorption to municipal solid waste. Environ. Sci. Technol. 35, 4569–4576.
Wurl, O., Obbard, J.P., 2004. A review of pollutants in the sea-surface microlayer (SML): a unique habitat for marine organisms. Mar. Pollut. Bull. 48, 1016–1030.
Yeo, B.G., Takada, H., Taylor, H., Ito, M., Hosoda, J., Allinson, M., Connell, S., Greaves, L., McGrath, J., 2015. POPs monitoring in Australia and New Zealand using plastic resin pellets, and International Pellet Watch as a tool for education and raising public awareness on plastic debris and POPs. Mar. Pollut. Bull. 101, 137–145.
Zarfl, C., Matthies, M., 2010. Are marine plastic particles transport vectors for organic pollutants to the Arctic. Mar. Pollut. Bull. 60, 1810–1814.
Zbyszewski, M., Corcoran, P.L., 2011. Distribution and degradation of fresh water plastic particles along the beaches of Lake Huron, Canada. Water Air Soil Pollut. 220, 365–372.
Zbyszewski, M., Corcoran, P.L., Hockin, A., 2014. Comparison of the distribution and degradation of plastic debris along shorelines of the Great Lakes, North America. J. Great Lakes Res. 40, 288–299.
Zhang, W.W., Ma, X.D., Zhang, Z.F., Wang, Y., Wang, J.Y., Wang, J., Ma, D.Y., 2015. Persistent organic pollutants carried on plastic resin pellets from two beaches in China. Mar. Pollut. Bull. 99, 28–34.
Ziccardi, L.M., Edgington, A., Hentz, K., Kulacki, K.J., Driscoll, S.K., 2016. Microplastics as vectors for bioaccumulation of hydrophobic organic chemicals in the marine environment: a state-of-the-science review. Environ. Toxicol. Chem. 35, 1667–1776.
Zitko, V., Hanlon, M., 1991. Another source of pollution by plastics: skin cleaners with plastic scrubbers. Mar. Pollut. Bull. 22, 41–42.

FURTHER READING

ten Hulscher, T.E.M., Cornelissen, G., 1996. Effect of temperature on sorption equilibrium and sorption kinetics of organic micropollutants—a review. Chemosphere 32, 609–626.

CHAPTER 8

The Effects of Microplastic Pollution on Aquatic Organisms

S. Michele Harmon
University of South Carolina Aiken, Aiken, SC, United States

8.1 INTRODUCTION

The presence of small plastic pieces in the oceans was first noted by scientists in the early 1970s (Carpenter et al., 1972). Since that time, many scientists have studied the potential problems associated with what we now term "microplastics." Microplastic debris in aquatic ecosystems is currently considered one of the most important global pollution problems of our time.

The majority of synthetic plastics polluting the aquatic environment include polyethylene terephthalate (PET), low- and high-density polyethylene (PE), polypropylene (PP), polyvinyl chloride (PVC), and polystyrene (PS). Microplastics are categorized as primary or secondary and then further classified as fragments, pellets, fibers, film, or foam for further study. The term "microplastic" generally refers to plastic particles that are <5 mm, with the term "nanoplastic" being used to describe a plastic particle that is <1 μm in at least one of its dimensions (da Costa et al., 2016).

Primary microplastics are those plastic particles intentionally manufactured in sizes <5 mm for use in personal care products or industrial applications, such as blasting scrubbers. Plastic microbeads have become common components in consumer products such as toothpastes, body washes, and facial cleansers. As such, they frequently flush into municipal wastewater treatment facilities (Fendall and Sewell, 2009). While wastewater treatment processes remove much of this material, a certain portion bypasses the treatment process to be discharged into the aquatic environment (Carr et al., 2016; Talvitie et al., 2015). Mason et al. (2016a) estimate that an average of 13 billion microbeads is released each day into waterways of the United States alone. Staggering numbers, such as these, have led to federal legislation in the United States to ban the manufacture of plastic microbeads for personal care products beginning in 2017 through the Microbead-Free Waters Act of 2015, and while this should decrease new inputs of microplastic pollution in the United States, there will continue to be a significant global pollutant load from other primary microplastic sources and from other countries into the future.

Microplastic Contamination in Aquatic Environments
https://doi.org/10.1016/B978-0-12-813747-5.00008-4

Secondary microplastics are the degraded fragments of larger plastic debris that have made their way into the environment. In the environment, plastic items degrade through photooxidative pathways (Singh and Sharma, 2008) that make the plastic brittle enough to break into pieces that become increasingly smaller over time. The formation of secondary microplastics from plastic debris depends upon a number of exposure factors including ultraviolet exposure, oxygen concentrations, temperature, mechanical forces, biofouling, and the size and shape of the debris pieces (Andrady, 2011; Barnes et al., 2009; Browne et al., 2010; Corcoran et al., 2009; Pegram and Andrady, 1989; ter Halle et al., 2016; Singh and Sharma, 2008; Tosin et al., 2012). In a recent study, Weinstein et al. (2016) demonstrated that microplastic production from strips of polyethylene, polypropylene, and polystyrene placed in an open estuary can begin in as little as 8 weeks, producing both fragments and fibers. It has even been shown that isopod crustaceans (*Sphaeroma quoianum*) can add secondary microplastic fragments to marine waters by boring into polystyrene floats under docks (Davidson, 2012).

Other contributors to microplastic pollution are plastic fibers. Plastic fibers come from the degradation of larger debris (Weinstein et al., 2016), from the breakdown of geotextile liners (Wiewel and Lamoree, 2016), or from household washing of synthetic textiles (Browne et al., 2011; Napper and Thompson, 2016; Pirc et al., 2016). In the case of synthetic textiles, microplastic fibers are discharged in wash water and then, like microbeads, make their way into municipal wastewater treatment facilities where some fibers are ultimately released at the outfall (Browne et al., 2011). Agricultural application of sewage sludge also serves as an additional source of microplastic fiber pollution through runoff into the watershed (Zubris and Richards, 2005).

8.1.1 Prevalence of Microplastics in Water and Sediment

There are many published studies that measure the presence of microplastic pollutants in water and sediments around the world. Recent review articles have attempted to summarize and estimate the overall abundance of microplastic pollutants in both freshwater (Anderson et al., 2016; Eerkes-Medrano et al., 2015; Ivleva et al., 2017; Mason et al., 2016b) and marine systems (Enders et al., 2015; Ivleva et al., 2017; van Sebille et al., 2015). While the results of these analyses vary in specific numbers, all agree that microplastics have become a problem of great concern because of the enormous volume of these materials in the world's oceans, lakes, and rivers. The global nature of the predicament is further evident in the fact that microplastics have even been found in the atmospheric fallout (Dris et al., 2016), arctic waters and sea ice (Amelineau et al., 2016; Lusher et al., 2015; Obbard et al., 2014; Tekman et al., 2017), the waters of remote lakes (Free et al., 2014; Zhang et al., 2016), and the guts of organisms collected from deep-sea sediments (Taylor et al., 2016).

The ubiquitous presence of these materials leads to many questions related to how these particles affect aquatic organisms at individual and population levels.

8.1.2 The Concern Over Microplastics in the Aquatic Environment

In aquatic ecosystems, microplastics can be found in the sediments, floating on the surface, and suspended throughout the water column, thus leading to numerous pathways of potential problems for aquatic organisms and populations (Fig. 8.1). The plastics themselves cause direct physical or nutritional problems when ingested, and these problems may be exacerbated by the presence of plasticizers in the particles themselves or by the presence of other toxic pollutants that have adhered to the surface. There are also some indications that microplastics can affect primary producers at the base of aquatic food webs.

The abundance of microplastics, combined with their small size and close proximity to plankton in the water column, allows for direct ingestion by aquatic biota at different trophic levels. Microplastic ingestion by marine and estuarine organisms has been noted by many scientists over the years and from locations all over the world. Recent reviews on this topic (Anderson et al., 2016; do Sul and Costa, 2014; Ivleva et al., 2017; Wright et al., 2013b) report microplastic ingestion by a long list of marine organisms, including fish, seabirds, marine mammals, and a host of marine/estuarine invertebrates.

While microplastic pollution in freshwater systems has drawn less attention than marine systems, we know that freshwater organisms are also exposed and that they also

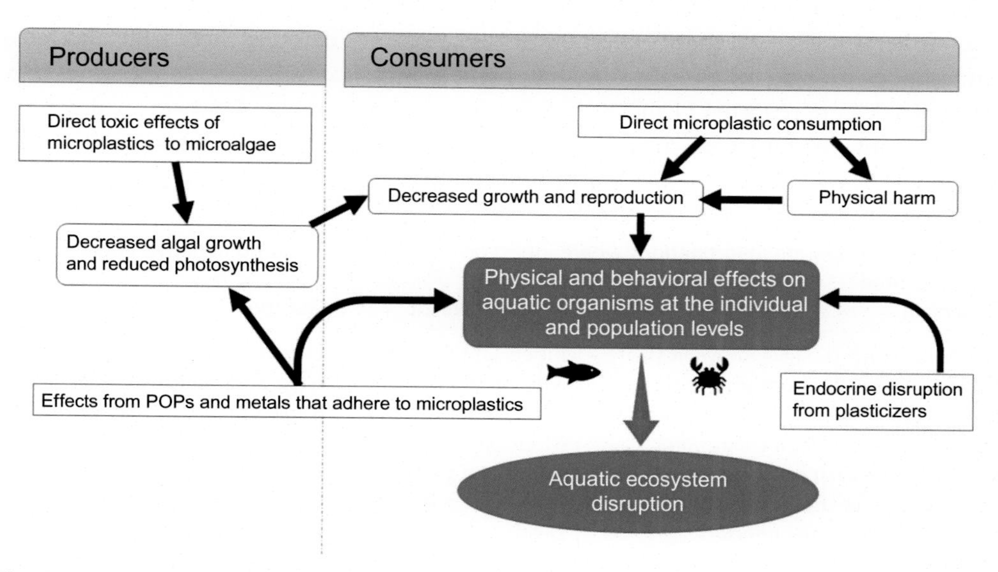

Fig. 8.1 Conceptual model of the pathways of potential effects from exposure to microplastics in the aquatic ecosystem.

ingest microplastic pollutants. This has been observed in several field studies (Biginagwa et al., 2016; Faure et al., 2015; McGoran et al., 2017; Peters and Bratton, 2016; Phillips and Bonner, 2015; Silva-Cavalcanti et al., 2017) and in many laboratory exposure settings [reviewed by Anderson et al. (2016) and by Eerkes-Medrano et al. (2015)].

Because there is no longer any doubt that microplastics are being incidentally and/or selectively ingested by aquatic organisms, the field of investigation has now turned toward elucidating the effects of constant assimilation of these pollutants into aquatic systems. This chapter will review current lines of investigation targeting the following four questions:

(1) What happens when microplastics are part of an aquatic organism's diet?
(2) Do microplastics affect primary producers?
(3) Is there potential for endocrine disruption and toxicity from plasticizers added to these plastic products during their manufacture?
(4) Are there effects to aquatic organisms from the pollutants that adhere to microplastics?

8.2 MICROPLASTICS AS PART OF AN AQUATIC ORGANISM'S DIET

While many studies have reported microplastic ingestion by fish (e.g., Bellas et al., 2016; Biginagwa et al., 2016; Boerger et al., 2010; Guven et al., 2017; Lusher et al., 2016; Ory et al., 2017), few of these report actual physical damage to the fish (Peda et al., 2016). Tadpoles of *Xenopus tropicalis* were shown to accumulate and eliminate polystyrene microspheres with no negative effects (Hu et al., 2016). Likewise, there have also been a number of investigations on invertebrates where ingestion was observed, but no negative effects were reported (Table 8.1). For a few recent examples, Devriese et al. (2015) observed uptake of microplastic particles in the digestive tract of brown shrimp (*Crangon*

Table 8.1 Invertebrate investigations where ingestion was observed, but no negative effects were reported

Organism	Ingested plastic	Citation
Brown shrimp (*Cragnon cragnon* (L.))	Fibers and particles	Devriese et al. (2015)
Pacific oyster (*Crassostrea gigas*) larvae	Microparticles	Cole and Galloway (2015)
Sea urchin (*Tripneustes gratilla*) larvae	Microspheres	Kaposi et al. (2014)
Marine isopod (*Idotea emarginata*)	Microparticles	Hamer et al. (2014)
Gooseneck barnacles (*Lepas* spp.)	Fibers and particles	Goldstein and Goodwin (2013)
Sea cucumbers (Echinodermata)	Fibers and particles	Graham and Thompson (2009)
Periwinkle (*Littorina littorea*)	Beads	Gutow et al. (2016)
Brine shrimp nauplii (*Artemia* sp.)	Particles	Batel et al. (2016)
Mud snails (*Potampoyrgus antipodarum*)	Particles	Imhof and Laforsch (2016)

crangon (L.)) from the southern shallow water habitats of the English Channel area and the southern part of the North Sea. While ingestion did occur, there was no indication of translocation into the tissues and no indication of physical harm. Imhof and Laforsch (2016) found that microplastic exposure caused no morphological changes or developmental effects in mud snails (*Potampoyrgus antipodarum*) after exposing to irregular shaped microplastic particles in their food. Kaposi et al. (2014) reported that the ingestion of microspheres by the larvae of sea urchins (*Tripneustes gratilla*) caused no measurable dose-dependent effects.

On the other hand, there are numerous publications showing physical effects to some lower-trophic-level aquatic species. The initial concerns related to the direct ingestion of nanosized and microplastic particles include physical damage to feeding structures or digestive organs, accumulation within organisms, and translocation of the particles from the digestive tract to other tissues. In the larger context, there are questions related to changes in foraging strategies, feeding behaviors, and ecosystem functioning.

8.2.1 Physical Damage

Nanosized particles, those with a size of 1 μm or less, are problematic for a number of species (reviewed by da Costa et al., 2016) and are of special concern because of their ability to pass through biological membranes (Mattsson et al., 2015b). Exposure to nanosized polystyrene particles resulted in reduced body size and reproduction in *Daphnia magna* exposed to concentrations of 0.22–103 mg nanopolystyrene/L and caused an increase in neonatal malformations when exposed to concentrations as low as 30 mg nanopolystyrene/L (Besseling et al., 2014). Bergami et al. (2016) reported that exposure to polystyrene nanoparticles (5–100 μg/mL) resulted in several sublethal effects to larvae of the brine shrimp *Artemia franciscana*, including adsorption to antennules and appendages resulting in an increased number of molting events. Likewise, Cole et al. (2013) reported microplastic particles adhered to the external carapace and appendages of copepods, potentially affecting swimming capabilities. Della Torre et al. (2014) investigated the disposition and toxicity of polystyrene nanoparticles in sea urchin embryos (*Paracentrotus lividus*) and observed the upregulation of genes involved in cellular stress responses and the facts that cationic (–NH_2) polystyrene nanoparticles caused severe developmental defects with EC_{50} values of 3.85 and 2.61 μg/mL for 24 and 48 h post-fertilization, respectively. Anionic (–COOH) polystyrene nanoparticles tended to accumulate inside the *P. lividus* embryo's digestive tract and caused no developmental effects at concentrations up to 50 μg/mL. Jeong et al. (2016) noted size-dependent effects after the ingestion of nano- and microsized (0.05, 0.5, and 6 μm) nonfunctionalized polystyrene microbeads by the monogonont rotifer (*Brachionus koreanus*). These effects included reduced growth rate, reduced fecundity, decreased life span, longer reproduction time, and elevated levels of enzymatic biomarkers of stress. When exposed at very high concentrations (12.5–400 mg/L), the ingestion of 1 μm polyethylene particles led to dose-dependent immobilization effects for *D. magna* (Rehse et al., 2016) with a 96 h EC_{50} of 57.43 mg/L.

8.2.2 Accumulation and Subsequent Effects on Growth and Fecundity

Perhaps the greatest effects are actually due to the accumulation of these particles in the digestive organs of these organisms. Accumulation within the gut has potential energetic costs to an organism, as this can affect feeding behavior and/or the ability to ingest or digest actual food. The results are often measured in terms of reduced growth or fecundity. These effects can move up trophic levels as predators consume organisms that are carrying a high load of these particles in the gut or body tissues.

Water-column-associated microplastics are similar in size and color to planktonic species, such as copepods, that are common prey items. Ingestion by marine zooplankton has been observed by Desforges et al. (2015), who reported microplastic ingestion by calanoid copepods (*Neocalanus cristatus*) and euphausiids (*Euphausia pacifica*) collected from the open ocean, and by Cole et al. (2013) who noted ingestion by several marine zooplankton species. It is hypothesized that the similarities in size and color explain the frequent ingestion of these particles by planktivorous fish species such as amberstripe scad (*Decapterus muroadsi*) (Ory et al., 2017) or Japanese anchovy (*Engraulis japonicus*) (Tanaka and Takada, 2016).

Once ingested, these microplastic particles have energetic costs to the organism. Cole et al. (2015) demonstrated that the ingestion of polystyrene microbeads could alter the feeding capacity of the pelagic copepod, *Calanus helgolandicus*, thus leading to decreased reproductive output due to energetic depletion. Polystyrene microbead exposure also reduced the fecundity of another group of copepods, *Tigriopus japonicus*, after laboratory exposure (Lee et al., 2013). A 2-month exposure of adult Pacific oysters (*Crassostrea gigas*) to polystyrene microplastics resulted in feeding modifications and reproductive disruption in the oysters, with significant impacts on offspring (Sussarellu et al., 2016). Daily exposure of Asian green mussels (*Perna viridis*) to suspended polyvinylchloride particles affected survival, filtration, and respiration rates, as well as byssus production (Rist et al., 2016). In a mesocosm exposure of polylactic acid microparticles, Green (2016) reported elevated respiration and assumed this to be an indicator of stress to European flat oysters (*Ostrea edulis*).

Fibers tend to be the microplastic material found most commonly in the stomach contents of field-collected fish (Nadal et al., 2016; Neves et al., 2015; Silva-Cavalcanti et al., 2017) and zooplankton (Sun et al., 2017). In the laboratory, exposure to polyethylene terephthalate microplastic fibers in the absence of algae resulted in mortality to *D. magna*, but no effect was found when daphnids were fed before the experiments (Jemec et al., 2016). It is still unclear if the presence of microplastic fibers cause actual physical damage or if the effects are due to their extended presence in the gut in place of actual food items. Chronic exposure of freshwater amphipods, *Hyalella azteca*, to polyethylene microplastic particles resulted in reduced growth and fecundity, and authors attributed this to a decrease in the ingestion and processing of actual food items (Au et al., 2015). These authors also noticed that retention in the gut was influenced by particle shape, as they reported a higher retention time for polypropylene microfibers in the gut compared with polyethylene microplastic particles. In 10-day testing, Au et al. (2015) also found that amphipods exposed to polypropylene microplastic fibers (0–90 microplastics/mL) demonstrated a dose-dependent decrease in growth. Similarly, the amphipod *Gammarus fossarum* ingested both polyamide fibers (500 × 20 μm) and polystyrene beads (1.6 μm) in a 28-day feeding experiment, but the authors reported that the fibers significantly reduced the assimilation efficiency of the animals (Blarer and Burkhardt-Holm, 2016). High levels of environmental microplastic pollution reduced body condition in Norway lobsters (*Nephrops norvegicus*), a commercially important species in Europe (Welden and Cowie, 2016). As with other species, the predominant microplastic found in field caught *N. norvegicus* was in the form of fibers/filaments, and dietary uptake of plastic filaments along with fish to *N. norvegicus* has been reported (Murray and Cowie, 2011).

Where fish are concerned, de Sa et al. (2015) found that common goby (*Pomatoschistus microps*) juveniles can ingest microplastics when mixed with prey items and that the ingestion of these particles may reduce predatory performance and efficiency. In freshwater, Grigorakis et al. (2017) conducted a feeding study and reported that neither microbeads nor microfibers were likely to accumulate within the gut contents of goldfish (*Carassius auratus*). Similarly, Mazurais et al. (2015) found no significant effects to growth or inflammatory response when larvae of the European sea bass (*Dicentrarchus labrax*) ingested fluorescent polyethylene microbeads.

Physical effects have also been noted for benthic organisms. When maintained in sediments spiked from 0% to 5% with microscopic polyvinylchloride, the polychaete worm *Arenicola marina* (lugworm) demonstrated a significant depletion in energy reserves due to a combination of reduced feeding activity, accumulation of microplastics in the gut, and associated inflammation (Wright et al., 2013a). These benthic animals provide important ecosystem services, as they constantly move sediment during feeding and burrowing, and they produce fecal casts that help to drive nitrogen cycling. Green et al. (2016) reported that after a 31-day mesocosm exposure with polyvinyl chloride, high-density polyethylene, and polyacetic acid microplastics in sediments, *A. marina* produced less casts. This had a subsequent effect on the microalgal biomass in the sediments, leading the authors to conclude that sediment-associated microplastic pollution has the potential to affect nitrogen cycling and primary productivity in benthic habitats. These exposures were conducted at 0%, 0.02%, 0.2%, and 2% by wet weight of sediment. Additional studies in a different habitat (Green et al., 2017) found that after 50 days of exposure, the filtration rates of oysters, *O. edulis*, increased by 25 μg/L of high-density polyethylene or polyacetic acid microplastics. The biomass of cyanobacteria was also reduced in the surrounding sediment. On the contrary, filtration rates of *Mytilus edulis* were reduced, and there were no other effects to primary producers or the benthic community.

8.2.3 Translocation in Tissues and Resulting Effects on Energetics

The problem with translocation of micro- and nanosized particles into other tissues is from the overall cost to the organism in terms of energetics. Translocation may result in an immune response that is costly, affecting growth and/or fecundity. For example, Browne et al. (2008) demonstrated that polystyrene microplastics (3.0 or 9.6 μm) could move from the gut of the mussel (*M. edulis*) into the circulatory system, where some of the particles persisted in the hemolymph for 48 days, eliciting an immune response. When von Moos et al. (2012) exposed *M. edulis* to additive-free high-density polyethylene microplastic particles (0–80 μm), the particles were drawn into the gills, taken up into the stomach, transported into the digestive gland, and then accumulated in the lysosomal system resulting in a strong inflammatory response.

In fish, Avio et al. (2015b) observed translocation of polystyrene and polyethylene microplastics from the stomach to the liver in the mullet, *Mugil cephalus*, after a laboratory exposure. Similarly, Lu et al. (2016) observed tissue accumulation of polystyrene microplastics in the gills, liver, and gut of zebra fish (*Danio rerio*) resulting in inflammation, lipid accumulation in the liver, and elevated enzymatic biomarkers of oxidative stress.

8.2.4 Trophic Transfer

Effects in upper trophic levels due to food chain transfer was demonstrated by Mattsson et al. (2015a) who administered polystyrene nanoparticles to fish through an aquatic food chain of algae (*Scenedesmus* sp.) to *D. magna* to fish (*Crassius crassius*). Effects on fish behavior included reduced feeding rate, changes in feeding behavior, and altered metabolism. These observed effects were likely linked to morphological changes in the brain and a general disturbance of cellular function. This demonstrated that nanoplastic exposure at lower trophic levels can result in effects to higher predatory consumers. On the other hand, Tosetto et al. (2017) reported no changes in fish behavior after tropic transfer of microplastics from beach hoppers (*Platorchestia smithi*), to Krefft's frillgobies (*Bathygobius krefftii*).

A study by Farrell and Nelson (2013) demonstrated that fluorescently labeled polystyrene microspheres could be fed to mussels (*M. edulis*) and then tropically transferred to crabs (*Carcinus maenas*). Furthermore, translocation to hemolymph and other tissues of the crab was also observed. Others not only observed this same pathway of trophic transfer but also reported inspiration across the gills in addition to ingestion through *M. edulis*, thus identifying ventilation as an additional route of microplastic uptake (Watts et al., 2014). Once inhaled into the gill chamber, microplastics caused a significant, but temporary, dose-dependent effect on oxygen consumption and osmoregulation (Watts et al., 2016). As with other aquatic species, the consumption of microplastic fibers by *C. maenas* resulted in reduced food consumption and a subsequent reduction in the energy available for growth (Watts et al., 2015).

Obviously, the idea of bioaccumulation by aquatic organisms raises concerns when humans are the ultimate consumer. Microplastics have been found in fish and shellfish being sold for human consumption in the United States and Indonesia (Rochman et al., 2015), Europe (Van Cauwenberghe and Janssen, 2014), British Columbia (Davidson and Dudas, 2016), and China (Li et al., 2015).

8.2.5 Microplastic Egestion

Once ingested by invertebrates, microplastics have even further-reaching effects when they are egested. Egestion of microplastic particles in fecal pellets can be effectively transferred to coprophagous biota (Cole et al., 2016) as a part of the natural system of nutrient cycling in aquatic ecosystems. Cole et al. (2016) showed that microplastics egested within fecal material of the copepod, *Centropages typicus*, could be transferred to *C. helgolandicus* via coprophagy. Furthermore, incorporation into fecal pellets has been shown to serve as a mechanism of transport from the water column to the benthic environment (Cole et al., 2013, 2016) further suggesting that interactions with marine life play an important role in the distribution of microplastics in the marine environment (Clark et al., 2016).

8.3 MICROPLASTICS AND PRIMARY PRODUCERS

Much less is known about the effects of microplastics on primary producers, despite the facts that their autotrophic processes support the aquatic food web and that algal photosynthesis is a major source of earth's oxygen. A number of laboratory studies have attempted to assess the effects of nanosized and microplastics on various algal species (Table 8.2). At this point, results are mixed and inconclusive. While these laboratory studies have shown that microplastics can affect microalgal cells, especially in high exposure concentrations (Sjollema et al., 2016), the environmental relevance and exact mechanisms of toxicity are still unclear.

The most straightforward hypotheses regarding the mechanisms of toxicity are based on simple physical adsorption of micro- and nanosized plastics to algal cell surfaces. It is thought that an algal cell can be covered with so many particles that photosynthesis is hampered through the physical blockage of light and gas exchange. This has been proposed by Bhattacharya et al. (2010) who exposed *Chlorella* and *Scenedesmus* to positively charged plastic nanoparticles and observed a reduction in photosynthetic activity for both species after particles adhered to cell surfaces. Negative effects on growth, photosynthetic efficiency, and chlorophyll content were reported for *Skeletonema costatum* exposed to polyvinyl chloride microspheres by Zhang et al. (2017) who used a combination of

Table 8.2 Summary of laboratory toxicity studies to assess the effects of nanosized and microplastics on algae

Algal species	Type of microplastic[a]	Result	Source
Dunaliella tertiolecta	Uncharged PS beads	No effect on photosynthesis Reduced growth after exposure at 25 and 250 mg/L	Sjollema et al. (2016)
Dunaliella tertiolecta	Negatively charged PS beads	No effect on photosynthesis No effect on growth after exposure to 25 and 250 mg/L	Sjollema et al. (2016)
Thalassiosira pseudonana	Negatively charged PS beads	No effect on photosynthesis after exposure to 25 and 250 mg/L	Sjollema et al. (2016)
Chlorella vulgaris	Negatively charged PS beads	No effect on photosynthesis after exposure to 25 and 250 mg/L	Sjollema et al. (2016)
Tetraselmis chuii	PE spheres	No effect on growth after exposure to 0.046–1.5 mg/L for 96 h	Davarpanah and Guilhermino (2015)
Chlorella sp.	Positively charged PS nanobeads	Reduced photosynthesis and increased ROS	Bhattacharya et al. (2010)
Scenedesmus spp.	Positively charged PS nanobeads	Reduced photosynthesis and increased ROS	Bhattacharya et al. (2010)
Skeletonema costatum	PVC microspheres	Inhibited growth, reduced chlorophyll, and reduced photosynthesis after exposure to 0–50 mg/L for 96 h	Zhang et al. (2017)
Scenedesmus obliquus	PS nanobeads	Inhibited growth and reduced chlorophyll after exposure to 44–1100 mg/L for 72 h	Besseling et al. (2014)
Chlamydomonas reinhardtii	High-density PE and PP fragments	No effect on growth or overexpression of stress genes	Lagarde et al. (2016)
Pseudokirchneriella subcapitata	Polyethyleneimine PS nanoparticles	Inhibited growth (EC_{50} values = 0.54–0.58 mg/L) after exposure to 0.1–1.0 mg/L for 72 h	Casado et al. (2013)

[a] *PS*, polystyrene; *PE*, polyethylene; *PVC*, polyvinyl chloride; *PP*, polypropylene.

scanning electron microscopy and experiments with larger plastic fragments to support their hypothesis that the negative effects were due to adsorption and aggregation of the microspheres on cell surfaces.

In a more in-depth study of the adsorption of microplastics to algal cell walls, Nolte et al. (2017) showed that adsorption of polystyrene nanoparticles to the cell walls of *Pseudokirchneriella subcapitata* was a function of water hardness and surface charge on the particle. These authors suggested that the growth inhibition seen in other studies (e.g., Bhattacharya et al., 2010) may have been due to increased turbidity in test solutions, changes in ion levels, or immobilization of algae due to agglomeration of the cell culture. Others have shown that the presence of microplastics encourages colonization on the surface that results in aggregates that trap algal cells (Lagarde et al., 2016; Long et al., 2015), pulling them from the water column. So, while growth is not specifically affected, the presence of actively photosynthesizing cells in the water column is reduced, perhaps leading to false positives in some toxicity tests.

8.4 POTENTIAL ENDOCRINE DISRUPTION AND TOXICITY FROM PLASTICIZERS AND OTHER ADDITIVES

During the manufacture of plastics, certain additives, such as phthalates or alkylphenols, are combined with the polymers for the purpose of achieving plasticity, color, or other desirable properties for the end product (Harper and Petrie, 2003). Halogenated compounds, like polybrominated diphenyl ethers (PBDEs) and hexabromocyclododecanes (HBCDs), may also be added during plastic manufacture because of their flame-retardant properties (Sun et al., 2016). Plastic additives are problematic from an environmental standpoint, as they are known to easily leach from the products, and some of them can be toxic at high concentrations. Furthermore, a number of these additives have long been suspected of possessing endocrine-disrupting capabilities at low concentrations. We know plastic leachates can affect aquatic invertebrates, such as barnacles (*Amphibalanus amphitrite*) (Li et al., 2016), *D. magna* (Lithner et al., 2012), and copepods (*Nitocra spinipes*) (Bejgarn et al., 2015) when exposed in laboratory settings. While these studies have been conducted using leachates from pieces of new consumer products, at least two studies have been conducted on the leachate from virgin plastic pellets in the microplastic size class. Silva et al. (2016) reported effects to embryo development of the brown mussel (*P. perna*) after exposure to leachate from virgin polypropylene pellets. Nobre et al. (2015) observed that the development of sea urchin (*Lytechinus variegatus*) larvae was also impaired by leachates from virgin polyethylene pellets. Both of these authors hypothesized that the effects were due to plastic additives, although there was no analytic identification of the additives in question.

There is still much speculation regarding the effects of these additives when organisms are exposed through the ingestion of microplastic particles, and models predict little potential risk from additives (Koelmans et al., 2014). However, laboratory studies, such as Browne et al. (2013), have reported that additives to PVC microplastics (Triclosan and PBDE-47) were taken up by *A. marina*, reducing their feeding rates and ability to engineer sediments. Rochman et al. (2014) have suggested the potential for endocrine disruption in Japanese medaka (*Oryzias latipes*) after ingestion of both virgin and marine-exposed microplastic particles. In a study of field-collected mussels (*M. galloprovincialis*) attached to Styrofoam substrates, Jang et al. (2016) found elevated concentrations of the brominated flame-retardant HBCD in the tissues along with the presence of Styrofoam microparticles inside the mussels, suggesting accumulation of HBCD along with ingested pieces of the substrate.

8.5 POLLUTANTS ADHERED TO MICROPLASTICS

The straightforward, direct problems caused by microplastic ingestion and exposure are further exacerbated by the fact that pollutant compounds (organics and metals) have the potential to sorb to these materials and increase the risk of exposure when in contact with algal cells or when ingested by aquatic organisms. When discussing this topic, the term "sorption" typically includes absorption of pollutants *into* the microplastics and adsorption *onto* the surfaces of these particles.

8.5.1 Persistent Organic Pollutants Sorbed to Microplastics

An abundance of published works has now documented the sorption of hydrophobic persistent organic pollutants to the surfaces of microplastics collected directly from the marine environment (Table 8.3). Furthermore, a number of scientific publications have been devoted to determining partitioning coefficients for these pollutants in the presence of various combinations of plastics (reviewed by Ziccardi et al., 2016). The pollutants that have garnered the most attention include polycyclic aromatic hydrocarbons (PAHs), polychlorinated biphenyls (PCBs), hexachlorocyclohexanes (HCHs), halogenated hydrocarbons, and organochlorine pesticides. Much less attention has been given to the sorptive properties of pharmaceuticals and personal care products (Wu et al., 2016).

The obvious concern is the desorption of these compounds and subsequent bioaccumulation once the particles have been ingested by an aquatic organism. Laboratory simulations have suggested that conditions in the gut will enhance chemical desorption from plastics (Bakir et al., 2014). Once desorbed, bioaccumulation is dependent upon the properties of the chemical and the conditions present in the animal in question. Several studies have attempted to link the concentrations of persistent organic pollutants sorbed to field-collected microplastics to concentrations of these contaminants in the tissues of aquatic organisms from the same location (Ziccardi et al., 2016), but these studies are still inconclusive, as a number of pathways potentially contribute to the body burdens of animals collected from the environment.

Table 8.3 Published works documenting sorption of persistent organic pollutants to the surfaces of microplastics collected from the marine environment

Pollutant(s)[a]	Location collected	Citation
PCBs, DDE, nonylphenols	Japan	Mato et al. (2001)
PAH, PCB, organochlorine pesticides	Portugal	Antunes et al. (2013)
PAHs, PCBs, HCHs, organochlorine pesticides	China	Zhang et al. (2015)
PCBs	Tokyo	Endo et al. (2005)
PAHs, PCBs, organochlorine pesticides	Portugal	Frias et al. (2010)
PAHs, PCBs, organochlorine pesticides	California	Van et al. (2012)
PAHs, PCBs, PBDEs, organochlorine pesticides	Brazil	Taniguchi et al. (2016)
PCBs, HCHs, organochlorine pesticides	Pacific Islands	Heskett et al. (2012)
PAHs, PCBs, HCHs, organochlorine pesticides	Greece	Karapanagioti et al. (2011)
PAHs, PCBs, HCHs, organochlorine pesticides	Vietnam	Le et al. (2016)
PCBs, HCHs, PAHs, hopanes, organochlorine pesticides	Portugal	Mizukawa et al. (2013)
PCB, HCH, organochlorine pesticides	South Africa	Ryan et al. (2012)
PAHs, PCBs	Uruguay	Lozoya et al. (2016)

[a]*PCBs*, polychlorinated biphenyls; *HCHs*, hexachlorocyclohexanes; *PAHs*, polycyclic aromatic hydrocarbons; *PBDEs*, polybrominated diphenylethers.

Modeling studies based on thermodynamics and equilibrium partitioning have suggested that the microplastics may not function as a major vector for the transfer of persistent organic pollutants to aquatic organisms (Beckingham and Ghosh, 2017: Gouin et al., 2011; Koelmans et al., 2013, 2016). Still, other models indicate that desorption and uptake of persistent organic pollutants from ingested microplastics is a significant risk (Bakir et al., 2016; Teuten et al., 2007).

Experimental studies of persistent organic pollutant bioaccumulation from aquatic animal exposed via microplastics in the laboratory show that desorption from microplastics and subsequent bioaccumulation by organisms after ingestion can occur. This has been observed in lugworms (Besseling et al., 2013; Browne et al., 2013), amphipods (Chua et al., 2014), mussels (Avio et al., 2015a), and fish (Rochman et al., 2013; Wardrop et al., 2016). A visual fluorescence tracking study has even shown that benzo[*a*]-pyrene sorbed to microplastics could be passed through trophic transfer from brine shrimp (*Artemia* sp.) to zebra fish (*D. rerio*) (Batel et al., 2016). While we now know that bioaccumulation from microplastic-associated contaminants is possible, the degree and significance of this transfer in the environment is unclear, especially when compared with other exposure pathways (Ziccardi et al., 2016).

In addition to bioaccumulation, some researches have attempted to link toxic effects to this transfer of persistent organic pollutants from the microplastics to the animal.

In aquatic invertebrates, the uptake of nonylphenol and phenanthrene from ingested polyvinyl chloride microplastics into the gut tissues of lugworms (*A. marina*) caused a reduction in the ability of coelomocytes to remove pathogenic bacteria by >60% (Browne et al., 2013). Avio et al. (2015a) demonstrated that pyrene-sorbed polyethylene and polystyrene microplastics were bioavailable to laboratory-exposed mussels (*M. galloprovincialis*) and that accumulation of pyrene was observed in the tissues where the microplastics localized (hemolymph, gills, and digestive tissues). Using DNA microarrays and biomarkers of cellular response, the authors noted changes in immunologic responses, lysosomal compartment, peroxisomal proliferation, antioxidant system, neurotoxic effects, and the onset of genotoxicity.

In studies with fish, Japanese medaka (*O. latipes*) accumulated PCBs and PBDEs from microplastics resulting in hepatic stress in the form of glycogen depletion and fatty vacuoles (Rochman et al., 2013).

The easy ability of these hydrophobic persistent organic pollutants to sorb to plastics not only has the potential to affect organisms directly but also has the potential to change the overall fate of these contaminants in the environment.

8.5.2 Metals Sorbed to Microplastics

Copper and other metals are known to adhere to the surface of microplastic particles from environmental samples (Ashton et al., 2010; Holmes et al., 2012, 2014; Turner and Holmes, 2015; Wang et al., 2017). In the laboratory, Brennecke et al., 2016 demonstrated that copper and zinc released from antifouling paint adsorbed to virgin polystyrene beads and aged polyvinyl chloride fragments in seawater. This has led to questions of enhanced metal toxicity in the presence of microplastics.

Because of the algicidal properties of copper, the concept that copper adheres to the surface of microplastics raises questions about synergistic toxic effects when microplastics are combined with metals. However, Davarpanah and Guilhermino (2015) did not see where the effects of copper and microplastics were synergistic after they investigated the effects of polyethylene microspheres, alone and in mixture with copper, on the population growth of the marine microalgae *Tetraselmis chuii*.

Khan et al. (2015) reported that the presence of polyethylene microplastic beads had no effect on the uptake or localization of silver by zebra fish when both were administered directly to test solutions. When the microbeads were incubated with silver to allow adsorption prior to fish exposure, silver uptake into fish tissues was reduced, and silver was localized in the intestine, demonstrating that microplastics can alter the bioavailability of a metal contaminant making it less likely to be taken up. On the other hand, Luis et al. (2015) observed toxicological interactions between polyethylene microplastic spheres and chromium(VI) in short-term (96 h) toxicity exposures of the common goby (*P. microps*).

8.6 CONCLUSION

Despite the vast amount of research that has been conducted, there are still many unanswered questions related to the effects of microplastics on aquatic organisms in the environment. Most studies are still conducted in the laboratory over short periods of time, typically days. They are conducted with a limited number of model organisms and only a few polymer types and shapes. It is still difficult to elucidate the actual effects on a population of aquatic organisms that are exposed over a lifetime to a variety of microplastics of differing composition, sizes, and shapes. There is still much work to be done to determine how trophic transfer of accumulated microplastic pollution affects nutrition, foraging behavior, and the health of organisms at upper trophic levels.

Uncertainty persists regarding the effects of these particles on the survival, growth, and photosynthetic capabilities of microalgae. Concern over the effects of these pollutants to algal primary producers is well placed, as hampering the growth and function of aquatic algae threatens aquatic ecosystem balance.

Additionally, there is still much contradictory discussion regarding the importance of microplastics as a true vector of hydrophobic persistent organic pollutants to aquatic animals. Part of the difficulty in quantifying the actual risks associated with exposure to and bioaccumulation of persistent organic pollutants sorbed to microplastics is in the ubiquitous nature of these pollutants in aquatic systems. While it is clear from the lab studies that transfer from plastics to tissues can occur, it is difficult to distinguish how much of an organism's body burden is from plastic desorption in the gut and how much is from other exposure pathways, such as water and food (Ziccardi et al., 2016). While effects associated with persistent organic pollutants sorbed to microplastics have been hypothesized and even supported by some studies, the differences in methodologies and other questions related to environmental relevance show that there is still much work to be done before this can be completely elucidated.

REFERENCES

Amelineau, F., Bonnet, D., Heitz, O., Mortreux, V., Harding, A.M.A., Karnovsky, N., et al., 2016. Microplastic pollution in the Greenland Sea: background levels and selective contamination of planktivorous diving seabirds. Environ. Pollut. 219, 1131–1139.

Anderson, J.C., Park, B.J., Palace, V.P., 2016. Microplastics in aquatic environments: implications for Canadian ecosystems. Environ. Pollut. 218, 269–280.

Andrady, A.L., 2011. Microplastics in the marine environment. Mar. Pollut. Bull. 62 (8), 1596–1605.

Antunes, J.C., Frias, J.G.L., Micaelo, A.C., Sobral, P., 2013. Resin pellets from beaches of the Portuguese coast and adsorbed persistent organic pollutants. Estuar. Coast. Shelf Sci. 130, 62–69.

Ashton, K., Holmes, L., Turner, A., 2010. Association of metals with plastic production pellets in the marine environment. Mar. Pollut. Bull. 60 (11), 2050–2055.

Au, S.Y., Bruce, T.F., Bridges, W.C., Klaine, S.J., 2015. Responses of *Hyalella azteca* to acute and chronic microplastic exposures. Environ. Toxicol. Chem. 34 (11), 2564–2572.

Avio, C.G., Gorbi, S., Milan, M., Benedetti, M., Fattorini, D., d'Errico, G., et al., 2015a. Pollutants bioavailability and toxicological risk from microplastics to marine mussels. Environ. Pollut. 198, 211–222.

Avio, C.G., Gorbi, S., Regoli, F., 2015b. Experimental development of a new protocol for extraction and characterization of microplastics in fish tissues: first observations in commercial species from Adriatic Sea. Mar. Environ. Res. 111, 18–26.
Bakir, A., Rowland, S.J., Thompson, R.C., 2014. Enhanced desorption of persistent organic pollutants from microplastics under simulated physiological conditions. Environ. Pollut. 185, 16–23.
Bakir, A., O'Connor, I.A., Rowland, S.J., Hendriks, A.J., Thompson, R.C., 2016. Relative importance of microplastics as a pathway for the transfer of hydrophobic organic chemicals to marine life. Environ. Pollut. 219, 56–65.
Barnes, D.K.A., Galgani, F., Thompson, R.C., Barlaz, M., 2009. Accumulation and fragmentation of plastic debris in global environments. Philos. Trans. R. Soc. Lond. B Biol. Sci. 364 (1526), 1985–1998.
Batel, A., Linti, F., Scherer, M., Erdinger, L., Braunbeck, T., 2016. Transfer of benzo[a] pyrene from microplastics to *Artemia* nauplii and further to zebrafish via a trophic food web experiment: CYP1A induction and visual tracking of persistent organic pollutants. Environ. Toxicol. Chem. 35 (7), 1656–1666.
Beckingham, B., Ghosh, U., 2017. Differential bioavailability of polychlorinated biphenyls associated with environmental particles: microplastic in comparison to wood, coal and biochar. Environ. Pollut. 220, 150–158.
Bejgarn, S., MacLeod, M., Bogdal, C., Breitholtz, M., 2015. Toxicity of leachate from weathering plastics: an exploratory screening study with *Nitocra spinipes*. Chemosphere 132, 114–119.
Bellas, J., Martinez-Armental, J., Martinez-Camara, A., Besada, V., Martinez-Gomez, C., 2016. Ingestion of microplastics by demersal fish from the Spanish Atlantic and Mediterranean coasts. Mar. Pollut. Bull. 109 (1), 55–60.
Bergami, E., Bocci, E., Vannuccini, M.L., Monopoli, M., Salvati, A., Dawson, K.A., et al., 2016. Nano-sized polystyrene affects feeding, behavior and physiology of brine shrimp *Artemia franciscana* larvae. Ecotoxicol. Environ. Saf. 123, 18–25.
Besseling, E., Wegner, A., Foekema, E.M., van den Heuvel-Greve, M.J., Koelmans, A.A., 2013. Effects of microplastic on fitness and PCB bioaccumulation by the Lugworm *Arenicola marina* (L.). Environ. Sci. Technol. 47 (1), 593–600.
Besseling, E., Wang, B., Lurling, M., Koelmans, A.A., 2014. Nanoplastic affects growth of *S. obliquus* and reproduction of *D. magna*. Environ. Sci. Technol. 48 (20), 12336–12343.
Bhattacharya, P., Lin, S.J., Turner, J.P., Ke, P.C., 2010. Physical adsorption of charged plastic nanoparticles affects algal photosynthesis. J. Phys. Chem. C 114 (39), 16556–16561.
Biginagwa, F.J., Mayoma, B.S., Shashoua, Y., Syberg, K., Khan, F.R., 2016. First evidence of microplastics in the African Great Lakes: recovery from Lake Victoria Nile perch and Nile tilapia. J. Great Lakes Res. 42 (1), 146–149.
Blarer, P., Burkhardt-Holm, P., 2016. Microplastics affect assimilation efficiency in the freshwater amphipod *Gammarus fossarum*. Environ. Sci. Pollut. Res. 23 (23), 23522–23532.
Boerger, C.M., Lattin, G.L., Moore, S.L., Moore, C.J., 2010. Plastic ingestion by planktivorous fishes in the North Pacific Central Gyre. Mar. Pollut. Bull. 60 (12), 2275–2278.
Brennecke, D., Duarte, B., Paiva, F., Cacador, I., Canning-Clode, J., 2016. Microplastics as vector for heavy metal contamination from the marine environment. Estuar. Coast. Shelf Sci. 178, 189–195.
Browne, M.A., Dissanayake, A., Galloway, T.S., Lowe, D.M., Thompson, R.C., 2008. Ingested microscopic plastic translocates to the circulatory system of the mussel, *Mytilus edulis* (L.). Environ. Sci. Technol. 42 (13), 5026–5031.
Browne, M.A., Galloway, T.S., Thompson, R.C., 2010. Spatial patterns of plastic debris along estuarine shorelines. Environ. Sci. Technol. 44 (9), 3404–3409.
Browne, M.A., Crump, P., Niven, S.J., Teuten, E., Tonkin, A., Galloway, T., et al., 2011. Accumulation of microplastic on shorelines worldwide: sources and sinks. Environ. Sci. Technol. 45 (21), 9175–9179.
Browne, M.A., Niven, S.J., Galloway, T.S., Rowland, S.J., Thompson, R.C., 2013. Microplastic moves pollutants and additives to worms, reducing functions linked to health and biodiversity. Curr. Biol. 23 (23), 2388–2392.
Carpenter, E.J., Anderson, S.J., Harvey, G.R., Miklas, H.P., Peck, B.B., 1972. Polystyrene spherules in coastal waters. Science 178, 749.
Carr, S.A., Liu, J., Tesoro, A.G., 2016. Transport and fate of microplastic particles in wastewater treatment plants. Water Res. 91, 174–182.

Casado, M.P., Macken, A., Byrne, H.J., 2013. Ecotoxicological assessment of silica and polystyrene nanoparticles assessed by a multitrophic test battery. Environ. Int. 51, 97–105.
Chua, E.M., Shimeta, J., Nugegoda, D., Morrison, P.D., Clarke, B.O., 2014. Assimilation of polybrominated diphenyl ethers from microplastics by the marine amphipod, *Allorchestes compressa*. Environ. Sci. Technol. 48 (14), 8127–8134.
Clark, J.R., Cole, M., Lindeque, P.K., Fileman, E., Blackford, J., Lewis, C., et al., 2016. Marine microplastic debris: a targeted plan for understanding and quantifying interactions with marine life. Front. Ecol. Environ. 14 (6), 317–324.
Cole, M., Galloway, T.S., 2015. Ingestion of nanoplastics and microplastics by Pacific oyster larvae. Environ. Sci. Technol. 49 (24), 14625–14632.
Cole, M., Lindeque, P., Fileman, E., Halsband, C., Goodhead, R., Moger, J., et al., 2013. Microplastic ingestion by zooplankton. Environ. Sci. Technol. 47 (12), 6646–6655.
Cole, M., Lindeque, P., Fileman, E., Halsband, C., Galloway, T.S., 2015. The impact of polystyrene microplastics on feeding, function and fecundity in the marine copepod *Calanus helgolandicus*. Environ. Sci. Technol. 49 (2), 1130–1137.
Cole, M., Lindeque, P.K., Fileman, E., Clark, J., Lewis, C., Halsband, C., et al., 2016. Microplastics alter the properties and sinking rates of zooplankton faecal pellets. Environ. Sci. Technol. 50 (6), 3239–3246.
Corcoran, P.L., Biesinger, M.C., Grifi, M., 2009. Plastics and beaches: a degrading relationship. Mar. Pollut. Bull. 58 (1), 80–84.
da Costa, J.P., Santos, P.S.M., Duarte, A.C., Rocha-Santos, T., 2016. (Nano)plastics in the environment—sources, fates and effects. Sci. Total Environ. 566, 15–26.
Davarpanah, E., Guilhermino, L., 2015. Single and combined effects of microplastics and copper on the population growth of the marine microalgae *Tetraselmis chuii*. Estuar. Coast. Shelf Sci. 167, 269–275.
Davidson, K., Dudas, S.E., 2016. Microplastic ingestion by wild and cultured Manila clams (*Venerupis philippinarum*) from Baynes Sound, British Columbia. Arch. Environ. Contam. Toxicol. 71 (2), 147–156.
Davidson, T.M., 2012. Boring crustaceans damage polystyrene floats under docks polluting marine waters with microplastic. Mar. Pollut. Bull. 64 (9), 1821–1828.
de Sa, L.C., Luis, L.G., Guilhermino, L., 2015. Effects of microplastics on juveniles of the common goby (*Pomatoschistus microps*): confusion with prey, reduction of the predatory performance and efficiency, and possible influence of developmental conditions. Environ. Pollut. 196, 359–362.
Della Torre, C., Bergami, E., Salvati, A., Faleri, C., Cirino, P., Dawson, K.A., et al., 2014. Accumulation and embryotoxicity of polystyrene nanoparticles at early stage of development of sea urchin embryos *Paracentrotus lividus*. Environ. Sci. Technol. 48 (20), 12302–12311.
Desforges, J.P.W., Galbraith, M., Ross, P.S., 2015. Ingestion of microplastics by zooplankton in the northeast Pacific Ocean. Arch. Environ. Contam. Toxicol. 69 (3), 320–330.
Devriese, L.I., van der Meulen, M.D., Maes, T., Bekaert, K., Paul-Pont, I., Frere, L., et al., 2015. Microplastic contamination in brown shrimp (*Crangon crangon*, Linnaeus 1758) from coastal waters of the southern North Sea and Channel area. Mar. Pollut. Bull. 98 (1–2), 179–187.
do Sul, J.A.I., Costa, M.F., 2014. The present and future of microplastic pollution in the marine environment. Environ. Pollut. 185, 352–364.
Dris, R., Gasperi, J., Saad, M., Mirande, C., Tassin, B., 2016. Synthetic fibers in atmospheric fallout: a source of microplastics in the environment? Mar. Pollut. Bull. 104 (1–2), 290–293.
Eerkes-Medrano, D., Thompson, R.C., Aldridge, D.C., 2015. Microplastics in freshwater systems: a review of the emerging threats, identification of knowledge gaps and prioritisation of research needs. Water Res. 75, 63–82.
Enders, K., Lenz, R., Stedmon, C.A., Nielsen, T.G., 2015. Abundance, size and polymer composition of marine microplastics $\geq$10 μm in the Atlantic Ocean and their modelled vertical distribution. Mar. Pollut. Bull. 100 (1), 70–81.
Endo, S., Takizawa, R., Okuda, K., Takada, H., Chiba, K., Kanehiro, H., et al., 2005. Concentration of polychlorinated biphenyls (PCBs) in beached resin pellets: variability among individual particles and regional differences. Mar. Pollut. Bull. 50 (10), 1103–1114.
Farrell, P., Nelson, K., 2013. Trophic level transfer of microplastic: *Mytilus edulis* (L.) to *Carcinus maenas* (L.). Environ. Pollut. 177, 1–3.

Faure, F., Demars, C., Wieser, O., Kunz, M., de Alencastro, L.F., 2015. Plastic pollution in Swiss surface waters: nature and concentrations, interaction with pollutants. Environ. Chem. 12 (5), 582–591.

Fendall, L.S., Sewell, M.A., 2009. Contributing to marine pollution by washing your face: microplastics in facial cleansers. Mar. Pollut. Bull. 58 (8), 1225–1228.

Free, C.M., Jensen, O.P., Mason, S.A., Eriksen, M., Williamson, N.J., Boldgiv, B., 2014. High-levels of microplastic pollution in a large, remote, mountain lake. Mar. Pollut. Bull. 85 (1), 156–163.

Frias, J., Sobral, P., Ferreira, A.M., 2010. Organic pollutants in microplastics from two beaches of the Portuguese coast. Mar. Pollut. Bull. 60 (11), 1988–1992.

Goldstein, M.C., Goodwin, D.S., 2013. Gooseneck barnacles (*Lepas* spp.) ingest microplastic debris in the North Pacific Subtropical Gyre. PeerJ 1, e184. https://doi.org/10.7717/peerj.184.

Gouin, T., Roche, N., Lohmann, R., Hodges, G., 2011. A thermodynamic approach for assessing the environmental exposure of chemicals absorbed to microplastic. Environ. Sci. Technol. 45 (4), 1466–1472.

Graham, E.R., Thompson, J.T., 2009. Deposit- and suspension-feeding sea cucumbers (Echinodermata) ingest plastic fragments. J. Exp. Mar. Biol. Ecol. 368 (1), 22–29.

Green, D.S., 2016. Effects of microplastics on European flat oysters, *Ostrea edulis*, and their associated benthic communities. Environ. Pollut. 216, 95–103.

Green, D.S., Boots, B., Sigwart, J., Jiang, S., Rocha, C., 2016. Effects of conventional and biodegradable microplastics on a marine ecosystem engineer (*Arenicola marina*) and sediment nutrient cycling. Environ. Pollut. 208, 426–434.

Green, D.S., Boots, B., O'Connor, N.E., Thompson, R., 2017. Microplastics affect the ecological functioning of an important biogenic habitat. Environ. Sci. Technol. 51 (1), 68–77.

Grigorakis, S., Mason, S.A., Drouillard, K.G., 2017. Determination of the gut retention of plastic microbeads and microfibers in goldfish (*Carassius auratus*). Chemosphere 169, 233–238.

Gutow, L., Eckerlebe, A., Gimenez, L., Saborowski, R., 2016. Experimental evaluation of seaweeds as a vector for microplastics into marine food webs. Environ. Sci. Technol. 50 (2), 915–923.

Guven, O., Gokdag, K., Jovanovic, B., Kideys, A.E., 2017. Microplastic litter composition of the Turkish territorial waters of the Mediterranean Sea, and its occurrence in the gastrointestinal tract of fish. Environ. Pollut. 223, 286–294.

Hamer, J., Gutow, L., Kohler, A., Saborowski, R., 2014. Fate of microplastics in the marine isopod *Idotea emarginata*. Environ. Sci. Technol. 48 (22), 13451–13458.

Harper, C.A., Petrie, E.M., 2003. Plastics Materials and Processes: A Concise Encyclopedia. John Wiley & Sons, Hoboken, New Jersey.

Heskett, M., Takada, H., Yamashita, R., Yuyama, M., Ito, M., Geok, Y.B., et al., 2012. Measurement of persistent organic pollutants (POPs) in plastic resin pellets from remote islands: toward establishment of background concentrations for International Pellet Watch. Mar. Pollut. Bull. 64 (2), 445–448.

Holmes, L.A., Turner, A., Thompson, R.C., 2012. Adsorption of trace metals to plastic resin pellets in the marine environment. Environ. Pollut. 160, 42–48.

Holmes, L.A., Turner, A., Thompson, R.C., 2014. Interactions between trace metals and plastic production pellets under estuarine conditions. Mar. Chem. 167, 25–32.

Hu, L.L., Su, L., Xue, Y.G., Mu, J.L., Zhu, J.M., Xu, J., et al., 2016. Uptake, accumulation and elimination of polystyrene microspheres in tadpoles of *Xenopus tropicalis*. Chemosphere 164, 611–617.

Imhof, H.K., Laforsch, C., 2016. Hazardous or not are adult and juvenile individuals of *Potamopyrgus antipodarum* affected by non-buoyant microplastic particles? Environ. Pollut. 218, 383–391.

Ivleva, N.P., Wiesheu, A.C., Niessner, R., 2017. Microplastic in aquatic ecosystems. Angew. Chem. Int. Ed. 56 (7), 1720–1739.

Jang, M., Shim, W.J., Han, G.M., Rani, M., Song, Y.K., Hong, S.H., 2016. Styrofoam debris as a source of hazardous additives for marine organisms. Environ. Sci. Technol. 50 (10), 4951–4960.

Jemec, A., Horvat, P., Kunej, U., Bele, M., Krzan, A., 2016. Uptake and effects of microplastic textile fibers on freshwater crustacean *Daphnia magna*. Environ. Pollut. 219, 201–209.

Jeong, C.B., Won, E.J., Kang, H.M., Lee, M.C., Hwang, D.S., Hwang, U.K., et al., 2016. Microplastic size-dependent toxicity, oxidative stress induction, and p-JNK and p-p38 activation in the monogonont rotifer (*Brachionus koreanus*). Environ. Sci. Technol. 50 (16), 8849–8857.

Kaposi, K.L., Mos, B., Kelaher, B.P., Dworjanyn, S.A., 2014. Ingestion of microplastic has limited impact on a marine larva. Environ. Sci. Technol. 48 (3), 1638–1645.

Karapanagioti, H.K., Endo, S., Ogata, Y., Takada, H., 2011. Diffuse pollution by persistent organic pollutants as measured in plastic pellets sampled from various beaches in Greece. Mar. Pollut. Bull. 62 (2), 312–317.

Khan, F.R., Syberg, K., Shashoua, Y., Bury, N.R., 2015. Influence of polyethylene microplastic beads on the uptake and localization of silver in zebrafish (*Danio rerio*). Environ. Pollut. 206, 73–79.

Koelmans, A.A., Besseling, E., Wegner, A., Foekema, E.M., 2013. Plastic as a carrier of POPs to aquatic organisms: a model analysis. Environ. Sci. Technol. 47 (14), 7812–7820.

Koelmans, A.A., Besseling, E., Foekema, E.M., 2014. Leaching of plastic additives to marine organisms. Environ. Pollut. 187, 49–54.

Koelmans, A.A., Bakir, A., Burton, G.A., Janssen, C.R., 2016. Microplastic as a vector for chemicals in the aquatic environment: critical review and model-supported reinterpretation of empirical studies. Environ. Sci. Technol. 50 (7), 3315–3326.

Lagarde, F., Olivier, O., Zanella, M., Daniel, P., Hiard, S., Caruso, A., 2016. Microplastic interactions with freshwater microalgae: hetero-aggregation and changes in plastic density appear strongly dependent on polymer type. Environ. Pollut. 215, 331–339.

Le, D.Q., Takada, H., Yamashita, R., Mizukawa, K., Hosoda, J., Tuyet, D.A., 2016. Temporal and spatial changes in persistent organic pollutants in Vietnamese coastal waters detected from plastic resin pellets. Mar. Pollut. Bull. 109 (1), 320–324.

Lee, K.W., Shim, W.J., Kwon, O.Y., Kang, J.H., 2013. Size-dependent effects of micro polystyrene particles in the marine copepod *Tigriopus japonicus*. Environ. Sci. Technol. 47 (19), 11278–11283.

Li, H.X., Getzinger, G.J., Ferguson, P.L., Orihuela, B., Zhu, M., Rittschof, D., 2016. Effects of toxic leachate from commercial plastics on larval survival and settlement of the barnacle *Amphibalanus amphitrite*. Environ. Sci. Technol. 50 (2), 924–931.

Li, J.N., Yang, D.Q., Li, L., Jabeen, K., Shi, H.H., 2015. Microplastics in commercial bivalves from China. Environ. Pollut. 207, 190–195.

Lithner, D., Nordensvan, I., Dave, G., 2012. Comparative acute toxicity of leachates from plastic products made of polypropylene, polyethylene, PVC, acrylonitrile-butadiene-styrene, and epoxy to *Daphnia magna*. Environ. Sci. Pollut. Res. 19 (5), 1763–1772.

Long, M., Moriceau, B., Gallinari, M., Lambert, C., Huvet, A., Raffray, J., et al., 2015. Interactions between microplastics and phytoplankton aggregates: impact on their respective fates. Mar. Chem. 175, 39–46.

Lozoya, J.P., de Mello, F.T., Carrizo, D., Weinstein, F., Olivera, Y., Cedres, F., et al., 2016. Plastics and microplastics on recreational beaches in Punta del Este (Uruguay): unseen critical residents? Environ. Pollut. 218, 931–941.

Lu, Y.F., Zhang, Y., Deng, Y.F., Jiang, W., Zhao, Y.P., Geng, J.J., et al., 2016. Uptake and accumulation of polystyrene microplastics in zebrafish (*Danio rerio*) and toxic effects in liver. Environ. Sci. Technol. 50 (7), 4054–4060.

Luis, L.G., Ferreira, P., Fonte, E., Oliveira, M., Guilhermino, L., 2015. Does the presence of microplastics influence the acute toxicity of chromium(VI) to early juveniles of the common goby (*Pomatoschistus microps*)? A study with juveniles from two wild estuarine populations. Aquat. Toxicol. 164, 163–174.

Lusher, A.L., Tirelli, V., O'Connor, I., Officer, R., 2015. Microplastics in Arctic polar waters: the first reported values of particles in surface and sub-surface samples. Sci. Rep. 5, 14947. https://doi.org/10.1038/srep14947.

Lusher, A.L., O'Donnell, C., Officer, R., O'Connor, I., 2016. Microplastic interactions with North Atlantic mesopelagic fish. ICES J. Mar. Sci. 73 (4), 1214–1225.

Mason, S.A., Garneau, D., Sutton, R., Chu, Y., Ehmann, K., Barnes, J., et al., 2016a. Microplastic pollution is widely detected in US municipal wastewater treatment plant effluent. Environ. Pollut. 218, 1045–1054.

Mason, S.A., Kammin, L., Eriksen, M., Aleid, G., Wilson, S., Box, C., et al., 2016b. Pelagic plastic pollution within the surface waters of Lake Michigan, USA. J. Great Lakes Res. 42 (4), 753–759.

Mato, Y., Isobe, T., Takada, H., Kanehiro, H., Ohtake, C., Kaminuma, T., 2001. Plastic resin pellets as a transport medium for toxic chemicals in the marine environment. Environ. Sci. Technol. 35 (2), 318–324.

Mattsson, K., Ekvall, M.T., Hansson, L.A., Linse, S., Malmendal, A., Cedervall, T., 2015a. Altered behavior, physiology, and metabolism in fish exposed to polystyrene nanoparticles. Environ. Sci. Technol. 49 (1), 553–561.

Mattsson, K., Hansson, L.A., Cedervall, T., 2015b. Nano-plastics in the aquatic environment. Environ. Sci. Process Impacts 17 (10), 1712–1721.

Mazurais, D., Ernande, B., Quazuguel, P., Severe, A., Huelvan, C., Madec, L., et al., 2015. Evaluation of the impact of polyethylene microbeads ingestion in European sea bass (*Dicentrarchus labrax*) larvae. Mar. Environ. Res. 112, 78–85.

McGoran, A.R., Clark, P.F., Morritt, D., 2017. Presence of microplastic in the digestive tracts of European flounder, *Platichthys flesus*, and European smelt, *Osmerus eperlanus*, from the River Thames. Environ. Pollut. 220, 744–751.

Mizukawa, K., Takada, H., Ito, M., Geok, Y.B., Hosoda, J., Yamashita, R., et al., 2013. Monitoring of a wide range of organic micropollutants on the Portuguese coast using plastic resin pellets. Mar. Pollut. Bull. 70 (1–2), 296–302.

Murray, F., Cowie, P.R., 2011. Plastic contamination in the decapod crustacean *Nephrops norvegicus* (Linnaeus, 1758). Mar. Pollut. Bull. 62 (6), 1207–1217.

Nadal, M.A., Alomar, C., Deudero, S., 2016. High levels of microplastic ingestion by the semipelagic fish bogue *Boops boops* (L.) around the Balearic Islands. Environ. Pollut. 214, 517–523.

Napper, I.E., Thompson, R.C., 2016. Release of synthetic microplastic plastic fibres from domestic washing machines: effects of fabric type and washing conditions. Mar. Pollut. Bull. 112 (1–2), 39–45.

Neves, D., Sobral, P., Ferreira, J.L., Pereira, T., 2015. Ingestion of microplastics by commercial fish off the Portuguese coast. Mar. Pollut. Bull. 101 (1), 119–126.

Nobre, C.R., Santana, M.F.M., Maluf, A., Cortez, F.S., Cesar, A., Pereira, C.D.S., et al., 2015. Assessment of microplastic toxicity to embryonic development of the sea urchin *Lytechinus variegatus* (Echinodermata: Echinoidea). Mar. Pollut. Bull. 92 (1–2), 99–104.

Nolte, T.M., Hartmann, N.B., Kleijn, J.M., Garnaes, J., van de Meent, D., Hendriks, A.J., et al., 2017. The toxicity of plastic nanoparticles to green algae as influenced by surface modification, medium hardness and cellular adsorption. Aquat. Toxicol. 183, 11–20.

Obbard, R.W., Sadri, S., Wong, Y.Q., Khitun, A.A., Baker, I., Thompson, R.C., 2014. Global warming releases microplastic legacy frozen in Arctic Sea ice. Earths Future 2 (6), 315–320.

Ory, N.C., Sobral, P., Ferreira, J.L., Thiel, M., 2017. Amberstripe scad *Decapterus muroadsi* (Carangidae) fish ingest blue microplastics resembling their copepod prey along the coast of Rapa Nui (Easter Island) in the South Pacific subtropical gyre. Sci. Total Environ. 586, 430–437.

Peda, C., Caccamo, L., Fossi, M.C., Gai, F., Andaloro, F., Genovese, L., et al., 2016. Intestinal alterations in European sea bass *Dicentrarchus labrax* (Linnaeus, 1758) exposed to microplastics: preliminary results. Environ. Pollut. 212, 251–256.

Pegram, J.E., Andrady, A.L., 1989. Outdoor weathering of selected polymeric materials under marine exposure conditions. Polym. Degrad. Stab. 26 (4), 333–345.

Peters, C.A., Bratton, S.P., 2016. Urbanization is a major influence on microplastic ingestion by sunfish in the Brazos River Basin, Central Texas, USA. Environ. Pollut. 210, 380–387.

Phillips, M.B., Bonner, T.H., 2015. Occurrence and amount of microplastic ingested by fishes in watersheds of the Gulf of Mexico. Mar. Pollut. Bull. 100 (1), 264–269.

Pirc, U., Vidmar, M., Mozer, A., Krzan, A., 2016. Emissions of microplastic fibers from microfiber fleece during domestic washing. Environ. Sci. Pollut. Res. 23 (21), 22206–22211.

Rehse, S., Kloas, W., Zarfl, C., 2016. Short-term exposure with high concentrations of pristine microplastic particles leads to immobilisation of *Daphnia magna*. Chemosphere 153, 91–99.

Rist, S.E., Assidqi, K., Zamani, N.P., Appel, D., Perschke, M., Huhn, M., et al., 2016. Suspended micro-sized PVC particles impair the performance and decrease survival in the Asian green mussel *Perna viridis*. Mar. Pollut. Bull. 111 (1–2), 213–220.

Rochman, C.M., Manzano, C., Hentschel, B.T., Simonich, S.L.M., Hoh, E., 2013. Polystyrene plastic: a source and sink for polycyclic aromatic hydrocarbons in the marine environment. Environ. Sci. Technol. 47 (24), 13976–13984.

Rochman, C.M., Kurobe, T., Flores, I., Teh, S.J., 2014. Early warning signs of endocrine disruption in adult fish from the ingestion of polyethylene with and without sorbed chemical pollutants from the marine environment. Sci. Total Environ. 493, 656–661.

Rochman, C.M., Tahir, A., Williams, S.L., Baxa, D.V., Lam, R., Miller, J.T., et al., 2015. Anthropogenic debris in seafood: plastic debris and fibers from textiles in fish and bivalves sold for human consumption. Sci. Rep. 5, 14340. https://doi.org/10.1038/srep14340.

Ryan, P.G., Bouwman, H., Moloney, C.L., Yuyama, M., Takada, H., 2012. Long-term decreases in persistent organic pollutants in South African coastal waters detected from beached polyethylene pellets. Mar. Pollut. Bull. 64 (12), 2756–2760.

Silva, P., Nobre, C.R., Resaffe, P., Pereira, C.D.S., Gusmao, F., 2016. Leachate from microplastics impairs larval development in brown mussels. Water Res. 106, 364–370.

Silva-Cavalcanti, J.S., Silva, J.D.B., de Franca, E.J., de Araujo, M.C.B., Gusmao, F., 2017. Microplastics ingestion by a common tropical freshwater fishing resource. Environ. Pollut. 221, 218–226.

Singh, B., Sharma, N., 2008. Mechanistic implications of plastic degradation. Polym. Degrad. Stab. 93 (3), 561–584.

Sjollema, S.B., Redondo-Hasselerharm, P., Leslie, H.A., Kraak, M.H.S., Vethaak, A.D., 2016. Do plastic particles affect microalgal photosynthesis and growth? Aquat. Toxicol. 170, 259–261.

Sun, B.B., Hu, Y.N., Cheng, H.F., Tao, S., 2016. Kinetics of brominated flame retardant (BFR) releases from granules of waste plastics. Environ. Sci. Technol. 50 (24), 13419–13427.

Sun, X.X., Li, Q.J., Zhu, M.L., Liang, J.H., Zheng, S., Zhao, Y.F., 2017. Ingestion of microplastics by natural zooplankton groups in the northern South China Sea. Mar. Pollut. Bull. 115 (1–2), 217–224.

Sussarellu, R., Suquet, M., Thomas, Y., Lambert, C., Fabioux, C., Pernet, M.E.J., et al., 2016. Oyster reproduction is affected by exposure to polystyrene microplastics. Proc. Natl. Acad. Sci. U. S. A. 113 (9), 2430–2435.

Talvitie, J., Heinonen, M., Paakkonen, J.P., Vahtera, E., Mikola, A., Setala, O., et al., 2015. Do wastewater treatment plants act as a potential point source of microplastics? Preliminary study in the coastal Gulf of Finland, Baltic Sea. Water Sci. Technol. 72 (9), 1495–1504.

Tanaka, K., Takada, H., 2016. Microplastic fragments and microbeads in digestive tracts of planktivorous fish from urban coastal waters. Sci. Rep. 6, 34351. https://doi.org/10.1038/srep34351.

Taniguchi, S., Colabuono, F.I., Dias, P.S., Oliveira, R., Fisner, M., Turra, A., et al., 2016. Spatial variability in persistent organic pollutants and polycyclic aromatic hydrocarbons found in beach-stranded pellets along the coast of the state of Sao Paulo, southeastern Brazil. Mar. Pollut. Bull. 106 (1–2), 87–94.

Taylor, M.L., Gwinnett, C., Robinson, L.F., Woodall, L.C., 2016. Plastic microfibre ingestion by deep-sea organisms. Sci. Rep. 6, 33997. https://doi.org/10.1038/srep33997.

Tekman, M.B., Krumpen, T., Bergmann, M., 2017. Marine litter on deep Arctic seafloor continues to increase and spreads to the North at the HAUSGARTEN observatory. Deep-Sea Res. I Oceanogr. Res. Pap. 120, 88–99.

ter Halle, A., Ladirat, L., Gendre, X., Goudouneche, D., Pusineri, C., Routaboul, C., et al., 2016. Understanding the fragmentation pattern of marine plastic debris. Environ. Sci. Technol. 50 (11), 5668–5675.

Teuten, E.L., Rowland, S.J., Galloway, T.S., Thompson, R.C., 2007. Potential for plastics to transport hydrophobic contaminants. Environ. Sci. Technol. 41 (22), 7759–7764.

Tosetto, L., Williamson, J.E., Brown, C., 2017. Trophic transfer of microplastics does not affect fish personality. Anim. Behav. 123, 159–167.

Tosin, M., Weber, M., Siotto, M., Lott, C., Degli Innocenti, F., 2012. Laboratory test methods to determine the degradation of plastics in marine environmental conditions. Front. Microbiol. 3, 225. https://doi.org/10.3389/fmicb.2012.00225.

Turner, A., Holmes, L.A., 2015. Adsorption of trace metals by microplastic pellets in fresh water. Environ. Chem. 12 (5), 600–610.

Van, A., Rochman, C.M., Flores, E.M., Hill, K.L., Vargas, E., Vargas, S.A., et al., 2012. Persistent organic pollutants in plastic marine debris found on beaches in San Diego, California. Chemosphere 86 (3), 258–263.

Van Cauwenberghe, L., Janssen, C.R., 2014. Microplastics in bivalves cultured for human consumption. Environ. Pollut. 193, 65–70.

van Sebille, E., Wilcox, C., Lebreton, L., Maximenko, N., Hardesty, B.D., van Franeker, J.A., et al., 2015. A global inventory of small floating plastic debris. Environ. Res. Lett. 10 (12), 124006. https://doi.org/10.1088/1748-9326.

von Moos, N., Burkhardt-Holm, P., Kohler, A., 2012. Uptake and effects of microplastics on cells and tissue of the blue mussel *Mytilus edulis* L. after an experimental exposure. Environ. Sci. Technol. 46 (20), 11327–11335.

Wang, J.D., Peng, J.P., Tan, Z., Gao, Y.F., Zhan, Z.W., Chen, Q.Q., et al., 2017. Microplastics in the surface sediments from the Beijiang River littoral zone: composition, abundance, surface textures and interaction with heavy metals. Chemosphere 171, 248–258.

Wardrop, P., Shimeta, J., Nugegoda, D., Morrison, P.D., Miranda, A., Tang, M., et al., 2016. Chemical pollutants sorbed to ingested microbeads from personal care products accumulate in fish. Environ. Sci. Technol. 50 (7), 4037–4044.

Watts, A.J.R., Lewis, C., Goodhead, R.M., Beckett, S.J., Moger, J., Tyler, C.R., et al., 2014. Uptake and retention of microplastics by the shore crab Carcinus maenas. Environ. Sci. Technol. 48 (15), 8823–8830.

Watts, A.J.R., Urbina, M.A., Corr, S., Lewis, C., Galloway, T.S., 2015. Ingestion of plastic microfibers by the crab *Carcinus maenas* and its effect on food consumption and energy balance. Environ. Sci. Technol. 49 (24), 14597–14604.

Watts, A.J.R., Urbina, M.A., Goodhead, R., Moger, J., Lewis, C., Galloway, T.S., 2016. Effect of microplastic on the gills of the shore crab *Carcinus maenas*. Environ. Sci. Technol. 50 (10), 5364–5369.

Weinstein, J.E., Crocker, B.K., Gray, A.D., 2016. From macroplastic to microplastic: degradation of high-density polyethylene, polypropylene, and polystyrene in a salt marsh habitat. Environ. Toxicol. Chem. 35 (7), 1632–1640.

Welden, N.A.C., Cowie, P.R., 2016. Long-term microplastic retention causes reduced body condition in the langoustine, *Nephrops norvegicus*. Environ. Pollut. 218, 895–900.

Wiewel, B.V., Lamoree, M., 2016. Geotextile composition, application and ecotoxicology: a review. J. Hazard. Mater. 317, 640–655.

Wright, S.L., Rowe, D., Thompson, R.C., Galloway, T.S., 2013a. Microplastic ingestion decreases energy reserves in marine worms. Curr. Biol. 23 (23), R1031–R1033.

Wright, S.L., Thompson, R.C., Galloway, T.S., 2013b. The physical impacts of microplastics on marine organisms: a review. Environ. Pollut. 178, 483–492.

Wu, C.X., Zhang, K., Huang, X.L., Liu, J.T., 2016. Sorption of pharmaceuticals and personal care products to polyethylene debris. Environ. Sci. Pollut. Res. 23 (9), 8819–8826.

Zhang, W.W., Ma, X.D., Zhang, Z.F., Wang, Y., Wang, J.Y., Wang, J., et al., 2015. Persistent organic pollutants carried on plastic resin pellets from two beaches in China. Mar. Pollut. Bull. 99 (1–2), 28–34.

Zhang, K., Su, J., Xiong, X., Wu, X., Wu, C.X., Liu, J.T., 2016. Microplastic pollution of lakeshore sediments from remote lakes in Tibet plateau, China. Environ. Pollut. 219, 450–455.

Zhang, C., Chen, X.H., Wang, J.T., Tan, L.J., 2017. Toxic effects of microplastic on marine microalgae *Skeletonema costatum*: interactions between microplastic and algae. Environ. Pollut. 220, 1282–1288.

Ziccardi, L.M., Edgington, A., Hentz, K., Kulacki, K.J., Driscoll, S.K., 2016. Microplastics as vectors for bioaccumulation of hydrophobic organic chemicals in the marine environment: a state-of-the-science review. Environ. Toxicol. Chem. 35 (7), 1667–1676.

Zubris, K.A.V., Richards, B.K., 2005. Synthetic fibers as an indicator of land application of sludge. Environ. Pollut. 138 (2), 201–211.

CHAPTER 9

Chemicals Associated With Marine Plastic Debris and Microplastics: Analyses and Contaminant Levels

Sang Hee Hong, Won Joon Shim, Mi Jang
Korean Institute of Ocean Science and Technology, Geoje, South Korea
Korea University of Science and Technology, Daejeon, South Korea

ABBREVIATIONS

Brominated flame retardants (BFRs)

BTBPE 1,2-bis(2,4,6-tribromphenoxy)ethan
HBCD hexabromocyclododecane
PBDE polybrominated diphenyl ethers
TBBPA tetrabromobisphenol A
TDBP-TAZTO tris(2,3-dibromopropyl)isocyanurate

Phosphorus flame retardants (PFRs)

DOPO 3,4:5,6-dibenzo-2*H*-1,2-oxaphosphorin-2-oxide
EHDP 2-ethylhexyldiphenyl phosphate
TCEP tris(2-chloroethyl) phosphate
TCIPP tris(2-chloroisopropyl)phosphate
TDCIPP tris(1,3-dichloroisopropyl)phosphate
TMPP tris(methylphenyl)phosphate
TPHP tris(phenyl) phosphate

Plasticizers

DEHA diethylhexyl adipate
DEHP di-(2-ethylhexyl) phthalate
DEP diethyl phthalate
DIBP diisobutyl phthalate
BBP butyl benzyl phthalate
BEHP bis(2-ethylhexyl) ester
DINA diisononyl adipate

Stabilizers

UV320 2-(2H-benzotriazol-2-yl)-4,6-bis(2-methyl-2-propanyl)phenol
UV326 2-(5-chlor-2*H*-benzotriazol-2-yl)-4-methyl-6-(2-methyl-2-propanyl)phenol
UV327 2,4-di-tert-butyl-6-(5-chlorobenzotriazol-2-yl)phenol
UV328 2-(2H-benzotriazol-2-yl)-4,6-bis(1,1-dimethylpropyl)phenol

Microplastic Contamination in Aquatic Environments
https://doi.org/10.1016/B978-0-12-813747-5.00009-6

Antioxidants

2,4-DTBP	2,4-di-tert-butylphenol
Irganox 1010	pentaerythritol tetrakis (3-(3′,5′-di-tert-butyl-4′-hydroxyphenyl) propionate)
Irganox 1076	octadecyl-3-(3,5-di-tert-butyl-4-hydroxyphenyl)propionate
Irganox 1098	*N,N*′-hexane-1,6-diylbis(3-(3,5-di-tert-butyl-4-hydroxyphenyl)propionamide)
Irganox 1019	*N,N*′-propane-1,3-diylbis(3-(3,5-di-tert-butyl-4-hydroxyphenyl)propionamide)
Irganox 1024	2′,3-bis(3-(3,5-di-tert-butyl-4-hydroxyphenyl)propionyl)propionohydrazide
Irgafos 168	tris (2,4-di-tert-butylphenyl)phosphite

9.1 INTRODUCTION

The mass production and mass consumption of plastics have led to the substantial accumulation of plastic debris in natural habitats, with multiple adverse effects on terrestrial and marine ecosystems. Currently, plastics are the most abundant items in marine debris and the most frequently reported materials encountered by marine organisms (Thompson et al., 2009; CBD, 2012). Over 80% of the adverse impacts of marine debris on marine species are associated with plastic debris (CBD, 2012). There is growing concern over marine plastic debris because of the increased bioavailability of their breakdown products, including microplastics. As larger pieces of plastic debris become smaller through environmental weathering and degradation processes, their potential encounters with or ingestion by marine animals could greatly expand from large marine animals (e.g., marine mammals, sea turtles, and seabirds) to small marine species (e.g., zooplankton, crustaceans, bivalves, polychaetes, and fish). It is well known that entanglement and ingestion of plastic debris have adverse physical effects on marine life, such as internal and/or external abrasions and blockage of digestive organs, which results in satiation, starvation, and physical deterioration. However, the chemical hazards of plastic debris have recently attracted attention from scientists.

Plastic products intrinsically contain various chemicals, such as nonreacted monomers or oligomers, additives, and by-products. Additionally, plastics have a high potential for sorbing (adsorbing or absorbing) and concentrating hydrophobic organic chemicals dissolved in surrounding waters. Consequently, plastic debris can be a "chemical cocktail" and may influence the surrounding environments (Fig. 9.1). A key question related to plastic debris and microplastics is whether they act as sources and/or vectors of hazardous chemicals in the marine environment and to marine wildlife. However, there are still significant gaps in our knowledge. This chapter focuses on chemicals that are found or are likely to exist in plastic debris and microplastics. We analyze and synthesize research approaches and current findings, particularly on the levels of chemicals measured in plastic samples collected from the marine environment (covering coasts, open oceans, and marine organisms). Based on these studies, we discuss the limitations of current information and future perspectives for microplastic-associated chemicals.

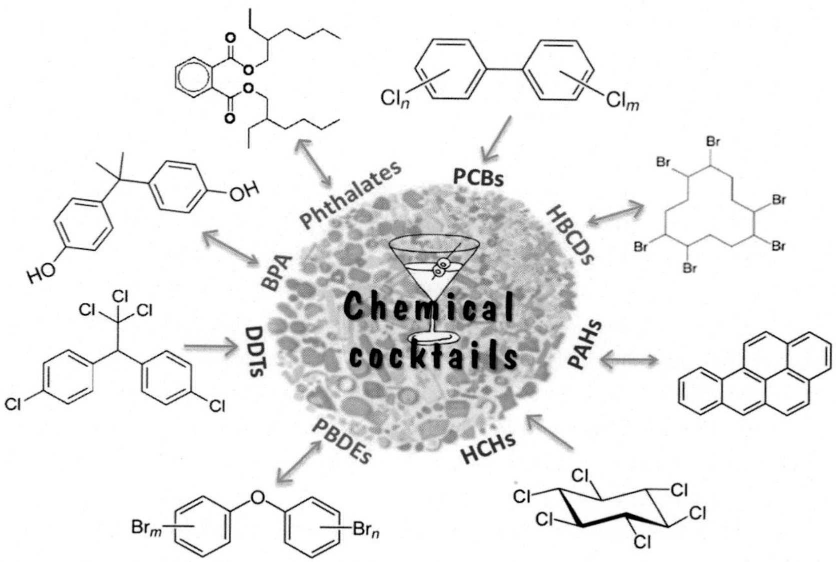

Fig. 9.1 Microplastics and their associated sorbed *(blue)* and additive *(red)* chemicals. *Arrows* indicate the direction of chemical movement from or to microplastics in water.

9.1.1 Chemicals in Marine Plastic Debris and Its Fragments

The chemicals included in marine plastic debris and microplastics can be divided into two categories based on their origins: (i) chemicals that originated from plastic materials (referred to here as "chemical ingredients of plastics") and (ii) chemicals sorbed from the environment. This section describes chemical categories and chemicals of concern found in marine plastics.

9.1.1.1 Chemical Ingredients of Plastics

Plastics are synthetic organic polymers made by polymerizing monomers (e.g., ethylene, propylene, and styrene) extracted from crude oil and natural gas. Because polymerization is rarely complete, unreacted monomers or oligomers (short chains of polymers) are commonly present in polymers. A modeling study ranked the families of polyurethane, polyacrylonitrile, polyvinyl chloride (PVC), epoxy resins, and styrenic copolymers as the most hazardous polymer groups for the environment and human health based on monomer hazard classification and determined that a considerable number of polymers (31 out of 55) are made of monomers that belong to the two worst of five hazard levels (Lithner et al., 2011). Aside from polymeric materials, various chemical additives are added to plastics to improve or modify the properties and processability of parent polymers. Examples of plastic additives include plasticizers, flame retardants (FRs), antioxidants, heat stabilizers, ultraviolet (UV) stabilizers, heat stabilizers, biocides, colorants, fillers, blowing agents, lubricants, and processing aids. Some additives are also used as monomers (e.g., bisphenol A (BPA) for polycarbonates and nonylphenol (NP) and octylphenol (OP) for phenol formaldehyde resins). In nearly all cases, additives in plastics are not chemically bound to polymers, with some exceptions. As a result, final plastic products frequently contain various chemical substances, including unreacted monomers, oligomers, additives, and other polymerization impurities (e.g., by-products, catalyst remnants, and polymerization solvents) in addition to base polymers. Among these substances, noncovalently bound low-molecular-weight substances have a great potential to migrate from plastics to the surrounding environment and other contact media during their production and use and after their disposal and can have adverse effects on human health and the environment. The typical additives used in plastics and their functions are briefly described below (Hansen et al., 2013; Ambrogi et al., 2017):

- *Plasticizers*: Plasticizers are organic substances that are added to polymers to improve their flexibility, extensibility, and processability. They are commonly used in rigid polymers that are in a glassy state at room temperature and become flexible by decreasing the glass transition temperature (T_g), melting temperature (T_m), and elastic modulus of polymers.
- *FRs*: FRs are used to prevent or delay fires by raising the ignition temperature of polymers and reducing the rates of burning and flame spread (Morose, 2006). They can be classified into four types based on the technology used to make them: halogen-based (e.g., brominated, chlorinated, and fluorinated FRs), phosphorus-based (e.g., ammonium polyphosphates and other phosphorus-based FRs), nitrogen-based (e.g., melamine and melamine derivatives), and inorganic salts (e.g., magnesium hydroxide and aluminum trihydroxide FRs).
- *Antioxidants*: Antioxidants are added to most hydrocarbon polymers (up to 2% *w/w*), including polyethylene (PE), polypropylene (PP), polystyrene (PS), and acrylonitrile butadiene styrene, to avoid thermal degradation and photooxidation, which may cause physical-mechanical changes in polymers and the loss of their original properties (Lau and Wong, 2000). Synthetic antioxidants are important ingredients in polyolefin polymers. Based on chemical structure, antioxidants are classified as amines, hindered phenols, phosphites, thioesters, and natural-based compounds. The types and amounts of antioxidants vary depending on the oxidizability of polymers, the processing temperature during production, and the target performance of end-use application.
- *Photostabilizers (UV or light stabilizers)*: Photostabilizers are added to plastics to prevent photochemical destruction processes and reactions caused by UV radiation (primarily 300–400 nm) and therefore ensure the long-term stability of polymers, particularly for

outdoor applications. They are classified into UV absorbers and quenchers. UV absorbers preferentially absorb more UV radiation than the polymer and convert it into harmless infrared radiation or heat, and quenchers deactivate the excited polymer molecules by energy transfer mechanisms.

- *Heat stabilizers*: Like UV light, heat can oxidize polymers during application and processing. Heat stabilizers improve the heat stability of polymers by stopping thermal oxidation or by attacking the decomposition products of oxidation. They are largely used in PVC products and recycled plastics as well. Metallic salts such as organometallic compounds (e.g., organotins) are major types of heat stabilizers.
- *Biocides (or antimicrobial agents)*: Biocides are used to protect polymers from attacks by bacteria, fungi, algae, etc. The main chemical groups are fungicides, bactericides, algaecides, and antifouling agents. Soft PVC and foamed polyurethanes are the major polymer types with applied biocides.
- *Colorants*: Colorants are added to mask undesirable color characteristics and enhance the aesthetic value of polymers without seriously altering their properties or performance.

Among these additives, the substances that are not chemically bound to the polymer matrix and have toxicological potential are chemicals of concern. Plastic additives that are not chemically bound to polymers and have been adopted on the REACH candidate list of substances of very high concern (https://echa.europa.eu/candidate-list-table), the OSPAR list of chemicals for priority action (OSPAR, 2013), the Norwegian priority list of hazardous substances (Norwegian Environment Agency, 2015), and the US EPA priority pollutant list (US EPA, 2014) are presented in Table 9.1 (adopted from Hansen et al., 2013).

9.1.1.2 Chemicals Sorbed From the Environment

Polymer materials can sorb and concentrate hydrophobic organic contaminants in the water column because of their lipophilic surface structure. This is the reason polymeric materials such as low-density PE, polyoxymethylene, polydimethylsiloxane, silicones, and poly(ethylene-*co*-vinylacetate) have been utilized as time-integrative passive samplers for monitoring nonpolar organic contaminants in the aquatic environment (Booij et al., 2016). Furthermore, the large surface-area-to-volume ratio of microplastics may enhance the sorption of contaminants dissolved in the water column. Field experiments using plastic pellets showed significant enrichment of chemicals in plastic pellets up to 10^6-fold higher than those in seawater (Mato et al., 2001; Rochman et al., 2013b). The most common sorbed chemicals found in microplastic field samples are persistent, bioaccumulative, and toxic substances (PBTs) such as polychlorinated biphenyls (PCBs), organochlorine pesticides (OCPs), and polycyclic aromatic hydrocarbons (PAHs). PBTs are hydrophobic (or lipophilic) and resistant to environmental degradation, resulting in their ubiquitous

Table 9.1 Priority substances having the potential for leaching from plastics

Substances[a–d]	Function	Relevant type of plastics	Binding in plastics	Concentration in plastics
Phthalates[a–d]	Plasticizer	Mainly PVC, PMMA, PP, polyamide, polyester, ABS, cellulose	A	Up to 50%
Bisphenol A[c]	Monomer	PC, epoxy resin, polyester, PUR, polyamide	B	Polycarbonate, 0.0003%–0.0141%
	Processing aid	Phenoplast cast resin		0.2% in phenoplast resin
	Antioxidant	PVC		0.2%
	Cross-linking agent	Rigid PUR foam		
Short- and medium-chain chlorinated paraffins[a–c]	Secondary plasticizer	PUR, acrylic	A	10%–15% (sealants)
	Flame retardant	Cellulosic textile		
Brominated flame retardants (PBDE, HBCD, TBBPA)[b,c]	Flame retardant	ABS, EPS, XPS, HIPS, polyamides, PBT, PE, PP, epoxy, unsaturated polyester, PUR; HBCD, mainly for EPS, XPS, and HIPS	A (except for TBBPA)	2%–28%, 0.1% (penta-BDE and octa-BDE) *HBCD, 1%–3% in XPS product, 0.5%–3% and 6%–25% in textile backcoating
Mercury and mercury compounds[b–d]	Catalyst	PUR	A, B	0.1%–0.3%
2-Methoxyethanol[a]	Solvent, intermediate	Epoxy resins, PVA	A	<0.1%
Nonylphenol and its etoxylates[b,c]	Monomer	Phenol/formaldehyde plastic	A	–
	Catalyst	Epoxy resins		
	Antioxidant	Vinyl, polyolefins, polystyrenics		
	Heat stabilizer	PVC		
Octylphenol and its ethoxylates[b,c]	Monomer	Phenol/formaldehyde resin	B	~3%–4% in resin
	Antioxidant (minor)	PVC cable		
	Emulsifier	Styrene-butadiene copolymers, PTFE		

Organic tin compounds (TBT, DBT, TPT)[b,c]	Biocide, impurity (stabilizers in PVC, catalyst for PUR foams)	PUR foam, PVC	A	1%–25% of TBT in antimicrobial products, 0.001%–1% of DBT in stabilizer, 0.05%–0.3% of DBT in PVC
Perfluorooctanoic acid (PFOA) and similar compounds[b,c]	Dispersing agent	PTFE, FEP, PVDF	A	Trace amount
Polycyclic aromatic hydrocarbons[b–d]	Impurity in plasticizer and carbon black	Soft plasticized plastics, ABS, PP, Black-colored plastics	A	–
Trichloroethylene[a,c,d]	Intermediate or chain transfer agent	PVC	C	–
Tris(2-chloroethyl) phosphate[a,c]	Plasticizer	Unsaturated polyester (mainly), PVC, PMMA, epoxy, polyamide, PC, PUR	A	Up to 20%
	Flame retardant	PVA		

ABS, acrylonitrile butadiene styrene; *EPS*, expanded polystyrene; *FEP*, fluoroethylene propylene; *PC*, polycarbonate; *PE*, polyethylene; *PMMA*, polymethyl methacrylate; *PP*, polypropylene; *PS*, polystyrene; *PTFE*, polytetrafluoroethylene (teflon); *PUR*, polyurethane; *PVA*, polyvinylacetate; *PVC*, polyvinyl chloride; *PVDF*, polyvinylidene fluoride; *XPS*, extruded polystyrene.

A, not chemically bound; *B*, unreacted monomer or oligomer; *C*, volatile.

[a]The REACH candidate list of substance of very high concern (SVHC) substances (https://echa.europa.eu/candidate-list-table).

[b]The OSPAR list of chemicals for priority action (OSPAR, 2013).

[c]Norwegian priority list of hazardous substances (Norwegian Environment Agency, 2015).

[d]US EPA priority pollutant list (US EPA, 2014).

Adapted from Hansen, E., Nilsson, N.H., Lithner, D., Lassen, C., 2013. Hazardous substances in plastic materials. Report number TA-3017/2013. Danish Technological Institute, Vejle, Denmark (Prepared by COWI incooperation with Danish Technological Institute).

presence in the environment (in water, sediment, soil, and air) and bioaccumulation and biomagnification in food webs. Similar to organic matter and biological tissues, plastic particles can rapidly sorb hydrophobic organic chemicals, not only PBTs but also organic additives dissolved in water.

9.1.1.3 Chemicals of Concern in Field Samples

To evaluate the environmental risks of microplastics, the level of exposure and effects of not only the particles themselves but also their associated chemicals in the environment must be analyzed. To meet this need, there have been increased efforts to identify the chemical concentrations and profiles in plastic debris and microplastics. The effort began by focusing on well-known PBT chemicals such as PCBs, OCPs, and PAHs because of their high toxic potential and widespread occurrence in the environment and has now been expanded to include emerging contaminants and additive chemicals. The chemicals that have been frequently found in macro- and microplastic samples from the marine environment are briefly introduced below. Their levels in field microplastic samples are described in detail in Section 9.2.3.

(i) *PCBs* are synthetic organochlorine chemicals with 1–10 chlorine atoms attached to a biphenyl backbone. By virtue of their physicochemical stability, they were used in a wide variety of applications, such as dielectric fluids in capacitors and transformers, plasticizers in paint and plastics, pigments, dyes, and other industrial applications. Due to their persistence, bioaccumulation, long-range transport potential, and high toxicity, PCBs have been banned in industrialized countries since the 1970s and have been designated as persistent organic pollutants (POPs) by the Stockholm Convention since 2001. The decline in the use of PCBs has led to a decline in PCB levels in the environment. However, they are still emitted from a few remaining sources such as old buildings and widely detected in both marine and terrestrial environments. PCBs are one of the classes of sorbed chemicals frequently reported in microplastics collected from the marine environment.

(ii) *OCPs* are synthetic pesticides widely used for agricultural and public health purposes (e.g., mosquito control) worldwide. Many OCPs (including aldrin, chlordane, chlordecone, dichlorodiphenyltrichloroethane (DDT), dieldrin, endrin, heptachlor, hexachlorobenzene (HCB), hexachlorocyclohexane (HCH), mirex, and toxaphene) have also been listed as priority POPs under the Stockholm Convention (see http://chm.pops.int/TheConvention/ThePOPs/AllPOPs/tabid/2509/Default.aspx). Although restrictions on the use and production of OCPs have been implemented worldwide, they are still detected in various environmental matrices, including water, sediment, soil, and air, and are still used in several developing countries (Ali et al., 2014). Among the OCP chemicals, DDT and HCHs are the most frequently reported in microplastics (Hong et al., 2017).

(iii) *PAHs* are ubiquitously distributed organic contaminants composed of two or more fused aromatic (benzene) rings. They are categorized into three groups based on their source: biogenic PAHs produced biologically, such as via diagenesis; petrogenic PAHs, derived from petroleum (e.g., oil spills, use of crude oil, and crude oil products); and pyrogenic PAHs, formed during the incomplete combustion of organic materials such as coal, crude oil, and wood (Zeng and Vista, 1997). PAHs are potentially toxic to living organisms (e.g., carcinogenic, mutagenic, and genotoxic) and accumulative in aquatic organisms. Owing to their potential toxicity, PAHs are included in the priority pollutant (or substances) lists of the US EPA (2014) and European Commission (2008).

(iv) *Brominated FRs (BFRs)* are a group of synthetic additives that have been applied in a wide variety of plastic products (e.g., computers, televisions, cars, and construction materials) and textiles to inhibit or delay the spread of fire. With some exceptions, such as tetrabromobisphenol A (TBBPA), most BFRs are not chemically bound to, but rather physically combined with, polymer materials. They therefore have a marked tendency to leach out of polymer materials into the surrounding environment, leading to widespread environmental contamination (Rani et al., 2017b). Among BFRs, hexabrominated biphenyls (hexa-BBs), pentabrominated diphenyl ethers (BDEs), and octa-BDEs were included in the Stockholm Convention on POPs in 2009, and hexabromocyclododecane (HBCD) was added in 2013, with a 5-year exemption for building insulation (see http://chm.pops.int/default.aspx). Restrictions on the use of PBBs, PBDEs, and HBCDs have led to the increased use of alternative FRs, such as other BFRs and nonbrominated FRs (e.g., phosphorus FRs (PFRs)).

(v) *Phthalates (or phthalic acid esters)* have been widely used as plasticizers added to polymers, mainly PVC, to increase both plasticity and durability since the 1920s. They can be found in flooring, furniture, medical devices, food packaging, pharmaceuticals, cosmetics, personal care products, cleaning products, toys, and more. Phthalates are not covalently bound to polymers and are therefore readily released into the environment during production, use, and disposal. To date, phthalates are widely found in all environmental compartments including air, water, sediment, and biota (Net et al., 2015). They have endocrine-disrupting potential and can cause adverse reproductive and developmental effects in aquatic organisms and humans (Staples et al., 1997; Katsikantami et al., 2016). Despite their relatively low potential for bioaccumulation and environmental persistence compared with PCBs, OCPs, and BFRs, their extensive use and widespread presence in the environment and toxic potential have drawn increasing attention to their adverse impacts on the environment and human health. Consequently, six phthalates have been listed as priority pollutants by the US EPA (2014) and European Commission (2008).

(vi) *BPA* is an organic compound with two phenol groups. It is used as a monomer for producing polymer materials, primarily PC plastics and epoxy resins, and thus commonly appears in a diverse range of products such as water bottles, sport equipment, water pipes, internal protective coating in food and beverage cans, toys, and thermal paper (Michałowicz, 2014). BPA is also used as an antioxidant and plasticizer in PVC plastics. BPA is biodegradable and has a short half-life compared with POPs, but it is frequently detected in the environment due to its abundant use and subsequent environmental release through various routes, including wastewater treatment plants, landfill leachates, and BPA-based products (Im and Loffler, 2016). It is known that BPA has endocrine-disrupting, oxidative, and mutagenic potential; has negative effects on reproduction and development; and is also related to diabetes, obesity, and cardiovascular diseases in humans (Michałowicz, 2014). Some countries in Europe, North America, and Asia have banned the production and sale of baby bottles made with BPA. However, it is widely applied to produce a variety of plastic products, including materials that come into contact with food. Several bisphenol analogues, such as 4,4′-methylenediphenol (BPF), 4-hydroxyphenyl sulfone (BPS), and 4,4′-(hexafluoroisopropylidene)diphenol (BPAF), have already been applied as BPA replacements in various applications, but their toxicity and environmental fate are still unknown.

It is interesting to note that 78% of priority pollutants listed by the US EPA and 61% of those listed by the European Union are associated with plastic debris (Rochman et al., 2013a). Two nontarget screening studies qualitatively confirmed that numerous chemicals existed in plastic marine debris (Gauquie et al., 2015; Rani et al., 2015). However, to date, there is limited information available on chemicals associated with plastic debris and microplastics, especially on additive chemicals, metals, and emerging contaminants.

9.1.2 Levels of Additive Chemicals in Plastic Consumer Products

Different types of plastic products contain different types and content of chemical ingredients (Chen et al., 2016). This section describes the levels of additives in consumer plastics reported in the literature. Tables 9.2 and 9.3 present the concentrations of BFRs, plasticizers, antioxidants, and UV stabilizers measured in various plastic products. The tables show that additives are used in a wide range of applications, from disposable goods to construction materials, in the concentration range from not detected (n.d.) to 236,000 mg/kg or greater. Most of the studies were conducted not to measure chemical levels in plastics but to assess human exposure to hazardous chemicals released from plastic products (Pezo et al., 2007; Weschler and Nazaroff, 2012). Overall, FRs were mostly monitored in indoor plastics (e.g., electronic goods, printed circuit boards, furniture, fabric, and car interiors), while plasticizers, stabilizers, antioxidants, BPA, and NP were measured in food contact items. The additive contents are diverse, not only among

Table 9.2 Concentrations of flame-retardant plastic additives in commercial products in the literature

Chemicals (no. of congener)	Product	Polymer	Sample number	Detection (%)	Concentration (μg/g)[a]		References
					Min-max	Mean ± STD	
Brominated flame retardants (BFRs)							
PBDEs (10)	Printed circuit board		2	100	281–4790		Ballesteros-Gómez et al. (2013)
PBDEs (3)	Television		12	83	85–89,200		Gallen et al. (2014)
	Household appliances		17	41	83–7200		Gallen et al. (2014)
	Electronic materials		10	50	2025–21,600		Gallen et al. (2014)
	Computer		2	50		1350	Gallen et al. (2014)
PBDEs (10)	Toy figurines		1	100		14.9	Ionas et al. (2016)
	Toy car		1	100		19.3	Ionas et al. (2016)
PBDEs (8)	Electronic equipment		4		<1.5–20,040		Abdallah et al. (2017)
	Car seat		2		22,000–30,000		Abdallah et al. (2017)
	Soft furnishing		5		<0.3–30,000		Abdallah et al. (2017)
PBDEs (11)	Furniture foam	PUF[b]	115	16	0.0009–0.04		Hammel et al. (2017)
HBCDs (3)	Ice box	EPS[c]	1	100		960 ± 29	Rani et al. (2014)
	Electronic appliance	EPS	2	100	0.65–602		Rani et al. (2014)
	Construction materials	EPS	5	100	709–905		Rani et al. (2014)
	Aquaculture farm buoy	EPS	2	100	0.1–53		Rani et al. (2014)
	Packaging material/ lab	EPS	4	100	3.3–591		Rani et al. (2014)

Continued

Table 9.2 Concentrations of flame-retardant plastic additives in commercial products in the literature—cont'd

Chemicals (no. of congener)	Product	Polymer	Sample number	Detection (%)	Concentration (μg/g)		References
					Min-max	Mean ± STD	
	Food packaging	XPS[d]	10	100	0.03–8.4		Rani et al. (2014)
	Food container	PS	8	100	0.02–0.12		Rani et al. (2014)
	Construction material	EPS	32	100	2500–24,000		Jeannerat et al. (2016)
	Construction material	XPS	54	100	2000–22,900		Jeannerat et al. (2016)
	Construction material	EPS	13			5800	Abdallah et al. (2017)
	Electronic equipment		4		<0.3–1600		Abdallah et al. (2017)
TBBPA	Printed circuit board		2	50	n.d.[e]–16,200		Ballesteros-Gómez et al. (2013)
	Car interior		2	100	7–15		Ballesteros-Gómez et al. (2013)
	Television		12	83	110–150,000		Gallen et al. (2014)
	Household appliances		17	59	26–160,000		Gallen et al. (2014)
	Electronic materials		10	70	59,000–164,000		Gallen et al. (2014)
	Toy plastic		2	50	150–143,000		Gallen et al. (2014)
	Computer		2	50		150	Gallen et al. (2014)
BTBPE	Printed circuit board		2	100	41–1700		Ballesteros-Gómez et al. (2013)
TDBP-TAZTO	Curtain (fabric)	Polyester	40	25	5350–23,700		Miyake et al. (2017)

Phosphorus flame retardants (PFRs)							
TCEP	Printed circuit board		2	100	0.8–1.5		Ballesteros-Gómez et al. (2013)
DOPO	Printed circuit board		2	100	35–90		Ballesteros-Gómez et al. (2013)
TPHP	Printed circuit board		2	100	590–3600		Ballesteros-Gómez et al. (2013)
	Car interior		2	100	9–2.5		Ballesteros-Gómez et al. (2013)
	Curtain (fabric)	Polyester	40	23	40–199	72.7 ± 48	Miyake et al. (2017)
EHDP	Printed circuit board		2	100	9–10		Ballesteros-Gómez et al. (2013)
TMPP	Printed circuit board		2	100	12–1800		Ballesteros-Gómez et al. (2013)
TCIPP	Car interior		2	100	200–16,400		Ballesteros-Gómez et al. (2013)
	Furniture foam	PUF	115	6	0.0008–0.05		Hammel et al. (2017)
	Curtain (fabric)	Polyester	40	5	3900–4310	1410 ± 289	Miyake et al. (2017)
TDCIPP	Furniture foam	PUF	115	19	0.0005–0.09		Hammel et al. (2017)

[a]On a plastic weight basis.
[b]PUF, polyurethane foam.
[c]Expanded polystyrene foam.
[d]Extruded polystyrene foam.
[e]n.d., not detected.

Table 9.3 Concentrations of plastic additives such as plasticizer, stabilizer, antioxidant, bisphenol A, and nonylphenol in commercial plastic products in the literature

Chemicals	Product	Polymer	Sample number	Detection (%)	Concentration (μg/g)[a]		References
					Min-max	Mean ± STD	
Plasticizer							
DEHP	Food container	PVC	17	12	17,900–18,100		Kao (2012)
	GVP film[b]		27	100	3.7–15.9	5.15	Li et al. (2016)
DEP	Food package	PP	1	100		11.4 ± 0.22	Li et al. (2014)
		PE	1	100		16.6 ± 2.15	Li et al. (2014)
DEHA	Wrapping films	PVC	23	43	162,000–204,000		Kao (2012)
	Packaging composite		3	100	0.6–19.3		Lin et al. (2015)
	Packaging film	PVC	1	100		45,500 ± 795	Lin et al. (2015)
DIBP	Food package	PP	1	100		49.5 ± 3.02	Li et al. (2014)
		PE	1	100		60.6 ± 2.44	Li et al. (2014)
	GVP film		27	100	2.2–16.7	3.65	Li et al. (2016)
BEHP	Food wrap film	PE	1	100		3.57 ± 0.12	Yang et al. (2017)
BBP	Food wrap film	PE	1	100		0.8 ± 0.09	Yang et al. (2017)
DINA	Wrapping films	PVC	23	57	208,000–236,000		Kao (2012)
	Food container	PVC	17	12	23,800–25,500		Kao (2012)
UV stabilizer							
UV320	Food-related items	PE, PP, PET	15	93	n.d.[c]–0.16		Rani et al. (2017a)
	Fishing float and net	PE, PP, acrylic	7	71	n.d.–56.5		Rani et al. (2017a)
	Packing string, fire cracker	PE, PP	5	100	0.03–0.23		Rani et al. (2017a)

UV326	Milk packing	PE	1	100		7.75	Zhang et al. (2016)
	Snack packing	PE	1	100		12	Zhang et al. (2016)
	Food-related items	PE, PP, PET	15	100	0.002–180		Rani et al. (2017a)
	Fishing float and net	PE, PP, acrylic	7	100	0.009–6.3		Rani et al. (2017a)
	Packing string, fire cracker	PE, PP	5	100	0.02–1.8		Rani et al. (2017a)
UV327	Food-related items	PE, PP, PET	15	100	0.002–36.9		Rani et al. (2017a)
	Fishing float and net	PE, PP, acrylic	7	100	0.002–3.7		Rani et al. (2017a)
	Packing string, fire cracker	PE, PP	5	100	0.02–1.75		Rani et al. (2017a)
UV328	Milk packing	PE	1	100		24.8	Zhang et al. (2016)
	Snack packing	PE	1	100		30.5	Zhang et al. (2016)
	Food-related items	PE, PP, PET	15	100	0.003–0.097		Rani et al. (2017a)
	Fishing float and net	PE, PP, acrylic	7	100	0.004–0.4		Rani et al. (2017a)
	Packing string, fire cracker	PE, PP	5	100	0.003–0.77		Rani et al. (2017a)
Antioxidant							
Irganox 1010	Packaging film	PE	1	100		11.2±0.6	Moreta and Tena (2015)
	Multilayer packaging	PE	8	100	7.2–208		Moreta and Tena (2015)

Continued

Table 9.3 Concentrations of plastic additives such as plasticizer, stabilizer, antioxidant, bisphenol A, and nonylphenol in commercial plastic products in the literature—cont'd

Chemicals	Product	Polymer	Sample number	Detection (%)	Concentration (μg/g)		References
					Min-max	Mean ± STD	
	Food-related items	PE, PP, PET	15	100	0.32–700		Rani et al. (2017a)
	Fishing float and net	PE, PP, acrylic	7	100	0.28–86		Rani et al. (2017a)
	Packing string, fire cracker	PE, PP	5	100	0.03–68		Rani et al. (2017a)
Irganox 1076	Packaging film	PE	1	100		99.5 ± 2.2	Moreta and Tena (2015)
	Multilayer packaging	PE	8	100	38.2–353		Moreta and Tena (2015)
	Food-related items	PE, PP, PET	15	100	0.06–786		Rani et al. (2017a)
	Fishing float and net	PE, PP, acrylic	7	86	n.d.–2130		Rani et al. (2017a)
	Packing string, fire cracker	PE, PP	5	100	0.03–68		Rani et al. (2017a)
Irgafos 168	Packaging film	PE	1	100		123 ± 6	Moreta and Tena (2015)
	Multilayer packaging	PE	8	88	n.d.–123		Moreta and Tena (2015)
Irganox 1098	Commercial plastic	PP	5	100	0.12–1.1	0.51 ± 0.37	Wang and Yuan (2016)
Irganox 1019	Commercial plastic	PP	5	100	0.11–0.54	0.37 ± 0.18	Wang and Yuan (2016)
Irganox 1024	Commercial plastic	PP	5	40	0.09–0.81	0.45 ± 0.51	Wang and Yuan (2016)

2,4-DTPB	Milk packing	PE	1	100		8.75	Zhang et al. (2016)
	Snack packing	PE	1	100		22	Zhang et al. (2016)
Bisphenol A (BPA) and nonylphenol (NP)							
BPA	Feeding bottle	PC[d]	30		0.7–70		Kao (2012)
	Plate	PC	3	100	4–9		Mercea (2009)
	Container	PC	3	100	2.5–70		Mercea (2009)
	Bottle	PC	2	100	2.1–25		Mercea (2009)
NP	Food containers, package	PE	12	58	n.d.–25.7		Kao (2012)
		PP	7	29	n.d.–2.9		Kao (2012)
		PS	4	25	n.d.–3.4		Kao (2012)
		PVC	3	100	78.4–2020		Kao (2012)

[a]On a plastic weight basis.
[b]GVP film, greenhouse vegetable production plastic film.
[c]n.d., not detected.
[d]PC, polycarbonate.

applications but also within the same applications. The content of FRs was relatively high in their applications compared with that of other additives, which might be due the fire safety requirements of those products. The highest levels of PBDE (up to 9%) and TBBPA (up to 16%) were detected in electronic appliances (Gallen et al., 2014). HBCD was mostly monitored in PS foam products, and a high concentration was found in construction materials (up to 2.4%) (Rani et al., 2014; Jeannerat et al., 2016). Much lower (but still detectable) amounts of HBCD were detected in ice boxes, floating buoys, and packing materials made of PS foam, indicating a lack of proper controls for the addition of HBCD to PS products (Rani et al., 2014). Global regulations on the use of PBDE and HBCD have led to an increasing market demand for alternative FRs. As a result, novel BFRs and PFRs are currently detected in plastic products and house dust (Saini et al., 2016; Mitro et al., 2016; Hammel et al., 2017; Miyake et al., 2017). However, plasticizers, stabilizers, and antioxidants (unlike FRs) have been monitored mostly in food containers and packing materials. Table 9.3 illustrates that disposable plastic items have considerable amounts of additives. PVC-based plastics tend to contain higher amounts of additive chemicals than other polymers.

9.2 ENVIRONMENTAL LEVELS AND PROFILES OF CHEMICALS IN MARINE PLASTIC SAMPLES

In the early 1970s, Carpenter et al. (1972) first reported that an environmental contaminant, PCB, was present in small PS spherules floating on the sea surface and first described its potential hazards to marine organisms. Microplastic-associated chemicals were readdressed in a paper from Mato et al. (2001) in the early 2000s. They found that plastic resin pellets had great potential for accumulating organic contaminants in marine waters with adsorption coefficients of 10^5–10^6 and represented the contamination levels of the surrounding environment. Subsequently, Takada (2006) started the International Pellet Watch program (IPW, a volunteer-based global monitoring program for POPs) in 2005, using PE pellets stranded along the coasts as a monitoring tool, and has posted global monitoring results on the website http://www.pelletwatch.org. The IPW program is providing useful information about chemical concentrations in microplastics distributed in the marine environment, particularly on sorbed chemicals. The assessment of chemical levels in microplastics has become more active since the 2010s, with increasing concern about microplastic pollution; the majority of related publications (~89%) were published from 2010 onward (Hong et al., 2017). Despite a recent increase, the overall number of studies that have measured plastic-associated chemicals in plastic marine debris and microplastics is still limited. This section summarized the chemical concentrations measured in plastic samples from the marine environment and monitoring approaches.

9.2.1 Characteristics of Plastic Samples

A recent article reviewed methods for analyzing chemicals associated with microplastics (Hong et al., 2017). This section briefly summarizes the monitoring approaches that have been applied for investigating microplastic-associated chemicals by referring to the review article and discusses the limitations of current information.

9.2.1.1 Sampling Locations

Chemical analysis has been conducted on macro- and microplastics mainly collected from beaches, with limited ocean surface samples and very limited samples from marine organisms (all from seabird stomachs) (Hong et al., 2017; Tables 9.4–9.10). No attempt has yet been made to analyze samples from seabeds. This bias may be due to the easy accessibility of beaches for microplastic sampling compared with the ocean and biota.

9.2.1.2 Shape

Microplastics have various shapes, including fragments, fibers/filaments, beads/spheres, films/sheets, and pellets. Among these, pellets (∼70%) and fragments (∼25%) have been commonly used for chemical analysis (Hong et al., 2017; Tables 9.4–9.10). Pellets and fragments were the dominant types of samples studied on beaches and from sea surface/biota, respectively. Foamed-type fragments were used in a small number of studies measuring a BFR, HBCD, and mercury (Graca et al., 2014; Jang et al., 2016, 2017). However, no data are available for fibers or films. Microplastic fibers are a dominant type of plastic found in marine organisms and abiotic matrices; therefore, assessing chemical composition and concentration in fibers is critical for understanding the role of plastics in the transfer of chemicals to the environment and marine organisms.

9.2.1.3 Polymer Type

Many studies conducted chemical analyses without classifying polymer types, particularly those on fragment samples (Hong et al., 2017). The studies that classified polymers mainly investigated less dense polymers, including PE, PP, and PS foams (expanded PS (EPS) and extruded PS (XPS)) because sampling was conducted mainly on beaches and on the ocean surface (Tables 9.4–9.10). Their lower specific densities (PE, 0.9–0.97 g/cm^3; PP, 0.94 g/cm^3; and EPS, 0.01–0.05 g/cm^3) than that of seawater (1.02 g/cm^3) led to these fragments floating on the ocean surface and washing ashore. Their highly frequent appearance may also be related to their high production volumes, accounting for ∼50% of global plastic demand (PlasticsEurope, 2015). Few studies have analyzed other polymer types such as polyurethane, nylon, acrylic/styrene, PC, and PVC (Nakashima et al., 2012; Mendoza and Jones, 2015; Rani et al., 2017a). Among resin pellet samples, yellowing PE pellets were frequently studied due to their utility for monitoring PBT contamination (Ogata et al., 2009;

Table 9.4 Concentrations of polychlorinated biphenyls (PCBs) in marine plastic debris in the literature

Chemicals (no. of congener)	Location	Year	Polymer	Shape and size (mm)	Conc. (ng/g)[a] Min-max	References
Beached plastic						
PCBs[b]	New Zealand (4 beaches)	1972–76	PE, PS	Pellet (~5)	<100–4000	Gregory (1978)
PCBs (18)	Japan (47 beaches)	2001	PE, PP	Pellet (1–5)	<28–2300	Endo et al. (2005)
PCBs (13)	Worldwide (32 beaches)	2001–08	PE	Pellet	5–605	Ogata et al. (2009)
PCBs	Japan (coastal region)	–	PE	Fragment (~35)	12–254	Teuten et al. (2009)
PCBs (18)	Portugal (2 beaches)	2008–09	PE, PP, PS-butadiene	Pellet	7–36	Frias et al. (2010)
PCBs (19)	Greece (4 beaches)	2008	PE	Pellet (1–5)	5–270[c]	Karapanagioti et al. (2011)
PCBs (39)	Worldwide (Japan, United States, Vietnam, and Costa Rica beaches)	2007–09	PE, PP	Fragment (~10)	1–436	Hirai et al. (2011)
PCBs (7)	Belgium (4 beaches)	2014	–[d]	Pellet	7.1–226[c]	Gauquie et al. (2015)
PCBs (13)	Eight remote island (Spain, United Kingdom, Australia, United States)	2008–11	PE	Pellet	0.1–9	Heskett et al. (2012)
PCBs	United States (8 beaches)	2010	–	(<5–50)	n.d.[e]–47	Van et al. (2012)
PCBs (13)	South Africa (3 beaches)	1984–2008	PE	Pellet (1–5)	16–113	Ryan et al. (2012)
PCBs (18)	Portugal (10 beaches)	2012	–	Pellet (3–6)	2–223	Antunes et al. (2013)

PCBs (13)	Portugal (8 beaches)	2008–12	PE	Pellet	10–310[c]	Mizukawa et al. (2013)
PCBs (13)	Ghana (11 beaches)	2009, 2013	PE	Pellet (1–5)	1–69	Hosoda et al. (2014)
PCBs (13)	Ghana (11 beaches)	2009, 2013	PE	Pellet	1–69	Hosoda et al. (2014)
PCBs (13)	Australia, New Zealand (23 beaches)	2012	PE	Pellet	n.d.–294	Yeo et al. (2015)
PCBs (20)	China (2 beach)	2012	–	Pellet	22–323	Zhang et al. (2015)
PCBs (7)	Mumbai (4 beaches)	2011–12	–	Pellet (2–5)	n.d.–210	Jayasiri et al. (2015)
PCBs (13)	Vietnam (2 beaches)	2007, 2017	–	Pellet	4–24	Le et al. (2016)
PCBs (51)	Brazil (41 beaches)	2010	–	Pellet	3–7550	Taniguchi et al. (2016)
PCBs (36)	Pacific (Hawaii and California beaches)	2003–04	PE, PP	Pellet, fragment	n.d.–980	Rios et al. (2007)
Floating plastic						
PCBs (Aroclor1254)	United States (Niantic Bay)	1971	PS	Sphere (0.1–2)	~5000	Carpenter et al. (1972)
PCBs (23)	Japan (4 regions)	1997–98	PP	Pellet (1–5)	4–117	Mato et al. (2001)
PCBs	Pacific (NPCG[f])	–	PE	Fragment (~35)	1–23	Teuten et al. (2009)
PCBs (39)	Pacific (including NPCG)	2005, 2008	PE, PP	Fragment (~10)	1–78	Hirai et al. (2011)
PCBs (39)	Atlantic (Caribbean Sea)	2008	PE, PP	Fragment (~10)	1–29	Hirai et al. (2011)

Continued

Table 9.4 Concentrations of polychlorinated biphenyls (PCBs) in marine plastic debris in the literature—cont'd

Chemicals (no. of congener)	Location	Year	Polymer	Shape and size (mm)	Conc. (ng/g) Min-max	References
PCBs (28)	South Atlantic (3 locations)	2010	–	Fragment	16–589	Rochman et al. (2014)
PCBs (34)	North Pacific	2007	PE, PP, PU, Nylon	(0.33–10)	n.d.–223	Mendoza and Jones (2015)
PCBs (12)	Mediterranean (2 coasts)	–	–	Fragment	356–500	Iñiguez et al. (2017)
Plastic ingested by organisms						
PCBs	Brazil (ingested plastics by eight species of seabirds)	1991–2008	PE, PP+	Pellet, fragment	243–491	Colabuono et al. (2010)
PCBs (14)	Norway (northern fulmar stomach)	2012–13	–	Fragment	0.08–64	Herzke et al. (2016)
PCBs (36)	Mexico (Albatross)	2003–04	PE, PP	Fragment	n.d.	Rios et al. (2007)

[a]On a plastic weight basis.
[b]Aroclor1221, 1232, 1242, and 1248.
[c]Median value.
[d]No information.
[e]Not detected.
[f]North Pacific central gyre.

Table 9.5 Concentrations of organochlorine pesticides (OCPs) in marine plastic debris in the literature

Chemicals	Location	Year	Polymer	Shape and size (mm)	Conc. (ng/g)[a] Min-max	References
Beached plastic						
Total DDTs	Worldwide (32 beaches)	2001–08	PE	Pellet	1.69–267	Ogata et al. (2009)
DDT					0.62–78	
DDD					0.25–32	
DDE					0.14–128	
DDE	Japan (coastal region)	–	PE	Fragment (∼35)	0.2–276	Teuten et al. (2009)
Total DDTs	Worldwide (Japan, United States, Vietnam, Costa Rica)	2007–09	PE, PP	Fragment (∼10)	0.2–198	Hirai et al. (2011)
DDT					n.d.[b]–98	
DDD					n.d.–16	
DDE					n.d.–198	
Total DDTs	Greece (4 beaches)	2008	PE	Pellet (1–5)	1.1–42	Karapanagioti et al. (2011)
DDT					0.56–25	
DDD					0.32–2.2	
DDE					0.11–15	
Total DDTs	South Africa (3 beaches)	1984–2008	PE	Pellet (1–5)	11–1281	Ryan et al. (2012)
DDT					4–762	
DDD					2.8–278	
DDE					0.3–241	
Total DDTs	Australia, New Zealand (23 beaches)	2012	PE	Pellet	0.5–422	Yeo et al. (2015)
DDT					0.94–305	
DDD					0.05–96	
Total DDTs	China (2 beaches)	2012	–[c]	Pellet	1.2–127	Zhang et al. (2015)
DDT					0.05–87	
DDD					0.3–26	
DDE					0.19–21	
Total DDTs	Pacific (Hawaii and California beaches)	2003–04	PE, PP	Pellet, fragment	n.d.–140	Rios et al. (2007)

Continued

Table 9.5 Concentrations of organochlorine pesticides (OCPs) in marine plastic debris in the literature—cont'd

Chemicals	Location	Year	Polymer	Shape and size (mm)	Conc. (ng/g) Min-max	References
Total DDTs	Portugal (2 beaches)	2008–09	PE, PP, PS-butadiene	Pellet	0.6–4.4	Frias et al. (2010)
DDT					n.d.–4.05	
DDD					n.d.–1.54	
DDE					n.d.–0.39	
Total DDTs	United States (8 beaches)	2010	–	Fragment (<5–50)	n.d.–76	Van et al. (2012)
Total DDTs	Eight remote islands (Spain, United Kingdom, Australia, United States)	2008–11	PE	Pellet	0.7–4.1	Heskett et al. (2012)
DDT					n.d.–2.8	
DDD					n.d.–1.5	
DDE					n.d.–2.5	
Total DDTs	Portugal (10 beaches)	2012	–	Pellet (3–6)	0.4–41	Antunes et al. (2013)
Total DDTs	Portugal (8 beaches)	2008–12	PE	Pellet	2–49[d]	Mizukawa et al. (2013)
DDT					1.12–31.8	
DDD					0.34–4.89	
DDE					0.05–13	
Total DDTs	Mumbai (4 beaches)	2011–12	–	Pellet (2–5)	0.3–110[d]	Jayasiri et al. (2015)
Total DDTs	Brazil (41 beaches)	2010	–	Pellet	<0.11–840	Taniguchi et al. (2016)
Total DDTs	Vietnam (2 beaches)	2007, 2017	–	Pellet	12–558	Le et al. (2016)
DDT					7.8–357	
DDD					3.4–187	
DDE					5.8–14.4	

HCHs	Worldwide (32 beaches)	2001–08	PE	Pellet	0.14–37.1	Ogata et al. (2009)
	Greece (4 beaches)	2008	PE	Pellet (1–5)	1.05–3.5	Karapanagioti et al. (2011)
	South Africa (3 beaches)	1984–2008	PE	Pellet (1–5)	2–112	Ryan et al. (2012)
	Eight remote islands (Spain, United Kingdom)	2008–11	PE	Pellet	0.4–19	Heskett et al. (2012)
	Australia, New Zealand (23 beaches)	2012	PE	Pellet	n.d.–29	Yeo et al. (2015)
	China (2 beaches)	2012	–	Pellet	n.d.–1.9	Zhang et al. (2015)
	Portugal (8 beaches)	2008–12	PE	Pellet	n.d.–3.3 [d]	Mizukawa et al. (2013)
	India (4 beaches in Mumbai)	2011–12	–	Pellet (2–5)	12–251[d]	Jayasiri et al. (2015)
	Brazil (41 beaches)	2010	–	Pellet	<0.24–4.1	Taniguchi et al. (2016)
	Vietnam (2 beaches)	2007, 2017	–	Pellet	0.4–1.4	Le et al. (2016)
Chlordanes	China (2 beaches)	2012	–	Pellet	n.d.–6.6	Zhang et al. (2015)
	United States (8 beaches)	2010	–		1.8–60	Van et al. (2012)
	Brazil (41 beaches)	2010	–	Pellet	<0.08–64	Taniguchi et al. (2016)
Heptachlor Aldrin Dieldrin Endrin	China (2 beaches)	2012	–	Pellet	n.d.–6.4 n.d.–0.54 n.d. n.d.–3.8	Zhang et al. (2015)
Drins Mirex	Brazil (41 beaches)	2010	–	Pellet	<0.44–38 <0.74–105	Taniguchi et al. (2016)
Cyclodienes	India (4 beaches in Mumbai)	2011–12	–	Pellet (2–5)	2.7–614[d]	Jayasiri et al. (2015)

Continued

Table 9.5 Concentrations of organochlorine pesticides (OCPs) in marine plastic debris in the literature—cont'd

Chemicals	Location	Year	Polymer	Shape and size (mm)	Conc. (ng/g) Min-max	References
Floating plastic						
Total DDTs	Pacific (including NPCG[e])	2005, 2008	PE, PP	Fragment (~10)	n.d.–2.0	Hirai et al. (2011)
DDT					n.d.–1	
DDD					n.d.	
DDE					n.d.–2	
Total DDTs	Atlantic (Caribbean Sea)	2008	PE, PP	Fragment	0.4–4.8	Hirai et al. (2011)
DDT					n.d.–3	
DDD					n.d.	
DDE					n.d.–2	
DDE	Pacific (NPCG)	–	PE	Fragment (~35)	0.1–4.7	Teuten et al. (2009)
Plastic ingested by organisms						
Total DDTs	Brazil (8 seabird species)	1991–2008	PE, PP	Pellet, fragment	64.4–88	Colabuono et al. (2010)
Chlordanes					4.29–14.2	
Mirex					6.48–14.6	
Total DDTs	Norway (northern fulmar)	2012–13	–	Fragment	n.d.–823	Herzke et al. (2016)
Total DDTs	Mexico (Albatross)	2003–04	PE, PP	Fragment	n.d.	Rios et al. (2007)

[a]On a plastic weight basis.
[b]Not detected.
[c]No information.
[d]Median value.
[e]North Pacific central gyre.

Table 9.6 Concentrations of polycyclic aromatic hydrocarbons (PAHs) in marine plastic debris in the literature

Chemicals (no. of congener)	Location	Year	Polymer	Shape and size (mm)	Conc. (ng/g)[a] Min-max	References
Beached plastic						
PAHs (16)	Pacific (Hawaii and California beaches)	2003–04	PE, PP	Pellet, fragment	n.d.[b]–1700	Rios et al. (2007)
PAHs	Japan (Coastal region)	–	PE	Fragment (~35)	<60–9370	Teuten et al. (2009)
PAHs (16)	Portugal (2 beaches)	2008–09	PE, PP, PS-butadiene	Pellet	75–1335	Frias et al. (2010)
PAHs (6)	Greece (2 beaches)	–	PE, PP	Pellet (1–5)	44–200	Karapanagioti et al. (2010)
PAHs (15)	Worldwide (Japan, United States, Vietnam, Costa Rica beaches)	2007–09	PE, PP	Fragment (~10)	0.26–8.45	Hirai et al. (2011)
PAHs (18)	Greece (4 beaches)	2008	PE	Pellet (1–5)	100–500	Karapanagioti et al. (2011)
PAHs	United States (8 beaches)	2010	–	Fragment (<5–50)	30–1900	Van et al. (2012)
PAHs (35)	Brazil (beach)	2010	–	Pellet	386–1996	Fisner et al. (2013b)
PAHs (12)	Brazil (beach)	2010	–	Pellet	130–27,735	Fisner et al. (2013a)
PAHs (17)	Portugal (10 beaches)	2012	–	Pellet (3–6)	53–44,800	Antunes et al. (2013)
PAHs (33)	Portugal (8 beaches)	2008–12	PE	Pellet	50–24,000[c]	Mizukawa et al. (2013)
PAHs (16)	China (2 beaches)	2012	–	Pellet	136–2384	Zhang et al. (2015)
PAHs (16)	Belgium (4 beaches)	2014	–	Pellet	105–7895	Gauquie et al. (2015)

Continued

Table 9.6 Concentrations of polycyclic aromatic hydrocarbons (PAHs) in marine plastic debris in the literature—cont'd

Chemicals (no. of congener)	Location	Year	Polymer	Shape and size (mm)	Conc. (ng/g) Min-max	References
PAHs (16)	Brazil (41 beaches)	2010	–	Pellet	192–13,708	Taniguchi et al. (2016)
Floating plastic						
PAHs	Pacific (NPCG[c])	–[d]	PE	Fragment (~35)	<100–959	Teuten et al. (2009)
PAHs (15)	Pacific (including NPCG)	2005, 2008	PE, PP	Fragment (~10)	12–868	Hirai et al. (2011)
PAHs (15)	Atlantic (Caribbean Sea)	2008	PE, PP	Fragment (~10)	88–105	Hirai et al. (2011)
PAHs (16)	North Pacific	2007	PE, PP, PU, Nylon	Fragment (0.33–10)	n.d.–249	Mendoza and Jones (2015)
PAHs (16)	Mediterranean (2 coasts)	–	–	Macrodebris	356–500	Iñiguez et al. (2017)
Plastic ingested by organisms						
PAHs (16)	Mexico (Albatross)	2003–04	PE, PP	Fragment	640	Rios et al. (2007)

[a]On a plastic weight basis.
[b]Not detected.
[c]North Pacific central gyre.
[d]No information.

Table 9.7 Concentrations of brominated and fluorinated compounds in marine plastic debris in the literature

Chemicals (no. of congener)	Location	Year	Polymer	Shape and size (mm)	Conc. (ng/g)[a] Min-max	References
Beached plastic						
PBDEs (20)	Worldwide (Japan, United States, Vietnam, and Costa Rica beaches)	2007–09	PE, PP	Fragment (~10)	0.3–412	Hirai et al. (2011)
PBDEs (7)	Brazil (41 beaches)	2010	–	Pellet	<0.26–2.2	Taniguchi et al. (2016)
HBCDs (3)	South Korea (12 beaches)	2015	EPS	Fragment (~5)	751,000–2700,000	Jang et al. (2017)
HBCDs (3)	South Korea (coast)	2013–14 and 2015	EPS	Fragment (>20 cm)	150–5,220,000	Jang et al. (2016, 2017)
HBCDs (3)	Asia-Pacific coastal regions (11 countries)	2014	EPS	Fragment (>20 cm)	10–4810	Jang et al. (2017)
HBCDs (3)	United States (Alaskan beach)	2014	EPS, XPS	Fragment (>20 cm)	980–14,500,000	Jang et al. (2017)
PFAAs (18)	Greece (5 beaches)	2011		Pellet (2–6)	0.01–0.18	Llorca et al. (2014)
Floating plastic						
PBDEs	Pacific (NPCG[b])	–	PE	Fragment (~35)	0.4–57	Teuten et al. (2009)
PBDEs	Japan (Coastal region)	–	PE	Fragment (~35)	0.9–2.1	Teuten et al. (2009)
PBDEs (20)	Pacific (including NPCG)	2005, 2008	PE, PP	Fragment (~10)	0.3–9909	Hirai et al. (2011)
PBDEs (20)	Atlantic (Caribbean Sea)	2008	PE, PP	Fragment (~10)	9.1–15.7	Hirai et al. (2011)
PBDEs (13)	South Atlantic (3 locations)	2010	–	Fragment	0.1–8	Rochman et al. (2014)

Continued

Table 9.7 Concentrations of brominated and fluorinated compounds in marine plastic debris in the literature—cont'd

Chemicals (no. of congener)	Location	Year	Polymer	Shape and size (mm)	Conc. (ng/g) Min-max	References
Plastic ingested by organisms						
PBDEs (49)	North Pacific (short-tailed shearwater)	2005	–[c]	Fragment (∼15)	n.d.[d]–16,444	Tanaka et al. (2013)
PBDEs (49)	North Pacific (short-tailed shearwater)	2008–10	–	Fragment (∼10)	0.16–4728	Tanaka et al. (2015)
PBDEs (10)	Norway (northern fulmar)	2012–13	–	Fragment	n.d.–17	Herzke et al. (2016)

[a]On a plastic weight basis.
[b]North Pacific central gyre.
[c]No information.
[d]Not detected.

Table 9.8 Concentrations of phenolic compounds in marine plastic debris in the literature

Chemicals (no. of congener)	Location	Year	Polymer	Shape and size (mm)	Conc. (ng/g)[a] Min-max	References
Beached plastic						
BPA	Worldwide (Japan, United States, Vietnam, and Costa Rica beaches)	2007–09	PE, PP	Fragment (~10)	n.d.[b]–730	Hirai et al. (2011)
NP					n.d.–3936	Hirai et al. (2011)
OP					n.d.–154	Hirai et al. (2011)
Floating plastic						
NP	Japan (4 regions)	1997–98	PP	Pellet (1–5)	13–16,000	Mato et al. (2001)
	Pacific (NPCG)	–[c]	PE	Fragment (~35)	24.9–2660	Teuten et al. (2009)
	Pacific (including NPCG)	2005, 2008	PE, PP	Fragment (~10)	5.8–997	Hirai et al. (2011)
	Atlantic (Caribbean Sea)	2008	PE, PP	Fragment (~10)	58.1–159	Hirai et al. (2011)
BPA	Pacific (NPCG)	–	PE	Fragment (~35)	5–284	Teuten et al. (2009)
	Pacific (including NPCG)	2005, 2008	PE, PP	Fragment (~10)	n.d.–283	Hirai et al. (2011)
	Atlantic (Caribbean Sea)	2008	PE, PP	Fragment (~10)	1–3.3	Hirai et al. (2011)
	South Atlantic (3 locations)	2010	–	Fragment	n.d.–5	Rochman et al. (2014)
OP	Pacific (including NPCG)	2005, 2008	PE, PP	Fragment (~10)	0.1–40.4	Hirai et al. (2011)
	Atlantic (Caribbean Sea)	2008	PE, PP	Fragment (~10)	8.5–26.4	Hirai et al. (2011)
Alkylphenols	South Atlantic (3 locations)	2010	–	Fragment	22–342	Rochman et al. (2014)
Alkylphenol ethoxylates	South Atlantic (3 locations)	2010	–	Fragment	0.8–98	Rochman et al. (2014)
Bromophenols(17)	Mediterranean (2 coasts)	–	–	Macrodebris	58–77	Iñiguez et al. (2017)
Chlorophenols (17)	Mediterranean (2 coasts)	–	–	Macrodebris	104–106	Iñiguez et al. (2017)

[a]On a plastic weight basis.
[b]Not detected.
[c]No information.

Table 9.9 Concentrations of antioxidants and benzotriazole-type ultraviolet stabilizers in marine plastic debris in the literature

Chemicals	Location	Year	Polymer	Shape and size (mm)	Conc. (μg/g)[a] Min-max	References
Beached plastic						
Antioxidants	Korea (coast)	2014	PE, PP, PET, PC, acrylic/styrene	Macrodebris	0.2–1620	Rani et al. (2017a)
Irganox 1076					n.d.[b]–1580	
Irganox 1010					0.02–155	
BHT					n.d.–29	
2,4-DTBP					0.03–16	
UV stabilizers					0.003–82	
UV326					n.d.–82	
UV320					n.d.–26	
UV327					n.d.–20	
UV328					n.d.–1.6	

[a]On a plastic weight basis.
[b]Not detected.

Karapanagioti et al., 2011; Heskett et al., 2012; Hosoda et al., 2014; Yeo et al., 2015; Le et al., 2016). It is known that PE has a high sorption capacity for hydrophobic contaminants, and the yellowness of the pellet is indicative of long residence time at sea (Endo et al., 2005; Rochman et al., 2013b). Some studies sorted samples according to color or weathering condition (Antunes et al., 2013; Fisner et al., 2013a; Gauquie et al., 2015; Taniguchi et al., 2016).

9.2.1.4 Sample Amount

An important parameter that must be considered prior to chemical analysis is the sample amount, as it must be sufficient to exceed the limits of detection of the target analytes. A suitable amount of a sample depends on the concentration of the target substances in the sample and the amount of interfering materials in the sample, as well as the sensitivity of the instrument. Most studies presented sample quantities used for extraction as the number of pellets or weights of pellets and fragments (Hong et al., 2017). Microplastic sample weights (including fragments and pellets) used per extraction ranged from 0.15 to 5 g, frequently $\leq$1 g. For pellet analysis, five pellets (equivalent to ~0.15 g) were commonly used per extraction. For EPS, 0.15–0.3 g samples were applied for chemical analysis (Graca et al., 2014; Jang et al., 2016, 2017). In field surveys, obtaining sufficient amounts of samples for chemical analysis increases in difficulty as the size of the plastics grows smaller, resulting in high detection limits and low detection frequencies for target chemicals. This is why there are limited data available on chemicals associated with microplastics.

Table 9.10 Concentrations of metals in marine plastic debris in the literature

Element	Location	Year	Polymer	Shape and size (mm)	Conc. (ng/g)[a] Min-max	References
Beached plastic						
Lead (Pb)	Japan (1 beach)	2009	PVC	PVC float (135)	~78[b]	Nakashima et al. (2012)
Aluminum (Al)	Malta (1 beach)	–[c]	–	Pellet	3.03–6.72[d]	Turner and Holmes (2011)
Iron (Fe)					6.37–16.8[d]	
Copper (Cu)					0.19–0.25[d]	
Aluminum (Al)	United Kingdom (4 beaches)	2009	–	Pellet (3–5)	0.007–0.05[b]	Ashton et al. (2010)
Iron (Fe)					0.03–0.07[b]	
Manganese (Mn)					0.001–0.008[b]	
Copper (Cu)					0.0001–0.0006[b]	
Zinc (Zn)					0.0004–0.002[b]	
Silver (Ag)					0.002–0.03[b]	
Cobalt (Co)					0.025–0.1[b]	
Cadmium (Cd)					0.002–0.01[b]	
Chromium (Cr)					0.02–0.15[b]	
Molybdenum (Mo)					0.007–0.015[b]	
Antimony (Sb)					0.006–0.017[b]	
Tin (Sn)					0.018–0.11[b]	
Uranium (U)					0.003–0.01[b]	
Lead (Pb)					0.0002–0.001[b]	
Mercury (Hg)	Poland (13 beaches)	2010, 2011	EPS	Fragmen t(~2 cm)	0.02–3863	Graca et al. (2014)
Floating plastic						
Copper (Cu)	China (river 8 sites)	2015	PE, PP, Copolymer	Fragment	0.08–0.5	Wang et al. (2017)
Lead (Pb)					0.04–0.13	
Zinc (Zn)					0.002–0.015	
Cadmium (Cd)					0.002–0.02	
Nickel (Ni)					0.0005–0.002	
Titanium (Ti)					13.6–39	

[a]On a plastic weight basis.
[b]Mean value.
[c]No information.
[d]Mean value of four weathered categories.

9.2.1.5 Sample Dimension

The size of plastic particles used for chemical analysis ranged from a few millimeters to a few centimeters. However, no studies have examined plastic particles <1 mm in size. The plastic particles most abundantly found in marine organisms such as bivalves and fish are a few hundred micrometers in size; therefore, there is a knowledge gap in estimating chemical exposure to marine organisms via ingestion of microplastics.

Microplastics collected from the environment have various sizes, shapes, colors, polymer compositions, and weathering conditions. Further studies are needed that examine microplastic samples with shapes, sizes, and polymer compositions frequently found in the environment and in marine organisms (e.g., fibers and fragments <1 mm). The overall number of studies on plastic- and microplastic-associated chemicals is limited. In particular, more studies on microplastics from the digestive tracts of marine organisms are needed to understand the contribution of microplastics in the transfer of hazardous chemicals to marine organisms, along with chemical concentrations in organisms, to clarify chemical transfer from ingested plastics to biota. To overcome the practical limitations of collecting a sufficient amount of microplastic samples from the environment and marine organisms, sensitive multiresidual analytic methods that are cost-effective and time-efficient are needed.

9.2.2 Non-Target Screening of Chemicals

Targeted analysis has been successfully applied to provide quantitative information (down to ultratrace levels) for a number of chemicals in plastic samples and environmental samples. However, it might miss useful information about unknown or untargeted substances in the samples that could be at high levels or have high potential toxicity. To fill this knowledge gap, a nontarget screening technique was recently applied to marine plastic debris (Gauquie et al., 2015; Rani et al., 2015). The nontarget analysis revealed a diversity of chemicals associated with plastic marine debris, identifying a variety of chemical ingredients of plastics (e.g., antioxidants, UV stabilizers, lubricants, intermediates, FRs, and degradation products), environmental pollutants (e.g., PCBs and PAHs), and biofilm and algal compounds. Previously, the majority of targeted analyses focused on a few sorbed chemicals (e.g., PCBs, PAHs, and OCPs) and a few additive chemicals (e.g., PBDEs and HBCDs). There is little information on emerging contaminants, metals, additive chemicals, and monomer/oligomer raw materials of plastics. To understand the potential harms and ecological risks of plastic debris and their fragments, comprehensive information on a wide range of chemicals is needed.

9.2.3 Levels of Chemicals in Environmental Samples

The target chemicals and their concentration ranges that have been detected in macro- and microplastic particles collected from the ocean surface, beaches, and marine organisms are summarized in Tables 9.4–9.10. Most of the studies analyzed PCBs, DDT and its degradation products (DDD and DDE), and PAHs. Some studies included brominated or fluorinated compounds such as PBDEs, HBCD, and perfluoroalkyl acids (PFAAs) as emerging contaminants and phenolic compounds such as BPA, NP, and OP as endocrine-disrupting substances. A few studies reported metals such as aluminum, iron, copper, and lead. One study measured plastic additives, such as phthalates, antioxidants, and UV stabilizers. Some other legacy OCPs, such as HCHs, HCB, chlordanes, and mirex, were also reported. In general, PCBs and DDT were present at high concentrations compared with other compounds. However, high concentrations of additive chemicals were sporadically found in some plastics sampled from coastal regions, remote islands, and open ocean (e.g., PBDEs in PP fragments from the Pacific central gyre and HBCDs in EPS debris from Korean and Alaskan beaches) (Hirai et al., 2011; Jang et al., 2017).

9.2.3.1 Polychlorinated Biphenyls

The number of congeners for PCB analysis varies from 7 to >100 (Aroclor mixture) among studies, which produce different PCB concentrations (generally the more congeners, the higher the reported levels) (Table 9.4). The concentrations of PCBs ranged from n.d. to 7550 ng/g plastics (*hereafter ng/g*). The highest PCB concentrations were reported in pellet samples (7550 ng/g) from Brazilian beaches (Taniguchi et al., 2016); PS spheres from Niantic Bay, United States (5000 ng/g, Carpenter et al., 1972); and pellets from New Zealand (4000 ng/g) and Japanese beaches (2300 ng/g) (Gregory, 1978; Endo et al., 2005). Most PCB concentrations detected in marine plastic particles were lower than 500 ng/g. In most studies, large differences were observed among pieces, even among those from the same location, which may be due to different residence times in the ocean, polymer composition, sizes, and shapes. Relatively higher PCB concentrations were measured in plastic fragments from urban beaches than from remote beaches and open ocean (Endo et al., 2005; Hirai et al., 2011), implying that the risks associated with these hydrophobic pollutants may be higher in urban areas. However, sporadic high concentrations of PCBs were also measured in pellets from remote islands, suggesting that plastic pellets may be an exposure route of the contaminants to remote sites. Three studies measured PCB concentrations (range, n.d.–491 ng/g) in plastic samples from seabird stomachs (Rios et al., 2007; Colabuono et al., 2010; Herzke et al., 2016).

9.2.3.2 Organochlorine Pesticides

In the case of DDT and its degradation products, the concentration of each compound was reported separately, and/or the total concentration of three compounds (expressed as DDTs) was provided (Table 9.5). The overall level of DDT compounds was higher than those of other OCPs, including HCHs, chlordanes, aldrin, endrin, and mirex, indicating their abundant use or high persistence in the environment. The concentrations of DDT,

DDD, DDE, and total DDTs ranged from n.d. to 762, n.d. to 278, n.d. to 276, and n.d. to 1281 ng/g, respectively. The highest concentrations of DDTs were detected in resin pellets stranded on beaches in South Africa (1281 ng/g) and Brazil (840 ng/g) (Ryan et al., 2012; Taniguchi et al., 2016). Technically, DDT was banned in the 1970s and the 1980s in many countries. However, DDT is still found in most microplastic samples, even in high proportions in some samples from beaches in Australia, China, Portugal, South Africa, and Vietnam (Ryan et al., 2012; Mizukawa et al., 2013; Zhang et al., 2015; Le et al., 2016). Among the degradation products, DDE was generally dominant over DDD. The overall level of total DDTs in floating microplastics from the open ocean (n.d.–4.8 ng/g; Teuten et al. 2009; Hirai et al., 2011) was much lower than those in samples from coastal beaches. Total DDT concentrations in pellets and fragments from seabird stomachs were in the range of n.d.–823 ng/g (Rios et al. 2007; Colabuono et al., 2010; Herzke et al., 2016). The highest concentration was detected in plastic fragments ingested by northern fulmars (Herzke et al., 2016).

Compared with DDTs, information on other OCPs is very limited. The concentration ranges of HCHs, chlordanes, and drins (aldrin, dieldrin, and endrin) detected in plastic pellets were 0.14–251 ng/g, n.d.–64 ng/g, and n.d.–38 ng/g, respectively (Table 9.5). The highest concentrations of HCHs and chlordanes were measured in resin pellets from the coasts of India and Brazil, respectively (Jayasiri et al., 2015; Taniguchi et al., 2016). Mirex was detected in pellet samples from Brazil (Taniguchi et al., 2016).

9.2.3.3 Polycyclic Aromatic Hydrocarbons

PAHs are ubiquitous in the environment and have multiple origins. PAHs were widely detected in marine plastic particles at concentrations up to 44,800 ng/g (Table 9.6). The highest level, 44,800 ng/g, was reported in pellets collected near industrial areas and port facilities along the Portuguese coast (Antunes et al., 2013). As with PCBs and DDTs, a wide variation in PAH concentrations was observed among the same polymer types at the same locations. Fragments form urban beaches had higher PAH concentrations than those from remote beaches and open ocean (Teuten et al., 2009; Hirai et al., 2011). However, PAHs were prevalent in plastic fragments from the ocean, with concentrations ranging from n.d. to 959 ng/g. Van et al. (2012) detected high concentrations of PAHs in PS foam samples from beaches (300–1900 ng/g) and PS foam products (240–1700 ng/g) that were not exposed to the environment and raised the possibility of potential PAH formation during PS foam production. PAHs were also detected in PE and PP fragments from Albatross boluses from Guadalupe Island, Mexico, at a concentration of 640 ng/g (Rios et al., 2007).

9.2.3.4 Brominated and Fluorinated Chemicals

The three primary brominated or fluorinated chemicals are considered as emerging POPs, and some of them are regulated under the Stockholm Convention. Among these,

PBDE, a BFR, was reported in several studies at concentrations ranging from n.d. to 16,444 ng/g (Table 9.7). Decabrominated diphenyl ether (BDE209) added to plastics as a FR contributed to the exceptionally high concentrations of PBDEs in some plastic fragments found in the North Pacific central gyre (167 and 9907 ng/g) and in seabird stomachs (9909 and 16,444 ng/g) (Hirai et al., 2011; Tanaka et al., 2013, 2015), which implies that microplastics can be a carrier of pollutants in the marine environment. The overall PBDE levels detected in plastic fragments from beaches, ocean surface, and marine organisms are much lower than those in plastic products (see Table 9.2).

Recent studies reported HBCD concentrations of up to 2700 μg/g in EPS microplastics (2–5 mm in diameter) stranded along the Korean coast (Al-Odaini et al., 2015; Jang et al., 2017; Table 9.7). HBCD, one of the widely used BFRs with PBDEs, is the main additive chemical applied to PS products (e.g., expanded or extruded PS) for construction materials and electronic housings (Alaee et al., 2003). HBCD is also commonly found in EPS debris collected from the Asia-Pacific coastal region (Bangladesh, Brunei, Canada, Hong Kong, Japan, Peru, Singapore, South Korea, Sri Lanka, Taiwan, Thailand, United States (California, Hawaii, and Alaska), and Vietnam), indicating chemical dispersion via EPS pollution in the marine environment (Table 9.7). The highest concentration of HBCD was detected at up to 14,500 μg/g in EPS debris from Alaskan beaches that was strongly suspected to be debris from the 2011 tsunami in Japan, as the HBCS levels were comparable with those reported in EPS and XPS construction materials (Table 9.2; Rani et al., 2014; Jeannerat et al., 2016). This finding implies that EPS debris has the potential to act as a moving source and as a medium for transporting hazardous chemicals in the marine environment.

Relatively low levels (0.01–0.18 ng/g) of PFAAs were detected in beached pellets compared with those of PBDEs and HBCDs (Table 9.7; Llorca et al., 2014). The similar chemical profiles of plastic pellets and sediments from the same locations indicate that PFAA compounds in pellets were sorbed from the surrounding waters.

9.2.3.5 Phenolic Compounds

Three phenolic compounds classified as endocrine-disrupting substances were also detected in the plastic fragments and pellets (Table 9.8). The concentration levels were in the ranges of n.d.–730 ng/g for BPA, n.d.–16,000 ng/g for NP, and n.d.–154 ng/g for OP. These compounds in plastic particles can be ascribed to plastic additives and/or sorption from seawater. The extraordinarily high levels of NP in resin pellets from remote beaches and open ocean, which is higher than those in food-related plastic products and comparable with those in PVC-based plastics, indicate that it is applied as an additive chemical (Hirai et al., 2011; see Table 9.3; Mercea, 2009; Kao, 2012). NP itself can be used as an antioxidant, or other antioxidant (tri(nonylphenyl) phosphite) and antistatic agent (NP polyethoxylates) additives in plastics may contain NP as nonreacting remnants or can be degraded to NP by oxidation (Mato et al., 2001). Sporadic high concentrations of BPA have also been measured in fragments from remote coasts (730 ng/g) and open ocean (~280 ng/g), which may be related to its use as a constituent monomer of PC plastics and/or other additive applications (Teuten et al., 2009; Hirai et al., 2011).

9.2.3.6 *Other Additives and Metals*

A recent study measured UV stabilizers and antioxidants in macroplastic debris commonly found on beaches, including plastic bowls, beverage bottles, snack packs, fishing floats and nets, ropes, and fire crackers, at concentrations of 0.2–1620 μg/g and 0.003–82 μg/g, respectively (Rani et al., 2017a; Table 9.9). Antioxidants were present at higher concentrations than UV stabilizers in both plastic debris and corresponding new plastics, indicating their high use over UV stabilizers (Table 9.3). Most antioxidants and UV stabilizers were relatively high in new plastics compared with corresponding plastic marine debris, implying their potential leaching or degradation during use or after disposal.

There have been a series of studies monitoring additive and sorbed chemicals and, to a lesser extent, metals, in floating, beached, or ingested plastic particles. Some metals are used as catalysts and additives in the plastic industry. Additionally, the oxidation of plastics by weathering changes their surface chemistry and facilitates the binding of metals from the surrounding water. The concentrations of aluminum, iron, and copper in beached pellets in Malta ranged 3.03–6.72, 6.37–16.8, and 0.19–0.25 g/g, respectively (Turner and Holmes, 2011; Table 9.10). A high level of lead, 78 g/g, was reported in one type of beached PVC fishing float in Japan (Nakashima et al., 2012).

9.3 ENVIRONMENTAL IMPLICATIONS: POTENTIAL SOURCES AND/OR VECTORS OF CHEMICALS IN THE MARINE ENVIRONMENT

A key question regarding microplastic pollution is the role of microplastics as carriers or vectors of hazardous chemicals to marine environments and marine organisms. In other words, plastic debris and microplastics can transfer chemicals from contaminated regions to remote and pristine areas, or if marine organisms ingest plastic particles, sorbed chemicals and additives can be released from plastics into gut fluids and subsequently bioaccumulate. Ryan et al. (1988) first raised the possibility that seabirds (shearwaters) assimilate toxic chemicals, PCBs, from ingested plastic particles, based on a positive correlation between the mass of ingested plastics and the body burden of PCBs. Thereafter, a series of controlled experiments proved the transfer of chemicals from ingested plastics to marine organisms, resulting in adverse health effects (Browne et al., 2013; Rochman et al., 2013b; Chua et al., 2014; Avio et al., 2015). Attempts have also been made to find evidence in the field. For example, Fossi et al. (2012) and Rochman et al. (2014) reported positive relationships between the density of plastics in seawater and the levels and profiles of additives (phthalates and PBDEs) in whales and fish, respectively. Tanaka et al. (2013) reported the presence of more highly brominated PBDE congeners (e.g., BDE-209 and BDE-183) that are not present in the natural prey items, in seabird tissues and ingested plastics. Recently, Jang et al. (2016) reported the strong enrichment of additive HBCDs in mussels inhabiting EPS debris compared with those living on other substrates (such as HDPE, metal, and rock), and the HBCD levels in mussels living on EPS closely reflected those of corresponding EPS substrates. On the other hand, recent model analyses and controlled exposure studies suggest that the relative contribution of ingested plastics to the bioaccumulation of hydrophobic contaminants would be negligible compared with those of water or prey and could even reduce the uptake of chemicals (Besseling et al., 2013, 2017; Bakir et al., 2016; Devriese et al., 2017; Koelmans et al., 2013, 2014, 2016). Koelmans et al. (2016) demonstrated that the majority of plastics (particularly 1–5 mm-sized microplastics) in the ocean have reached sorption equilibrium for most hydrophobic organic chemicals, including both additives and sorbed chemicals, and concluded that overall, the flux of organic chemicals bioaccumulated from natural prey overwhelms the flux from ingested microplastics in most habitats. However, macroplastic particles containing high amounts of additives would take longer to reach equilibrium in the ocean. For example, a high concentration of HBCD (up to 14,500 μg/g, Table 9.7) was found in EPS tsunami debris stranded along Alaskan beaches in 2014 that may have drifted across the Pacific Ocean over 2 years (Jang et al., 2017). The macrosized EPS debris can generate numerous nonequilibrated EPS fragments where it moves. Although their contribution as sources and/or vectors of hazardous chemicals to the marine environment or marine organisms is likely small compared with that of natural prey, plastic debris and their fragments are likely to have the potential to influence surrounding environments or marine organisms. Fragmentation of plastics in the environment leads to the leaching of chemicals from their newly exposed surfaces. Plastic debris and microplastics can transfer chemicals to marine organisms in two ways: indirect intake via water and direct uptake by ingestion. Chemicals leached out of plastics into water can be adsorbed onto organic matter or dissolved in the water column and subsequently taken up by organisms. Direct uptake takes place mainly in the digestive tract of organisms after the ingestion of plastics. Their relative importance in the fate and transfer of chemicals in the marine environment and the bioaccumulation of these chemicals by marine organisms remain unknown. The important point is that the widespread use of plastics and the resulting marine plastic pollution have become sources of hazardous chemicals in the marine environment. As aforementioned, plastic products and their debris may contain a complex mixture of chemicals, but our knowledge of the chemicals associated with plastic debris and microplastics is still very limited. Therefore, to answer the ultimate questions about the role of plastic debris and microplastics as sources and vectors of hazardous chemicals in the environment, further studies regarding their chemical constituents, sorption-desorption properties in the environment and digestive tracts of organisms, environmental fate, and biological effects are required.

REFERENCES

Abdallah, M.A., Drage, D.S., Sharkey, M., Berresheim, H., Harrad, S., 2017. A rapid method for the determination of brominated flame retardant concentrations in plastics and textiles entering the waste stream. J. Sep. Sci. 40, 1–9.

Alaee, M., Arias, P., Sjodin, A., Bergman, Å., 2003. An overview of commercially used brominated flame retardants, their applications, their use patterns in different countries/regions and possible modes of release. Environ. Int. 29, 683–689.

Ali, U., Syed, J.H., Malik, R.N., Katsoyiannis, A., Li, J., Zhang, G., Jones, K.C., 2014. Organochlorine pesticides (OCPs) in South Asian region: a review. Sci. Total Environ. 476, 705–717.

Al-Odaini, N.A., Shim, W.J., Han, G.M., Jang, M., Hong, S.H., 2015. Enrichment of hexabromocyclododecanes in coastal sediments near aquaculture areas and a wastewater treatment plant in a semi-enclosed bay in South Korea. Sci. Total Environ. 505, 290–298.

Ambrogi, V., Carfagna, C., Cerruti, P., Marturano, V., 2017. In: Jasso-Gastinel, C.F., Kenny, J.M. (Eds.), Additives in Polymers in Modification of Polymer Properties. William Andrew Publishing, pp. 87–108. ISBN 978-0-323-44353-1.

Antunes, J.C., Frias, J.G.L., Micaelo, A.C., Sobral, P., 2013. Resin pellets from beaches of the Portuguese coast and adsorbed persistent organic pollutants. Estuar. Coast. Shelf Sci. 130, 62–69.

Ashton, K., Holmes, L., Turner, A., 2010. Association of metals with plastic production pellets in the marine environment. Mar. Pollut. Bull. 60, 2050–2055.

Avio, C.G., Gorbi, S., Milan, M., Benedetti, M., Fattorini, D., d'Errico, G., Pauletto, M., Bargelloni, L., Regoli, F., 2015. Pollutants bioavailability and toxicological risk from microplastics to marine mussels. Environ. Pollut. 198, 211–222.

Bakir, A., O'Connor, I.A., Rowland, S.J., Hendriks, A.J., Thompson, R.C., 2016. Relative importance of microplastics as a pathway for the transfer of hydrophobic organic chemicals to marine life. Environ. Pollut. 219, 56–65.

Ballesteros-Gómez, A., de Boer, J., Leonards, P.E.G., 2013. Novel analytical methods for flame retardants and plasticizers based on gas chromatography, comprehensive two-dimensional gas chromatography, and direct probe coupled to atmospheric pressure chemical ionization-high resolution time-of-flight-mass spectrometry. Anal. Chem. 85, 9572–9580.

Besseling, E., Wegner, A., Foekema, E.M., van den Heuvel-Greve, M.J., Koelmans, A.A., 2013. Effects of microplastic on fitness and PCB bioaccumulation by the lugworm *Arenicola marina (L.)*. Environ. Sci. Technol. 47 (1), 593–600.

Besseling, E., Foekema, E.M., van den Heuvel-Greve, M.J., Koelmans, A.A., 2017. The effect of microplastic on the uptake of chemicals by the lugworm *Arenicola marina (L.)* under environmentally relevant exposure conditions. Environ. Sci. Technol. 51, 8795–8804.

Booij, K., Robinson, C.D., Burgess, R.M., Mayer, P., Roberts, C.A., Ahrens, L., Allan, I.J., Brant, J., Jones, L., Kraus, U.R., Larsen, M.M., Lepom, P., Petersen, J., Profrock, D., Roose, P., Schafer, S., Smedes, F., Tixier, C., Vorkamp, K., Whitehouse, P., 2016. Passive sampling in regulatory chemical monitoring of nonpolar organic compounds in the aquatic environment. Environ. Sci. Technol. 50, 3–17.

Browne, M.A., Niven, S.J., Galloway, T.S., Rowland, S.J., Thompson, R.C., 2013. Microplastic moves pollutants and additives to worms, reducing functions linked to health and biodiversity. Curr. Biol. 23, 2388–2392.

Carpenter, E., Anderson, S.J., Harvey, G.R., Miklas, H.P., Peck, B.B., 1972. Polystyrene spherules in coastal waters. Science 178, 749–750.

CBD (Convention on Biological Diversity)—GEF, 2012. Impacts of Marine Debris on Biodiversity: Current Status and Potential Solutions, Montreal, Technical Series No. 67. http://www.cbd.int/doc/publications/cbd-ts-67-en.pdf (Retrieved 12 November 2012).

Chen, D., Kannan, K., Tan, H.L., Zheng, Z.G., Feng, Y.L., Wu, Y., Widelka, M., 2016. Bisphenol analogues other than BPA: environmental occurrence, human exposure, and toxicity—a review. Environ. Sci. Technol. 50, 5438–5453.

Chua, E.M., Shimeta, J., Nugegoda, D., Morrison, P.D., Clarke, B.O., 2014. Assimilation of polybrominated diphenyl ethers from microplastics by the marine amphipod, *Allorchestes compressa*. Environ. Sci. Technol. 48 (14), 8127–8134.

Colabuono, F.I., Taniguchi, S., Montone, R.C., 2010. Polychlorinated biphenyls and organochlorine pesticides in plastics ingested by seabirds. Mar. Pollut. Bull. 60, 630–634.

Devriese, L.I., De Witte, B., Vethaak, A.D., Hostens, K., Leslie, H.A., 2017. Bioaccumulation of PCBs from microplastics in Norway lobster (*Nephrops norvegicus*): an experimental study. Chemosphere 186, 10–16.

Endo, S., Takizawa, R., Okuda, K., Takada, H., Chiba, K., Kanehiro, H., Ogi, H., Yamashita, R., Date, T., 2005. Concentration of polychlorinated biphenyls (PCBs) in beached resin pellets: variability among individual particles and regional differences. Mar. Pollut. Bull. 50, 1103–1114.

European Commission, 2008. Priority Substances and Certain Other Pollutants (According to Annex II of the Directive 2008/105/EC). http://ec.europa.eu/environment/water/water-framework/priority_substances.htm.

Fisner, M., Taniguchi, S., Majer, A.P., Bicego, M., Turra, A., 2013a. Concentration and composition of polycyclic aromatic hydrocarbons (PAHs) in plastic pellets: implications for small-scale diagnostic and environmental monitoring. Mar. Pollut. Bull. 76, 349–354.

Fisner, M., Taniguchi, S., Moreira, F., Bicego, M., Turra, A., 2013b. Polycyclic aromatic hydrocarbons (PAHs) in plastic pellets: variability in the concentration and composition at different sediment depths in a sandy beach. Mar. Pollut. Bull. 70, 219–226.

Fossi, M.C., Panti, C., Guerranti, C., Coppola, D., Giannetti, M., Marsili, L., Minutoli, R., 2012. Are baleen whales exposed to the threat of microplastics? A case study of the Mediterranean fin whale (*Balaenoptera physalus*). Mar. Pollut. Bull. 64 (11), 2374–2379.

Frias, J.P.G.L., Sobral, P., Ferreira, A.M., 2010. Organic pollutants in microplastics from two beaches of the Portuguese coast. Mar. Pollut. Bull. 60, 1988–1992.

Gallen, C., Banks, A., Brandsma, S., Baduel, C., Thai, P., Eaglesham, G., Heffernan, A., Leonards, P., Bainton, P., Mueller, J.F., 2014. Towards development of a rapid and effective non-destructive testing strategy to identify brominated flame retardants in the plastics of consumer products. Sci. Total Environ. 491, 255–265.

Gauquie, J., Devriese, L., Robbens, J., De Witte, B., 2015. A qualitative screening and quantitative measurement of organic contaminants on different types of marine plastic debris. Chemosphere 138, 348–356.

Graca, B., Beldowska, M., Wrzesien, P., Zgrundo, A., 2014. Styrofoam debris as a potential carrier of mercury within ecosystems. Environ. Sci. Pollut. Res. 21, 2263–2271.

Gregory, R., 1978. Accumulation and distribution of virgin plastic granules on New Zealand beaches. N. Z. J. Mar. Freshw. Res. 12, 399–414.

Hammel, S.C., Hoffman, K., Lorenzo, A.M., Chen, A., Phillips, A.L., Butt, C.M., Sosa, J.A., Webster, T.F., Stapleton, H.M., 2017. Associations between flame retardant applications in furniture foam, house dust levels, and residents' serum levels. Environ. Int. 107, 181–189.

Hansen, E., Nilsson, N.H., Lithner, D., Lassen, C., 2013. Hazardous substances in plastic materials (Report number TA-3017/2013). Danish Technological Institute, Vejle, Denmark (Prepared by COWI incooperation with Danish Technological Institute).

Herzke, D., Anker-Nilssen, T., Nost, T.H., Gotsch, A., Christensen-Dalsgaard, S., Langest, M., Fangel, K., Koelmans, A.A., 2016. Negligible impact of ingested microplastics on tissue concentrations of persistent organic pollutants in northern fulmars off coastal Norway. Environ. Sci. Technol. 50, 1924–1933.

Heskett, M., Takada, H., Yamashita, R., Yuyama, M., Ito, M., Geok, Y.B., Ogata, Y., Kwan, C., Heckhausen, A., Taylor, H., Powell, T., Morishige, C., Young, D., Patterson, H., Robertson, B., Bailey, E., Mermoz, J., 2012. Measurement of persistent organic pollutants (POPs) in plastic resin pellets from remote islands: toward establishment of background concentrations for International Pellet Watch. Mar. Pollut. Bull. 64, 445–448.

Hirai, H., Takada, H., Ogata, Y., Yamashita, R., Mizukawa, K., Saha, M., Kwan, C., Moore, C., Gray, H., Laursen, D., Zettler, E.R., Farrington, J.W., Reddy, C.M., Peacock, E.E., Ward, M.W., 2011. Organic micropollutants in marine plastics debris from the open ocean and remote and urban beaches. Mar. Pollut. Bull. 62, 1683–1692.

Hong, S.H., Shim, W.J., Hong, L., 2017. Methods of analysing chemicals associated with microplastics: a review. Anal. Methods 9, 1361–1368.

Hosoda, J., Ofosu-Anim, J., Sabi, E.B., Akita, L.G., Onwona-Agyeman, S., Yamashita, R., Takada, H., 2014. Monitoring of organic micropollutants in Ghana by combination of pellet watch with sediment analysis: E-waste as a source of PCBs. Mar. Pollut. Bull. 86, 575–581.

Im, J., Loffler, F.E., 2016. Fate of bisphenol A in terrestrial and aquatic environments. Environ. Sci. Technol. 50, 8403–8416.

Iñiguez, M.E., Conesa, J.A., Fullana, A., 2017. Pollutant content in marine debris and characterization by thermal decomposition. Mar. Pollut. Bull. 117, 359–365.

Ionas, A.C., Ulevicus, J., Gomez, A.B., Brandsma, S.H., Leonards, P.E.G., van de Bor, M., Covaci, A., 2016. Children's exposure to polybrominated diphenyl ethers (PBDEs) through mouthing toys. Environ. Int. 87, 101–107.

Jang, M., Shim, W.J., Han, G.M., Rani, M., Song, Y.K., Hong, S.H., 2016. Styrofoam debris as a source of hazardous additives for marine organisms. Environ. Sci. Technol. 50, 4951–4960.

Jang, M., Shim, W.J., Han, G.M., Rani, M., Song, Y.K., Hong, S.H., 2017. Widespread detection of a brominated flame retardant, hexabromocyclododecane, in expanded polystyrene marine debris and microplastics from South Korea and the Asia-Pacific coastal region. Environ. Pollut. 232, 785–794.

Jayasiri, H.B., Purushothaman, C.S., Vennila, A., 2015. Bimonthly variability of persistent organochlorines in plastic pellets from four beaches in Mumbai coast, India. Environ. Monit. Assess. 187, 469.

Jeannerat, D., Pupier, M., Schweizer, S., Mitrev, Y.N., Favreau, P., Kohler, M., 2016. Discrimination of hexabromocyclododecane from new polymeric brominated flame retardant in polystyrene foam by nuclear magnetic resonance. Chemosphere 144, 1391–1397.

Kao, Y.M., 2012. A review on safety inspection and research of plastic food packaging materials in Taiwan. J. Food Drug Anal. 20, 734–743.

Karapanagioti, H.K., Ogata, Y., Takada, H., 2010. Eroded plastic pellets as monitoring tools for polycyclic aromatic hydrocarbons (PAH): laboratory and field studies. Global NEST J. 12, 327–334.

Karapanagioti, H.K., Endo, S., Ogata, Y., Takada, H., 2011. Diffuse pollution by persistent organic pollutants as measured in plastic pellets sampled from various beaches in Greece. Mar. Pollut. Bull. 62, 312–327.

Katsikantami, I., Sifakis, S., Tzatzarakis, M.N., Vakonaki, E., Kalantzi, O.I., Tsatsakis, A.M., Rizos, A.K., 2016. A global assessment of phthalates burden and related links to health effects. Environ. Int. 97, 212–236.

Koelmans, A.A., Besseling, E., Wegner, A., Foekema, E.M., 2013. Plastic as a carrier of POPs to aquatic organisms: a model analysis. Environ. Sci. Technol. 47 (14), 7812–7820.

Koelmans, A.A., Besseling, E., Foekema, E.M., 2014. Leaching of plastic additives to marine organisms. Environ. Pollut. 187, 49–54.

Koelmans, A.A., Bakir, A., Burton, G.A., Janssen, C.R., 2016. Microplastic as a vector for chemicals in the aquatic environment: critical review and model-supported reinterpretation of empirical studies. Environ. Sci. Technol. 50, 3315–3326.

Lau, O.W., Wong, S.K., 2000. Contamination in food from packaging material. J. Chromatogr. A 882, 255–270.

Le, D.Q., Takada, H., Yamashita, R., Mizukawa, K., Hosoda, J., Tuyet, D.A., 2016. Temporal and spatial changes in persistent organic pollutants in Vietnamese coastal waters detected from plastic resin pellets. Mar. Pollut. Bull. 109, 320–324.

Li, B., Wang, Z., Lin, Q., Hu, C., Su, Q., Wu, Y., 2014. Determination of polymer additives-antioxidants, ultraviolet stabilizers, plasticizers and photoinitiators in plastic food package by accelerated solvent extraction coupled with high-performance liquid chromatography. J. Chromatogr. Sci. 53, 1026–1035.

Li, C., Chen, J.Y., Wang, J.H., Han, P., Luan, Y.X., Ma, X.P., Lu, A.X., 2016. Phthalate esters in soil, plastic film, and vegetable from greenhouse vegetable production bases in Beijing, China: concentrations, sources, and risk assessment. Sci. Total Environ. 568, 1037–1043.

Lin, Q.B., Cai, L.F., Wu, S.J., Yang, X., Chen, Z.N., Zhou, S.H., Wang, Z.W., 2015. Determination of four types of hazardous chemicals in food contact materials by UHPLC-MS/MS. Packag. Technol. Sci. 28, 461–474.

Lithner, D., Larsson, A., Dave, G., 2011. Environmental and health hazard ranking and assessment of plastic polymers based on chemical composition. Sci. Total Environ. 409, 3309–3324.

Llorca, M., Farre, M., Karapanagioti, H.K., Barcelo, D., 2014. Levels and fate of perfluoroalkyl substances in beached plastic pellets and sediments collected from Greece. Mar. Pollut. Bull. 87, 286–291.
Mato, Y., Isobe, T., Takada, H., Kanehiro, H., Ohtake, C., Kaminuma, T., 2001. Plastic resin pellets as a transport medium for toxic chemicals in the marine environment. Environ. Sci. Technol. 35, 318–324.
Mendoza, L.M.R., Jones, P.R., 2015. Characterisation of microplastics and toxic chemicals extracted from microplastic samples from the North Pacific Gyre. Environ. Chem. 12, 611–617.
Mercea, P., 2009. Physicochemical processes involved in migration of bisphenol A from polycarbonate. J. Appl. Polym. Sci. 112, 579–593.
Michałowicz, J., 2014. Bisphenol A—sources, toxicity and biotransformation. Environ. Toxicol. Pharmacol. 37, 738–758.
Mitro, S.D., Dodson, R.E., Singla, V., Adamkiewicz, G., Elmi, A.F., Tilly, M.K., Zota, A.R., 2016. Consumer product chemicals in indoor dust: a quantitative meta-analysis of U.S. studies. Environ. Sci. Technol. 50, 10661–10672.
Miyake, Y., Tokumura, M., Nakayama, H., Wang, Q., Amagai, T., Ogo, S., Kume, K., Kobayashi, T., Takasum, S., Ogawa, K., Kannan, K., 2017. Simultaneous determination of brominated and phosphate flame retardants in flame-retarded polyester curtains by a novel extraction method. Sci. Total Environ. 601–602, 1333–1339.
Mizukawa, K., Takada, H., Ito, M., Geok, Y.B., Hosada, J., Yamashita, R., Saha, M., Suzuki, S., Miguez, C., Frias, J., Antunes, J.C., Sobral, P., Santos, I., Micaelo, C., Ferreira, A.M., 2013. Monitoring of a wide range of organic micropollutants on the Portuguese coast using plastic resin pellets. Mar. Pollut. Bull. 70, 296–302.
Moreta, C., Tena, M.T., 2015. Determination of plastic additives in packaging by liquid chromatography coupled to high resolution mass spectrometry. J. Chromatogr. A 1414, 77–87.
Morose, G., 2006. An Overview of Alternatives to Tetrabromobisphenol A (TBBPA) and Hexabromocyclododecane (HBCD). Lowell Center for Sustainable Production, University of Massachusetts Lowell, Lowell, MA.
Nakashima, E., Isobe, A., Kako, S., Itai, T., Takahashi, S., 2012. Quantification of toxic metals derived from macroplastic litter on Ookushi Beach, Japan. Environ. Sci. Technol. 46, 10099–10105.
Net, S., Sempere, R., Delmont, A., Paluselli, A., Ouddane, B., 2015. Occurrence, fate, behavior and ecotoxicological state of phthalates in different environmental matrices. Environ. Sci. Technol. 49, 4019–4035.
Norwegian Environment Agency, 2015. List of Priority Substances. http://www.environment.no/topics/hazardous-chemicals/lists-of-hazardous-substances/list-of-priority-substances/.
Ogata, Y., Takada, H., Mizukawa, K., Hirai, H., Iwasa, S., Endo, S., Mato, Y., Saha, M., Okuda, K., Nakashima, A., Murakami, M., Zurcher, N., Booyatumanondo, R., Zakaria, M.P., Dung, L.Q., Gordon, M., Miguez, C., Suzuki, S., Moore, C., Karapanagioti, H.K., Weerts, S., McClurg, T., Burres, E., Smith, W., Van Velkenburg, M., Lang, J.S., Lang, R.C., Laursen, D., Danner, B., Stewardson, N., Thompson, R.C., 2009. International pellet watch: global monitoring of persistent organic pollutants (POPs) in coastal waters. 1. Initial phase data on PCBs, DDTs, and HCHs. Mar. Pollut. Bull. 58, 1437–1446.
OSPAR, 2013. OSPAR List of Chemicals for Priority Action. Reference number 2004-12, https://www.ospar.org/work-areas/hasec/chemicals/priority-action.
Pezo, D., Salafranca, J., Nerin, C., 2007. Development of an automatic multiple dynamic hollow fibre liquid-phase microextraction procedure for specific migration analysis of new active food packagings containing essential oils. J. Chromatogr. A 1174, 85–97.
PlasticsEurope 2015. Plastics—the Facts 2014/2015: an analysis of European plastics production, demand and waste data. Plastics Europe, Association of Plastic Manufacturers, Brussels, 2015. http://www.plasticseurope.org/Document/plastics—the-facts-2015.aspx?Page=DOCUMENT&FolID=2 (Accessed April 1, 2016).
Rani, M., Shim, W.J., Han, G.M., Jang, M., Song, Y.K., Hong, S.H., 2014. Hexabromocyclododecane in polystyrene based consumer products: an evidence of unregulated use. Chemosphere 110, 111–119.
Rani, M., Shim, W.J., Han, G.M., Jang, M., Al-Odaini, N.A., Song, Y.K., Hong, S.H., 2015. Qualitative analysis of additives in plastic marine debris and its new products. Arch. Environ. Contam. Toxicol. 69, 352–366.

Rani, M., Shim, W.J., Han, G.M., Jang, M., Song, Y.K., Hong, S.H., 2017a. Benzotriazole-type ultraviolet stabilizers and antioxidants in plastic marine debris and their new products. Sci. Total Environ. 579, 745–754.

Rani, M., Shim, W.J., Jang, M., Han, G.M., Hong, S.H., 2017b. Releasing of hexabromocyclododecanes from expanded polystyrenes in seawater—field and laboratory experiments. Chemosphere 185, 798–805.

Rios, L.M., Moore, C., Jones, P.R., 2007. Persistent organic pollutants carried by synthetic polymers in the ocean environment. Mar. Pollut. Bull. 54, 1230–1237.

Rochman, C.M., Browne, M.A., Halpern, B.S., Hentschel, B.T., Hoh, E., Karapanagioti, H.K., Rios-Mendoza, L.M., Takada, H., Teh, S., Thompson, R.C., 2013a. Classify plastic waste as hazardous. Nature 494, 169–171.

Rochman, C.M., Hoh, E., Hentschel, B.T., Kaye, S., 2013b. Long-term field measurement of sorption of organic contaminants to five types of plastic pellets: implications for plastic marine debris. Environ. Sci. Technol. 47, 1646–1654.

Rochman, C.M., Lewison, R.L., Eriksen, M., Allen, H., Cook, A.M., The, S.J., 2014. Polybrominated diphenyl ethers (PBDEs) in fish tissue may be an indicator of plastic contamination in marine habitats. Sci. Total Environ. 476–477, 622–633.

Ryan, P.G., Connell, A.D., Gardner, B.D., 1988. Plastic ingestion and PCBs in seabirds: is there a relationship? Mar. Pollut. Bull. 19, 174–176.

Ryan, P.G., Bouwman, H., Moloney, C.L., Yuyama, M., Takada, H., 2012. Long-term decreases in persistent organic pollutants in South African coastal waters detected from beached polyethylene pellets. Mar. Pollut. Bull. 64, 2756–2760.

Saini, A., Thysen, C., Jantunen, L., McQueen, R.H., Diamond, M.L., 2016. From clothing to laundry water: investigating the fate of phthalates, brominated flame retardants, and organophosphate esters. Environ. Sci. Technol. 50, 9289–9297.

Staples, A., Adams, W.J., Parkerton, T.F., Gorsuch, J.W., Biddinger, G.R., Reinert, K.H., 1997. Aquatic toxicity of eighteen phthalate esters. Environ. Toxicol. Chem. 16, 875–891.

Takada, H., 2006. Call for pellets! International Pellet Watch Global Monitoring of POPs using beached plastic resin pellets. Mar. Pollut. Bull. 52, 1547–1548.

Tanaka, K., Takada, H., Yamashita, R., Mizukawa, K., Fukuwaka, M., Watanuki, Y., 2013. Accumulation of plastic-derived chemicals in tissues of seabirds ingesting marine plastics. Mar. Pollut. Bull. 69, 219–222.

Tanaka, K., Takada, H., Yamashita, R., Mizukawa, K., Fukuwaka, M., Watanuki, Y., 2015. Facilitated leaching of additive-derived PBDEs from plastic by seabirds' stomach oil and accumulation in tissues. Environ. Sci. Technol. 49, 11799–11807.

Taniguchi, S., Colabuono, F.I., Dias, P., Oliveira, R., Fisner, M., Turra, A., Izar, G.M., Abessa, D.M.S., Saha, M., Hosodab, J., Yamashita, R., Takada, H., Lourenço, R.A., Magalhães, C.A., Bícegoa, M.C., Montone, R.C., 2016. Spatial variability in persistent organic pollutants and polycyclic aromatic hydrocarbons found in beach-stranded pellets along the coast of the state of São Paulo, southeastern Brazil. Mar. Pollut. Bull. (109), 87–94.

Teuten, E.L., Saquing, J.M., Knappe, D.R.U., Barlaz, M.A., Jonsson, S., Bjorn, A., Rowland, S.J., Thompson, R.C., Galloway, T.S., Yamashita, R., Ochi, D., Watanuki, Y., Moore, C., Viet, P.H., Tana, T.S., Prudente, M., Boonyatumanond, R., Zakaria, M.P., Akkhavong, K., Ogata, Y., Hirai, H., Iwasa, S., Mizukawa, K., Hagino, Y., Imamura, A., Saha, M., Takada, H., 2009. Transport and release of chemicals from plastics to the environment and to wildlife. Philos. Trans. R. Soc. Lond. Ser. B 364, 2027–2045.

Thompson, R.C., Moore, C.J., vom Saal, F.S., Swan, S.H., 2009. Plastics, the environment and human health: current consensus and future trends. Philos. Trans. R. Soc. B 364, 2153–2166.

Turner, A., Holmes, L., 2011. Occurrence, distribution and characteristics of beached plastic production pellets on the island of Malta (central Mediterranean). Mar. Pollut. Bull. 62, 377–381.

U.S. EPA, 2014. Priority Pollutant List (40 CFR Part 423, Appendix A). https://www.epa.gov/sites/production/files/2015-09/documents/priority-pollutant-list-epa.pdf.

Van, L., Rochman, C.M., Flores, E.M., Hill, K.L., Vargas, E., Vargas, S.A., Hoh, E., 2012. Persistent organic pollutants in plastic marine debris found on beaches in San Diego, California. Chemosphere 86, 258–263.

Wang, H., Yuan, J., 2016. Identification and quantification of unknown antioxidants in plastic materials by ultrasonic extraction and ultra-performance liquid chromatography coupled with quadrupole time-of-flight mass spectrometry. Eur. J. Mass Spectrom. 22 (1), 19–29.

Wang, J., Peng, J., Tan, Z., Gao, Y., Zhan, Z., Chen, Q., Cai, L., 2017. Microplastics in the surface sediments from the Beijiang River littoral zone: composition, abundance, surface textures and interaction with heavy metals. Chemosphere 171, 248–258.

Weschler, C.J., Nazaroff, W.W., 2012. SVOC exposure indoors: fresh look at dermal pathways. Indoor Air 22 (5), 356–377.

Yang, J.L., Li, Y.X., Wu, X., Ren, L., Zhang, J., Wang, Y., Zhang, Y.J., Sun, C.J., 2017. Gas chromatography-triple quadrupole tandem mass spectrometry for successive single-surface migration study of phthalate esters from polythene film. Food Control 73, 1134–1143.

Yeo, B.G., Takada, H., Taylor, H., Ito, M., Hsoda, J., Allinson, M., Connell, S., Greaves, L., McGrath, J., 2015. POPs monitoring in Australia and New Zealand using plastic resin pellets, and International Pellet Watch as a tool for education and raising public awareness on plastic debris and POPs. Mar. Pollut. Bull. 101, 137–145.

Zeng, E.Y., Vista, C.L., 1997. Organic pollutants in the coastal environment off San Diego, California. 1. Source identification and assessment by compositional indices of polycyclic aromatic hydrocarbons. Environ. Toxicol. Chem. 16, 179–188.

Zhang, W., Ma, X., Zhang, Z., Wang, Y., Wang, J., Wang, J., Ma, D., 2015. Persistent organic pollutants carried on plastic resin pellets from two beaches in China. Mar. Pollut. Bull. 99, 28–34.

Zhang, D., Liu, C., Yang, Y., 2016. Determination of UV absorbers and light stabilizers in food packing bags by magnetic solid phase extraction followed by high performance liquid chromatography. Chromatographia 79, 45–52.

CHAPTER 10

Occurrence and Fate of Microplastics in Wastewater Treatment Plants

Huase Ou, Eddy Y. Zeng
Jinan University, Guangzhou, China

10.1 INTRODUCTION

Microplastic (MP) contamination is a growing issue that is prevalent in global soils, air, and water. MP contamination occurs throughout the entire water cycle (hydrologic cycle) on earth, as MPs are ubiquitous in both the natural water cycle and the anthropogenic water cycle. In terms of the natural water cycle, considerable amount of MPs is detected in atmospheric precipitation, in wastewater and treated water, and in surface water, including oceans, lakes, and rivers (Andrady, 2011; Dris et al., 2015b; Lambert and Wagner, 2016). Even in the Southern Ocean around Antarctica, MPs are extracted from deep-sea sediments and surface waters (Sul et al., 2011). Microplastics pose an environmental concern due to their potential to physically and chemically harm a variety of exposed aquatic organisms ranging from zooplankton to mammals (Cole et al., 2015) and their potential threat to human health through ingesting contaminated seafood (Rochman et al., 2015). In the anthropogenic water cycle, MPs are found in municipal sewage (Tagg et al., 2015) and in the influent and effluent of wastewater treatment plants (WWTPs) (Carr et al., 2016). Studies that examined the source, concentration, and composition of MPs in wastewater have confirmed that the dominant MPs found in the influent of WWTPs consist of microfibers from domestic clothing washing (Baldwin et al., 2011) and microbeads from personal care products (van Wezel et al., 2016).

Since WWTPs are the final step of the anthropogenic water cycle, they are receptors for cumulative loadings of contaminates derived from industry, landfill, domestic wastewater, and storm water. Furthermore, WWTPs also act as the link between the natural and anthropogenic water cycles and thus act as the ultimate barriers against these undesired contaminants before MPs are released into the natural aquatic environment. Therefore, effective WWTPs are essential in reducing MP contamination and associated threats to ecosystems and human health.

The wastewater treatment process includes pretreatment, primary treatment, secondary treatment, and tertiary treatment. Bar screening, degreasing, air flotation, and primary sedimentation are typical pretreatment and primary treatments designed to remove large-size particles and oil in influent. In the secondary treatment, biofilm/activated sludge

Microplastic Contamination in Aquatic Environments
https://doi.org/10.1016/B978-0-12-813747-5.00010-2

processes and secondary sedimentation are employed for the removal of organic materials enriched in carbon, nitrogen, and phosphorus. Tertiary treatment usually applies sand filtration, chemical oxidation, and membrane separation (Le-Clech et al., 2006; Meng et al., 2010; Ternes et al., 2003) so to further improve wastewater quality. Although some of these treatment processes remove MPs by trapping them in the sludge, there is no treatment method that is specially designed for eliminating MPs. Considering the particulate feature of MPs, mechanical methods may be efficient. However, MPs are still detected in the final effluent from tertiary WWTPs that apply membrane filtration—indicating that existing treatment methods are still unable to completely eliminate MPs in wastewater (Ziajahromi et al., 2017). Since WWTPs are inefficient to completely remove MPs from wastewater, MPs are then discharged into ambient waters (Lasee et al., 2017; Mason et al., 2016).

Removal efficiencies of MPs may be variable in different water treatment processes, and this hypothesis has been tested by researchers that assessed the transport of MPs in full-scale WWTPs (Carr et al., 2016; Dris et al., 2015a; Talvitie et al., 2017). In a series of studies that evaluated residual MPs in the final effluent from WWTPs, in regard to morphology and composition, one study showed that substantial amounts (10^6–10^8 particles day^{-1}) of MPs were released from just one WWTP (Talvitie et al., 2017). These initial studies provided important information about the fate of MPs during wastewater treatment, but further research is still needed for more comparison and comprehensive investigation.

In this chapter, sampling, pretreatment, and analytic methods for MPs in wastewater are summarized. The latest knowledge about the classification, source, and transport of MPs in WWTPs is covered. Discharge of MPs from WWTPs into receiving waters is also evaluated based on existing literature. Data summarized in this chapter are useful as references for cultivating policies that prevent MPs from entering wastewater treatment system and for guiding the improvements of monitoring methods and wastewater treatment technologies.

10.2 SAMPLING, PRETREATMENT, AND ANALYTICAL METHOD

Wastewater contains high concentrations of various contaminants. To determine the sizes, shapes, and constituents of MPs in wastewater, methods for sampling, pretreatment, and analytic methods with careful and robust designs are critical.

10.2.1 Sampling Devices and Strategies

Sampling is the first step toward the identification and quantification of MPs. Samples include influent, effluent, and sludge. The use of appropriate technology and methods is important to obtain robust data and minimize random errors. Though there is still no standardize sampling method for MPs in wastewater, there are several effective methods that can be employed. For example, these days, MPs in wastewater samples are collected using various customized separation devices composed of metallic mesh/sieve, which are designed based on mechanical screening. For example, mesh screens with pore sizes from 25 to 500 μm were stacked on top of each other, with the largest pore size at the top (Fig. 10.1) (Ziajahromi et al., 2017). With this arrangement, size-segregated particles can be separated from wastewater continuously at high efficiency.

Commonly applied sampling strategies include grab sampling and continuous sampling. Grab sampling is a sampling technique in which a single sample or measurement is taken at a certain point in time (Lasee et al., 2017; Talvitie et al., 2017). With this sampling technique, wastewater or activated sludge samples with high water content are generally acquired using amber glass jars or steel buckets and then pumped through separation devices to obtain particles (Fig. 10.2). One grab sampling usually lasts 1–2 h (Ziajahromi et al., 2017). The grab sampling method is simple and efficient, but the obtained sample only represents one cross profile data point during a given period. Due to the variability and heterogeneity of wastewater, it should be noted that random errors are inevitable when using the grab sampling technique.

Continuous sampling provides a dataset in temporal dimension. For example, Dyachenko et al. (2017) collected the effluent from a WWTP over a period of 24 h at 2 h interval to calculate the average concentration of MPs. In another study, 24 samples were taken at 1 h intervals within a day to evaluate the removal efficiency of MPs in a WWTP (Talvitie et al., 2017). Unlike grab sampling, results from continuous sampling

Fig. 10.1 Separating devices based on metallic mesh/sieve. (A) A simple stack of four 8 in-diameter stainless steel sieves of various mesh sizes at 5000, 1000, 355, and 125 μm (Dyachenko et al., 2017) and (B) an enclosed sampling device based on sieve column with various mesh sizes at of 500, 190, 100, and 25 μm (Ziajahromi et al., 2017).

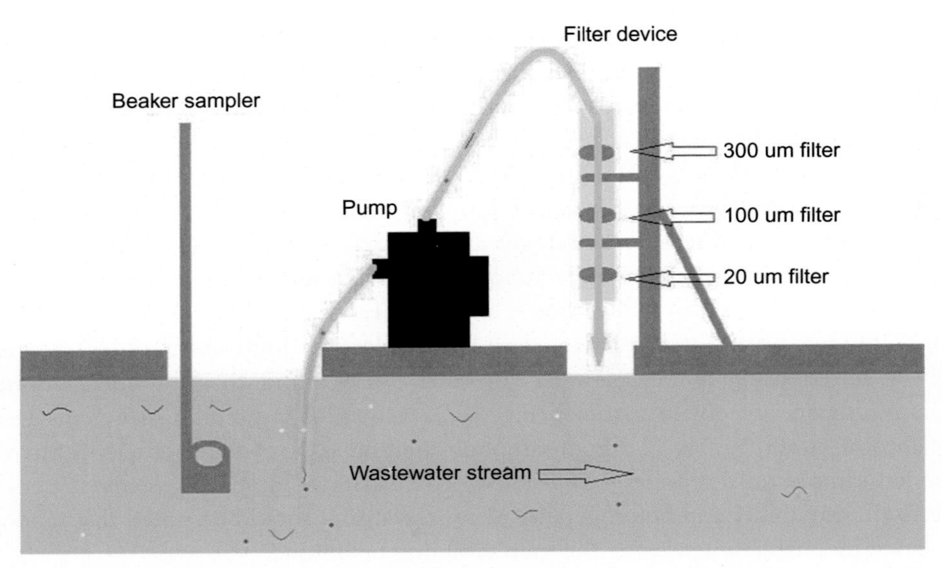

Fig. 10.2 Schematic picture of the grab sampling methods, beaker sampler and pump, and filtering assembly (Talvitie et al., 2017).

were able to identify a lower concentration of MPs at night than in the day and thus able to suggest an uneven temporal distribution of MPs.

Sludge samples in WWTPs are often collected with specially made samplers, such as the Van Veen grab sampler (Arp et al., 2011). Sludge samples generally contain more solid materials than wastewater, which can include microorganisms, organic materials, inorganic materials, and MPs. To separate MPs from colloidal solid materials, sludge samples need to be softened and washed by deionized water through a series of mesh screens (Mahon et al., 2017). Elutriation is also used to extract MPs from sludge, which is able to separate lighter particles from heavier ones through an upward flow of liquid and/or gas (Mahon et al., 2017). With these procedures, visible MP particles can be separated from wastewater and sludge.

10.2.2 Pretreatments Using H_2O_2 and Fenton Oxidation

Municipal sewage contains abundant particulates, organic matter, ions, and other trace impurities (e.g., chemicals). Organic matter, especially oil and colloids, can adhere to the surface of MPs. Furthermore, during the wastewater treatment processes, coagulant, flocculant, and microorganisms can also accumulate on MPs. Microplastics coated with organics cause significant interferences during the detection and identification of MPs using Fourier transform infrared (FTIR) or Raman spectroscopy (Tagg et al., 2017). Thus, sample pretreatment is recommended before analysis to effectively clean MP particles without breaking polymer backbone or altering particle size.

A commonly applied pretreatment method is preoxidation using ~30% hydrogen peroxide (H_2O_2), which can remove surface biogenic materials from the polymer (Nuelle et al., 2014). This method is widely used to eliminate attached biogenic materials on MPs obtained from wastewater and sludge samples (Leslie et al., 2017; Majewsky et al., 2016; Ziajahromi et al., 2017). The effect of H_2O_2 on FTIR of MPs was evaluated by Tagg et al. (2015). Polyethylene, polypropylene, polyvinyl chloride, polystyrene, nylon 6, and polyethylene terephthalate were tested. The MP pellets were firstly immersed in 15 mL of 30% H_2O_2, briefly shaken and stored for 3, 5, or 7 days. A 7-day exposure to H_2O_2 had no observable impact on the FTIR spectra of these polymers. The efficiency of sample filtration was also dramatically improved after H_2O_2 pretreatment. Heating of suspended sample matrix with H_2O_2 further improved the oxidation efficiency, and so, the pretreatment time was reduced to 30 min (Dyachenko et al., 2017). However, Munno et al. (2017) observed that preoxidation using H_2O_2 combined with heating induced melting of some MPs, especially microbeads. Therefore, additional heating treatment is not recommended for MP pretreatment using H_2O_2.

Fenton treatment using H_2O_2 and Fe(II) serves as an alternative pretreatment method. It is a fast and environmentally friendly oxidation method that can degrade most natural and artificial organic materials (Oturan and Aaron, 2014). National Oceanic and Atmospheric Administration recently recommended a wet peroxide oxidation (WPO) pretreatment method based on Fenton to remove organic impurities on MPs in water and sludge samples (Masura et al., 2015). This WPO method stated that 20 mL of 30% H_2O_2 and 20 mL of aqueous 0.05 M Fe(II) should be added to the beaker containing 0.3 mm size fraction from water samples and then subjected to procedural agitation and heating. This procedure has now been widely applied to improve the efficiency of sample pretreatment (Baldwin et al., 2016; Dyachenko et al., 2017; Tagg et al., 2017; Yonkos et al., 2014). Tagg et al. (2017) tested the potential degradation and fragmentation of MPs after using this WPO method. A comparison of treated and control samples suggested no evident variation of particle sizes or FTIR spectra. This WPO method can reduce the exposure time from days or hours to several minutes, which is an appropriate pretreatment method for rapid extraction and detection of MPs from field wastewater and sludge samples.

After oxidation pretreatments, water samples containing MPs need to be transferred to a density separator for further separation of particles from suspension (Masura et al., 2015).

10.2.3 Analytical Methods

Various analytic methods, such as scanning electron microscopy, gas chromatography-mass spectrometry (GC-MS), FTIR, and Raman spectroscopy, have been applied for the identification of MPs. Microscopic observation is the most common method

(Free et al., 2014; Lechner et al., 2014; Mason et al., 2016), but it is vulnerable to potential errors due to the interference of MPs by other particles, such as sand and grit. Substantial overestimation or underestimation (up to 50%) is frequently reported when extracted samples contain large amounts of particle mixtures (Dekiff et al., 2014; Lenz et al., 2015).

To obtain more precise data, GC/MS and related techniques may be applied. A modified pyrolysis/gas chromatography/mass spectrometry (py/GC/MS) can be used to analyze MP particles in sediment samples (Fabbri, 2001; Fries et al., 2013). In general, sample mass for py/GC/MS is small; for example, a single particle weighs 10–350 μg (Fries et al., 2013). To gain sufficient sample mass, a thermal desorption system-gas chromatography-mass spectrometry (TDS-GC-MS) can be exploited after thermal extraction in thermogravimetric analysis (Dumichen et al., 2015). This TDS-GC-MS technique can analyze nonhomogeneous high sample mass (about 200 times higher than that used in py-GC-MS) on a small scale. It can also identify and quantify the decomposition products of MPs. However, GC-/MS-based methods are destructive and require complicated equipment/operation.

Vibrational spectroscopic techniques, for example, FTIR and Raman spectroscopy, are traditional analytic methods for organic polymers and can determine the size, number, and type of MPs in aquatic samples. The principles of both FTIR and Raman spectroscopy involve the excitation of molecules and detection of their vibrations. Formation of FTIR depends on the changes of the permanent dipole moment of a chemical bond (e.g., C═O bond), while Raman spectroscopy is generated by the change of the polarizability of a chemical bond (e.g., C═C, C—H, and aromatic bonds). The existing reference spectra databases further improve accuracy and rate of FTIR and Raman spectroscopy. Thus, specific fingerprints of MPs can be characterized (Fig. 10.3). These two methods have been used in the detection of MPs in wastewater samples from WWTPs (Murphy et al., 2016).

Kappler et al. (2016) compared the advantages and disadvantages of FTIR and Raman spectroscopy by measuring the number, size, and type of MPs. FTIR can detect large particles (50–500 μm) quickly and reliably (32 particles within 20 min) but may underestimate small particles (1–50 μm). On the other hand, Raman spectroscopy can identify small MP particles (1–50 μm) precisely but is a time-consuming process (49 particles in 38 h). Therefore, FTIR and Raman spectroscopy are complementary techniques, and a synergistic application of these two methods should be considered for the analysis of MPs in heterogeneous wastewater samples.

Some modifications of FTIR and Raman spectroscopy have been made to extend their applications for complex samples. Combination of microscopes onto FTIR or Raman spectroscopy enables selective quantitative analyses of multiple MP particles in samples containing other farraginous particles (Ziajahromi et al., 2017). This method

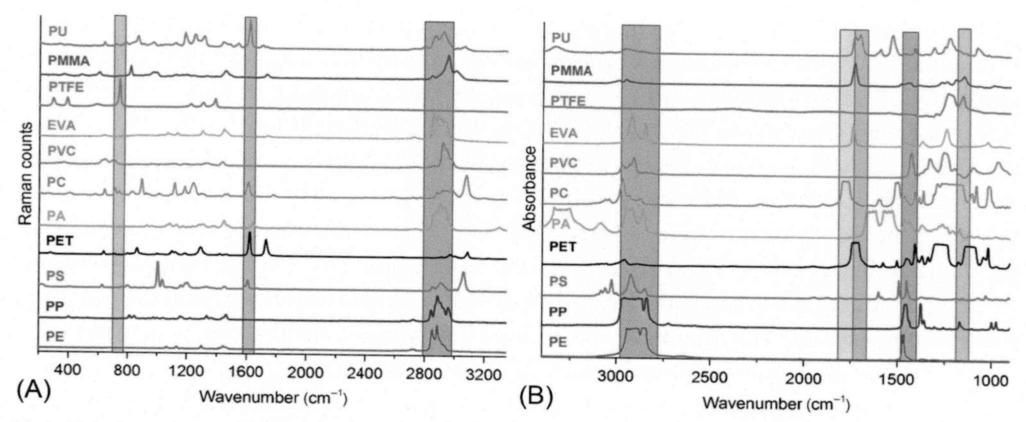

Fig. 10.3 Raman spectra (A) and FTIR spectra (B) of several common synthetic polymers. Different color tapes can be used to identify different polymers. For example, in Raman (A), characteristic peaks in blue tape can identify PE, PP, PA, PVC, EVA, PMMA, and PU, while green tape can be used to identify PS, PET, PC, and PU (Kappler et al., 2016).

produces visual images so that the dimension of MPs can be determined. Tagg et al. (2015) developed a focal plane array (FPA)-based reflectance micro-FTIR, which can produce images and identify different MP types (polyethylene, polypropylene, nylon 6, polyvinyl chloride, and polystyrene) in wastewater samples that have undergone pretreatment by 30% H_2O_2. This is effective for accurate and semiautomated detection of MPs in wastewater, even in the presence of abundant biological organic matter. A recent study that used FPA-FTIR to detect MPs in wastewater (Mintenig et al., 2017) extended the detection size limit to $<20\,\mu m$. However, the application of this method may be limited because the analysis time is relatively long (several hours for one sample).

It should be noted that FTIR or Raman spectroscopy is very time-consuming, requires skilled personnel, and is difficult to automate. Furthermore, pretreatment is required for MPs from wastewater and sludge samples before FTIR or Raman spectroscopy analyses.

10.3 CLASSIFICATION AND SOURCE OF MICROPLASTICS IN WASTEWATER TREATMENT PLANTS

The composition of MPs in influent of wastewater treatment plants is complex (Dyachenko et al., 2017; Ziajahromi et al., 2016, 2017), and it is difficult to fully clarify the composition and source of MPs in WWTPs due to knowledge limitations. This is because WWTPs receive domestic wastewater, partial industrial wastewater, and storm water runoff. This results from the fact that when separate sewage systems are applied, municipal WWTPs only receive domestic wastewater from inhabitants; however, when combined sewage systems or intercepted sewage systems are used, both domestic wastewater and fractional storm water are converged and discharged into WWTPs.

Similar to MPs in natural water bodies, MPs in WWTPs can be classified into two different types, primary MPs and secondary MPs. Primary MPs are industrially created particles, such as those microbeads that are added to several cosmetic and personal care products (Chang, 2015; Fendall and Sewell, 2009) and air-blasting media (Gregory, 2009). Secondary MPs are particles and fibers formed from the fragmentation of large plastic debris in aqueous medium via sunlight irradiation, wind blow, water stir, and other external environmental disturbances (Waller et al., 2017).

The dominating primary MPs in WWTPs are plastic microbeads from personal care products and cosmetics used as exfoliating agent. These microbeads have been found in marine water since the 1990s (Gregory, 1996; Zitko and Hanlon, 1991). They are spherical or irregularly shaped (Fig. 10.4), with a majority being smaller than 0.5 mm in size (Fendall and Sewell, 2009). After use, plastic microbeads are disposed down the drain and into the sewage. It has been estimated that ~94,500 microbeads can be released after a single wash (Napper et al., 2015). Due to their small sizes and large quantity, plastic microbeads are considered to be tough impurities for WWTPs. Conventional treatment methods in modern WWTPs transfer most of these microbeads from wastewater into the sludge, and the residual microbeads are discharged into receiving waters (Mason et al., 2016; Murphy et al., 2016). This fact was verified by the homologous comparison between microbeads in personal care products and those in field samples with FTIR (Cheung and Fok, 2016).

Synthetic fiber clothing is also a concern as these fabrics can shed small fibers and particles into the sewage during laundering. Theoretically, these small fiber debris from clothing can be classified as secondary MPs. A single wash of one garment by domestic washing machine can produce >1900 fibers per wash (Browne et al., 2011). Hartline et al. (2016) evaluated the mass releases of polyester apparel and found that approximately 0.3% of the unwashed garment mass was released during a single wash. It is believed that laundry discharges released to the sewage are the primary sources of MP fibers in WWTPs (Eriksen et al., 2013; Mason et al., 2016). It should be noted that abundant MP fibers found in the marine environment are derived from direct discharged sewage without WWTP treatment as a consequence of washing clothes (Browne et al., 2011), suggesting that WWTPs are not the only sources of fiber debris.

Of course, there are other types of MPs in wastewater besides microbeads and microfibers. For example, the debris from tires that spread across highways and roads, the debris from rubber surfaces, and the atmospheric deposition particles can all be carried by the rainwater runoff from surface ground into the combined or intercepted sewage systems, which finally flow into WWTPs (Table 10.1).

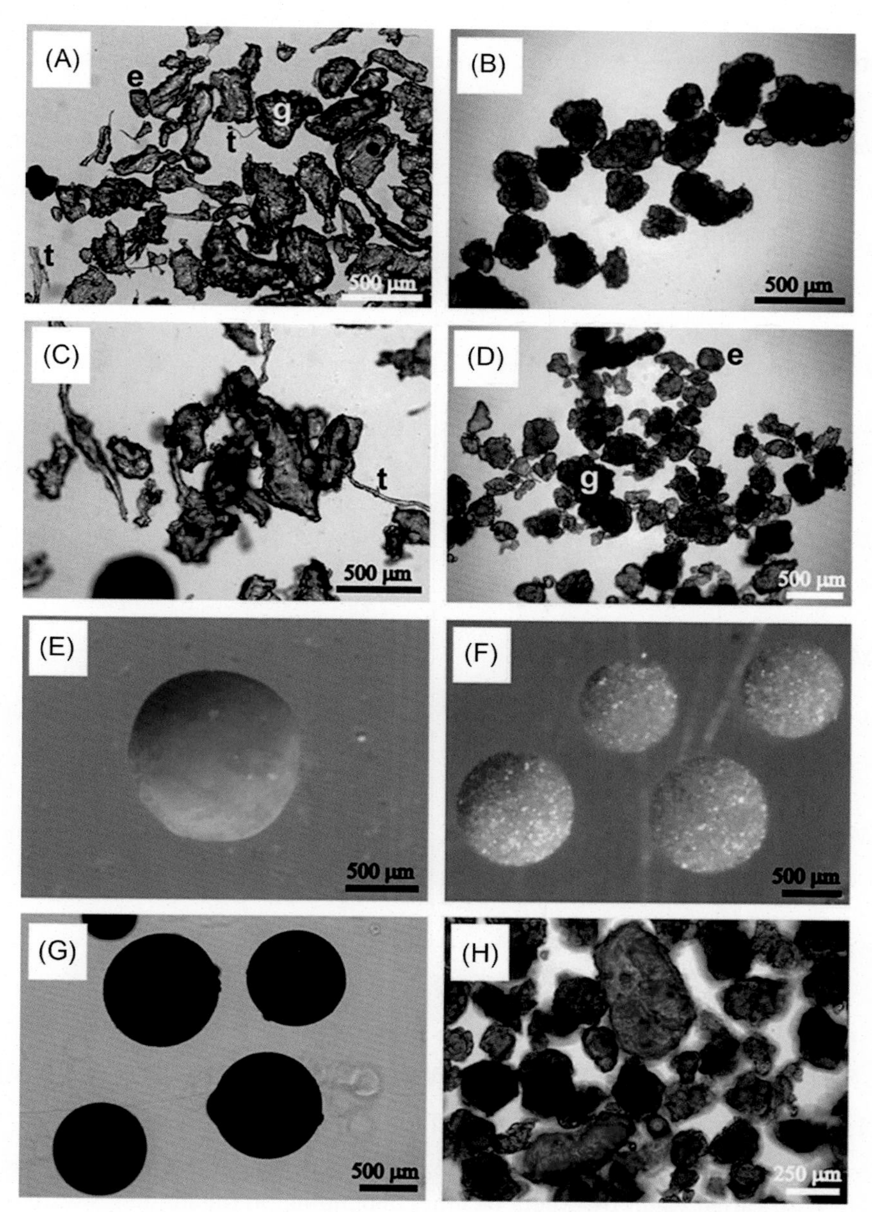

Fig. 10.4 Photomicrographs of the microplastics and colored inclusions in facial cleanser brands A–D. Scale bar in all panels except H 500 lm. (A) Microplastics from brand A include variable irregular shapes that include granular particles (g), ellipses (e), and threads (t). (B) Microplastics from brand B are uniform and granular in shape. (C) Microplastics from brand C include variable irregular shapes that are rounded or threadlike (t). (D) Microplastics from brand D are uniform and elliptical (e) or slightly granular (g) in shape. (E) Blue-colored material from brand A. Product labeling refers to these as "pore cleansing power beads" that contain lactic acid to "help open clogged pores." (F) Orange-colored material from brand B. Chemical composition unknown. (G) Blue-colored material from brand C. Chemical composition unknown. (H) Blue-colored material from brand D. Chemical composition unknown (Fendall and Sewell, 2009).

Table 10.1 Steel mesh or steel sieves used for microplastic separation from wastewater

No.	Water type	Size (μm)	Column/ stack	Strategy	US standard sieve	References
1	WWTP effluent	125, 335	No	Grab sampling	No	Sutton et al. (2016)
3	Wastewater in different stages of WWTP	65	No	–	No	Murphy et al. (2016)
4	WWTP effluent	20, 106, 300, 850, 4750	Yes	Grab sampling	Yes	Michielssen et al. (2016)
5	Wastewater in different stages of WWTP	125, 355	No	Composite sampling, continuous sampling	Yes	Mason et al. (2016)
6	Wastewater in different stages of WWTP	20–400	Yes	Continuous sampling	No	Carr et al. (2016)
7	WWTP effluent	125, 355, 1000, 5000	Yes	Composite sampling	Yes	Dyachenko et al. (2017)
8	Wastewater in different stages of WWTP	20, 100, 300	Yes	Grab sampling	No	Talvitie et al. (2017)
9	Wastewater in different stages of WWTP	500, 190, 100, 25	Yes	Composite sampling	No	Ziajahromi et al. (2017)

10.4 FATE OF MICROPLASTICS IN WASTEWATER TREATMENT PLANTS

10.4.1 Microplastics in Wastewater Treatment Processes

Wastewater treatment plants are designed to have distinct water treatment process combinations with varied water treatment facilities depending on the influent's water quality and the effluent discharge standard. Conventional wastewater treatment includes pre-treatment, primary treatment, and secondary treatment. A series of treatment processes, for example, bar screening, degreasing, air flotation, primary sedimentation, biofilm process/activated sludge process, and secondary sedimentation, are applied. To further improve the effluent quality, tertiary treatment with (sand) filtration, advanced oxidation process, and membrane filtration are used. So far, no treatment method is specially designed to remove MPs, and only a few studies have investigated the detailed removal efficiencies of MPs at different stages of WWTPs.

Removal efficiencies of MPs differ from one another in different treatment processes (Table 10.2). Generally, the removal efficiencies of MPs in different stages follow an order: primary treatment > secondary treatment > tertiary treatment. However, comparison between the detailed removal efficiencies is made difficult because of the variable treatment processes and sampling/identification techniques.

Dris et al. (2015a) conducted the first investigation about the fate of MPs in a WWTP through the analysis of wastewater influent and effluent. The WWTP applied screening and grit and oil removal as primary treatment, which was then followed by a primary settling tank and biological treatment. Biofilters were used in the tertiary stage, where the total removal rate of MPs into the sludge reached ~90%. In the influent, 1000–5000 μm MPs contributed to 45% of the total amount, which were completely removed after tertiary treatment. On the other hand, only small MPs (100–1000 μm) were found in the final effluent. It should be noted that fibers, but not fragments, were the predominant MPs in this WWTP. One shortcoming of this study was that there was no detailed investigation of MP morphology.

Around the same time, Talvitie et al. (2015) also studied a WWTP in Helsinki, Finland, which applied a conventional tertiary treatment procedure. Influent of this WWTP contained approximately 180 textile fibers and 430 synthetic particles per liter. Microplastic fibers were mostly removed by primary sedimentation, while MP particles were mostly settled in secondary sedimentation. Biological filtration in tertiary treatment further improved the removal efficiency of MPs. After the treatment process, an average of 4.9 (±1.4) fibers and 8.6 (±2.5) particles per liter were found in final effluent. Artificial textile fibers and synthetic plastic particles were identified as the dominating MPs following a similar pattern in the WWTP effluent and receiving sea water, verifying the role of WWTP as a route for MPs entering the sea.

Carr et al. (2016) investigated the transport of MPs in a tertiary wastewater reclamation plant, but only limited information was provided in regard to the concentration of MPs. The study also confirmed that pretreatment and primary treatment were effective for removing MPs. The majority of MPs in this WWTP had a profile (color, shape, and size) similar to the blue polyethylene particles in toothpaste formulations, implying that the additives in cosmetic and personal care products were the main sources of MPs in WWTPs. It should be noted that the concentration of MPs in return activated sludge reached ~50 particles L^{-1}, indicating a transport of MPs from wastewater to activated sludge during biological treatment.

Murphy et al. (2016) investigated a WWTP in England serving 6.5×10^5 populations, which used secondary treatment facilities with average treatment capacity of 2.6×10^6 m^3 day^{-1}. Only grab sampling was applied, and microscope combined with FTIR was used for determining the concentration and composition of MPs.

Table 10.2 Removal efficiencies of microplastics in various wastewater treatment processes

No. of WWTP	Location	Inhabitant population (thousand)	Treatment capacity (m^3 day^{-1})	Influent concentration (particles L^{-1})	Pretreatment/ primary treatment (particles L^{-1})	Secondary treatment (particles L^{-1})	Tertiary treatment (particles L^{-1})	Total discharge amounts (particles L^{-1})	References
1	Australia	1227	30,800	–	1.50	–	–	4.6×10^7	Ziajahromi et al. (2017)
2	Australia	67	1700	–	1.44	0.48	–	8.2×10^5	Ziajahromi et al. (2017)
3	Australia	151	1300	–	2.20	–	0.21	2.7×10^5	Ziajahromi et al. (2017)
4	Finland	800	270,000	380 (±52)–686 (±155)	9.9 (±1.0)–14.2 (±4.0)	1.0 (±0.6)–2.0 (±0.2)	0.7 (±0.6)–3.5 (±1.3)	1.9×10^8–9.5×10^8	Talvitie et al. (2017)
5	England	650	260,954	15.70 (±5.23)	3.40 (±0.28)	0.25 (±0.04)	–	6.5×10^7	Murphy et al. (2016)
6	France	–	240,000	293 (260–320)	90 (50–120)	–	35 (14–50)	–	Dris et al. (2015a)

The WWTP influent contained on average 15.70 (±5.20) particles L^{-1}, which was reduced to 0.25 (±0.04) particles L^{-1} in the final effluent (removal rate reached 98.4%). Approximately 45% of MPs were removed by pretreatment with coarse screening. Subsequent fine screening, grit sedimentation, degreasing, and primary sedimentation removed an additional ~34%. The secondary treatment stage handled the other ~20% MPs, implying that traditional biological treatment followed by a second sedimentation is also effective for MP elimination. Despite the high removal rate, it was calculated that 65 million pieces of MPs are still discharged into receiving water every day from this WWTP. This suggests that modern WWTPs transport a huge amount of MPs, especially small-size particles, into ambient waters.

In addition, this study also investigated MPs in grit and grease samples and in the sludge cake from sludge treatment. The grease sample contained an average of 19.67 (±4.51) particles per 2.5 g, which was significantly higher than the grit and sludge cake samples. The study also found that polyethylene microbeads from cosmetic and personal care products were dominant in grease samples. Fortunately, because of their lightweight nature and hydrophobicity, polyethylene microbeads are buoyant on the surface of wastewater and can therefore be easily skimmed off during the degreasing treatment. The final effluent of this WWTP contained no intact microbeads, which may be destructed into some smaller fragments with irregular shapes after the treatment processes.

Two full-scale WWTPs, which employed traditional secondary treatment and tertiary treatment, were examined by Michielssen et al. (2016). Removal efficiency of MPs in a novel pilot-scale WWTP with a microfiltration membrane bioreactor system was also evaluated. A mesh sieve stack was used for sampling, but only stereomicroscope was employed for MP identification. The total removal efficiency reached 95.6% and 97.2%, respectively, after the secondary treatment and tertiary treatment. The membrane bioreactor system removed 99.4% MPs, discharging 0.5 particles L^{-1} MPs. Fibers but not microbeads were identified as the major part in the effluent from two full-scale WWTPs.

A later study performed by Talvitie et al. (2017) evaluated the stepwise removal of MPs in a Finnish WWTP, which received municipal wastewater from 8.0×10^5 inhabitants and had a treatment capacity of 2.7×10^6 m^3 day^{-1}. Compared with Murphy's study (2016), the WWTP in Talvitie's (2017) study used similar pretreatment, primary treatment, and secondary treatment, but a tertiary treatment with a biologically active filter was added. Grab sampling and sequential sampling with mesh screen separating devices were used at different points along the treatment processes in this WWTP (Fig. 10.5). Stereomicroscope and FTIR were used to analyze MPs. Results showed that 97.4%–98.4% MPs were removed after pretreatment and primary treatment. Large particles (≥300 μm) were intercepted primarily in the pretreatment stages, while smaller ones (100–300 μm) were removed in the second and tertiary treatments. The smallest MPs (20–100 μm) were still able to bypass all the treatment facilities, including the tertiary treatment, and were discharged into the final effluent.

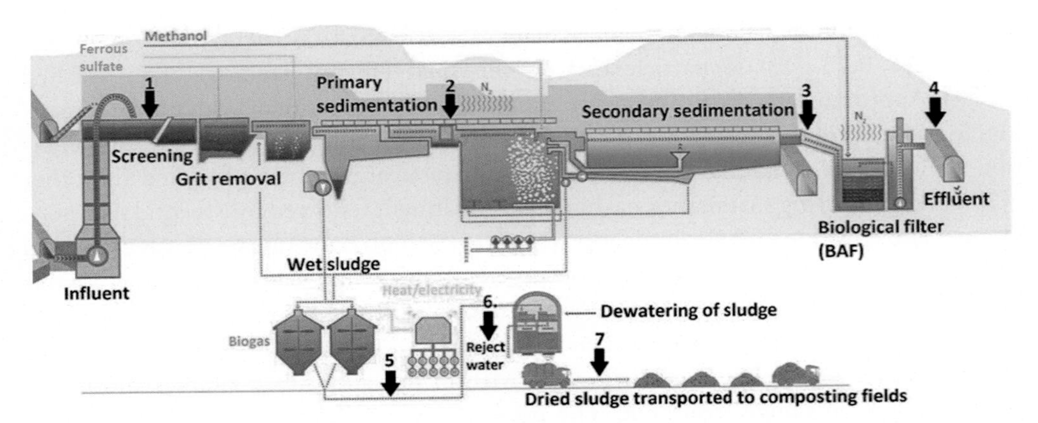

Fig. 10.5 The sampling sites in a WWTP: (1) influent, (2) after pretreatment, (3) after the activated sludge process, (4) plant effluent, (5) excess sludge, (6) reject water, and (7) dried sludge (Talvitie et al., 2017).

More recently, Ziajahromi et al. (2017) compared the removal efficiencies of MPs in three WWTPs. Grab sampling was performed using a customized separating devices (Fig. 10.1), and FTIR was used for the identification of MP particles. After primary treatment and secondary treatment, the concentration of MPs decreased to 1.44–2.20 and 0.48 particles L^{-1}, respectively, while the tertiary treatment only yielded a slight improvement. Ziajahromi et al. (2017) investigated the size distribution of MPs in effluent from different stages, and results demonstrated that large particles ($\geq$190 μm) were removed in mechanical primary treatment. However, smaller MPs (25–190 μm) were still abundant in the secondary and tertiary effluent. It should be noted that fibers from domestic laundry discharges were dominant in the effluent of all three WWTPs.

In modern WWTPs, primary treatments and secondary treatments have a high removal efficiency of MPs, especially for large particles with low density, which is transported into inorganic sludge (from primary sedimentation or flotation) and organic sludge (from biological treatments), implying that WWTPs are not the final ends of most MPs in wastewater (see Section 11.4.2). Surprisingly, advanced methods in the tertiary stage only slightly improve the ability to intercept smaller residual MPs. For example, effluent of reverse osmosis (Ziajahromi et al., 2017) or microfiltration (Michielssen et al., 2016) still contained MP particles (0.21 and 0.50 particles L^{-1}, respectively), indicating that most of the existing treatment methods are inefficient for completely removing MPs from wastewater. MPs of small sizes (<0.5 mm) in the shape of fiber and microbeads are ubiquitous in the final effluent. The concentrations of MPs in the final effluent of most WWTPs are relatively low (<1 particles L^{-1}), but the discharge volume of normal WWTPs generally reaches 10^8 L day^{-1} level. This means that a large amount of MPs can enter the receiving water on a daily basis.

10.4.2 Microplastics in Sludge Treatment Processes

Sludge assembly and disposition are important processes in WWTPs. In these steps, particulates, excess activated sludge, and organic materials are separated from wastewater and transported into sludge from different treatment stages. For example, inorganic particles with abundant large-scale MPs are settled in grit removal or primary sedimentation. In biological treatment stages, excess activated sludge is formed by organic materials, inorganic materials, microorganisms, and their derivatives, which also contains MPs. Thus, sewage sludge in WWTPs becomes another vector for MPs to reach the environment.

Sludge treatment is performed prior to sludge utilization, and it focuses on reducing sludge weight, volume, and potential health risks. A common sludge treatment includes thickening, dewatering, digesting, composting, incinerating, and drying. After sludge treatment, sewage sludge is transformed into biosolids. Biosolids can then be reused, for example, as a soil conditioner for agricultural usage. So far, there is limited information surrounding the changes in quantity and morphology of MPs during sludge treatments.

Mahon et al. (2017) extracted and analyzed MPs in sludge samples from seven WWTPs in Ireland and found that mean abundances of MPs ranged from 4196 to 15,385 particles kg^{-1} (dry weight) in sludge samples. Anaerobic digestion (AD), thermal drying (TD), and lime stabilization (LS) treatment processes were employed in these WWTPs, which resulted in different transformations of MPs. Abundant amounts of smaller sized MPs with fracture and flaking were observed in LS sludge samples, which may have resulted from the shearing of larger MP particles. Melting and blistering were observed in TD samples. This indicates that sludge treatment processes had disruptive effects on MPs, as it induced alterations to the amount, size, and shape. These changes may aggravate the photodegradation and thermooxidative degradation processes of MPs in biosolids applied to farming or other reusages. In another study, high concentration of MPs (e.g., 510–760 particles kg^{-1} wet weight (ww) and particle sizes between 10 and 5000 μm) in sewage sludge also indicated that WWTPs were able to transport MPs from wastewater to biosolids. This means that the final receptor can be changed from water bodies to land soil, such as farmland and forestry (Leslie et al., 2017). Therefore, more attention should be paid to the potential risks of MPs after sludge treatment.

10.5 DISCHARGE OF MICROPLASTICS FROM WASTEWATER TREATMENT PLANTS

10.5.1 Rivers

As most large urban agglomerations are located near long rivers and large lakes, discharge of MPs from WWTPs is relatively centralized into these drainage basins. Rivers are the most common receptors for WWTP effluent. Additionally, since small-scale MPs are dominant in effluent, concentrations of these MPs significantly increase in the downstream of WWTPs (Estahbanati and Fahrenfeld, 2016). McCormick et al. (2014) surveyed the concentrations of MPs upstream and downstream of a WWTP in Chicago. It was found that MPs of different sizes, shapes, and polymers all increased significantly in the downstream river water. The nutrient concentrations also increased in downstream samples with a similar pattern, thus indicating that the MPs and nutrient were all sourced from the same WWTP. A large-scale investigation of MP discharge from 17 WWTPs in the United States was conducted by Mason et al. (2016). Average concentration of MPs in the effluent was 0.05 ± 0.024 particles L^{-1}. Based on the statistical data of effluent volume and MP concentrations, it was estimated that an average of 13 billion (3–23 billion) MPs would be released into receiving waters every day via municipal wastewater in the United States.

In another study, Estahbanati and Fahrenfeld (2016) also found that WWTP effluent brought an abundant amount of small-size MPs to receiving waters. The study investigated the concentrations of MPs with different sizes (500–2000, 250–500, and 125–250 μm) upstream and downstream of four major municipal WWTPs on Raritan River, New Jersey. The concentrations of 250–500 and 125–250 μm-sized MPs increased significantly downstream. A morphology comparison between MPs collected in the field and those extracted from personal care products was also conducted (Fig. 10.6), and a

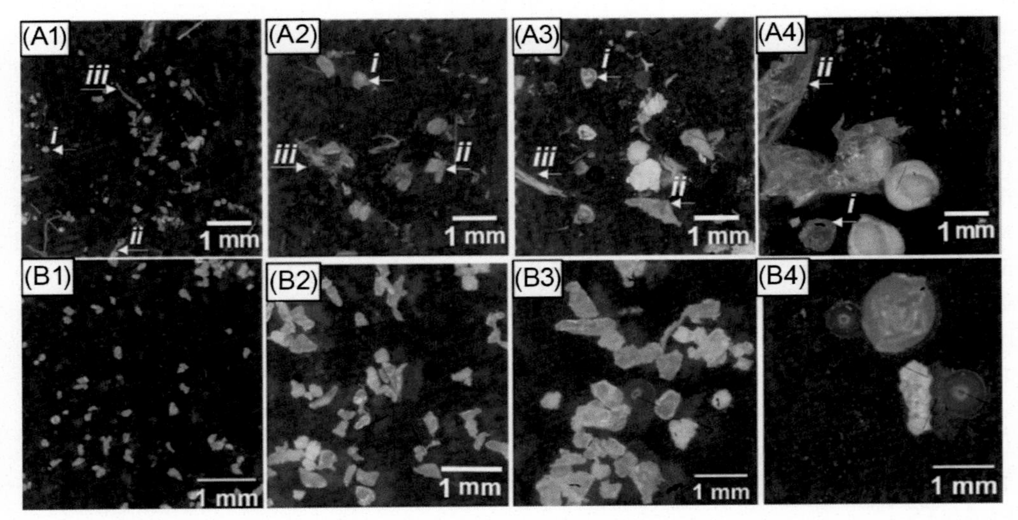

Fig. 10.6 Microplastics recovered from wastewater and personal care products. (A) Microplastics recovered in samples in the (1) 63–125 μm size category, (2) 125–250 μm size category, (3) 250–500 μm size category, and (4) 500–2000 μm size category. (B) Microplastics recovered from personal care products in the (1) 63–125 μm size category, (2) 125–250 μm size category, (3) 250–500 μm size category, and (4) 500–2000 μm size category. Examples of different particle classifications are labeled (i) primary microplastic, (ii) secondary microplastic, and (iii) nonmicroplastic particles excluded during the counting step (Estahbanati and Fahrenfeld, 2016).

moderate correlation was observed. Thus, WWTPs contribute to the increase of small MPs in receiving river waters.

Microplastics in WWTP effluent have distinct characteristics compared with those from other sources and are usually dominated by microfibers and microbeads. Gallagher et al. (2016) evaluated the concentrations of MPs in four rivers in Solent estuarine complex, the United Kingdom. Through microscope and FTIR analysis, fibers/lines and pellets/beads were identified as the dominating MPs in river waters, thus confirming that both local WWTPs and plastic industry were the main sources. In another study, the concentrations of MPs upstream and downstream of WWTPs at nine rivers in Illinois were measured (Mason et al., 2016). Pellets, fibers, and fragments were confirmed as the dominant shapes, while polypropylene, polyethylene, and polystyrene were identified as the dominating polymers. However, through morphology analysis, other studies have demonstrated that fiber, not microbeads, was the most common type of MPs in effluent from several WWTPs (Dris et al., 2015a; Mason et al., 2016; Michielssen et al., 2016; Ziajahromi et al., 2017). Either way, the take-home message is that WWTPs are considered to be important point sources of MPs into receiving rivers.

Discharge of MPs from WWTP affects the pollutant composition of river water, and it also significantly alters microorganism distribution (Hoellein et al., 2014), as MPs can provide habitat for specific microorganisms. McCormick et al. (2014) used high-throughput sequencing to analyze the bacteria species on MPs. They found that *Pseudomonas* spp., which is abundant in wastewater biological treatment processes but not in natural water bodies, was the main species on MP surface (contributed ~14% of the abundance). Furthermore, the assemblages colonizing on MPs were different in taxonomic composition compared with those on other suspended particulates in the same field (Fig. 10.7).

10.5.2 Lakes

Lakes are important receptors of WWTP effluent. One of the first studies to find microbeads in the Great Lakes water (the United States) was conducted by Eriksen et al. (2013). They attributed the source of microbeads to the application and direct discharge of personal care products through the sewage. They also found coexistence of significant amounts of coal ash and coal fly ash in lake water samples, suggesting that elemental analysis (such as SEM) or chemical analysis (such as FTIR) should be used to avoid misidentification of MPs.

Most lakes receive tributaries covering a large watershed, including multiple inlets, such as WWTPs, agricultural irrigation, and surface runoff. Therefore, WWTPs may not be the only sources of MPs for lake waters. Baldwin et al. (2016) characterized the quantity and morphology of floating MPs in wastewater effluent from several

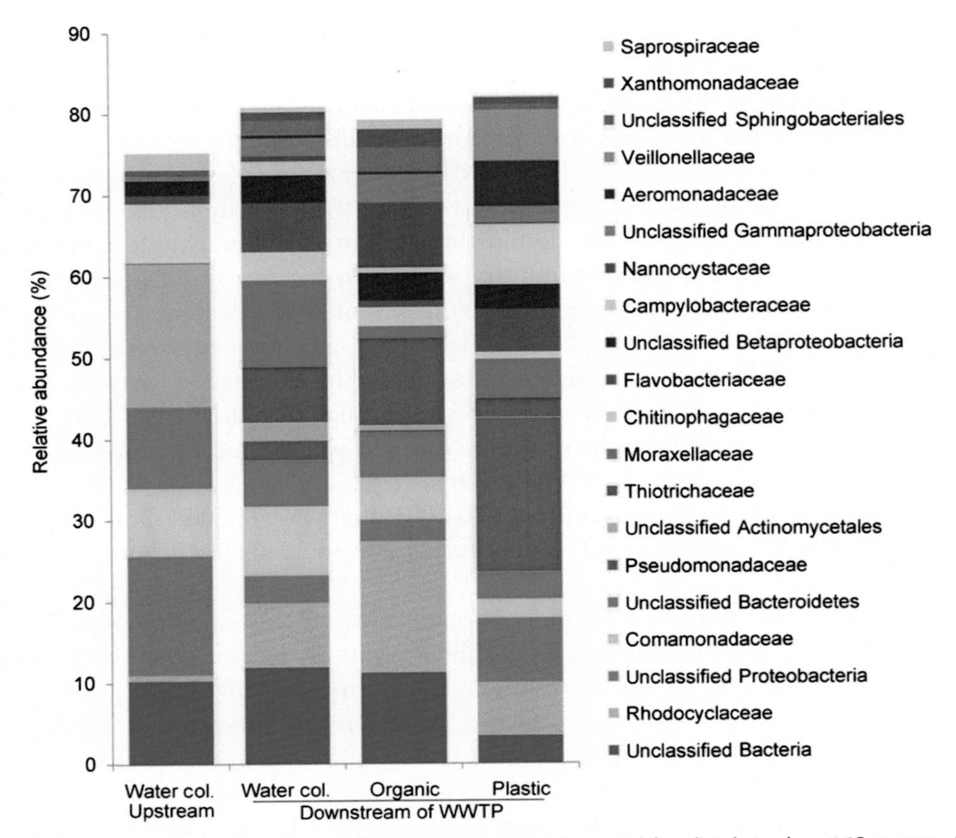

Fig. 10.7 Relative mean abundance of 20 most abundant bacterial families based on 16S sequencing data for samples collected in the North Shore Channel (McCormick et al., 2014).

WWTPs and in the receiving water in the Great Lakes tributaries. They found that the most common MP type was fibers/lines, which made up ~71% of all samples. They also confirmed that wastewater effluent contributed to part of the fibers/lines and pellets/beads found in receiving lake waters, suggesting that tributaries were presumably a substantial source of MPs for large water bodies such as lakes or oceans.

In another study, the seasonal variations of MP concentrations in a lake system were investigated (Lasee et al., 2017). Microplastic concentrations were generally higher in warmer months (form May to July) but decreased from August to December. This phenomenon was partly attributed to large rainfall in warmer months, which brought subsequent washing of secondary MPs onto the land surface. These results also confirmed that WWTPs were not the sole contributors of MPs in lake systems.

10.5.3 Oceans

Aside from rivers and lakes, oceans are the largest receiving water bodies of WWTP effluent, and WWTP-sourced MPs along the coastline have been reported worldwide. In a research that analyzed MPs from sediments on shoreline habitats, morphology analysis showed that the proportions of polyester and acrylic fibers in sewage effluent, receiving sea water, and clothing were similar, confirming that fibers from washing clothes were important sources of MPs in the sea (Browne et al., 2011).

Concentrations of MPs in San Francisco Bay and the final effluent from eight WWTPs in this watershed were investigated (Sutton et al., 2016). They demonstrated that the bay surface water contained the highest levels of MPs out of North America's urban water bodies. These WWTPs discharged considerable MPs into the bay. In one study, van Wezel et al. (2016) estimated the release of primary MPs from consumer products via WWTPs into marine environment. Targeted products included cosmetics, personal care products, cleaning agents, paint, and coatings. Total concentrations of primary MPs from these consumer products were in the range of 0.2–66 $\mu g L^{-1}$ in WWTP effluent.

10.6 CONCLUSIONS

Various sampling, pretreatment, and analytic methods for MPs in wastewater and sludge have been developed. Common sampling and separating devices are designed based on simple mechanical screening, while sampling strategies include grab sampling and continuous sampling. Before analysis, chemical digestion and density separation are used as pretreatment methods. Synergistic method that combines microscope, FTIR, and Raman spectroscopy is widely applied for MP identification. Two categories of MPs, including the microbeads from personal care products and cosmetics and the small fiber debris from domestic clothing washing, are identified as the dominating MPs in WWTPs. Primary and secondary treatments in WWTPs remove most MPs, and the next step includes transportation to sewage sludge. The residual MPs in WWTP effluent and sludge are ultimately discharged into receiving waters and/or lands. Therefore, WWTPs have been confirmed as considerable point sources of MPs to the environment.

ACKNOWLEDGMENT

This project was supported by the National Natural Science Foundation of China (No. 51778270).

REFERENCES

Andrady, A.L., 2011. Microplastics in the marine environment. Mar. Pollut. Bull. 62, 1596–1605.

Arp, H.P.H., Moskeland, T., Andersson, P.L., Nyholm, J.R., 2011. Presence and partitioning properties of the flame retardants pentabromotoluene, pentabromoethylbenzene and hexabromobenzene near suspected source zones in Norway. J. Environ. Monit. 13, 505–513.

Baldwin, V., Bhatia, M., Luckey, M., 2011. Folding studies of purified LamB protein, the maltoporin from the *Escherichia coli* outer membrane: trimer dissociation can be separated from unfolding. Biochim. Biophys. Acta 1808, 2206–2213.

Baldwin, A.K., Corsi, S.R., Mason, S.A., 2016. Plastic debris in 29 Great Lakes tributaries: relations to watershed attributes and hydrology. Environ. Sci. Technol. 50, 10377–10385.

Browne, M.A., Crump, P., Niven, S.J., Teuten, E., Tonkin, A., Galloway, T., Thompson, R., 2011. Accumulation of microplastic on shorelines woldwide: sources and sinks. Environ. Sci. Technol. 45, 9175–9179.

Carr, S.A., Liu, J., Tesoro, A.G., 2016. Transport and fate of microplastic particles in wastewater treatment plants. Water Res. 91, 174–182.

Chang, M., 2015. Reducing microplastics from facial exfoliating cleansers in wastewater through treatment versus consumer product decisions. Mar. Pollut. Bull. 101, 330–333.

Cheung, P.K., Fok, L., 2016. Evidence of microbeads from personal care product contaminating the sea. Mar. Pollut. Bull. 109, 582–585.

Cole, M., Lindeque, P., Fileman, E., Halsband, C., Galloway, T.S., 2015. The impact of polystyrene microplastics on feeding, function and fecundity in the marine copepod *Calanus helgolandicus*. Environ. Sci. Technol. 49, 1130–1137.

Dekiff, J.H., Remy, D., Klasmeier, J., Fries, E., 2014. Occurrence and spatial distribution of microplastics in sediments from Norderney. Environ. Pollut. 186, 248–256.

Dris, R., Gasperi, J., Rocher, V., Saad, M., Renault, N., Tassin, B., 2015a. Microplastic contamination in an urban area: a case study in Greater Paris. Environ. Chem. 12, 592–599.

Dris, R., Imhof, H., Sanchez, W., Gasperi, J., Galgani, F., Tassin, B., Laforsch, C., 2015b. Beyond the ocean: contamination of freshwater ecosystems with (micro-)plastic particles. Environ. Chem. 12, 539–550.

Dumichen, E., Barthel, A.K., Braun, U., Bannick, C.G., Brand, K., Jekel, M., Senz, R., 2015. Analysis of polyethylene microplastics in environmental samples, using a thermal decomposition method. Water Res. 85, 451–457.

Dyachenko, A., Mitchell, J., Arsem, N., 2017. Extraction and identification of microplastic particles from secondary wastewater treatment plant (WWTP) effluent. Anal. Methods 9, 1412–1418.

Eriksen, M., Mason, S., Wilson, S., Box, C., Zellers, A., Edwards, W., Farley, H., Amato, S., 2013. Microplastic pollution in the surface waters of the Laurentian Great Lakes. Mar. Pollut. Bull. 77, 177–182.

Estahbanati, S., Fahrenfeld, N.L., 2016. Influence of wastewater treatment plant discharges on microplastic concentrations in surface water. Chemosphere 162, 277–284.

Fabbri, D., 2001. Use of pyrolysis-gas chromatography/mass spectrometry to study environmental pollution caused by synthetic polymers: a case study: the Ravenna Lagoon. J. Anal. Appl. Pyrolysis 58, 361–370.

Fendall, L.S., Sewell, M.A., 2009. Contributing to marine pollution by washing your face: microplastics in facial cleansers. Mar. Pollut. Bull. 58, 1225–1228.

Free, C.M., Jensen, O.P., Mason, S.A., Eriksen, M., Williamson, N.J., Boldgiv, B., 2014. High-levels of microplastic pollution in a large, remote, mountain lake. Mar. Pollut. Bull. 85, 156–163.

Fries, E., Dekiff, J.H., Willmeyer, J., Nuelle, M.T., Ebert, M., Remy, D., 2013. Identification of polymer types and additives in marine microplastic particles using pyrolysis-GC/MS and scanning electron microscopy. Environ. Sci. Process Impacts 15, 1949–1956.

Gallagher, A., Rees, A., Rowe, R., Stevens, J., Wright, P., 2016. Microplastics in the Solent estuarine complex, UK: an initial assessment. Mar. Pollut. Bull. 102, 243–249.

Gregory, M.R., 1996. Plastic 'scrubbers' in hand cleansers: a further (and minor) source for marine pollution identified. Mar. Pollut. Bull. 32, 867–871.

Gregory, M.R., 2009. Environmental implications of plastic debris in marine settings-entanglement, ingestion, smothering, hangers-on, hitch-hiking and alien invasions. Philos. Trans. R. Soc. B 364, 2013–2025.

Hartline, N.L., Bruce, N.J., Karba, S.N., Ruff, E.O., Sonar, S.U., Holden, P.A., 2016. Microfiber masses recovered from conventional machine washing of new or aged garments. Environ. Sci. Technol. 50, 11532–11538.

Hoellein, T., Rojas, M., Pink, A., Gasior, J., Kelly, J., 2014. Anthropogenic litter in urban freshwater ecosystems: distribution and microbial interactions. PLoS One 9, e98485.

Kappler, A., Fischer, D., Oberbeckmann, S., Schernewski, G., Labrenz, M., Eichhorn, K.J., Voit, B., 2016. Analysis of environmental microplastics by vibrational microspectroscopy: FTIR, Raman or both? Anal. Bioanal. Chem. 408, 8377–8391.

Lambert, S., Wagner, M., 2016. Exploring the effects of microplastics in freshwater environments. Integr. Environ. Assess. Manag. 12, 404–405.

Lasee, S., Mauricio, J., Thompson, W.A., Karnjanapiboonwong, A., Kasumba, J., Subbiah, S., Morse, A.N., Anderson, T.A., 2017. Microplastics in a freshwater environment receiving treated wastewater effluent. Integr. Environ. Assess. Manag. 13, 528–532.

Lechner, A., Keckeis, H., Lumesberger-Loisl, F., Zens, B., Krusch, R., Tritthart, M., Glas, M., Schludermann, E., 2014. The Danube so colourful: a potpourri of plastic litter outnumbers fish larvae in Europe's second largest river. Environ. Pollut. 188, 177–181.

Le-Clech, P., Chen, V., Fane, T.A.G., 2006. Fouling in membrane bioreactors used in wastewater treatment. J. Membr. Sci. 284, 17–53.

Lenz, R., Enders, K., Stedmon, C.A., Mackenzie, D.M.A., Nielsen, T.G., 2015. A critical assessment of visual identification of marine microplastic using Raman spectroscopy for analysis improvement. Mar. Pollut. Bull. 100, 82–91.

Leslie, H.A., Brandsma, S.H., van Velzen, M.J.M., Vethaak, A.D., 2017. Microplastics en route: field measurements in the Dutch river delta and Amsterdam canals, wastewater treatment plants, North Sea sediments and biota. Environ. Int. 101, 133–142.

Mahon, A.M., O'Connell, B., Healy, M.G., O'Connor, I., Officer, R., Nash, R., Morrison, L., 2017. Microplastics in sewage sludge: effects of treatment. Environ. Sci. Technol. 51, 810–818.

Majewsky, M., Bitter, H., Eiche, E., Horn, H., 2016. Determination of microplastic polyethylene (PE) and polypropylene (PP) in environmental samples using thermal analysis (TGA-DSC). Sci. Total Environ. 568, 507–511.

Mason, S.A., Garneau, D., Sutton, R., Chu, Y., Ehmann, K., Barnes, J., Fink, P., Papazissimos, D., Rogers, D.L., 2016. Microplastic pollution is widely detected in US municipal wastewater treatment plant effluent. Environ. Pollut. 218, 1045–1054.

Masura, J., Baker, J., Foster, G., Arthur, C., 2015. Laboratory Methods for the Analysis of Microplastics in the Marine Environment: Recommendations for Quantifying Synthetic Particles in Waters and Sediments. National Oceanic and Atmospheric Administration, Silver Spring, MD.

McCormick, A., Hoellein, T.J., Mason, S.A., Schluep, J., Kelly, J.J., 2014. Microplastic is an abundant and distinct microbial habitat in an urban river. Environ. Sci. Technol. 48, 11863–11871.

Meng, N.C., Bo, J., Chow, C.W.K., Saint, C., 2010. Recent developments in photocatalytic water treatment technology: a review. Water Res. 44, 2997–3027.

Michielssen, M.R., Michielssen, E.R., Ni, J., Duhaime, M.B., 2016. Fate of microplastics and other small anthropogenic litter (SAL) in wastewater treatment plants depends on unit processes employed. Environ. Sci.: Water Res. Technol. 2, 1064–1073.

Mintenig, S.M., Int-Veen, I., Loder, M.G.J., Primpke, S., Gerdts, G., 2017. Identification of microplastic in effluents of waste water treatment plants using focal plane array-based micro-Fourier-transform infrared imaging. Water Res. 108, 365–372.

Munno, K., Helm, P.A., Jackson, D.A., Rochman, C., Sims, A., 2017. Impacts of temperature and selected chemical digestion methods on microplastic particles. Environ. Toxicol. Chem. 37 (1), 91–98.

Murphy, F., Ewins, C., Carbonnier, F., Quinn, B., 2016. Wastewater treatment works (WwTW) as a source of microplastics in the aquatic environment. Environ. Sci. Technol. 50, 5800–5808.

Napper, I.E., Bakir, A., Rowland, S.J., Thompson, R.C., 2015. Characterisation, quantity and sorptive properties of microplastics extracted from cosmetics. Mar. Pollut. Bull. 99, 178–185.

Nuelle, M.T., Dekiff, J.H., Remy, D., Fries, E., 2014. A new analytical approach for monitoring microplastics in marine sediments. Environ. Pollut. 184, 161–169.

Oturan, M.A., Aaron, J.J., 2014. Advanced oxidation processes in water/wastewater treatment: principles and applications. A review. Crit. Rev. Environ. Sci. Technol. 44, 2577–2641.

Rochman, C.M., Tahir, A., Williams, S.L., Baxa, D.V., Lam, R., Miller, J.T., Teh, F.C., Werorilangi, S., Teh, S.J., 2015. Anthropogenic debris in seafood: plastic debris and fibers from textiles in fish and bivalves sold for human consumption. Sci. Rep. 5, 14340–14349

Sul, J.A.I.D., Barnes, D.K.A., Costa, M.F., Convey, P., Costa, E.S., Campos, L.S., 2011. Plastics in the Antarctic environment: are we looking only at the tip of the iceberg? Oecologia Australis 15, 150–170.

Sutton, R., Mason, S.A., Stanek, S.K., Willis-Norton, E., Wren, I.F., Box, C., 2016. Microplastic contamination in the San Francisco Bay, California, USA. Mar. Pollut. Bull. 109, 230–235.

Tagg, A.S., Sapp, M., Harrison, J.P., Ojeda, J.J., 2015. Identification and quantification of microplastics in wastewater using focal plane array-based reflectance micro-FT-IR imaging. Anal. Chem. 87, 6032–6040.

Tagg, A.S., Harrison, J.P., Ju-Nam, Y., Sapp, M., Bradley, E.L., Sinclair, C.J., Ojeda, J.J., 2017. Fenton's reagent for the rapid and efficient isolation of microplastics from wastewater. Chem. Commun. 53, 372–375.

Talvitie, J., Heinonen, M., Paakkonen, J.P., Vahtera, E., Mikola, A., Setala, O., Vahala, R., 2015. Do wastewater treatment plants act as a potential point source of microplastics? Preliminary study in the coastal Gulf of Finland, Baltic Sea. Water Sci. Technol. 72, 1495–1504.

Talvitie, J., Mikola, A., Setala, O., Heinonen, M., Koistinen, A., 2017. How well is microlitter purified from wastewater? A detailed study on the stepwise removal of microlitter in a tertiary level wastewater treatment plant. Water Res. 109, 164–172.

Ternes, T.A., Stuber, J., Herrmann, N., McDowell, D., Ried, A., Kampmann, M., Teiser, B., 2003. Ozonation: a tool for removal of pharmaceuticals, contrast media and musk fragrances from wastewater? Water Res. 37, 1976–1982.

van Wezel, A., Caris, I., Kools, S.A.E., 2016. Release of primary microplastics from consumer products to wastewater in the Netherlands. Environ. Toxicol. Chem. 35, 1627–1631.

Waller, C.L., Griffiths, H.J., Waluda, C.M., Thorpe, S.E., Loaiza, I., Moreno, B., Pacherres, C.O., Hughes, K.A., 2017. Microplastics in the Antarctic marine system: an emerging area of research. Sci. Total Environ. 598, 220–227.

Yonkos, L.T., Friedel, E.A., Perez-Reyes, A.C., Ghosal, S., Arthur, C.D., 2014. Microplastics in four estuarine rivers in the Chesapeake Bay, USA. Environ. Sci. Technol. 48, 14195–14202.

Ziajahromi, S., Neale, P.A., Leusch, F.D.L., 2016. Wastewater treatment plant effluent as a source of microplastics: review of the fate, chemical interactions and potential risks to aquatic organisms. Water Sci. Technol. 74, 2253–2269.

Ziajahromi, S., Neale, P.A., Rintoul, L., Leusch, F.D.L., 2017. Wastewater treatment plants as a pathway for microplastics: development of a new approach to sample wastewater-based microplastics. Water Res. 112, 93–99.

Zitko, V., Hanlon, M., 1991. Another source of pollution by plastics: skin cleaners with plastic scrubbers. Mar. Pollut. Bull. 22, 41–42.

CHAPTER 11

Microplastics in Marine Food Webs

Outi Setälä*, Maiju Lehtiniemi*, Rachel Coppock[†,‡], Matthew Cole[†,‡]
*Finnish Environment Institute, Helsinki, Finland
[†]Plymouth Marine Laboratory, Plymouth, United Kingdom
[‡]University of Exeter, Exeter, United Kingdom

11.1 INTRODUCTION

A multitude of food webs exist in the world's oceans, are made up of a wide variety of organisms that occupy distinct niches, and possess different behavioral and feeding strategies. So far, only a small fraction of these taxa have been included in studies concerning microplastic debris in marine ecosystems. Microplastics (microscopic plastic debris, 100 nm to 5 mm diameter) are now widely recognized as a pollutant of international concern (Galgani et al., 2013; GESAMP, 2016). Understanding the potential impacts this prolific contaminant can have on marine life and food webs has become of intense interest, with an exponential increase in research being conducted in recent years. In this chapter, we explore how microplastics enter marine food webs and consider the complex, iterative relationship between microplastics, biota, and biologically mediated ecological processes. Microplastic ingestion has been documented in animals throughout the marine food web, including zooplankton (Desforges et al., 2014), fish (Bellas et al., 2016; Lusher et al., 2013), marine mammals (Lusher et al., 2015a; Bravo Rebolledo et al., 2013), turtles (Nelms et al., 2016), and seabirds (Tourinho et al., 2010). We explore the factors affecting microplastic consumption and infiltration into marine food webs, with consideration given to spatial overlap, predator-plastic ratios, the properties of microplastic debris, and the life history and feeding strategies of biota demonstrated to consume plastic. At the individual level, microplastics pose a risk to the health of the organism; indeed, a growing number of experimental studies have demonstrated that at critical concentrations, microplastics can adversely affect feeding, energetic reserves, reproduction, growth, and survival in invertebrate and vertebrate species, including calanoid copepods (Cole et al., 2015; Lee et al., 2013), polychaete worms (Wright et al., 2013b; Green et al., 2016), fish (Rochman et al., 2015), and oysters (Sussarellu et al., 2016). The latest evidence suggests that microplastics could also affect higher levels of biological organization, with population shifts and altered behavior impacting upon the ecological function of keystone species (Galloway et al., 2017). While the risks microplastics pose to individual biota are explored in greater detail in other chapters of this book, here, we focus on how plastics have the potential to affect food webs and marine

Microplastic Contamination in Aquatic Environments
https://doi.org/10.1016/B978-0-12-813747-5.00011-4

ecosystems as a whole. Furthermore, we consider how trophic interactions and ecological processes can change the microplastics themselves.

11.2 THE OVERLAP BETWEEN PLASTICS AND BIOTA

Perhaps the most important variable affecting the flux of microplastic particles into marine food webs is their abundance and distribution in the environment and physical overlap with biota.

11.2.1 Geographical Overlap

In recent years, there has been a concerted effort to identify the different habitats polluted with plastic debris and ascertain the concentrations of microplastics across a wide range of aquatic ecosystems. Microplastics are ubiquitous in the world's oceans, and their presence in remote locations, including the Arctic (Lusher et al., 2015b), Antarctic (Waller et al., 2017), mid-oceanic atolls (Do Sul et al., 2014), and oceanic depths (Woodall et al., 2014), have highlighted their widespread distribution. However, accurately determining the concentrations and type of microplastics present in seawater and sediments has proved a challenge. Adaptations to traditional sampling techniques (e.g., trawls and sediment grabs; see review by Hidalgo-Ruz et al., 2012) have proved invaluable for collecting samples; however, isolating and identifying microplastics have required a more novel approach (see Box 11.1). In recent years, a wide range of methodologies have been

BOX 11.1 Methodological Approach

Although microplastics are a relatively new topic in the environmental sciences, researchers have been able to learn from the experimental approaches and understanding gleaned from the fields of ecotoxicology, marine biology, and aquatic chemistry. Basic mechanisms of feeding and energy transfer in marine food webs are well understood, and this knowledge has been useful in understanding observed interactions between microplastics and biota. Lessons learnt from nanoparticle research have been of particular relevance to microplastic exposure studies, particularly with respect to uptake mechanisms and mechanisms underpinning observed health effects and developing sound ecological risk assessment (Syberg et al., 2015; Hüffer et al., 2017). In contrast, collecting field data on the distribution and quantity of microplastics in different ecological compartments (water surface, water column, seafloor habitats, and strandline) has turned out to be a significant challenge, requiring novel approaches, method development, and optimization (Hidalgo-Ruz et al., 2012; Lusher et al., 2017). An ongoing issue facing microplastic researchers is the absence of harmonized sampling or sample analysis protocols, and a forward challenge for the field is to work toward methodological standardization.

suggested for extracting and analyzing plastics (see reviews by Lusher et al., 2017; Miller et al., 2017); however, the variety of methods employed can often result in incomparable datasets. Analyzing such data is further confounded by the heterogeneous distribution and temporal variability in microplastic concentrations.

Global sampling efforts have helped to identify "hot spots" of plastic (Eriksen et al., 2014; Cózar et al., 2015; Van Sebille et al., 2015). For example, the North Pacific, South Pacific, and North Atlantic subtropical oceanic gyres, which amass flotsam from throughout the oceanic basins, have all been highlighted as accumulation zones for microplastic debris (Moore et al., 2001; Law et al., 2010; Eriksen et al., 2013). Oceanic gyres are largely oligotrophic and therefore relatively devoid of marine life; however, for biota that can survive in the gyres, interactions with microplastic will be commonplace. For example, in the North Pacific gyre, Moore et al. (2001) observed a 6:1 plastic-to-plankton ratio, and Goldstein and Goodwin (2013) identified that 33% of gooseneck barnacles (*Lepas* spp.) had consumed between 1 and 30 items of microplastic. However, our understanding of the numbers and distribution patterns of microplastics in marine environments is far from complete. This was pointed out already in the study dataset of >330 μm particles from surface water tows, which showed smallest particles to be most prevalent, but only down to a certain size group (1 mm) after which the concentrations decreased (Cozar et al., 2015). This absence of smaller plastic may result from difficulties in identifying very small particles or might be explained by biotic or abiotic degradation or movement of these plastics.

Enclosed and semienclosed seas like the Mediterranean Sea and the Baltic Sea have also been noted for their high microplastic concentrations (Collignon et al., 2012; Setälä et al., 2016b; Gewert et al., 2017) and thus have been proposed to accumulate plastic debris in greater amounts than open oceans (Fossi et al., 2016). As increasing concentrations inevitably increase the exposure of organisms at the base of the food webs, this may be the case also at higher trophic levels. In the Mediterranean Sea, stomach analyses from large pelagic predators (swordfish and tuna) revealed that 18.5% of the fish examined contained microplastics. The reported concentrations of microplastics from the surface waters of another highly polluted semienclosed sea basin, the Baltic Sea, show how the microplastic concentrations in surface waters may significantly differ spatially (Setälä et al., 2016b; Gewert et al., 2017) and may reach high concentrations (up to $4.7 \times 105\ km^{-2}$) close to highly populated urban areas with low water exchange, or as was found by Gorokhova (2015), in deep water layers separated by a halocline. In the Baltic Sea, the field observations of microplastics in the food web have mainly related to fish, herring being the most studied fish species. Bråte et al. (2017) analyzed the data from various studies on microplastics in fish from these Nordic waters; in the analyzed dataset consisting of 1425 individuals of Atlantic and Baltic herring, microplastic ingestion varied between 0% and 30%. Ogonowski et al. (2017) reported that approximately 50% of herring individuals had ingested plastics along the Swedish coast in the Baltic Sea, although the numbers of microplastics on individual fish were low (0–1 per fish), reflecting great variability between samples. In comparison, very low numbers of particulate microplastics (fibers were excluded) were also found in a recent study containing over 500 herring individuals from the open sea areas of the northern Baltic Sea (Budimir et al., 2018). The reported share of herring with ingested microplastic particles varies greatly between these studies and may at least partly be explained by spatial differences in the overlap of microplastics and herring. Differences in methods used for extracting microplastics from fish tissue makes comparisons between studies difficult and conclusions vague.

A recent study predicts the greatest overlap between microplastics and marine life will occur in coastal regions (Clark et al., 2016). Coastal waters and estuaries have relatively high biological productivity owing to their shallow, protected waters and fresh nutritional inputs from rivers, which are valued by aquaculture and fisheries, and encompass important nursery grounds for commercially exploited marine taxa. It is postulated that their proximity to sources of anthropogenic pollution (e.g., maritime industry, urban areas, and riverine inputs) puts them at high risk of microplastic pollution. Microplastic sampling in coastal regions is problematic owing to the density of organic material in these waters (Cole et al., 2014); nevertheless, recent studies have highlighted the overlap between plastics and biota in coastal waters. In the English Channel, a 36.5% incidence of microplastic ingestion in demersal and pelagic fish species has been observed (Lusher et al., 2013), while 70% of brown shrimps (*Crangon crangon*) sampled from the coastlines of European countries along the English Channel have been shown to consume microplastic (Devriese et al., 2015). More recently, Steer et al. (2017) identified the ratio of microplastics to fish larvae ranged from 27:1 nearest Plymouth (United Kingdom) to 1:1 35 km from the shoreline.

11.2.2 Habitats

Microplastics consist of a wide range of polymers which have their own special characteristics that affect their distribution in the water, and thereby which organisms and habitats are prone to plastic exposure. Local wind conditions, water currents, and geomorphology all affect the distribution of microplastics in water and their spatial accumulation (Barnes et al., 2009). The vast amounts of anthropogenic debris washing up on beaches across the globe (Browne et al., 2011) provide visual evidence of the efficiency with which floating plastic debris can be transported on the sea surface. Approximately half of marine plastic debris is initially buoyant (e.g., polystyrene, polyethylene, and polypropylene), while denser plastic (e.g., polyvinylchloride and nylon) readily sinks in seawater. As observed from numerous sampling campaigns, microplastics can permeate throughout the water column, with plastic and microplastic debris, including low-density polymer plastic, widely evident in benthic ecosystems (Miller et al., 2017).

Laboratory exposures have been used to demonstrate that biotic interactions including biofouling (Fazey and Ryan, 2016; Kaiser et al., 2017), egestion (Cole et al., 2013, 2016), and bioturbation (Näkki et al., 2017), as well as physical processes such as fragmentation (Andrady, 2017), can affect the properties and movement of plastics; it is hypothesized that these processes could result in changes to the distribution of microplastics within marine ecosystems where biota and plastics overlap (Fig. 11.1; Clark et al., 2016). In these waters, we might expect a downward flux of plastic debris, resulting in an accumulation of microplastics on the seafloor (Barnes et al., 2009; Woodall et al., 2014). However, it is important to recognize that vertical flux should be considered a redistribution of plastics, and not a "removal" mechanism. Benthic ecosystems can be highly biologically productive habitats, supporting a diverse array of life that play vital roles in the oceanic carbon pump (Turner, 2015), reef formation (Beck et al., 2011), and bioturbation (Cadée, 1976). Environmental sampling has identified plastic pollution in every benthic habitat investigated, including highly remote areas such as both Arctic (Bergmann et al., 2017) and Antarctic (Munari et al., 2017) polar regions and the deep sea (Woodall et al., 2014; Bergmann et al., 2017). Plastic concentrations in sediments are highly variable, due in part to different sampling and extraction methodologies and also to the natural heterogeneity of sediments. Concentrations of up to 6600 microplastics per

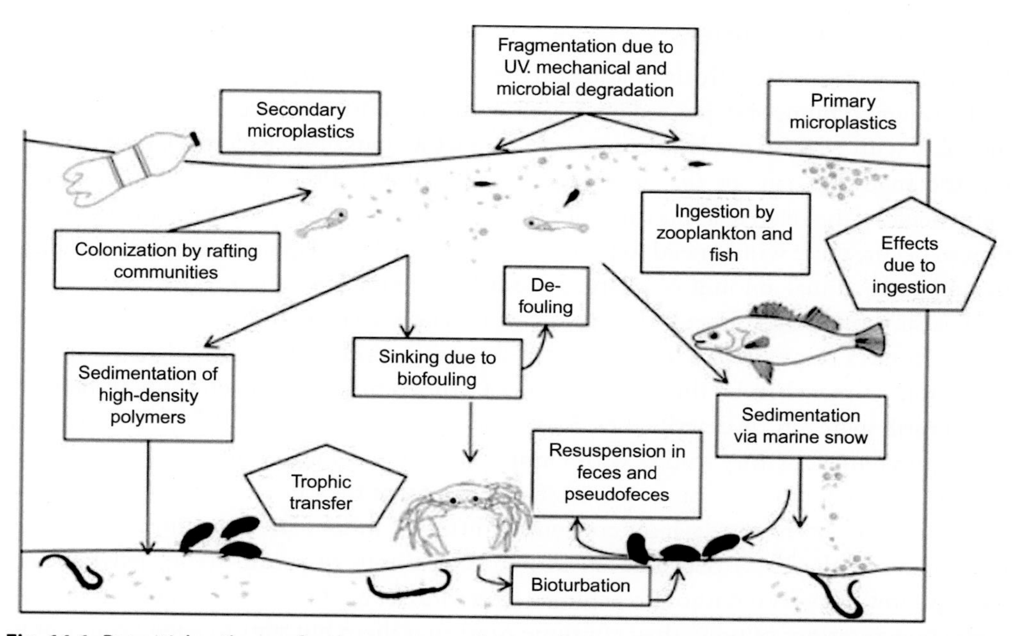

Fig. 11.1 Potential pathways for the transport of microplastics and its biological interactions. *Courtesy of Wright, S.L., Thompson, R.C., Galloway, T.S., 2013a. The physical impacts of microplastics on marine organisms: a review. Environ. Pollut. 178, 483–492.*

kg have been reported in Arctic sediments (Bergmann et al., 2017), and in a study of 42 sites around the Australian coastline (Ling et al., 2017), a regional average of 3400 microplastics per L was reported, with the highest individual sample yielding 12,500 plastics per L. Laboratory exposures have shown that benthic invertebrates readily consume plastic, and this can have a detrimental impact on their health and functionality. A reduction in energy reserves (Wright et al., 2013a), reproduction (Sussarellu et al., 2016), metabolism, and bioturbation activity (Green et al., 2016) has been reported in benthic organisms, with potential impacts to ecosystem functioning (Volkenborn et al., 2007).

11.3 ENCOUNTERING AND DETECTION OF MICROPLASTICS

Compared with the dynamic interactions between a predator/grazer and their natural prey, the relationships between an animal and microplastic are somewhat simplified. The feeding mode and life history of an organism will affect both its encounter and ingestion rate of microplastic. Organisms may actively select microplastics from the environment in search of prey, or they may ingest them accidentally while feeding on food particles or animals that contain plastic.

11.3.1 A Passive Particle

Microplastics are passive: freely floating on the water surface, suspended or slowly sinking in the water column, or deposited on or within the seabed. Encounter rate (i.e., the commonality with which a predator comes into contact with its prey) is a crucial factor affecting the ingestion rate of that prey (e.g., Evans, 1989). Primarily, encounter rate is influenced by the relative abundance of predator/grazer and prey; for microplastic ingestion to occur, there would need to be a significant spatial overlap between biota and plastic and a substantial amount of plastic present for a likely encounter to occur.

Classic work on feeding efficiencies has shown how changes in prey density affect the ingestion rates of predators. Ingestion increases with an increasing prey density up to a saturation point, whereby the predator cannot process more prey even though the prey density still increases, as described by Solomon (1949) and Holling (1959). This has also been shown in laboratory studies with virgin microplastics and various invertebrate taxa: the more particles the organisms were offered, the more they were ingested, even when working with the relatively high concentrations used in laboratory settings (Fig. 11.2) (e.g., Cole et al., 2013; Setälä et al., 2016a). Gelatinous organisms (e.g., jellyfish and ctenophores) may feed without reaching a saturation level. This means that even in very high concentrations of prey, they continue capturing them but start to egest/vomit prey that they are unable to process. However, it has been observed that jellyfish ingested relatively low numbers of microplastics compared with other filter feeders (e.g., copepods)

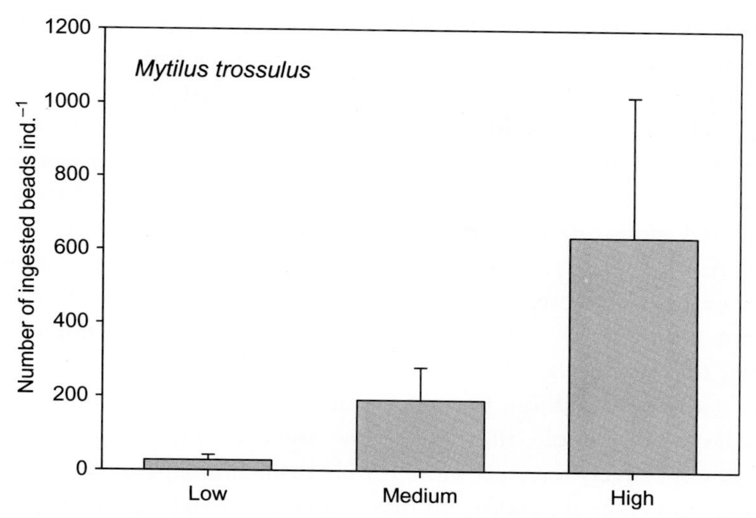

Fig. 11.2 The number of ingested 10 μm spheres (mean ± SD) in *blue* mussel (*Mytilus trossulus*) at three different bead concentrations (low, 5; medium, 50; and high, 250 beads mL^{-1}). *Data from Setälä, O., Norkko, J., Lehtiniemi, M., 2016. Feeding type affects microplastic ingestion in a coastal invertebrate community. Mar. Pollut. Bull. 102(1), 95–101.*

in the South China Sea (Sun et al., 2017). The classical Holling-type ingestion patterns may also be affected by clogging of feeding appendages. In such cases, a high concentration of microplastics (fibers) may decrease feeding activity, resulting in lower ingestion rates.

Active, motile predators (e.g., cruising predators) will encounter prey, and we therefore assume plastic, more readily as they move through the water or sediment. Nonmotile animals will encounter microplastics the same way they come into contact with suspended or deposited prey (i.e., water currents bringing particles close enough for capture or generating localized currents to draw suspended particles to the organism). Sessile organisms are also not able to avoid exposure to microplastics and are subjected to all particles present in the suspension they are feeding in. However, passively floating and sessile organisms and ambush predators can compensate for reduced encounter rates through high efficient filtering activity (Green et al., 2003).

11.3.2 Detecting Microplastics

Animals detect prey using visual or chemical cues or hydromechanical signals when identifying motile prey moving through the water. Organisms relying upon visual detection may mistake microplastics as prey. For example, ocean-foraging Fulmars travel vast distances across the North Atlantic, relying on visual cues to select prey floating near the ocean surface; dissections of Fulmars beached along European coastlines have routinely identified that the seabirds' stomachs are full of plastic (Van Franeker et al., 2011). Researchers often note that microplastic debris comes in a wide range of shapes, size, and color; however, it is currently unclear whether these attributes have any influence on its likelihood of being consumed by animals relying on visual detection.

The swimming activity and speed of motile prey affect their encounter rate, with numerous studies establishing that actively moving prey are detected more frequently and encountered more often (Gerritsen and Strickler, 1977; Gerritsen, 1984; Tiselius et al., 1993). As microplastics are passive particles, they cannot be detected using hydromechanical signals, and we would therefore expect them to be encountered less frequently than motile prey at similar concentrations. For example, in pelagic communities, the swimming activity of the predator is affecting the encounter rate of microplastic particles in addition to their density and overall distribution. However, as plastic particles are nonmotile, they make easy targets for predators and may therefore be ingested (if not actively rejected) more readily than natural prey that can incite escape responses (e.g., Green et al., 2003) and may require an active capturing process.

Chemical cues play a significant but variable role in the prey selection of marine organisms from invertebrates to mammals. For example, fish have diversely developed olfactory organs (Hara, 1975) for detecting signals related to reproduction and feeding. Some marine species possess highly developed chemosensory organs (e.g., sharks), while in some others, they may be poorly developed (e.g., visual predators like pike) (Hara, 1975). Crustaceans, such as copepods are generally considered to be selective feeders that display flexibility in their feeding behavior (Koehl and Stickler, 1981); discrimination between prey can be based on size (Frost, 1972), motility (Atkinson, 1995), or chemical signals (Cowles et al., 1988). Not all chemicals are sensed; what is important is that in order for an organism to receive chemical stimuli, the chemical itself should be soluble in water. Chemical signals can assist in the selection for high-quality food, determined by protein content (Cowles et al., 1988), or be used to avoid unsuitable prey (e.g., harmful algae containing toxic compounds like saxitoxin). However, active avoidance of unsuitable or toxic prey by copepods is most likely a result of a common history, that is, coevolution of the prey and predator (Colin and Dam, 2002).

Field-collected data and exposure experiments show that plastic particles floating in the water and embedded in the sediment are rapidly colonized by rich microbial communities comprising prokaryotic and eukaryotic organisms, like bacteria and algae (Oberbeckmann et al., 2014; Harrison et al., 2014). So far, there is very little information on how the formation of biofilm actually affects the ingestion of microplastics. Recent studies show that the effects of biofouling are most likely taxon- or even species-specific. Vroom et al. (2017) identified that biofouling of polystyrene beads promoted ingestion by planktonic crustaceans, although this was somewhat dependent on taxon, size, and stage of the grazers. For two of the three copepod species studied (*Acartia longiremis* and *Calanus finmarchicus*, excluding the adult females of the latter), it was shown that

BOX 11.2 Experimental Work

Most of the information that has so far been produced on the parameters affecting microplastic ingestion by marine organisms come from simplified laboratory experiments. Results from experimental work should not be directly applied to natural conditions where confounding factors exist. When conducting environmentally relevant experimental work on ingestion and effects of microplastics in food webs, the concentration, size, and type of the used particles should be adjusted to correspond to natural conditions. At the moment, there is still a mismatch between "reality" and laboratory experiments. So far, most experiments are run with microplastic concentrations higher than those commonly found in the environment and with virgin particles of uniform size and shape that fail to accurately represent the conditions in the field (Phuong et al., 2016). This inconsistency is likely to influence our understanding of the marine microplastic problem as Ogonowski et al. (2016) showed in laboratory experiments comparing the effects of primary and secondary microplastics. They showed that secondary microplastics have more negative effects on feeding in a cladoceran, *Daphnia magna*, compared with primary microplastics commonly used in the previous studies. The reason why experimental laboratory studies have not used microplastic concentrations commonly observed in marine environment is not only their "low" concentrations but also the uncertainty in assessing their concentrations. Microplastic concentrations found in marine environments vary significantly between areas and habitats but seem to be low when compared with the numbers of the real prey, which makes environmentally relevant exposure studies difficult. Long-lasting exposure experiments in mesocosms mimicking natural conditions would be needed to more accurately assess the relationships between microplastics and their potential predators.

in most cases, the fouled microplastics were ingested by more individuals and at higher rates than the unfouled plastics. However, one copepod species, *Pseudocalanus* spp., did not ingest any of the microplastic particles offered. Contradictory results were reported by Allen et al. (2017) who studied the ingestion of weathered, fouled, and unfouled preproduction pellets (polystyrene (PS), low-density polyethylene (LDPE), and high-density polyethylene (HDPE)), by a scleractinian coral species known to use chemosensory cues for feeding. Their results showed that the corals ingested different types of plastics, consuming significantly more unfouled than fouled microplastics that were taken up (Box 11.2).

11.4 INTO THE FOOD WEBS

The ingestion, entanglement, or inhalation of microplastic by marine organisms can be viewed as an entry point into marine food webs. Owing to their small size, microplastics are bioavailable to a wide range of marine organisms and can be both selectively and accidentally ingested (Schuyler et al., 2012). The ingestion of microplastic particles is affected by their concentration, size, shape, distribution, and chemical character (i.e., density and chemical signal) and the animal's feeding habits. In animals with developed organs for prey detection, plastic polymers may thus not be selected, or they may be rejected if they are recognized as being unfavorable or if a more preferable prey is available.

11.4.1 Filter Feeding

Filter-feeding organisms are prevalent throughout marine food webs, from small planktonic invertebrates and benthic taxa to megafauna, where they feed on suspended organic material, such as algae, zooplankton, fish larvae, and detritus. The size range of particles that can be ingested by a grazer depends on the feeding mode (e.g., filter feeding or raptorial), gape size, and specific feeding mechanisms of the grazer/predator. For filter feeders, the actual size limits for the ingested prey are set by the structure and function of the filtering apparatus used for trapping particles from the suspension (Riisgård and Larsen, 2010). Filtering devices in suspension-feeding organisms are not simple sieves that mechanically clean the water from suspended particles. The structures of filtering apparatus found in unicellular, invertebrate, or vertebrate organisms differ greatly, both between and among taxa, with varying levels of adaptability and sensory capability. Particle capture depends on particle type (e.g., shape, size, and density), particle concentration, water viscosity, the quantity of water that is filtered, and filtering efficiency. Besides direct contact, the capturing mechanisms may also involve other factors, such as chemo- and mechanoreception (Riisgård and Larsen, 2010). Moreover, experimentally measured clearance rates of plankton have been found to vary also depending on temperature, salinity, and the type of prey that has been offered (e.g., Kiørboe et al., 1982; Garrido et al., 2013). Daily clearance rates of marine invertebrates can vary from microliters (unicellular organisms, like ciliates) to milliliters (copepods), liters (bivalves), hundreds of liters (gelatinous zooplankton), or more (baleen whales).

Two parameters are commonly used to estimate the efficiency and outcome of filter feeding: ingestion and clearance rate. The ingestion rate denotes the number of prey particles ingested per predator in a time unit. Ingestion rate can be experimentally estimated directly, through observations of ingested prey particles inside the organism, or indirectly, as the disappearance of prey from the experimental media over time. In the past, inert plastic particles (spheres) have been used as surrogates for natural prey to estimate feeding parameters in planktonic organisms (Huntley et al., 1983; Borsheim, 1984; Nygaard et al., 1988). These historical studies with *Calanus* and related copepod genera have demonstrated a preference for algae over polystyrene beads, alongside size selectivity (Fernandez, 1979; Donaghay and Small, 1979; Huntley et al., 1983). However, observations for such preferences do not necessarily hold for all developmental stages, which further complicates things, that is, when exposure studies are being conducted. Clearance rate is a derivative of ingestion rate and is calculated by dividing the latter by prey concentration. The clearance rate thus measures the water volume that an individual organism can clear of food particles in a time unit. To understand the probability of any suspended particle to be ingested by a filter-feeding organism, both the clearance rate and the concentration of suitable prey should be taken into account.

From the viewpoint of a small filter-feeding organism under natural conditions, microplastic concentrations may be too low for routinely encountering a plastic particle. However, in waters containing high concentrations of microplastics, the situation is different even for a small organism with a relatively low clearance rate and efficiency, such as a copepod. As an example, the experimentally defined daily clearance rates of common copepods may vary between ~10 and <200 mL (Frost, 1975; Engström et al., 2000; Setälä et al., 2009). In theory, a copepod feeding, for example, with a high clearance rate of 144 mL/day (Frost, 1975), at a concentration of 9200 plastics per m^3 as has been observed from the Pacific Ocean (Desforges et al., 2014), a single microplastic would be ingested by every 0.7 copepods, assuming all particles are edible and the animals are solely undertaking passive ingestion without rejection of plastic. Assessments based on animals collected from the field have also confirmed the role of zooplankton as entry points for microplastics to food webs. The study of Desforges et al. (2015) which was based on the analysis of the number of ingested microplastics from subsurface-collected zooplankton and the overall distribution of these species from the Northeast Pacific Ocean, identified encounter of microplastics by zooplankton as 1 particle per every 34 copepods and 1 particle per every 17 euphausiid. The authors further estimated that both the juvenile salmon and adult returning fish would be affected daily with ingested microplastics through their zooplankton prey.

Invertebrates with a capacity for filtering larger quantities of water and with a longer life span (e.g., bivalves) or large filter feeders (such as whales) may encounter microplastics far more frequently than zooplankton. Bivalves are one of the key organisms when entry points of microplastics to marine food webs are assessed. They are efficient suspension-feeding animals that form links between the pelagic and benthic ecosystems and are a key source of prey for many marine fish, birds, and mammals. In the Baltic Sea, it has been assessed that within 1 year, the blue mussel beds would, in theory, filter a water volume equivalent to the whole sea basin (Kautsky and Kautsky, 2000). The numbers of microplastics found in bivalves vary significantly ranging from <0.5 particles (Eastern Atlantic and Baltic Sea) to over 100 particles (Western Atlantic) per animal (Mathalon and Hill, 2014; Vandermeersch et al., 2015; Railo, 2017). Exposure of large filter feeders to microplastics has been shown by Fossi et al. (2014) after examining concentrations of phthalates and organochlorine compounds of a basking shark and a baleen whale. The authors concluded that microlitter is ingested by these large filter feeders together with their neustonic prey. A comparative study carried out in two semienclosed basins, the Mediterranean Sea and the Sea of Cortez in the Gulf of California (Fossi et al., 2016), gives supporting information indicating that fin whales in highly polluted areas are

BOX 11.3 Microplastics, an Issue of Size

"Microplastic" is typically used to describe plastic particles smaller than 5 mm in diameter, with a lower size limit of 100 nm; plastics larger than 5 mm are considered "macroplastics," while plastics smaller than 100 nm in size are termed "nanoplastic" (Cole et al., 2011). Using these size classifications, the largest microplastic particles (5000 μm) have a diameter 50,000 times larger than the smallest microplastic (0.1 μm). Moreover, when we consider volume and surface area, these differences become even more apparent. Imagine a spherical shaped microplastic particle, like the ones used in experimental studies, or the plastic microbeads commonly used in exfoliating personal care products: a 5 mm-diameter bead is 1.25×10^{14} times greater in volume and 2.50×10^{9} larger by area than a 100 nm-diameter bead. Of course, most of the weathered microplastic particles that are found in the marine environment are not uniform in shape, with fibrous, planar, and irregularly shaped plastic being most prevalent. Nevertheless, differences in a particle's dimensions will have a significant impact on the risk they pose to marine life. For example, microplastics of different sizes may differ in their behavior under marine conditions (i.e., buoyancy), biological availability, and capacity to incite biological effects. Furthermore, the larger surface-area-to-volume ratios associated with smaller particles greatly increase the plastic's capacity for adsorbing (and potentially desorbing) waterborne pollutants (e.g., persistent organic pollutants and hydrophobic organic contaminants) (Koelmans et al., 2016), up to 1 million times greater than that found in the surrounding seawater (Mato et al., 2001).

exposed to major health hazards due to microplastics and their cocontaminants. Considering the vast amounts of water these animals filter (5893 m^3 day^{-1}; Fossi et al., 2014), this conclusion is more than relevant (Box 11.3).

11.4.2 Respiratory Intake

Ventilation has also been identified in exposure experiments as a means by which microplastics can be concentrated from the surrounding water. Watt et al. (2014) identified that the shore crab (*Carcinus maenas*) was able to respire polystyrene microbeads, which accumulated on the surface of their gills. Blue mussels (*Mytilus trossulus*) and Baltic clams (*Macoma balthica*) have also been shown to accumulate microplastic particles to their gills after 24 h incubations; however, the bead concentrations were much higher in the digestive tracts of the same animals (Setälä et al., 2016a).

11.4.3 Entanglement

Numerous organisms have been shown to entangle with fibers or larger plastics (e.g., Laist, 1997; Cole et al., 2013; NOAA, 2014; Taylor et al., 2016). They may be found in the swimming or feeding appendages of invertebrates and in the valve gapes of bivalves or entangled around larger animals. Entanglement with fibers in field-collected animals has been observed even in remote areas such as the deep seas, where fibers were found on sea pens and hermit crabs (Taylor et al., 2016). When these organisms are eaten by higher trophic level predators, the plastics adhered to external surfaces of the organisms will be eaten as well.

11.4.4 Trophic Transfer

Once ingested, microplastics will be either egested or retained by the organism. If a predator consumes an organism that has retained microplastic, the predator will be indirectly consuming this plastic as part of its diet, in a process referred to as "trophic transfer." The trophic transfer of plastic has been documented in predatory Norway lobsters (*Nephrops norvegicus*) that consumed polypropylene rope fibers embedded in fish (Murray and Cowie, 2011), shore crabs (*C. maenas*) that indirectly ingested fluorescent polystyrene 0.5 and 10 μm microspheres present in common mussels (*M. edulis*) (Farrell and Nelson, 2013; Watt et al., 2014), mysid shrimps (*Neomysis integer*) that consumed fluorescent polystyrene 10 μm spheres previously taken up by mesozooplankton (Setälä et al., 2014), and fish (*Gasterosteus aculeatus*) that consumed an insect larvae containing microbeads in a mesocosm experiment (Lehtiniemi and Setälä, unpublished). The trophic transfer of microplastics and associated POPs from *Artemia* nauplii to zebra fish (*Danio rerio*) was also verified in a laboratory experiment (Batel et al., 2016), and microplastic debris found in fecal pellets of predatory seabirds (great skuas, *Stercorarius skua*) was greatest when correlated with the remains of surface-feeding Northern fulmars (*Fulmarus glacialis*) (Hammer et al., 2016).

For trophic transfer to occur, microplastic must be consumed alongside the prey. This includes plastic adhered to algae (Bhattacharya et al., 2010; Gutow et al., 2015) or the external surfaces of an animal (e.g., entrapped in the setae of a copepods' appendages; Cole et al., 2013) or retained indefinitely within the organism itself. Plastics are commonly observed in the intestinal tract of marine animals, including seabirds (Van Franeker and Law, 2015), fish (Lusher et al., 2013), invertebrates (Murray and Cowie, 2011), and turtles (Nelms et al., 2016); this occurs where larger plastics or coalesced polymeric fibers cause a gut blockage, preventing the plastic from being shifted via peristaltic action. In the common shore crab (*C. maenas*), polystyrene microspheres have been observed to lodge between the microvilli that line the stomach, resulting in prolonged gut retention times. In copepods, starvation has been observed to increase gut retention times, with 10 μm polystyrene microspheres remaining in the intestinal tracts of *C. helgolandicus* for up to 7 days, far exceeding the typical gut passage times of just 2 h (Cole et al., 2013). In the common mussel (*M. edulis*), 3.0–9.6 μm polystyrene microspheres have been demonstrated to translocate into the circulatory fluid (hemolymph), where they can remain for in excess of 48 days (Browne et al., 2008; von Moos et al., 2012). Owing to their small size, nanoplastics (<100 nm diameter) have the capacity to cross epithelia and therefore have the capacity to enter tissues and circulatory fluids, for example, in dendritic cells that transport small particles (e.g., bacteria) across gut epithelial cell walls (Rescigno et al., 2001). Microplastic transfer has also

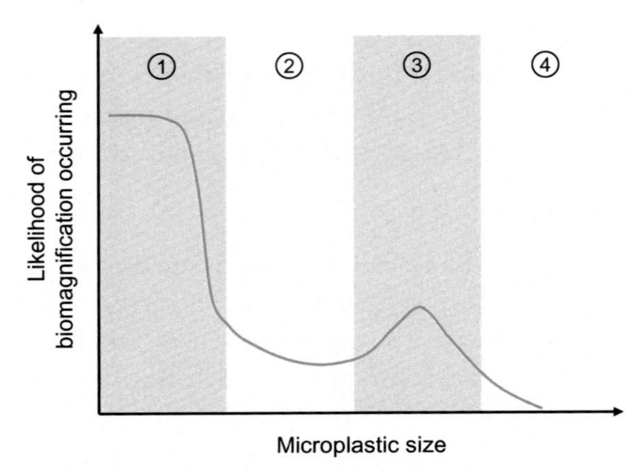

Fig. 11.3 Considering how microplastic size might influence the probability of biomagnification of plastics occurring in a food chain. (1) Very small (i.e., nano) plastics are readily absorbed by the gut and are retained within the circulatory fluid and/or tissues; (2) moderately sized plastics are ingested, are present within the organism during gut transit, and are then readily egested; (3) larger and fibrous plastics are ingested but, owing to their size, remain in the intestinal tract; (4) the largest microplastics are inedible to organisms at the base of the food chain.

been documented in a top marine predator, where the presence of microplastics in captive gray seal scats was attributed to trophic transfer from the wild-caught mackerel they were fed upon (Nelms et al., 2018).

In numerous aquatic ecosystems, persistent chemical pollutants (i.e., PCBs, PAHs, and methyl mercury) have been shown to biomagnify as they pass up the food chain (reviewed by Blais et al., 2007). The increasing body burdens of such pollutants in higher trophic organisms arise from the hydrophobicity of these chemicals, resulting in their accumulation within fatty tissues of prey species. So far, there have been no quantitative measures of microplastics passing up the food chain, and it therefore remains unclear whether plastics will biomagnify in marine food webs. Biomagnification will largely depend upon the transience of plastics in an organism, with biomagnification only occurring where plastics are readily ingested and retained. Retention of plastics can be influenced by food availability (Cole et al., 2013; Watt et al., 2014) and shape (Murray and Cowie, 2011) but will be predominantly governed by the size of the plastic (Galloway, 2015). In Fig. 11.3, we predict how the size of a plastic particle is likely to relate to the probability of that microplastic biomagnifying up the food chain.

11.5 ALTERATION, REPACKAGING AND TRANSPORT OF MICROPLASTICS WITHIN MARINE FOOD WEBS

In this section, we consider how marine organisms, trophic dynamics, and biologically mediated ecological processes can alter the fate of a microplastic and highlight how microplastics might impinge on biota, food webs, and marine ecosystems.

11.5.1 Biological Transport of Microplastic

Microplastics consumed, respired, or adhered by an organism will be subject to passive, biologically mediated transportation, with both vertical and lateral movement to be expected across a variety of habitats (e.g., water column and sediments). The distances by which microplastics can be transported via a biological vector will largely depend on the movement, migratory routes, and gut transit times of the individual organism (Fig. 11.4).

Diel vertical migrations, a synchronous daily migration of a wide range of taxa, have been highlighted as a potential route by which microplastics could be transported from the sea surface to deeper waters (Cole et al., 2016; Clark et al., 2016). Organisms may ingest plastics while feeding at the surface at night, which can then be egested hundreds of meters below the surface. For example, a large (2–3 mm) copepod swimming at speeds of between 30 and 90 m h^{-1} (Enright, 1977), with a gut evacuation time of approximately 2 h (Cole et al., 2013), could vertically transport microplastic to depths of 60–180 m. Lusher et al. (2016) identified that 11% of mesopelagic fish caught in the Northeast Atlantic had microplastics in their digestive tracts, and although it was unknown at what depth these plastics were consumed, the majority of the species identified undergo diel vertical migration and follow their zooplankton prey to the surface to feed; it is therefore plausible to suggest that ingestion of the microplastics may have occurred at the surface while feeding and egested at depth.

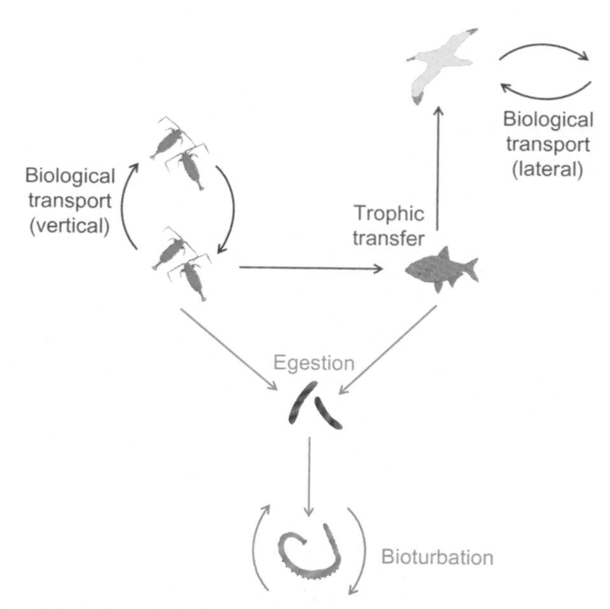

Fig. 11.4 How biota transport microplastics within marine ecosystems. *Image by Matthew Cole (original content).*

The geographic distribution of marine plastic has largely been considered from a physical perspective, with abiotic processes (i.e., wind, rivers, and oceanic currents) expected to be the dominant factors in distributing this pervasive pollutant (Sherman and Van Sebille, 2016). We consider that migratory species could also facilitate the transport of plastics. Migratory species have been widely demonstrated to play a vital role in the long-range transport of persistent pollutants (e.g., PCBs, DDT, and methyl mercury; Blais et al., 2007). For example, migratory fish (e.g., trout and salmon) have been shown to accumulate persistent organochlorines in their tissues while feeding in marine habitats, which are released in their eggs during spawning at otherwise pristine freshwater sites (Krümmel et al., 2003; Mu et al., 2004). Numerous migratory species, including turtles (Nelms et al., 2016), ocean-foraging seabirds (Van Franeker and Law, 2015), and cetaceans (Lusher et al., 2015a), are routinely sampled with plastics in their intestinal tracts. These animals undertake large-scale annual migrations; for example, the gray whale (*Eschrichtius robustus*) travels 6000 km annually from the coast of Mexico to the Chukchi Sea, and the Arctic tern (*Sterna paradisaea*) migrates 19,000 km from Greenland to the Antarctic each year (Alerstam et al., 2003). The egestion of plastic within feces, scat, or guano; the regurgitation of plastics by seabirds when feeding their young (Sileo et al., 1990); or the death of the animal will all contribute to the deposition of plastic in terrestrial, freshwater, or marine habitats far from the waters where such plastic was ingested.

11.5.2 Incorporation of Microplastics Into Biological Matrices

Within the marine environment, microplastics are rapidly colonized by "biofilms," made up of microorganisms, plants, and epibionts that attach and grow on substrates. The characteristics of the biofilm that forms on a plastic will be influenced by the polymer and the biological or ecological matrix through which it has passed; as such, the microbial complex that forms on the surface of plastics may act as a tracer of the journey of a microplastic within marine compartments (Galloway et al., 2017). The development of a biofilm can change the characteristics of the plastic polymer, for example, by increasing their mass (Lobelle and Cunliffe, 2011; Zettler et al., 2013; Rummel et al., 2017) and altering their chemical signal (see Section 11.3.2). It has been postulated that biofilm formation could be enough to cause otherwise buoyant plastics to sink or oscillate within the water column, depending on the size and density of the plastic (Ye and Andrady, 1991; Kooi et al., 2017).

In bivalves, feeding or rejection of particles that are suspended in the water is the outcome of passive and active selection. The size of the particles that may be ingested depends on the filtration apparatus of the particular species. In Pacific oyster (*Crassostrea gigas*) larvae, uptake of polystyrene microbeads was size-dependent, with microplastics larger than the oral groove unable to be ingested, while smaller plastics were readily consumed (Cole and Galloway, 2015). If the size is right and prey is directed to the specialized feeding organs (ctenidium), it may still be rejected as pseudofeces if considered unpalatable. Studies made with blue mussels have shown that the identification of unsuitable particles and their sorting in suspension-feeding bivalves take place in the lectin-containing mucus that covers feeding organs, where interaction with carbohydrates from suspension takes place (Espinosa et al., 2010). Mussels (*M. edulis*) have been visualized rejecting nanopolystyrene (Ward and Koch, 2009) and microplastic polyvinylchloride in their pseudofeces (personal observations of authors). The fate of microplastics incorporated into pseudofeces remains unclear.

Ingested microplastics will typically be passed along the intestinal tract through peristaltic action. Within the intestinal tract, microplastics will either be adsorbed across the gut lining, become entrapped in the gut (i.e., intestinal blockage causing retention of plastic), or become incorporated into the animal's feces and egested. Microplastics have been identified in the fecal pellets of copepods (Cole et al., 2013), and it is assumed that most animals that consume plastics will then egest them. Microplastics have been observed in commercially caught fish (e.g., Lusher et al., 2013), and while there are currently no data to explain the fate of plastic post ingestion, it could be assumed that the majority would pass through the gut and get packaged in fecal pellets. The repackaging of plastic into the feces of an animal will alter the properties (i.e., relative buoyancy) of the plastics within the water column (Cole et al., 2015) and represent an alternate route by which plastics can be transferred within marine ecosystems (Clark et al., 2016).

Sinking feces and marine aggregates play a vital role in the biological pump, whereby carbon and nutrients in the euphotic zone are repackaged and transported to the ocean depths (Turner, 2015). Feces from anchovies in the productive upwelling system off the coast of Peru were observed as a key contributor to downward flux in sediment traps, with fecal sinking rates averaging >1 km day^{-1} (Staresinic et al., 1983). In this scenario, any microplastics contained within these pellets may reach benthic sediments within a very short space of time. However, experimental work has documented that the incorporation of microplastics into fecal pellets (Cole et al., 2016) and marine aggregates (Long et al., 2015) will alter the buoyancy of the biological matrix. Many carbon flux studies have concluded that slowly sinking feces are unlikely to reach the seabed, instead becoming repackaged through coprophagy (i.e., the consumption of fecal matter) by larger zooplankton species (Turner, 2002), or broken down through microbial action. In feces containing microplastic, coprophagy would therefore represent a route by which plastics can reenter the marine food web. This has been demonstrated with copepods, when polystyrene microplastics ingested by the small copepod, *Centropages typicus*, were egested in their fecal pellets and subsequently ingested by the larger copepod, *C. helgolandicus* (Cole et al., 2016). The study further highlighted that microplastic-laden pellets were more prone to fragment, making them more bioavailable to detritivores during their descent through the water column.

11.5.3 The Fate of Microplastics in Benthic Ecosystems

Benthic sediments have been identified as an important sink for microplastics, including high-density plastics, which readily settle out of the water column, and lower-density plastics whose movement to the benthos is facilitated by biological matrices. Highly polluted coastal sediments may comprise 3% microplastics (Carson et al., 2011), while estimates of 4 billion bioplastic and polymer fibers per km^2 are reported in Indian seamount sediments (Woodall et al., 2014). Within sediments, microplastics become bioavailable to benthic dwelling fauna, including important commercial species such as Norwegian lobster, *N. norvegicus* (Murray and Cowie, 2011), and shellfish (Rochman et al., 2015). A number of papers having highlighted the capacity for benthic organisms, including bivalves (Sussarellu et al., 2016), echinoderms (Graham and Thompson, 2009), and polychaetes (Wright et al., 2013b; Besseling et al., 2012; Green et al., 2016) to ingest microplastics, with the potential to incite negative health effects with repercussions for their functionality (i.e., reduced bioturbation activity and reduced energetic reserves). As with pelagic organisms, it is hypothesized that benthic taxa can alter the properties of microplastics and through bioturbation move plastics from the sediment–water interface deeper into sediments. This has been evidenced in polychaetes and clams that transported microplastic fibers (polyethylene fishing line <1 mm) to depths of 1.7–5.1 cm during a 3-week mesocosm experiment (Näkki et al., 2017). However, determining the capacity for sediment-dwelling biota to redistribute plastic under natural conditions remains unknown, and it is unclear whether bioturbation can result in the permanent burial of this plastic.

11.6 CONCLUSIONS

Microplastics are under extensive research, and their complex interactions with marine food webs are becoming increasingly evident. Microplastics are pervasive, environmentally persistent particles, which have the potential to flux between the water column, seabed, and biota. Nano- and microplastics can enter marine food webs via a number of entry points and can subsequently be cycled through different biotic compartments; these biotic processes can result in changes to the properties and movement of the microplastic. Parameters governing the entrance of microplastics into food webs include the spatial overlap of microplastics and biota, the feeding strategy and motility of the organism, and the characteristics of the plastic. From the studies carried out so far, we have learned that different taxa, species, and developmental stage of a species will each process, handle, and react to microplastics in a myriad of ways. Some organisms have mechanisms that protect them from consuming anthropogenic contaminants, while others readily ingest large numbers of microplastic particles together with their natural prey. With microplastic pollution in the marine environment becoming a growing threat, the numbers of both primary and secondary microplastics are increasing. There may therefore come a time when the exposure experiments that are carried out today and that have been criticized because of their high microplastic concentrations will be considered as "historic" research with environmentally relevant concentrations.

FUNDING

OS and ML acknowledge Ministry of Environment and Academy of Finland (MIF 296169) for funding. RLC is funded through a Natural Environment Research Council GW4+ PhD studentship (NE/L002434/1). MC acknowledges funding from the Natural Environment Research Council discovery grant (NE/L007010).

REFERENCES

Alerstam, T., Hedenström, A., Åkesson, S., 2003. Long-distance migration: evolution and determinants. Oikos 103 (2), 247–260.

Allen, A.S., Seymour, A.C., Rittschof, D., 2017. Chemoreception drives plastic consumption in a hard coral. Mar. Pollut. Bull. 124 (1), 198–205.

Andrady, A.L., 2017. The plastic in microplastics: a review. Mar. Pollut. Bull. 119 (1), 12–22.

Atkinson, A., 1995. Omnivory and feeding selectivity in five copepod species during spring in the Bellingshausen Sea, Antarctica. ICES J. Mar. Sci. 52 (3–4), 385–396.

Barnes, D.K., Galgani, F., Thompson, R.C., Barlaz, M., 2009. Accumulation and fragmentation of plastic debris in global environments. Philos. Trans. R. Soc. Lond. Ser. B Biol. Sci. 364 (1526), 1985–1998.

Batel, A., Linti, F., Scherer, M., Erdinger, L., Braunbeck, T., 2016. The transfer of benzo [a] pyrene from microplastics to Artemia nauplii and further to zebrafish via a trophic food web experiment—CYP1A induction and visual tracking of persistent organic pollutants. Environ. Toxicol. Chem. 35 (7), 1656–1666.

Beck, M.W., Brumbaugh, R.D., Airoldi, L., Carranza, A., Coen, L.D., Crawford, C., Defeo, O., Edgar, G.J., Hancock, B., Kay, M.C., Lenihan, H.S., 2011. Oyster reefs at risk and recommendations for conservation, restoration, and management. Bioscience 61 (2), 107–116.

Bellas, J., Martinez-Armental, J., Martinez-Camara, A., Besada, V., Martinez-Gomez, C., 2016. Ingestion of microplastics by demersal fish from the Spanish Atlantic and Mediterranean coasts. Mar. Pollut. Bull. 109 (1), 55–60.

Bergmann, M., Wirzberger, V., Krumpen, T., Lorenz, C., Primpke, S., Tekman, M.B., Gerdts, G., 2017. High quantities of microplastic in Arctic deep-sea sediments from the HAUSGARTEN observatory. Environ. Sci. Technol. 51 (19), 11000–11010.

Besseling, E., Wegner, A., Foekema, E.M., Van Den Heuvel-Greve, M.J., Koelmans, A.A., 2012. Effects of microplastic on fitness and PCB bioaccumulation by the lugworm *Arenicola marina* (L.). Environ. Sci. Technol. 47 (1), 593–600.

Bhattacharya, P., Lin, S., Turner, J.P., Ke, P.C., 2010. Physical adsorption of charged plastic nanoparticles affects algal photosynthesis. J. Phys. Chem. C Nanomater. Interfaces 114 (39), 16556.

Blais, J.M., Macdonald, R.W., Mackay, D., Webster, E., Harvey, C., Smol, J.P., 2007. Biologically mediated transport of contaminants to aquatic systems. Environ. Sci. Technol. 41 (4), 1075–1084.

Borsheim, K.Y., 1984. Clearance rates of bacteria sized particles by freshwater ciliates measured with monodisperse fluorescent latex beads. Oecologia 63, 286–288.

Bråte, I.L.N., Huwer, B., Thomas, K.V., Eidsvoll, D.P., Halsband, C., Carney Almroth, B., Lusher, A., 2017. Micro-and macro-plastics in marine species from Nordic waters. Nordic Council of Ministers. 101 p, 48–49. https://doi.org/10.6027/TN2017-549

Bravo Rebolledo, E.L., Van Franeker, J.A., Jansen, O.E., Brasseur, S.M., 2013. Plastic ingestion by harbour seals (*Phoca vitulina*) in the Netherlands. Mar. Pollut. Bull. 67, 200–202.

Browne, M.A., Dissanayake, A., Galloway, T.S., Lowe, D.M., Thompson, R.C., 2008. Ingested microscopic plastic translocates to the circulatory system of the mussel, *Mytilus edulis* (L.). Environ. Sci. Technol. 42 (13), 5026–5031.

Browne, M.A., Crump, P., Niven, S.J., Teuten, E., Tonkin, A., Galloway, T.S., Thompson, R.C., 2011. Accumulation of microplastic on shorelines worldwide: sources and sinks. Environ. Sci. Technol. 45 (21), 9175–9179.

Budimir, S., Setälä, O., Lehtiniemi, M., 2018. Effective and easy to use extraction method shows low numbers of microplastics in offshore planktivorous fish from the northern Baltic Sea. Mar. Pollut. Bull. 127, 586–592. https://doi.org/10.1016/j.marpolbul.2017.12.054.

Cadée, G.C., 1976. Sediment reworking by *Arenicola marina* on tidal flats in the Dutch Wadden Sea. Neth. J. Sea Res. 10 (4), 440–460.

Carson, H.S., Colbert, S.L., Kaylor, M.J., McDermid, K.J., 2011. Small plastic debris changes water movement and heat transfer through beach sediments. Mar. Pollut. Bull. 62, 1708–1713.

Clark, J., Cole, M., Lindeque, P.K., Fileman, E., Blackford, J., Lewis, C., Lenton, T.M., Galloway, T.S., 2016. Marine microplastic debris: a targeted plan for understanding and quantifying interactions with marine life. Front. Ecol. Environ. 14 (6), 317–324.

Cole, M., Galloway, T.S., 2015. Ingestion of nanoplastics and microplastics by Pacific oyster larvae. Environ. Sci. Technol. 49, 14625–14632.

Cole, M., Lindeque, P., Halsband, C., Galloway, T.S., 2011. Microplastics as contaminants in the marine environment: a review. Mar. Pollut. Bull. 62 (12), 2588–2597.

Cole, M., Lindeque, P., Fileman, E., Halsband, C., Goodhead, R., Moger, J., Galloway, T.S., 2013. Microplastic ingestion by zooplankton. Environ. Sci. Technol. 47, 6646–6655.

Cole, M., Webb, H., Lindeque, P.K., Fileman, E.S., Halsband, C., Galloway, T.S., 2014. Isolation of microplastics in biota-rich seawater samples and marine organisms. Sci. Rep. 4 (4528), 1–8.

Cole, M., Lindeque, P., Fileman, E., Halsband, C., Galloway, T.S., 2015. The impact of polystyrene microplastics on feeding, function and fecundity in the marine copepod *Calanus helgolandicus*. Environ. Sci. Technol. 49 (2), 1130–1137.

Cole, M., Lindeque, P.K., Fileman, E., Clark, J., Lewis, C., Halsband, C., Galloway, T.S., 2016. Microplastics alter the properties and sinking rates of zooplankton faecal pellets. Environ. Sci. Technol. 50 (6), 3239–3246.

Colin, S.P., Dam, H.G., 2002. Effects of the toxic dinoflagellate *Alexandrium fundyense* on the copepod *Acartia hudsonica*: a test of the mechanisms that reduce ingestion rates. Mar. Ecol. Prog. Ser. 248, 55–65.

Collignon, A., Hecq, J.H., Galgani, F., Voisin, P., Collard, F., Goffart, A., 2012. Neustonic microplastic and zooplankton in the North Western Mediterranean Sea. Mar. Pollut. Bull. 64 (4), 861–864.

Cowles, T.J., Olson, R.J., Chisholm, S.W., 1988. Food selection by copepods: discrimination on the basis of food quality. Mar. Biol. 100, 41–49.

Cózar, A., Sanz-Martín, M., Martí, E., González-Gordillo, J.I., Ubeda, B., Gálvez, J.Á., Irigoien, X., Duarte, C.M., 2015. Plastic accumulation in the Mediterranean Sea. PLoS ONE 10(4) e0121762.

Desforges, J.P.W., Galbraith, M., Dangerfield, N., Ross, P.S., 2014. Widespread distribution of microplastics in subsurface seawater in the NE Pacific Ocean. Mar. Pollut. Bull. 79, 94–99.

Desforges, J.P., Galbraith, M., Ross, P.S., 2015. Ingestion of microplastics by zooplankton in the Northeast Pacific Ocean. Arch. Environ. Contam. Toxicol. 69 (3), 320–330.

Devriese, L.I., van der Meulen, M.D., Maes, T., Bekaert, K., Paul-Pont, I., Frère, L., Robbens, J., Vethaak, A.D., 2015. Microplastic contamination in brown shrimp (*Crangon crangon*, Linnaeus 1758) from coastal waters of the Southern North Sea and Channel area. Mar. Pollut. Bull. 98 (1), 179–187.

Do Sul, J.A.I., Costa, M.F., Fillmann, G., 2014. Microplastics in the pelagic environment around oceanic islands of the Western Tropical Atlantic Ocean. Water Air Soil Pollut. 225 (7), 2004. https://doi.org/10.1007/s11270-014-2004-z.

Donaghay, P.L., Small, L.F., 1979. Food selection capabilities of the estuarine copepod *Acartia clausi*. Mar. Biol. 52, 137–146.

Engström, J., Koski, M., Viitasalo, M., Reinikainen, M., Repka, S., Sivonen, K., 2000. Feeding interactions of the copepods *Eurytemora affinis* and *Acartia bifilosa* with the cyanobacteria *Nodularia* sp. J. Plankton Res. 22 (7), 1403–1409.

Enright, J.T., 1977. Copepods in a hurry: sustained high-speed upward migration. Limnol. Oceanogr. 22, 118–125.

Eriksen, M., Maximenko, N., Thiel, M., Cummins, A., Lattin, G., Wilson, S., Hafner, J., Zellers, A., Rifman, S., 2013. Plastic pollution in the South Pacific subtropical gyre. Mar. Pollut. Bull. 68 (1), 71–76.

Eriksen, M., Lebreton, L.C.M., Carson, H.S., Thiel, M., Moore, C.J., Borerro, J.C., Galgani, F., Ryan, P.G., Reisser, J., 2014. Plastic pollution in the world's oceans: 5 trillion plastic pieces weighing over 250.000 tons afloat at sea. PLoS One 9 (12), 1–15. https://doi.org/10.1371/journal.pone.0111913.

Espinosa, E.P., Hassan, D., Ward, J.E., Shumway, S., Allam, B., 2010. Role of epicellular molecules in the selection of particles by the blue mussel, *Mytilus edulis*. Biol. Bull. 219, 50–60.

Evans, G.T., 1989. The encounter speed of moving predator and prey. J. Plankton Res. 11, 415–417.

Farrell, P., Nelson, K., 2013. Trophic level transfer of microplastic: *Mytilus edulis* (L.) to *Carcinus maenas* (L.). Environ. Pollut. 177, 1–3.

Fazey, F.M., Ryan, P.G., 2016. Biofouling on buoyant marine plastics: an experimental study into the effect of size on surface longevity. Environ. Pollut. 210, 354–360.

Fernandez, F., 1979. Particle selection in the nauplius of *Calanus pacificus*. J. Plankton Res. 1 (4), 313–328.

Fossi, M.C., Coppola, D., Baini, M., Giannetti, M., Guerranti, C., Marsili, L., 2014. Large filter feeding marine organisms as indicators of microplastic in the pelagic environment: the case studies of the Mediterranean basking shark (*Cetorhinus maximus*) and fin whale (*Balaenoptera physalus*). Mar. Environ. Res. 100, 17–24.

Fossi, M.C., Marsili, L., Baini, M., Giannetti, M., Coppola, D., Guerranti, C., 2016. Fin whales and microplastics: the Mediterranean Sea and the Sea of Cortez scenarios. Environ. Pollut. 209, 68–78.

Frost, B.W., 1972. Effects of size and concentration of food particles on the feeding behaviour of the marine planktonic copepod *Calanus pacificus*. Limnol. Oceanogr. 17, 805–815.

Frost, B.W., 1975. A threshold feeding behavior in *Calanus pacificus*. Limnol. Oceanogr. 20, 263–266.

Galgani, F., Hanke, G., Werner, S.D.V.L., De Vrees, L., 2013. Marine litter within the European marine strategy framework directive. ICES J. Mar. Sci. 70 (6), 1055–1064.

Galloway, T.S., 2015. Micro- and nano-plastics and human health. In: Marine Anthropogenic Litter. Springer International Publishing, Cham, pp. 343–366.

Galloway, T.S., Cole, M., Lewis, C., 2017. Interactions of microplastic debris throughout the marine ecosystem. Nat. Ecol. Evol. 1, 0116.

Garrido, S., Cruz, J., Santo, M., Saiz, E., 2013. Effects of temperature, food type and food concentration on the grazing of the calanoid copepod *Centropages chierchiae*. J. Plankton Res. 35 (4), 843–854.

Gerritsen, J., 1984. Size efficiency reconsidered: a general foraging model for free-swimming aquatic animals. Am. Nat. 123, 450–467.

Gerritsen, J., Strickler, J.R., 1977. Encounter probabilities and community structure in zooplankton: a mathematical model. J. Fish. Res. Board Can. 34, 73–82.

GESAMP—Joint Group of Experts on the Scientific Aspects of Marine Environmental Protection, 2016. Kershaw, P.J., Rochman, C.M. (Eds.), Sources, fate and effects of microplastics in the marine environment: Part 2 of a global assessment. Report No. 93. Available from: www.gesamp.org.

Gewert, B., Ogonowski, M., Barth, A., MacLeod, M., 2017. Abundance and composition of near surface microplastics and plastic debris in the Stockholm Archipelago, Baltic Sea. Mar. Pollut. Bull. 120, 1–2.

Goldstein, M.C., Goodwin, D.S., 2013. Gooseneck barnacles (*Lepas* spp.) ingest microplastic debris in the North Pacific Subtropical Gyre. PeerJ. 1, e184.

Gorokhova, E., 2015. Screening for microplastic particles in plankton samples: how to integrate marine litter assessment into existing monitoring programs? Mar. Pollut. Bull. 99 (1–2), 271–275.

Graham, E.R., Thompson, J.T., 2009. Deposit-and suspension-feeding sea cucumbers (Echinodermata) ingest plastic fragments. J. Exp. Mar. Biol. Ecol. 368 (1), 22–29.

Green, S., Visser, A.W., Titelman, J., Kiørboe, T., 2003. Escape responses of copepod nauplii in the flow field of the blue mussel, *Mytilus edulis*. Mar. Biol. 142 (4), 727–733.
Green, D.S., Boots, B., Sigwart, J., Jiang, S., Rocha, C., 2016. Effects of conventional and biodegradable microplastics on a marine ecosystem engineer (*Arenicola marina*) and sediment nutrient cycling. Environ. Pollut. 208, 426–434.
Gutow, L., Eckerlebe, A., Gimenez, L., Saborowski, R., 2015. Experimental evaluation of seaweeds as vector for microplastics into marine food webs. Environ. Sci. Technol. 50, 915–923.
Hammer, S., Nager, R.G., Johnson, P.C.D., Furness, R.W., Provencher, J.F., 2016. Plastic debris in great skua (*Stercorarius skua*) pellets corresponds to seabird prey species. Mar. Pollut. Bull. 103, 206–210.
Hara, T.J., 1975. Olfaction in fish. Prog. Neurobiol. 5 (4), 271–335.
Harrison, J.P., Schratzberg, M., Sapp, M., Osborn, A.M., 2014. Rapid bacterial colonization of low-density polyethylene microplastics in coastal sediment microcosms. BMC Microbiol. 14 (1), 232.
Hidalgo-Ruz, V., Gutow, L., Thompson, R.C., Thiel, M., 2012. Microplastics in the marine environment: a review of the methods used for identification and quantification. Environ. Sci. Technol. 46, 3060–3307.
Holling, C.S., 1959. The components of predation as revealed by a study of small-mammal predation of the European pine sawfly. Can. Entomol. 91, 293–320.
Hüffer, T., Praetorius, A., Wagner, S., von der Kammer, F., Hofmann, T., 2017. Microplastic exposure assessment in aquatic environments: learning from similarities and differences to engineered nanoparticles. Environ. Sci. Technol. 51 (5), 2499–2507.
Huntley, M.E., Barthel, K.G., Star, J.L., 1983. Particle rejection by *Calanus pacificus*: discrimination between similarly sized particles. Mar. Biol. 74, 151–160.
Kaiser, D., Kowalski, N., Waniek, J.J., 2017. Effects of biofouling on the sinking behavior of microplastics. Environ. Res. Lett.
Kautsky, L., Kautsky, N., 2000. Baltic Sea, including Bothnian Sea and Bothnian Bay. In: Sheppard, C.R.C. (Ed.), Seas at the Millennium: An Environmental Evaluation. Elsevier Science Ltd., The Netherlands, pp. 1–14 (Chapter 8).
Kiørboe, T., Møhlenberg, F., Nicolajsen, H., 1982. Ingestion rate and gut clearance in the planktonic copepod *Centropages hamatus* (Liljeborg) in relation to food concentration and temperature. Ophelia 21, 181–194.
Koehl, M.A.R., Stickler, J.R., 1981. Copepod feeding currents: food capture at low Reynolds number. Limnol. Oceanogr. 26, 1062–1073.
Koelmans, A.A., Bakir, A., Burton, G.A., Janssen, C.R., 2016. Microplastic as a vector for chemicals in the aquatic environment: critical review and model-supported reinterpretation of empirical studies. Environ. Sci. Technol. 50 (7), 3315–3326.
Kooi, M., Van Nes, E.H., Scheffer, M., Koelmans, A.A., 2017. Ups and downs in the ocean: effects of biofouling on the vertical transport of microplastics. Environ. Sci. Technol. 51 (14), 7963–7971. https://doi.org/10.1021/acs.est.6b04702.
Krümmel, E., Macdonald, R.W., Kimpe, L.E., Gregory-Eaves, I., Demers, M., Smol, J.P., Finney, B., Blais, J.M., 2003. Delivery of pollutants by spawning salmon. Nature 425, 255–256.
Laist, D.W., 1997. Impacts of marine debris: entanglement of marine life in marine debris including a comprehensive list of species with entanglement and ingestion records. In: Marine Debris. Springer, New York, pp. 99–139.
Law, K.L., Morét-Ferguson, S., Maximenko, N.A., Proskurowski, G., Peacock, E.E., Hafner, J., Reddy, C.M., 2010. Plastic accumulation in the North Atlantic Subtropical Gyre. Science 329, 1185–1188.
Lee, K.W., Shim, W.J., Kwon, O.Y., Kang, J.H., 2013. Size-dependent effects of micro polystyrene particles in the marine copepod *Tigriopus japonicus*. Environ. Sci. Technol. 47, 11278–11283.
Ling, S.D., Sinclair, M., Levi, C.J., Reeves, S.E., Edgar, G.J., 2017. Ubiquity of microplastics in coastal seafloor sediments. Mar. Pollut. Bull. 121 (1–2), 104–110.
Lobelle, D., Cunliffe, M., 2011. Early microbial biofilm formation on marine plastic debris. Mar. Pollut. Bull. 62 (1), 197–200.

Long, M., Moriceau, B., Gallinari, M., Lambert, C., Huvet, A., Raffray, J., Soudant, P., 2015. Interactions between microplastics and phytoplankton aggregates: impact on their respective fates. Mar. Chem. 175, 39–46.

Lusher, A.L., McHugh, M., Thompson, R.C., 2013. Occurrence of microplastics in the gastrointestinal tract of pelagic and demersal fish from the English Channel. Mar. Pollut. Bull. 67 (1–2), 94–99.

Lusher, A.L., Hernandez-Milian, G., O'Brien, J., Berrow, S., O'Connor, I., Officer, R., 2015a. Microplastic and macroplastic ingestion by a deep diving, oceanic cetacean: the True's beaked whale *Mesoplodon mirus*. Environ. Pollut. 199, 185–191.

Lusher, A.L., Tirelli, V., O'Connor, I., Officer, R., 2015b. Microplastics in Arctic polar waters: the first reported values of particles in surface and sub-surface samples. Sci. Rep. 5, 14947. https://doi.org/10.1038/srep14947.

Lusher, A.L., O'Donnell, C., Officer, R., O'Connor, I., 2016. Microplastic interactions with North Atlantic mesopelagic fish. ICES J. Mar. Sci. 73 (4), 1214–1225.

Lusher, A.L., Welden, N.A., Sobral, P., Cole, M., 2017. Sampling, isolating and identifying microplastics ingested by fish and invertebrates. Anal. Methods 9, 1346–1360.

Mathalon, A., Hill, P., 2014. Microplastic fibers in the intertidal ecosystem surrounding Halifax Harbor, Nova Scotia. Mar. Pollut. Bull. 81 (1), 69–79.

Mato, Y., Isobe, T., Takada, H., Kanehiro, H., Ohtake, C., Kaminuma, T., 2001. Plastic resin pellets as a transport medium for toxic chemicals in the marine environment. Mar. Pollut. Bull. 35 (2), 318–324.

Miller, M.E., Kroon, F.J., Motti, C.A., 2017. Recovering microplastics from marine samples: a review of current practices. Mar. Pollut. Bull. 123 (1–2), 6–18. https://doi.org/10.1016/j.marpolbul.2017.08.058.

Moore, S.L., Leecaster, M.K., Weisberg, S.B., 2001. A comparison of plastic and plankton in the north Pacific Central Gyre. Mar. Pollut. Bull. 42, 1297–1300.

Mu, H., Ewald, G., Nilsson, E., Sundin, P., Wesén, C., 2004. Fate of chlorinated fatty acids in migrating sockeye salmon and their transfer to arctic grayling. Environ. Sci. Technol. 38, 5548–5554.

Munari, C., Infantini, V., Scoponi, M., Rastelli, E., Corinaldesi, C., Mistri, M., 2017. Microplastics in the sediments of Terra Nova Bay (Ross Sea, Antarctica). Mar. Pollut. Bull. 122, 161–165.

Murray, P.R., Cowie, P.R., 2011. Plastic contamination in the decapod crustacean *Nephrops norvegicus* (Linnaeus, 1758). Mar. Pollut. Bull. 62, 1207–1217.

Näkki, P., Setälä, O., Lehtiniemi, M., 2017. Bioturbation transports secondary microplastics to deeper layers in soft marine sediments of the northern Baltic Sea. Mar. Pollut. Bull. 119 (1), 255–261. https://doi.org/10.1016/j.marpolbul.2017.03.065.

Nelms, S.E., Duncan, E.M., Broderick, A.C., Galloway, T.S., Godfrey, M.H., Hamann, M., Lindeque, P.K., Godley, B.J., 2016. Plastic and marine turtles: a review and call for research. ICES J Mar. Sci. 73 (2), 165–181.

Nelms, S.E., Galloway, T.S., Godley, B.J., Jarvis, D.S., Lindeque, P.K., 2018. Investigating microplastic trophic transfer in marine top predators. Environ. Pollut. https://doi.org/10.1016/j.envpol.2018.02.016.

NOAA—National Oceanic and Atmospheric Administration Marine Debris Program (2014). Report on the Entanglement of Marine Species in Marine Debris With an Emphasis on Species in the United States, Silver Spring, MD. pp. 28.

Nygaard, K., Borsheim, K.Y., Thingstad, T.F., 1988. Grazing rates on bacteria by marine heterotrophic microflagellates compared to uptake rates of bacterial-sized monodisperse fluorescent latex beads. Mar. Ecol. Prog. Ser. 44, 159–165.

Oberbeckmann, S., Loeder, M.G.J., Gerdts, G., Osborn, M., 2014. Spatial and seasonal variation in diversity and structure of microbial biofilms on marine plastics in Northern European waters. FEMS Microbiol. Ecol. 90, 478–492.

Ogonowski, M., Schür, C., Jarsén, Å., Gorokhova, E., 2016. The effects of natural and anthropogenic microparticles on individual fitness in *Daphnia magna*. PLoS One 11(5) e0155063.

Ogonowski, M., Wenman, D., Gorokhova, E., 2017. In: Ingested microplastic is not correlated to HOC concentrations in Baltic Sea herring. *15th International Conference on Environmental Science and Technology*, Rhodes, Greece, 31 August to 2 September 2017.

Phuong, N.N., Zalouk-Vergnoux, A., Poirier, L., Kamari, A., Châtel, A., Mouneyrac, C., Lagarde, F., 2016. Is there any consistency between the microplastics found in the field and those used in laboratory experiments? Environ. Pollut. 211, 111–123.
Railo, S., 2017. Microlitter in *Mytilus trossulus* and Its Environment in the Northern Baltic Sea: Wastewater as Point Source Pollution. MSc thesis, University of Helsinki, Finland.
Rescigno, M., Urbano, M., Valzasina, B., Francolini, M., Rotta, G., Bonasio, R., Ricciardi-Castagnoli, P., 2001. Dendritic cells express tight junction proteins and penetrate gut epithelial monolayers to sample bacteria. Nat. Immunol. 2 (4), 361–367.
Riisgård, H.U., Larsen, P.S., 2010. Particle capture mechanisms in suspension-feeding invertebrates. Mar. Ecol. Prog. Ser. 418, 255–293. https://doi.org/10.3354/meps08755.
Rochman, C.M., Tahir, A., Williams, S.L., Baxa, D.V., Lam, R., Miller, J.T., Teh, F., Werorilangi, S., Teh, S.J., 2015. Anthropogenic debris in seafood: plastic debris and fibers from textiles in fish and bivalves sold for human consumption. Sci. Rep. 5, 14340.
Rummel, C.D., Jahnke, A., Gorokhova, E., Kühnel, D., Schmitt-Jansen, M., 2017. The impacts of biofilm formation on the fate and potential effects of microplastic in the aquatic environment. Environ. Sci. Technol. Lett. 4 (7), 258–267. https://doi.org/10.1021/acs.estlett.7b00164.
Schuyler, Q., Hardesty, B.D., Wilcox, C., Townsend, K., 2012. To eat or not to eat? Debris selectivity by marine turtles. PLoS One 7(7) e40884.
Setälä, O., Sopanen, S., Autio, R., Erler, K., 2009. Grazing and prey selection of the calanoid copepods *Eurytemora affinis* and *Acartia bifilosa* feeding on plankton assemblages containing *Dinophysis* spp. Boreal Environ. Res. 14, 837–849.
Setälä, O., Fleming-Lehtinen, V., Lehtiniemi, M., 2014. Ingestion and transfer of microplastics in the planktonic food web. Environ. Pollut. 185, 77–83.
Setälä, O., Norkko, J., Lehtiniemi, M., 2016a. Feeding type affects microplastic ingestion in a coastal invertebrate community. Mar. Pollut. Bull. 102 (1), 95–101.
Setälä, O., Magnusson, K., Lehtiniemi, M., Norén, F., 2016b. Distribution and abundance of surface water microlitter in the Baltic Sea: a comparison of two sampling methods. Mar. Pollut. Bull. 110 (1), 177–183.
Sherman, P., Van Sebille, E., 2016. Modeling marine surface microplastic transport to assess optimal removal locations. Environ. Res. Lett. 11 (1), 014006.
Sileo, L., Sievert, P.R., Samuel, M.D., Fefer, S.I., 1990. Prevalence and characteristics of plastic ingested by Hawaiian seabirds. *Proceedings of the Second International Conference on Marine Debris*. NOAA Technical Memo, Honolulu, pp. 665–681.
Solomon, M.E., 1949. The natural control of animal populations. J. Anim. Ecol. 18, 1–35.
Staresinic, N., Farrington, J., Gagosian, R.B., Clifford, C.H., Hulburt, E.M., 1983. Downward transport of particulate matter in the Peru coastal upwelling: role of the anchoveta, *Engraulis ringens*. In: Suess, E., Theide, J. (Eds.), Coastal Upwelling: Its Sediment Record. Part A. Responses of the Sedimentary Regime to Present Coastal Upwelling. Plenum, New York, pp. 225–240.
Steer, M., Cole, M., Thompson, R.C., Lindeque, P.K., 2017. Microplastic ingestion in fish larvae in the western English Channel. Environ. Pollut. 226, 250–259.
Sun, X., Li, Q., Zhu, M., Liang, J., Zheng, S., Zhao, Y., 2017. Ingestion of microplastics by natural zooplankton groups in the northern South China Sea. Mar. Pollut. Bull. 115 (1–2), 217–224.
Sussarellu, R., Suquet, M., Thomas, Y., Lambert, C., Fabioux, C., Pernet, M.E.J., Huvet, A., 2016. Oyster reproduction is affected by exposure to polystyrene microplastics. Proc. Natl. Acad. Sci. U. S. A. 113 (9), 2430–2435.
Syberg, K., Khan, F.R., Selck, H., Palmqvist, A., Banta, G.T., Daley, J., Sano, L., Duhaime, M.B., 2015. Microplastics: addressing ecological risk through lessons learned. Environ. Toxicol. Chem. 34 (5), 945–953.
Taylor, M.L., Gwinnett, C., Robinson, L.F., Woodall, L.C., 2016. Plastic microfibre ingestion by deep-sea organisms. Sci. Rep. 6, 33997.
Tiselius, P., Jonsson, P.R., Verity, P.G., 1993. A model evaluation of the impact of food patchiness on foraging strategy and predation risk in zooplankton. Bull. Mar. Sci. 53, 247–264.

Tourinho, P.S., Ivar do Sul, J.A., Fillmann, G., 2010. Is marine debris ingestion still a problem for the coastal marine biota of southern Brazil? Mar. Pollut. Bull. 60 (3), 396–401.
Turner, J.T., 2002. Zooplankton fecal pellets, marine snow and sinking phytoplankton blooms. Aquat. Microb. Ecol. 27, 57–102.
Turner, J.T., 2015. Zooplankton fecal pellets, marine snow, phytodetritus and the ocean's biological pump. Prog. Oceanogr. 130, 205–248.
Van Franeker, J.A., Law, K.L., 2015. Seabirds, gyres and global trends in plastic pollution. Environ. Pollut. 203, 89–96.
Van Franeker, J.A., Blaize, C., Danielsen, J., Fairclough, K., Gollan, J., Guse, N., Hansen, P.L., Heubeck, M., Jensen, J.K., Le Guillou, G., Olsen, B., 2011. Monitoring plastic ingestion by the northern fulmar *Fulmarus glacialis* in the North Sea. Environ. Pollut. 159 (10), 2609–2615.
Van Sebille, E., Wilcox, C., Lebreton, L., Maximenko, N., Hardesty, B.D., Van Franeker, J.A., Eriksen, M., Siegel, D., Galgani, F., Law, K.L., 2015. A global inventory of small floating plastic debris. Environ. Res. Lett. 10 (12), 124006.
Vandermeersch, G., Van Cauwenberghe, L., Janssen, C.R., Marques, A., Granby, K., Fait, G., Devriese, L., 2015. A critical view on microplastic quantification in aquatic organisms. Environ. Res. 143, 46–55.
Volkenborn, N., Hedtkamp, S.I.C., Van Beusekom, J.E.E., Reise, K., 2007. Effects of bioturbation and bioirrigation by lugworms (*Arenicola marina*) on physical and chemical sediment properties and implications for intertidal habitat succession. Estuar. Coast. Shelf Sci. 74, 331–343.
von Moos, N., Burkhardt-Holm, P., Köhler, A., 2012. Uptake and effects of microplastics on cells and tissue of the blue mussel *Mytilus edulis* L. after an experimental exposure. Environ. Sci. Technol. 46 (20), 11327–11335.
Vroom, R.J.E., Koelmans, A.A., Besseling, E., Halsband, C., 2017. Aging of microplastics promotes their ingestion by marine zooplankton. Environ. Pollut. 231 (1), 987–996.
Waller, C.L., Griffiths, H.J., Waluda, C.M., Thorpe, S.E., Loaiza, I., Moreno, B., Pacherres, C.O., Hughes, K.A., 2017. Microplastics in the Antarctic marine system: an emerging area of research. Sci. Total Environ. 598, 220–227.
Ward, J.E., Koch, D.J., 2009. Marine aggregates facilitate ingestion of nanoparticles by suspension-feeding bivalves. Mar. Environ. Res. 68, 137–142.
Watt, A.J., Lewis, C., Goodhead, R.M., Beckett, S.J., Moger, J., Tyler, C.R., Galloway, T.S., 2014. Uptake and retention of microplastics by the shore crab *Carcinus maenas*. Environ. Sci. Technol. 48 (15), 8823–8830.
Woodall, L.C., Sanchez-Vidal, A., Canals, M., Paterson, G.L.J., Coppock, R., Sleight, V., Calafat, A., Rogers, A.D., Narayanaswamy, B.E., Thompson, R.C., 2014. The deep sea is a major sink for microplastic debris. Royal Soc. Open Sci. 1 (4), 140317. https://doi.org/10.1098/rsos.140317.
Wright, S.L., Thompson, R.C., Galloway, T.S., 2013a. The physical impacts of microplastics on marine organisms: a review. Environ. Pollut. 178, 483–492.
Wright, S.L., Rowe, D., Thompson, R.C., Galloway, T.S., 2013b. Microplastic ingestion decreases energy reserves in marine worms. Curr. Biol. 23 (23), 1031–1033.
Ye, S., Andrady, A.L., 1991. Fouling of floating plastic debris under Biscayne Bay exposure conditions. Mar. Pollut. Bull. 22, 608–613.
Zettler, E.R., Mincer, T.J., Amaral-Zettler, L.A., 2013. Life in the "plastisphere": microbial communities on plastic marine debris. Environ. Sci. Technol. 47 (13), 7137–7146.

FURTHER READING

Goldstein, M.C., Titmus, A.J., Ford, M., 2013. Scales of spatial heterogeneity of plastic marine debris in the Northeast Pacific Ocean. PLoS One. https://doi.org/10.1371/journal.pone.0080020.
Green, D.S., 2016. Effects of microplastics on European flat oysters, *Ostrea edulis* and their associated benthic communities. Environ. Pollut. 216, 95–103.

CHAPTER 12

Microplastics in the Terrestrial Environment

Lei Mai*, Lian-Jun Bao*, Charles S. Wong*,†, Eddy Y. Zeng*
*Jinan University, Guangzhou, China
†University of Winnipeg, Winnipeg, MB, Canada

12.1 INTRODUCTION

A handful of reports on the occurrence of microplastics (MPs, <5 mm) in the marine environment have appeared in the literature in recent years. Since plastic contamination in the oceanic surface water was first reported in 1972 (Carpenter et al., 1972; Carpenter and Smith, 1972), the focus of MP studies has been placed on the marine environment. Jambeck et al. (2015) estimated that 275 million tons of plastic litter were produced in 2010, of which 4.8–12.7 million tons were ended up in the ocean. These numbers indicate that a large portion of plastic litter has remained in the terrestrial environment. In this age of plastics, our reliance on plastic products is undeniable (Thompson et al., 2009). Therefore, aside from the huge amount of plastics lost in the seas (Thompson et al., 2004), there is also a vast quantity of plastic wastes remaining in the terrestrial environment. Both are of great concern to human and ecosystem health and merit-intensive scrutiny, because the terrestrial environment is so "close to home," so to speak (Rillig, 2012).

As a main source of plastics to the oceans, attention has been paid to the occurrence, fate, and effects of terrestrial MPs in the form of anthropogenic litter. A recent review of MPs in both aquatic and terrestrial environments demonstrated that few studies have been conducted on MPs in the latter, because of the complexity of sample analysis (Duis and Coors, 2016). Despite the lack of data on terrestrial MPs, growing evidence suggests that MPs ubiquitously occur in the terrestrial environment, as they are likely generated from larger plastic materials through physical or chemical decomposition (Huerta Lwanga et al., 2016; Hodson et al., 2017). Eventually, they can be transported to the oceans through riverine runoff and/or atmospheric transport and deposition.

In the terrestrial environment, MPs have been found in soil, terrestrial biota, and sewage sludge, which may be further used as fertilizers in agriculture. Macroplastic particles (>5 mm in size) can be fragmented or degraded into MPs from soil erosion and from ingestion by terrestrial biota (e.g., earthworms) (Huerta Lwanga et al., 2017). MPs may also act as a transport media for metals and organic pollutants and increase the

Microplastic Contamination in Aquatic Environments
https://doi.org/10.1016/B978-0-12-813747-5.00012-6

likelihood for exposure of terrestrial biota to metals and other pollutants (Hodson et al., 2017). MPs in sewage sludge may enter the terrestrial environment via sludge application in agriculture, similar to the fate of silver nanoparticles reported by Pradas del Real et al. (2016). Besides, other sources may also contribute to MPs in the terrestrial environment, such as atmospheric fallout of fibers and fine particles during thermal cutting of plastic products (Zhang et al., 2012; Dris et al., 2016). Therefore, it is important to understand the occurrence of MPs in the terrestrial environment, to gauge the impacts of MPs outside the marine environment.

While abundant data have been accumulated on the occurrence of MPs in the marine environment (Eriksen et al., 2014; Song et al., 2014; Reisser et al., 2015), little is known about the state of MPs in the terrestrial environment. The difficulties and challenges on analytic methods of MP extraction and identification could be one of the important reasons for the limited data on the occurrence of MPs in the terrestrial environment (Rillig, 2012). In contrast to water, it is hard to identify MPs in a complex environmental matrix, such as soil and sludge. This chapter aims to assimilate and discuss existing information on the sources, fate, and effects of MPs in the terrestrial environment and to estimate the contribution of terrestrial plastics to aquatic environments. Given the limited literature to date on terrestrial MPs, this chapter also includes macroplastics, which have the potential to generate microsized plastic particles.

12.2 OCCURRENCE OF MPs IN THE TERRESTRIAL ENVIRONMENT

Although no survey to date concerning the global distribution of MPs in the terrestrial environment has been conducted to our knowledge, there are clearly many pathways for MPs to enter the terrestrial environment. A few examples are through disposing of litter from daily life, dumping of sewage sludge in landfills, and application of sewage sludge in agriculture as biosolid fertilizer (Zubris and Richards, 2005). Currently, few data are available in the open peer-reviewed literature regarding the abundance of MPs in the terrestrial environment. All such available peer-reviewed journal literature on terrestrial MPs, to our knowledge, is summarized in Table 12.1 In this section, plastic debris (including macro- and microplastics) in the soil, terrestrial organisms, and sewage sludge will be discussed. An overview of the current state of our "plastic planet" and strategies for plastic waste management will also be provided.

12.2.1 Plastic Debris in Soil

Plastic products have completely transformed our daily lives and standard of living. Most disposable products are made of plastics, as they are durable, lightweight, and cost-effective to produce (Barnes et al., 2009). Data on plastics in Europe in 2014 showed that only 29.7% of 25.8 million tons of plastic litter was recycled (PlasticsEurope, 2016). Not all of plastic waste is suitable to be recycled. For example, some plastics are not recycled

Table 12.1 Current literature data on the distribution of microplastics in the terrestrial environment

Matrix	Abundance	Plastic type	Plastic size	References
Soil	$3.0 \pm 1.9\,g/m^2$ soil	Polyethylene film	25 μm	Ramos et al. (2015)
	11 ± 10 pieces/m^2		100 μm	
	NA[a]	Polyethylene film	<2 mm	Feuilloley et al. (2005)
Biota	$0.45 \pm 0.25\%$ w/w earthworms	Polyethylene from plastic bag	<150 μm	Huerta Lwanga et al. (2016)
	364 Items from 16 birds	Natural fibers, plastic fibers, and fragmented plastics	0.5–8.5 mm	Zhao et al. (2016)
Sludge	0 – 2 pieces/g wet weight	Synthetic fiber	NA[a]	Zubris and Richards (2005)
	5 particles/g wet weight	Microplastics	<5 mm	Carr et al. (2016)
	4200–15,000 particles/kg dry weight	Polyethylene, polyester, acrylic, polyethylene terephthalate, polypropylene, and polyamide	250–5000 μm	Mahon et al. (2017)
Others	2–355 particles/m^2/day	Synthetic fibers (atmospheric fallout)	<5 mm	Dris et al. (2016)
	1×10^{12} particles/m^3	Polystyrene foam nanoparticles in the air through thermal cutting	10–10,000 nm	Zhang et al. (2012)

[a]NA means data not available.

because of technical limitations, such as requirements for purity, or for economic reasons. This can explain why more than 80% of recycled plastics are from plastic packaging and bottles (PlasticsEurope, 2016). Landfilling was always chosen for the treatment of plastic waste that could not be recycled by many countries, with 30.8% of plastics went to landfill in Europe in 2014 (PlasticsEurope, 2016). Throwing away used plastic as litter (e.g., Fig. 12.1) and landfill disposal of waste lead to plastic contamination of soil. Plastics are generally chemically resistant and consequently can persist in the environment for a long period of time (Andrady and Neal, 2009). Some plastics, when exposed to UV light or other crucial environmental conditions, can be physically fragmented or chemically and/or biologically degraded to fine plastic particles, defined as MPs (Thompson et al., 2004). Plastic wastes may be degraded to MPs and eventually enter the marine environment through riverine runoff or wash off from beaches.

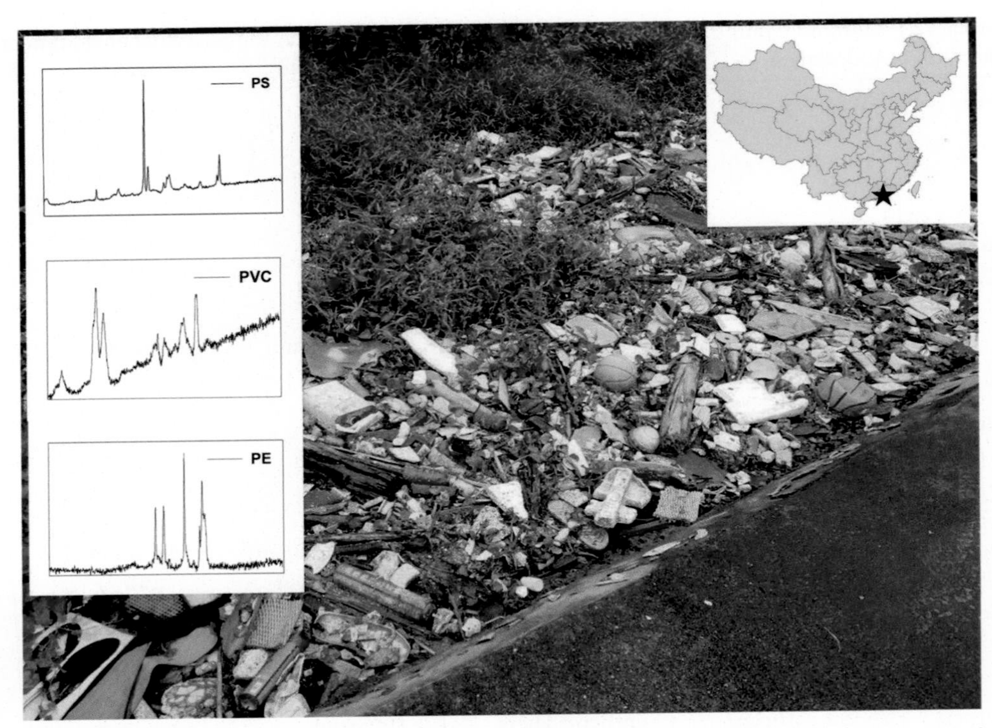

Fig. 12.1 Photo taken on 12 June 2017 of the terrestrial environment (*black star* on map) near the Humen estuary of the Pearl River in Guangzhou, China. Plastic wastes include polystyrene (PS), polyvinyl chloride (PVC) and polyethylene (PE) products, and other plastic litter. Absorption spectra of PS, PVC, and PE (top to bottom, respectively) are inset. *Photo credit: Jing Liu.*

There are many pathways for MPs to accumulate in soil, either from fragmentation of directly discarded plastic litter or from leaching of buried wastes in landfills. Upon disposal, plastic wastes (e.g., plastic bags, bottles and films for packaging, and soil protection) can be exposed to the natural environment. Terrestrial biota could ingest plastics by mistaking them as food. Bacteria or microorganisms may use it as a habitat, as they are also in coastal areas (Harrison et al., 2014). Polyethylene (PE) film used in horticulture was reported to be able to incorporate into soil and act as a vector for transport of pollutants such as pesticides (Ramos et al., 2015). Over time, plastics such as horticultural film left on fields without proper disposal could become fragmented through abrasion and erosion, resulting in numerous MPs in soil. Similarly, PE film is prevalently used in agriculture as mulch (Hablot et al., 2014). Although the mulch will be collected after use, it is nearly impossible to collect all of them without leaving some residues, which are derived from fragmentation and aging of thin films (Steinmetz et al., 2016). The degradation of PE film in soil to generate microsized plastic residues has been intensively reported

(Feuilloley et al., 2005; Kyrikou and Briassoulis, 2007; Briassoulis et al., 2015). However, the abundance, fate, and effects of MPs thus derived remain unknown, and their assessment is a challenge without sufficient analytic methods.

Until recently, Nizzetto et al. (2016b) estimated the annual input of MPs into farmland soil in Europe and North America to be 63–430 and 44–300 thousand tons, respectively. These numbers even outweigh the calculated amounts of MPs in global marine surface waters, estimated at 93–236 thousand tons (Van Sebille et al., 2015). Such astonishing values reflect a global concern on MP pollution in the terrestrial environment. As a result, studies are gradually emerging about MPs in soil, from occurrence to biological effects (Huerta Lwanga et al., 2016). However, most such reports have only demonstrated the existence of MPs in soil. One of the possible obstacles might be that the mineral matrices of soils make it difficult to process such samples for the extraction of MPs. Comprehensive and quantitative data on MP abundance in different regions are still lacking. Rillig (2012) strongly suggested that systematic examination of the MP occurrence in soil is necessary and that analytic technologies used to investigate MPs in the marine environment could also be used for detecting and measuring MPs in soil. As with beach sands, studies on MPs in soil should be promoted, as soil shares some features (e.g., living habitat for organisms and accumulation of chemical contaminants) of coastal beach sands in some extent (Rillig, 2012). Besides, Dümichen et al. (2017) suggested that a thermal degradation method could be used to extract and identify MPs from complex environmental matrices.

12.2.2 Interaction Between Terrestrial Biota and MPs

In both marine and terrestrial environments, a variety of biotic species can be affected by MPs. Although modes of feeding may differ between marine organisms affected by MPs (e.g., many are filter feeders) and terrestrial biota, exposure to MPs through dietary intake can lead to accumulation of plastic particles in digestive tracts. Compared with most marine biota, terrestrial biota (e.g., birds) are much more variable in dieting behavior and in use of living habitats (Sun et al., 2012). This finding indicates that terrestrial biota are more likely to ingest MPs as food or use MPs as a habitat (i.e., as a replacement for rocks or other natural shelters). Ingestion of MPs by biota in the marine environment has been widely studied from whales to fish and mussels and from seabirds to zooplankton (van Franeker et al., 2011; Foekema et al., 2013; Desforges et al., 2015; Lusher et al., 2015; Li et al., 2016a). However, information on the interactions of terrestrial biota with MPs (e.g., ingestion and translocation) is scarce.

Extremely few regional studies have been conducted to investigate the prevalence of MPs in the terrestrial biota. Zhao et al. (2016) surveyed anthropogenic plastic waste between 0.5 and 5 mm in size in terrestrial birds in China. They found that the majority of the pieces identified in the birds' digestive tracts were plastics, accounting for 62.6% of the total particles found in the birds. Further research was urgently suggested by these authors on the occurrence and effects of MPs in terrestrial birds. Aside from birds, other soil organisms also have the potential to interact with MPs in soil. For example, earthworms build burrows in soil. Bioturbation of MPs by earthworms was observed by Huerta Lwanga et al. (2017), who demonstrated that 73.5% of MPs were moved by earthworms from the soil surface downward into the bulk soil during burrow formation. In addition, Maaß et al. (2017) investigated the interaction of two collembolan species with MPs in soil, as these microarthropods also moved plastic particles downward. Pollutants carried by MPs (e.g., metals and organic contaminants) may affect soil and groundwater quality, because chemicals may leach from MPs in bulk soil through desorption. However, no study has been carried out so far to our knowledge on the bioaccumulation of MPs in the terrestrial biotic food web. Data on abundances (including quantitation, size fractions, and classifications) and retention of different types of plastics in terrestrial biota globally have remained scarce. Further efforts along these lines are needed to better describe the state of MP pollution in the terrestrial food web.

12.2.3 MPs in Sewage Sludge

Another source of MPs in the terrestrial environment is municipal wastewater treatment plant (WWTP) sludge; as noted by Nizzetto et al. (2016a), "Are agricultural soils dumps for microplastics of urban origin?" WWTPs generally receive wastewaters from households, hospitals, and industries, resulting in multiple pathways for MPs to end up in sewage sludge. Some commonly used personal care products, for example, facial cleansers (Fendall and Sewell, 2009) and toothpaste (Carr et al., 2016), contain abundant MPs. Increasing production and widespread use of synthetic textiles for clothing result in large amounts of synthetic fibers being drained into sewer systems through washing. These MPs are accumulated in sludge during sedimentation in WWTPs. Carr et al. (2016) estimated that a majority of MPs from WWTPs were retained in sludge. In particular, particles with higher density than water, such as polyvinyl chloride, are more likely to retain in sewage sludge.

The European Environmental Agency forecasted that the proportion of sewage sludge used as fertilizers would rise to 54% in 2005 (European Environmental Agency, n.d.). The reuse of sewage sludge for land application is one of the most economically friendly operations for sewage sludge disposal (Wang, 1997). The application of sewage sludge in agricultural land would potentially bring both benefits and risks (Singh and Agrawal, 2008). The association of sludge application with soil pollution has been widely recognized, such as increased concentrations of brominated diphenyl ethers in soil (Sellström et al., 2005), altered soil biological characteristics (Banerjee et al., 1997), and worsen metal contamination (Kidd et al., 2007; Pradas del Real et al., 2016). A large number of studies focused on metals and organic pollutants in sludge but few concerned MPs.

Synthetic fibers were still detectable in soil and sludge products, even after more than a decade since sewage sludge was applied. As a result, synthetic fibers may be used as an indicator of the old practice applying sewage sludge in agricultural soil (Zubris and Richards, 2005). Mahon et al. (2017) reported that the abundances of MPs in the sludge samples collected in Ireland ranged between 4200 and 15,000 particles/kg of dry weight. Nizzetto et al. (2016a) estimated that 125–850 ton of MPs per million residents were emitted to the terrestrial environment through sewage sludge application or biosolid processing in European countries. Although sludge containing toxic substances at concentrations higher than regulated concentrations is not allowed to be applied in agriculture, MP is not a concern by any regulations to date. Regulations concerning the agricultural application of sludge should better include MPs as harmful substances, because of potential harmful consequences on the reproduction of earthworms from exposure to MPs (Huerta Lwanga et al., 2016). To reduce MP pollution during sludge application, post-treatment measures to remove MPs from sludge should be taken before disposal or reuse of sludge.

12.2.4 Other Sources of MPs in the Terrestrial Environment

The use of plastic products is ubiquitous in human beings' daily life. All plastic products have the potential to generate MPs through fragmentation, aging, and deterioration. For example, MPs can be generated from the abrasion of tire wear, surface detachment of plastic coating, and blasting of plastics during industrial processes (e.g., thermal cutting). Zhang et al. (2012) observed the release of plastic particles during thermal cutting of plastic products. Synthetic fibers, also MPs, ubiquitously occurred in both indoor and outdoor environments (Dris et al., 2016, 2017). Polymer fragments were also found in biocomposting residues, most of which were identified as PE and polypropylene films and to a less extent as polystyrene and PE terephthalate particles (Dümichen et al., 2017). Nevertheless, the occurrence and fate of MPs in the terrestrial environment remain underinvestigated and are area for further studies.

12.3 BIOLOGICAL EFFECTS OF MPs ON TERRESTRIAL ORGANISMS

MPs are ubiquitously present in both the terrestrial and aquatic environments, where a variety of organisms inhabit. The effects of MPs on aquatic biota have been widely studied (Wright et al., 2013; Sigler, 2014). Some typical physical and chemical effects of MPs on marine organisms were reported by Yamashita et al. (2011). Such effects include entanglement and increased mortality. Furthermore, Erren et al. (2009) speculated that increased wildlife cancer might be a reflection of global plastic contamination, although little evidence currently exists to either support or refute this claim.

In contrast, few studies have investigated the effects of MPs on terrestrial organisms. The majority of the previous studies have been conducted at laboratory scale.

Earthworms, which are commonly found in soil, were used to investigate the effects of MPs on terrestrial organisms in the laboratory. Huerta Lwanga et al. (2016) studied the exposure of *Lumbricus terrestris* to PE particles with sizes $<150\,\mu m$ and a size distribution of $<50\,\mu m$ (50%), 50–100 μm (27%), and $>100\,\mu m$ (23%). They showed that mortality increased with increasing MP concentration in soil, while the growth rate exhibited an opposite trend. Earthworms also egested MPs in a size-selective manner, as 90% of MPs present in casts were $<50\,\mu m$. Although the exact damage to earthworms through ingestion of MPs is not yet clear, negative effects are inevitable. Zhao et al. (2016) also reported ingestion of MPs by terrestrial birds; however, no results on the effects of MPs were presented. In contrast, the effects of MPs on seabirds have been intensively studied (Tanaka et al., 2013; van Franeker and Law, 2015). Further studies are urgently desirable on the physical and chemical damage or effects on other terrestrial organisms, especially free-range poultries, where the intake of MPs may accidently occur during feeding. Given the importance of poultry as the human diet, MP contamination in poultry may pose a potential risk to human health.

The effects of MPs on terrestrial organisms are caused not only by original plastic particles but also by plasticizers, plastic additives, and other contaminants that may be sorbed on MPs. Because of the prevalent use of plastic films in agriculture and greenhouses, residual plastics may release phthalate esters (PAEs), which are widely used as plasticizers (Ma et al., 2015) and are weakly estrogenic. Organic matter containing PAEs may be ingested by soil organisms; thus, soil organisms may be affected by PAEs. Another common terrestrial contaminant is zinc, which can be adsorbed by MPs (such as PE) more than by soil particles with equal mass (Hodson et al., 2017). In this case, although no MPs were retained in the guts of earthworms, the ingestion process may have resulted in transport of zinc from MPs to earthworms (Hodson et al., 2017). Clearly, sorbed contaminants in terrestrial MPs may be released and desorbed to organisms during the ingestion process. This process is analogous to the sorption of persistent organic pollutants in MPs in the aquatic environment (Bakir et al., 2014). Strong sorption of MPs for some hydrophobic organic chemicals, such as PAHs and PCBs (Guo et al., 2012; Hüffer and Hofmann, 2016) and metals (Rochman et al., 2014; Turner and Holmes, 2015), is widely reported in the aquatic environment. However, the co-occurrence of MPs and organic pollutants or metals in the terrestrial environment has rarely been studied and requires further investigations, particularly its effects on organisms and the food web in the terrestrial environment.

12.4 CONTRIBUTION OF TERRESTRIAL PLASTICS TO MARINE MPs

The sources of marine MPs have been widely reviewed, and terrestrial inputs appeared to constitute a large portion of the total amount of marine MPs (Duis and Coors, 2016; Li et al., 2016b). In the early 1990s, the US Environmental Protection Agency

(USEPA, 1992) reported that plastic pellets released to the environment were mainly derived from the plastic industry. A previous study estimated that 44% of marine MPs in Korea were sourced from land (Jang et al., 2014). While sources may differ from region to region, a large number of marine MP sources are land-based. Terrestrially derived MPs can be transported to the marine environment through plastic waste disposal, sludge application, soil erosion, and runoff, as conceptualized in Fig. 12.2.

In conclusion, the contribution of terrestrial plastics to marine MP pollution is not negligible and may constitute a large proportion of the entire input. Actions should be taken to reduce the terrestrial inputs, such as recycling plastic materials to produce new products and generating energy in landfills from unrecyclable plastics. Legislation is also recommended for proper plastic waste management. In fact, 65 associations from 34 countries around the world have signed the *Global Declaration for Solutions on Marine Litter* since 2011 (PlasticsEurope, 2016), in an attempt to reduce plastic waste inputs. There are six commitments outlined in this declaration to reduce ocean plastic pollution: *raising awareness*, *promoting best policies*, *enhanced recovery*, *research for facts*, *sharing knowledge*,

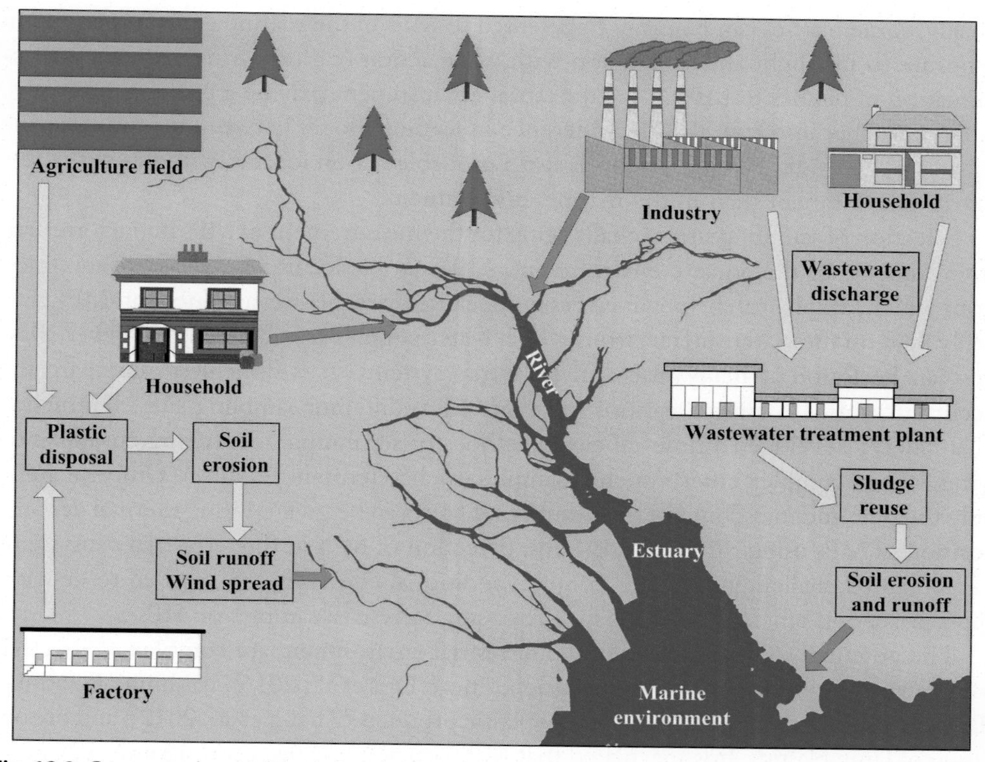

Fig. 12.2 Conceptual model depicting the transport of plastics from the terrestrial environment to the marine environment.

and preventing pellet loss (Marine Litter Solutions, 2016). However, these guidelines may not have any significant effects on plastic waste management in the terrestrial environment.

12.5 CHALLENGES AHEAD

At present, the main focus of MP studies has been placed on the marine environment. The occurrence and fate of MPs in the terrestrial environment have been inadequately addressed. There are several possible explanations for this. First, plastics are widespread and commonly found in terrestrial systems, so that they may have been overlooked as contaminants. On the other hand, the direct observation of plastics in the oceans came as a surprise to the general public (Thompson et al., 2004). Second, there are numerous challenges in the identification and quantitation of MPs in the terrestrial environment. The sample matrices in the terrestrial systems are more complex than those in the marine environment. The terrestrial settings vary substantially from region to region, making it extremely difficult to compare the abundances and compositions of MPs among regions. Finally, some plastics can be easily fragmented to MPs in the marine environment upon exposure to UV light and interaction with wave action (e.g., shearing), whereas transformation of plastics to MPs in the terrestrial environment may be a prolonged process. Rillig (2012) pointed out that the difference in feeding modes between organisms in the aquatic and terrestrial environments is also a probable reason for fewer studies in the terrestrial environment than in the marine environment.

Selection of sampling sites is challenging for the measurement of MPs in the terrestrial environment. In the aquatic environment, MPs can be accumulated along shorelines, from which samples can be obtained easily. Because many plastics are buoyant, MPs generally float on the water surface, from which water samples are obtained. However, plastics can be found at many places in terrestrial systems. It is difficult to compare the occurrence of MPs among various soil types from different sampling sites. Dümichen et al. (2017) developed a time-efficient method for separating, identifying, and quantifying MPs in complex environmental samples (such as ferment residues). Unfortunately, only the type and mass, but not the number, of MPs can be given during thermal decomposition of MPs using this method. Thus, detection of MPs in the terrestrial ecosystems has remained challenging. In this context, additional efforts are still needed to develop appropriate and efficient methods for processing terrestrial samples of MPs.

The potential sources of MPs in the terrestrial environment are complex. Terrestrial MPs can be inputted from atmospheric deposition (Dris et al., 2016), washing of clothing (Browne et al., 2011), thermal cutting of plastic products (Zhang et al., 2012), and breakdown of large plastics. It is speculated that MP particles may be constituents of $PM_{2.5}$ in the atmosphere.

Apparently, collecting samples from the terrestrial environment for investigating MPs remains a great challenge. The biological effects of terrestrial MPs are also complex and difficult to deal with, especially under field conditions. All the obstacles discussed above may have been the factors hindering research efforts concerning MPs in the terrestrial environment.

12.6 OUTLOOKS AND SUGGESTIONS

To date, limited data on the occurrence, fate, and biological effects of MPs in the terrestrial environment are available in the literature. Because terrestrial plastics are important precursors of marine environment MPs, further studies on MPs in the terrestrial environment must be encouraged. The limited studies on terrestrial MPs have focused on agricultural soil, earthworms, and sewage sludge. A broad range of terrestrial matrices therefore should be involved, such as poultry, birds, landfills, and atmosphere. To comprehensively understand the fate of MPs in the environment, all potential pathways should be quantified to better estimate the current and future contributions of anthropogenic plastic wastes to marine MPs.

To assess the biological risk of MPs, the toxicity of plastics and any attached or contained pollutants to terrestrial organisms should be evaluated. This should include new knowledge about the sorption and desorption of associated chemicals on plastics, which can be evaluated in the laboratory and then characterized in the field. The retention time of plastics in soil and the degradation rate of large plastics to form MPs are also worthy of additional investigations. It is also useful to assess correlations, if any, between the levels of terrestrial MP contamination and regional economic development, as this can be beneficial for management of plastic wastes. Such results are highly desirable for governments, environmental agencies, and other decision-makers to better manage plastic wastes.

REFERENCES

Andrady, A.L., Neal, M.A., 2009. Applications and societal benefits of plastics. Philos. Trans. R. Soc. B 364, 1977–1984.

Bakir, A., Rowland, S.J., Thompson, R.C., 2014. Enhanced desorption of persistent organic pollutants from microplastics under simulated physiological conditions. Environ. Pollut. 185, 16–23.

Banerjee, M.R., Burton, D.L., Depoe, S., 1997. Impact of sewage sludge application on soil biological characteristics. Agric. Ecosyst. Environ. 66, 241–249.

Barnes, D.K., Galgani, F., Thompson, R.C., Barlaz, M., 2009. Accumulation and fragmentation of plastic debris in global environments. Philos. Trans. R. Soc. B 364, 1985–1998.

Briassoulis, D., Babou, E., Hiskakis, M., Kyrikou, I., 2015. Degradation in soil behavior of artificially aged polyethylene films with pro-oxidants. J. Appl. Polym. Sci. 132, 3262–3271.

Browne, M.A., Crump, P., Niven, S.J., Teuten, E., Tonkin, A., Galloway, T., Thompson, R., 2011. Accumulation of microplastic on shorelines woldwide: sources and sinks. Environ. Sci. Technol. 45, 9175–9179.

Carpenter, E.J., Smith, K.L., 1972. Plastics on the sargasso sea surface. Science 175, 1240–1241.

Carpenter, E.J., Anderson, S.J., Harvey, G.R., Miklas, H.P., Peck, B.B., 1972. Polystyrene spherules in coastal waters. Science 178, 749–750.

Carr, S.A., Liu, J., Tesoro, A.G., 2016. Transport and fate of microplastic particles in wastewater treatment plants. Water Res. 91, 174–182.

Desforges, J.P., Galbraith, M., Ross, P.S., 2015. Ingestion of microplastics by zooplankton in the Northeast Pacific Ocean. Arch. Environ. Contam. Toxicol. 69, 320–330.

Dris, R., Gasperi, J., Saad, M., Mirande, C., Tassin, B., 2016. Synthetic fibers in atmospheric fallout: a source of microplastics in the environment? Mar. Pollut. Bull. 104, 290–293.

Dris, R., Gasperi, J., Mirande, C., Mandin, C., Guerrouache, M., Langlois, V., Tassin, B., 2017. A first overview of textile fibers, including microplastics, in indoor and outdoor environments. Environ. Pollut. 221, 453–458.

Duis, K., Coors, A., 2016. Microplastics in the aquatic and terrestrial environment: sources (with a specific focus on personal care products), fate and effects. Environ. Sci. Eur. 28, 1–25.

Dümichen, E., Eisentraut, P., Bannick, C.G., Barthel, A.-K., Senz, R., Braun, U., 2017. Fast identification of microplastics in complex environmental samples by a thermal degradation method. Chemosphere 174, 572–584.

Eriksen, M., Lebreton, L.C., Carson, H.S., Thiel, M., Moore, C.J., Borerro, J.C., Galgani, F., Ryan, P.G., Reisser, J., 2014. Plastic pollution in the world's oceans: more than 5 trillion plastic pieces weighing over 250,000 tons Afloat at sea. PLoS One 9, e111913.

Erren, T., Zeu, D., Steffany, F., Meyer-Rochow, B., 2009. Increase of wildlife cancer: an echo of plastic pollution? Nat. Rev. Cancer 9, 842.

European Environmental Agency.n.d. Generation and treatment of sewage sludge. https://www.eea.europa.eu/data-and-maps/indicators/generation-and-treatment-of-sewage-sludge#toc-0 (Accessed 20 October 2017).

Fendall, L.S., Sewell, M.A., 2009. Contributing to marine pollution by washing your face: microplastics in facial cleansers. Mar. Pollut. Bull. 58, 1225–1228.

Feuilloley, P., César, G., Benguigui, L., Grohens, Y., Pillin, I., Bewa, H., Lefaux, S., Jamal, M., 2005. Degradation of polyethylene designed for agricultural purposes. J. Polym. Environ. 13, 349–355.

Foekema, E.M., De, G.C., Mergia, M.T., van Franeker, J.A., Murk, A.J., Koelmans, A.A., 2013. Plastic in north sea fish. Environ. Sci. Technol. 47, 8818–8824.

Guo, X., Wang, X., Zhou, X., Kong, X., Tao, S., Xing, B., 2012. Sorption of four hydrophobic organic compounds by three chemically distinct polymers: role of chemical and physical composition. Environ. Sci. Technol. 46, 7252–7259.

Hablot, E., Dharmalingam, S., Hayes, D.G., Wadsworth, L.C., Blazy, C., Narayan, R., 2014. Effect of simulated weathering on physicochemical properties and inherent biodegradation of PLA/PHA Nonwoven Mulches. J. Polym. Environ. 22, 417–429.

Harrison, J.P., Schratzberger, M., Sapp, M., Osborn, A.M., 2014. Rapid bacterial colonization of low-density polyethylene microplastics in coastal sediment microcosms. BMC Microbiol. 14, 1–15.

Hodson, M.E., Duffus-Hodson, C., Clark, A., Prendergast-Miller, M., Thorpe, K.L., 2017. Plastic bag derived-microplastics as a vector for metal exposure in terrestrial invertebrates. Environ. Sci. Technol. 51, 4714–4721.

Huerta Lwanga, E., Gertsen, H., Gooren, H., Peters, P., Salánki, T., Van, d.P.M., Besseling, E., Koelmans, A.A., Geissen, V., 2016. Microplastics in the terrestrial ecosystem: implications for Lumbricus terrestris (Oligochaeta, Lumbricidae). Environ. Sci. Technol. 50, 2685–2691.

Huerta Lwanga, E., Gertsen, H., Gooren, H., Peters, P., Salánki, T., Ploeg, M.V.D., Besseling, E., Koelmans, A.A., Geissen, V., 2017. Incorporation of microplastics from litter into burrows of Lumbricus terrestris. Environ. Pollut. 220, 523–531.

Hüffer, T., Hofmann, T., 2016. Sorption of non-polar organic compounds by micro-sized plastic particles in aqueous solution. Environ. Pollut. 214, 194–201.

Jambeck, J.R., Geyer, R., Wilcox, C., Siegler, T.R., Perryman, M., Andrady, A., Narayan, R., Law, K.L., 2015. Plastic waste inputs from land into the ocean. Science 347, 768–771.

Jang, Y.C., Lee, J., Hong, S., Lee, J.S., Shim, W.J., Song, Y.K., 2014. Sources of plastic marine debris on beaches of Korea: more from the ocean than the land. Ocean Sci. 49, 151–162.

Kidd, P.S., Domínguezrodríguez, M.J., Díez, J., Monterroso, C., 2007. Bioavailability and plant accumulation of heavy metals and phosphorus in agricultural soils amended by long-term application of sewage sludge. Chemosphere 66, 1458–1467.

Kyrikou, I., Briassoulis, D., 2007. Biodegradation of agricultural plastic films: a critical review. J. Polym. Environ. 15, 125–150.

Li, J., Qu, X., Su, L., Zhang, W., Yang, D., Kolandhasamy, P., Li, D., Shi, H., 2016a. Microplastics in mussels along the coastal waters of China. Environ. Pollut. 214, 177–184.

Li, W.C., Tse, H.F., Fok, L., 2016b. Plastic waste in the marine environment: a review of sources, occurrence and effects. Sci. Total Environ. 566–567, 333–349.

Lusher, A.L., Hernandez-Milian, G., O'Brien, J., Berrow, S., O'Connor, I., Officer, R., 2015. Microplastic and macroplastic ingestion by a deep diving, oceanic cetacean: the True's beaked whale Mesoplodon mirus. Environ. Pollut. 199, 185–191.

Ma, T.T., Wu, L.H., Chen, L., Zhang, H.B., Teng, Y., Luo, Y.M., 2015. Phthalate esters contamination in soils and vegetables of plastic film greenhouses of suburb Nanjing, China and the potential human health risk. Environ. Sci. Pollut. R. 22, 12018–12028.

Maaß, S., Daphi, D., Lehmann, A., Rillig, M.C., 2017. Transport of microplastics by two collembolan species. Environ. Pollut. 225, 456–459.

Mahon, A.M., O'Connell, B., Healy, M.G., O'Connor, I., Officer, R., Nash, R., Morrison, L., 2017. Microplastics in sewage sludge: effects of treatment. Environ. Sci. Technol. 51, 810–818.

Marine Litter Solutions, 2016. The declaration of the global plastics associations for solutions on marine litter. https://www.marinelittersolutions.com (Accessed 26 June 2017).

Nizzetto, L., Futter, M., Langaas, S., 2016a. Are agricultural soils dumps for microplastics of urban origin? Environ. Sci. Technol. 50, 10777–10779.

Nizzetto, L., Langaas, S., Futter, M., 2016b. Pollution: do microplastics spill on to farm soils? Nature 537, 488.

PlasticsEurope, (2016) Plastics—the Facts 2016. An analysis of European plastics production, demand and waste data, http://www.plasticseurope.org/en/resources/publications/3-plastics-facts-2016, Accessed 12 August 2017.

Pradas del Real, A.E., Castillomichel, H.A., Kaegi, R., Sinnet, B., Magnin, V., Findling, N., Villanova, J., Carriere, M., Santaella, C., Fernandezmartinez, A., 2016. Fate of Ag-NPs in sewage sludge after application on agricultural soils. Environ. Sci. Technol. 50, 1759–1768.

Ramos, L., Berenstein, G., Hughes, E.A., Zalts, A., Montserrat, J.M., 2015. Polyethylene film incorporation into the horticultural soil of small periurban production units in Argentina. Sci. Total Environ. 523, 74–81.

Reisser, J., Slat, B., Noble, K., Plessis, K.D., Epp, M., Proietti, M., De Sonneville, J., Becker, T., Pattiaratchi, C., 2015. The vertical distribution of buoyant plastics at sea: an observational study in the North Atlantic Gyre. Biogeosciences 12, 1249–1256.

Rillig, M.C., 2012. Microplastic in terrestrial ecosystems and the soil? Environ. Sci. Technol. 46, 6453–6454.

Rochman, C.M., Hentschel, B.T., Teh, S.J., 2014. Long-term sorption of metals is similar among plastic types: implications for plastic debris in aquatic environments. PLoS One 9, e85433.

Sellström, U., de Wit, C.A., Lundgren, N., Tysklind, M., 2005. Effect of sewage-sludge application on concentrations of higher-brominated diphenyl ethers in soils and earthworms. Environ. Sci. Technol. 39, 9064–9070.

Sigler, M., 2014. The effects of plastic pollution on aquatic wildlife: current situations and future solutions. Water Air Soil Pollut. 225, 1–9.

Singh, R.P., Agrawal, M., 2008. Potential benefits and risks of land application of sewage sludge. Waste Manag. 28, 347–358.

Song, Y.K., Hong, S.H., Mi, J., Kang, J.H., Kwon, O.Y., Han, G.M., Shim, W.J., 2014. Large accumulation of micro-sized synthetic polymer particles in the sea surface microlayer. Environ. Sci. Technol. 48, 9014–9021.

Steinmetz, Z., Wollmann, C., Schaefer, M., Buchmann, C., David, J., Tröger, J., Muñoz, K., Frör, O., Schaumann, G.E., 2016. Plastic mulching in agriculture. Trading short-term agronomic benefits for long-term soil degradation? Sci. Total Environ. 550, 690–705.

Sun, Y., Luo, X., Wu, J., Mo, L., Chen, S., Zhang, Q., Zou, F., Mai, B., 2012. Species- and tissue-specific accumulation of Dechlorane Plus in three terrestrial passerine bird species from the Pearl River Delta, South China. Chemosphere 89, 445–451.

Tanaka, K., Takada, H., Yamashita, R., Mizukawa, K., Fukuwaka, M.-A., Watanuki, Y., 2013. Accumulation of plastic-derived chemicals in tissues of seabirds ingesting marine plastics. Mar. Pollut. Bull. 69, 219–222.

Thompson, R.C., Olsen, Y., Mitchell, R.P., Davis, A., Rowland, S.J., Anthony, W.G.J., McGonigle, D., Russell, A.E., 2004. Lost at sea: where is all the plastic? Science 304, 838.

Thompson, R.C., Swan, S.H., Moore, C.J., Saal, F.S.V., 2009. Introduction: our plastic age. Philos. Trans. Biol. Sci. 364, 1973–1976.

Turner, A., Holmes, L.A., 2015. Adsorption of trace metals by microplastic pellets in fresh water. Environ. Chem. 12, 600–610.

USEPA, 1992. Plastics pellets in the aquatic environment: sources and recommendations—final report, pp. 19238–19293.

van Franeker, J.A., Law, K.L., 2015. Seabirds, gyres and global trends in plastic pollution. Environ. Pollut. 203, 89–96.

van Franeker, J.A., Blaize, C., Danielsen, J., Fairclough, K., Gollan, J., Guse, N., Hansen, P.-L., Heubeck, M., Jensen, J.-K., Le Guillou, G., Olsen, B., Olsen, K.-O., Pedersen, J., Stienen, E.W.M., Turner, D.M., 2011. Monitoring plastic ingestion by the northern fulmar Fulmarus glacialis in the North Sea. Environ. Pollut. 159, 2609–2615.

Van Sebille, E., Wilcox, C., Lebreton, L., Maximenko, N., Hardesty, B.D., Van Franeker, J.A., Eriksen, M., Siegel, D., Galgani, F., Lavender Law, K., 2015. A global inventory of small floating plastic debris. Environ. Res. Lett. 10, 124006.

Wang, M.J., 1997. Land application of sewage sludge in China. Sci. Total Environ. 197, 149.

Wright, S.L., Thompson, R.C., Galloway, T.S., 2013. The physical impacts of microplastics on marine organisms: a review. Environ. Pollut. 178, 483–492.

Yamashita, R., Takada, H., Fukuwaka, M.-A., Watanuki, Y., 2011. Physical and chemical effects of ingested plastic debris on short-tailed shearwaters, Puffinus tenuirostris, in the North Pacific Ocean. Mar. Pollut. Bull. 62, 2845–2849.

Zhang, H., Kuo, Y.Y., Gerecke, A.C., Wang, J., 2012. Co-release of hexabromocyclododecane (HBCD) and nano- and microparticles from thermal cutting of polystyrene foams. Environ. Sci. Technol. 46, 10990–10996.

Zhao, S., Zhu, L., Li, D., 2016. Microscopic anthropogenic litter in terrestrial birds from Shanghai, China: not only plastics but also natural fibers. Sci. Total Environ. 550, 1110–1115.

Zubris, K.A., Richards, B.K., 2005. Synthetic fibers as an indicator of land application of sludge. Environ. Pollut. 138, 201–211.

CHAPTER 13

Nanoplastics in the Aquatic Environment

Karin Mattsson*[,†], Simonne Jocic[‡], Isa Doverbratt[†], Lars-Anders Hansson[†]
[*]University of Gothenburg, Gothenburg, Sweden
[†]Lund University, Lund, Sweden
[‡]Oak Crest Institute of Science, Monrovia, CA, United States

13.1 INTRODUCTION

Since the Industrial Revolution, humans have extensively been contributing to the accumulation of rubble in marine and freshwater ecosystems. Because the buildup of trash in water bodies was previously considered miniscule owing to its capacity to drift away from vantage points, the growing impact of plastic pollutants has historically been neglected. Today, however, pollution of aquatic systems is recognized as one of the biggest environmental threats to our planet. Ever since the mass production of plastic material in the 1940s, plastic has been statistically the largest contributor to marine pollution (Ryan et al., 2009). Concerns have been raised about the ecotoxicology of not only the macroform of plastic but also more recently plastic degradation products, namely micro- and nanosized particles. The following chapter highlights the ecological effects of differently sized plastics, with specific emphasis on nanosized plastic particles. The following text confines the definition of "nanoparticles" to mean all particles that have at least one dimension between 1 and 100 nm, that is, nanospheres, nanowires/nanotubes and nanofilms. To facilitate understanding, "nanoplastics" are defined as plastic nanoparticles, that is, any synthetic or semisynthetic organic polymers with at least one dimension between 1 and 100 nm.

13.2 SOURCES OF NANOPARTICLES IN THE AQUATIC ENVIRONMENT

Nanoplastics in the aquatic system are derived from primary and secondary particles. Primary particles are intentionally manufactured to a fixed size whereas secondary particles arise from fragmentation of larger matter.

Industrial operations and human activities contribute to the distribution of secondary particles, that is, plastic litter in freshwater and marine ecosystems. Human-originating sources include solid waste disposal from land and individual vessels at sea and coastal landfill operations (Pruter, 1987). Furthermore, accidental loss or spillage of plastics

Microplastic Contamination in Aquatic Environments
https://doi.org/10.1016/B978-0-12-813747-5.00013-8

during transportation and manufacturing practices in the plastics industry also contribute to the accumulation of secondary particles in aquatic environments (Derraik, 2002; Colton et al., 1974; Mato et al., 2001). Microplastic and nanoplastic particles used in consumer products from pharmaceutical and cosmetic industries (Lorenz et al., 2011) can reach the environment via wastewater or during consumer use (Sharma and Chatterjee, 2017). Most microplastics are removed in the wastewater treatment plants (Carr et al., 2016; Talvitie et al., 2017); however, not all particles are removed, and the plants may constitute to a considerable source of microplastics (Talvitie et al., 2017). Ecologically induced processes, such as tsunamis and storms, may also partially contribute to the widespread of plastic particles (Zettler et al., 2013).

It has been argued that the increase in the amount of primary particles in the aquatic environment is caused by the recent expansion in the manufacturing and use of nanoparticles (Biswas and Wu, 2005; Chow et al., 2005; Simonet and Valcarcel, 2009), and engineered nanoparticles released into the atmosphere inevitably end up in the aquatic environment or in soil (Nowack and Bucheli, 2007; Ryan et al., 2009; Simonet and Valcarcel, 2009). Nanoplastic particles produced from research and medical applications can also contribute as a source for nanoplastics in the environment. However, a more pronounced amount can arise from cosmetic consumer products that can, for example, enter the aquatic environment through the wastewater or during use. However, how small particles can escape through the wastewater plants is not known since most studies in wastewater plants do not detect particles with a size smaller than 20 μm (Carr et al., 2016; Talvitie et al., 2017).

The introduction and implementation of environmentally protective legislation has aided in reducing human-induced sources of maritime and freshwater pollution. For example, the Marpol 73/78 agreement assisted in limiting the accumulation of secondary plastic particles in marine environments by restricting waste disposal from ships. Since 2015, 99.2% of the world shipping tonnage, represented by 152 state parties, supported this convention. Moreover, in December 1988, the Ocean Dumping Ban Act was passed by the United States to forbid cities from depositing untreated sewage into the ocean.

13.3 FRAGMENTATION AND DEGRADATION OF PLASTICS

There are two main steps in the breakdown process of polymer particles: fragmentation and degradation. Fragmentation is the break down process of larger polymer chains into smaller, still polymeric, fragments (Thompson et al., 2005). Polymers are then susceptible to polymer degradation which is a bond-breaking process preceding chemical transformation that changes the polymer properties such as molecular weight. Degradation of plastics is chemically, physically, or biologically induced (Singh and Sharma, 2008; Andrady, 2011).

There are six processes involved in degradation of plastics: thermal degradation, hydrolysis, mechanical/physical degradation, thermooxidative degradation, photodegradation, and biodegradation (Andrady, 2011). However, owing to the moderate temperatures in aquatic environments, thermal degradation can be disregarded unlike hydrolysis, mechanical/physical degradation, thermooxidative degradation, photodegradation, and biodegradation, which are naturally occurring in saltwater and freshwater systems. Thermal degradation is a process employed in factories to induce bulk polymer breakdown of plastics following vigorous heat cycles with temperatures up to 430°C (Singh and Sharma, 2008).

- *Hydrolysis* is an environmentally induced bond-breaking reaction, introduced by the addition of water.
- *Mechanical* and/or physical degradation occurs to bulk materials when they are subject to mechanical stress from waves, rocks, sand, and other forces or substances that the polymer can interact with in aquatic environments. When the polymer is subject to this shearing force, the molecule breaks.
- *Thermooxidative* degradation is a slow oxidative breakdown that takes place at moderate temperatures.
- *Photodegradation*, a sunlight-induced degradation process which is more prominent on land and in air but slower in water due to the attenuation of light. Ultraviolet (UV) radiation has enough energy to cut carbon-carbon bonds, although the efficiency of this type of degradation is dependent on the wavelength of the light and the polymer chemical structure. During photodegradation, the polymer changes its physical and optical properties such that it experiences yellow discoloration and weakened mechanical properties, that is, the breakdown occurs only on the surface of the polymer.
- *Biodegradation* occurs when living organisms, usually microbes such as bacteria, break down organic substances.

Biodegradable plastics are plastics that can be broken down into water, carbon monoxide, and some biomaterials from microorganisms, such as fungi and bacteria. Some plastics are made from biomaterials or materials made from biological and renewable resources such as grains, corn, potatoes, beet sugar, sugar cane, or vegetable oils, but many biodegradable plastics are made from oil (PlasticsEurope, 2015). Nonetheless, most plastics are thermoplastics which means that they can be remelted into the liquid phase, and this type is generally considered nonbiodegradable (PlasticsEurope, 2015). The degree of biodegradability depends on the properties of the polymer and those of the biological environment. When a product is defined as biodegradable, it is considered to be biodegradable under specific conditions, such as when an industrial composter reaches a temperature of 70°C (UNEP, 2015). However, this is not the case in aquatic environments where the temperature, UV exposure and microbial colonizations are different to the preceding conditions (Moore, 2008).

In 2010, O'Brine and Thompson studied the degradation of four different plastic carrier bags in the marine environment—two oxo-biodegradable polyethylene (PE) bags, one biodegradable bag of polyester, and one standard PE bag produced from 33% of recycled materials. After 4 weeks of exposure, they saw the formation of biofilms on the samples and a decrease in tensile strength. Moreover, after 24 weeks, the compostable polyester was no longer detectable, whereas 98% of the other plastic remained intact even after 40 weeks (O'Brine and Thompson, 2010). Another study, made by Lambert and Wagner (2016), investigated the degradation of 1 cm pieces of polystyrene (PS) disposable coffee cup lids and found an increase in the number of nanoparticles over time. In other words, larger plastics were degraded into nanosized particles. The pieces were placed in a weathering chamber for 24 h together with 20 mL demineralized water at 30°C and exposed to both visible and ultraviolet radiation (Lambert and Wagner, 2016). According to nanoparticle tracking analysis (NTA), the remaining plastic pieces had decreased in particle size and had unquantifiable weights during the experiment. This is in agreement with a study made by Lambert et al. (2013) that studied the degradation of a latex polymer both in freshwater and in saltwater. Nanoparticles were detected with NTA, and the conclusion drawn from their experiment was that the degradation is dependent on light, that is, photooxidation was the primary degradation pathway in aquatic environments (Lambert et al., 2013). Moreover, the effect of UV radiation on the decrease in particle size, that is, the degradation of microplastic fragments (PE 90% and polypropylene (PP) 10%) collected in the North Atlantic gyre, has been studied using dynamic light scattering (DLS) instrumentation. This study showed that small nanoparticles are formed during the initial stage of degradation and that they then rearrange and form aggregates. Moreover, these newly formed nanoparticles also generate defects on the original material, which in turn causes degradation (Gigault et al., 2016).

In addition to physical effects, Zettler et al. (2013) observed bacterial growth on plastics (PE and PP) from collected plastic debris in seawater. They identified a higher diversity of bacteria on the plastics than in the sea water itself, suggesting that bacteria growing on plastic litter may be transported from the land to the sea (Zettler et al., 2013). In summary, plastic debris introduce organisms not native to the marine environment; some of which facilitate biodegradation.

In the aquatic environment there are particles of many different sizes and shapes (Cole et al., 2011). Ter Halle et al. (2016) suggested that rectangular pieces of plastic fragment differently than cubic particles owing to the fact that surface polymers of rectangular-shaped pieces will fragment from cracks created on the sun-exposed side. Cubic particles have more evenly distributed cracks owing to the fact that they roll at the surface. This difference in behavior will affect degradation rate such that cubic particles will degrade faster (Ter Halle et al., 2016).

Few studies have quantifiably detected nanoplastic particles from degradation processes; nonetheless, it is reasonable to assume that these processes produce nanosized materials of different sizes. Nanoparticles from degradation processes possess an extremely large surface area, and a common plastic shopping bag transformed into the form of 40 nm plastic particles has a surface area of 2600 m^2 (Mattsson et al., 2015b). Even though it is unlikely for an entire plastic shopping bag to be transformed into 40 nm plastic particles, the large amount of plastic bags disposed in the environment suggests that plastic litter eventually will generate an enormous surface area as degradation proceeds. The surface area is important for their biological impact (Brown et al., 2001). Nanoparticles exhibit properties both from solids (such as the fluorescence in quantum dots) and molecules (such as their ability to move). They also have novel physical features that make it possible for them to pass through biological barriers (Mattsson et al., 2017), penetrate tissue (Kashiwada, 2006) and accumulate in organs (von Moos et al., 2012) as well as affect the behavior and metabolism of organisms (Cedervall et al., 2012; Mattsson et al., 2015a). Moreover, nanoparticles can enter the base of a food chain more easily than larger particles.

13.4 TOXICITY OF NANOPARTICLES WHEN REACHING THE ENVIRONMENT

Nanoparticles have diverse chemical, mechanical, electric and physical properties that differ considerably from their corresponding bulk material owing to their broad distribution in size with at least one dimension between 1 and 100 nm (Biswas and Wu, 2005; Lowry et al., 2012). Hence, a material known to be nontoxic in bulk can be toxic at the nanometer scale due to its characteristic properties (Karlsson et al., 2009). It is not entirely clear which parameter is of most importance from a toxicity perspective and the potentially harmful effects of nanoparticles may be attributed to a large variety of properties. The most obvious are their chemical composition and the shape of the particles; however, due to their small size and the large surface area per unit mass, the surface energy cannot be neglected. The size and surface energy may affect, for example, surface functionalization, grafting, adsorption, homo- and heteroaggregation, reactivity, interaction between other nanoparticles, and their interaction with the environment (Lowry et al., 2012; Zoroddu et al., 2014).

Nanosized materials possess novel properties owing to their size. For example, quantum dots have size-dependent optical and electric properties related to their bandgap energy (Alivisatos, 1996). These properties are size-dependent because bandgap energy is intrinsically related to the fluorescence wavelength which is directly proportional to particle diameter. In addition, gold, which has historically been classified as an inert material, is catalytic in its nanoparticle form (Haruta, 2004).

Nanoparticles constitute larger surface areas than bulk materials; this large surface area is an important parameter when considering toxicity. Surface atoms have fewer nearest neighbors but have unsaturated bonds that make them more reactive than bulk material

(Oberdorster et al., 2005; Auffan et al., 2009). Surface area relates to surface energy that in turn affects the interaction between nanoparticles and biological components. Examples of parameters that affect the toxicity of nanoparticles once they reach the biological metabolism (e.g., inside an organism) are purity, doping, morphology, redox parameters, nature or composition of the shell of the particle or coating material, surface modifications and surfactants, solubility, duration of exposure, behavior under electromagnetic field exposure, size heterogeneity, chemical and colloidal stability and biodegradability (Lowry et al., 2012). Size and shape of nanoparticles are the most influential determinants of how far and where in the metabolism a specific nanoparticle reaches. The size decides the propensity of particles to pass through biological barriers such as the blood-brain barrier (Mattsson et al., 2017). From a toxicity point of view, also, dose is important since relevant and realistic dose regimes must be achieved to draw meaningful conclusions from an environmental perspective. However, studies using high doses can also be useful in order to assess which mechanisms and metabolic pathways are important. Conventional toxicity regulations for materials are generally not relevant for nanoparticles since they are often based on total mass, whereas for assessing effects of nanoparticles, the mass of each particle is more relevant. It is also important to characterize the materials to be able to assess their potential toxicity. Hence, a small change in one property will affect the fate of a specific type of nanoparticle when reaching the environment. Therefore, if the properties are not known, it is almost impossible to state what parameter has caused the effect and additionally, coatings make the fate of nanoparticles much more complicated (Oberdorster et al., 2005).

The biological fate, mobility and bioavailability of nanoparticles in aquatic systems depend on their size, shape and charge. Nanoparticles undergo different transformations such as macromolecular interactions and physical, chemical, and biological transformations (Lowry et al., 2012; Dale et al., 2015). Below is a list of the main transformations that occur to nanoparticles released in the environment:

- *Macromolecular interactions* define the interactions between particles and natural organic matter or biomolecules such as proteins in which a biomolecular corona is created around the particle (Cedervall et al., 2007; Lynch et al., 2006). The corona has a different surface structure and aggregation tendency than the naked particle. This determines how it affects organisms. Adsorption of proteins and biomolecules to the particles can occur in all environments and the corona can change over time by the replacement of biomolecules (Dell'Orco et al., 2010; Tenzer et al., 2013) and with changes in the environment (Lundqvist et al., 2011). This new surface will be important in the environment and in biological systems as it will influence the fate of the nanoparticles.
- *Physical transformation* or aggregation can occur as homoaggregation, between the same type of particles, or as heteroaggregation, between different types of nanoparticles.

When particles aggregate, the surface area to volume ratio is reduced and a new surface structure is created.

- *Biological transformations* occur when particles interact with both intracellular and extracellular living tissues. These reactions can cause changes both in the nanomaterial itself and in the potential coatings surrounding the particle, resulting in an alteration in, for example, surface charge, aggregation state, and reactivity. Biological transformations of carbon-based nanoparticles or coatings may reduce the concentration and affect the transport routes of the particles in the environment. However, it is still not known if these processes occur at high enough rates to be considered important.
- *Chemical transformation* occurs when the particles are reduced or oxidized. The redox reactions in water are induced by sunlight and can affect the coating, oxidation state, generation of reactive oxygen species (ROS), and persistence.

The fate of a particle will depend not only on the specific environment but also on the history of the particle as this will affect its properties, state, and number of future transitions (Lowry et al., 2012). The number of transitions that a particle will undergo is difficult to predict since some transitions take months or years. In 2016, Besseling et al. modeled the fate of nanoplastics and microplastic particles in a river, and according to their model, the nanoparticles and microparticles may remain in the river system instead of, as believed, continuing to the marine ecosystem (Besseling et al., 2017).

Although plastics vary in their general chemical composition, they generally have nonpolar surfaces that may absorb or attract other hydrophobic compounds. Plastic particles can absorb or adsorb persistent organic pollutants (POPs) (Ziccardi et al., 2016) such as PCB (Ryan et al., 1988). Absorption describes the scenario in which a particle imbibes a chemical, whereas adsorption describes that of which a chemical binds to the surface of particles. Both processes require a sufficient concentration gradient of POPs favoring plastic as opposed to the surrounding environment. Nevertheless, this context is not representative of aquatic systems that contain a concentration gradient of POPs favoring seawater over plastic particles, indicating that bioaccumulation of plastic particles cannot occur (Koelmans et al., 2016). Potential transfer calculations using worst-case parameters carried out by Koelmans et al. (2016) showed that the effects from plastics as carriers are irrelevant to the preceding conclusions. Moreover, microplastics act more as passive samplers instead of vectors for POPs even though chemical transfer still occurs (Herzke et al., 2016). Considering that the yearly plastics production rate is increasing exponentially, water bodies will still contain a higher level of the newly introduced chemicals, and thereby, a potential adsorption and absorption effect from plastics may be neglected (Koelmans et al., 2016). However, there are still many unknowns with regard to POP absorption and adsorption to plastics, which make predictions of plastic behavior difficult.

13.5 MEASUREMENT OF NANOPLASTICS IN THE ENVIRONMENT

There are five major methods for sampling plastic debris and marine anthropogenic particles in the environment, including beach combing, marine observational surveying, biological sampling, marine trawling, and sediment sampling (Cole et al., 2011; Hidalgo-Ruz et al., 2012). Neither beach combing nor marine observational surveying can be used for the monitoring of any nanoparticles in the environment since they are not visible. Biological sampling is used to detect litter consumed by marine and bird species. Marine trawling is a method used to collect litter from an aqueous environment through the use of filters on a moving boat. Most of these techniques filter surface water on a mesh, and the mesh size sets the particle-size cutoff. However, detection of nanoparticles is challenging considering that the most common mesh size for particle collection is around 300 μm (Hidalgo-Ruz et al., 2012). Sediment sampling combined with collection of benthic materials from beaches, estuaries, and the sea floor is also used for the sampling of plastic debris. Often, a density separation method is applied to separate the plastic particles from the sediment since low-density particles float up to the surface (Hidalgo-Ruz et al., 2012). Although it is possible to isolate low-density nanoparticles with this method, the denser particles that tend to sink will, obviously, not be isolated, which reduces the value of the method.

Available and frequently used techniques to quantify the amount and size of nanoparticles are, for example, DLS, NTA, UV-visible spectroscopy, differential sedimentation centrifugation (DSC), and electron microscopy. However, none of these techniques can identify the chemical identity of the particle. For larger particles, techniques such as Raman spectroscopy, energy-dispersive X-ray crystallography, near-infrared spectroscopy, and Fourier-transform infrared spectroscopy have been used (Hidalgo-Ruz et al., 2012). The general lack of suitable methods for nanoparticle collection from the aquatic environment, in combination with the lack of appropriate methods for polymer identification, limits the knowledge accumulated concerning nanoplastic concentrations in natural environments. However, one study estimated the total nanoparticle concentrations at nine locations in Sweden (Gallego-Urrea et al., 2010), where nanoparticle concentrations were measured in lakes, rivers, coastal areas, waste waters, and stormwater runoff. The majority of the particles had a diameter between 100 and 250 nm and occurred in concentrations of between 10^7 and 10^9 particles per mL (Gallego-Urrea et al., 2010).

13.6 NANOPLASTIC EFFECTS ON ORGANISMS IN THE AQUATIC ENVIRONMENT

Aquatic accumulation of primary and secondary nanoplastics poses a threat to the health status of exposed invertebrates and species from higher trophic levels that rely on these invertebrates as food. Studies on filter feeders, mussels, algae, oysters, and *Daphnia magna* have indicated that nanoplastic materials, owing to their small size, have the propensity to readily pass through biological barriers and accumulate in tissues and organs of exposed organisms (Kashiwada, 2006; von Moos et al., 2012; Mattsson et al., 2017), triggering physiological distress, metabolic disorders, growth impediments, hindered autotrophic capabilities, diminished reproductive fitness, and early mortality (von Moos et al., 2012). Toxicological harm toward a given species is dependent on the duration and route of exposure, as well as the nanoparticle surface charge, functionalization, aggregation status, and size (Oberdorster et al., 2005).

The interaction and behavior of nanoparticles with living organisms differ from those of larger pieces of bulk material because of their unique nanoscale properties. Due to their small size, they can more easily pass through biological barriers, penetrate through tissue (Kashiwada, 2006), and accumulate in organs (von Moos et al., 2012). The large surface area of nanoparticles, for instance, is responsible for the increased reactivity, which in turn increases its biotoxicity (Booth et al., 2016). Hence, particles with just a small difference in, for example, size, charge, or surface coating, can show completely different outcomes from a toxicological perspective.

13.6.1 Nanoplastic Effects on Algae

Algae are aquatic primary producers of different shapes and sizes. The smallest are a few micrometers in length and the largest are several meters long. Algae serve as a starting point for many food webs (Brönmark and Hansson, 2005; Mattsson et al., 2015a) and literature provides many examples of nanoparticles interacting with algae. However, few studies focus on plastic nanoparticles but instead focus on PS nanoparticles.

Bhattacharya et al. (2010) exposed single-celled alga genus *Chlorella* and multicelled *Scenedesmus* to differently charged PS nanoparticles in order to investigate the effect of particle surface charge and algal morphology on plastic adsorption (Bhattacharya et al., 2010). They used 20 nm PS particles with positively charged amidine PS and negatively charged carboxyl PS. To mimic natural conditions, the algae were exposed to the particles both in nonaggregated and aggregated forms. Overall, negatively charged particles showed a lower binding affinity than the positively charged particles. However, they varied in their adsorption levels between species. The single-cell alga, *Chlorella*, had a higher adsorption rate than the multicelled, *Scenedesmus*, for the negatively charged particles. Electrostatic interactions between the positively charged particles and the major component of cell walls in green plants, cellulose, are likely responsible for the higher affinity of positively charged particles. Furthermore, the adsorption of nanoplastic particles hindered algal photosynthesis and promoted the formation of ROS, which is a stress response that may affect algae viability. Their results also imply that cellulose in the cell wall plays an essential role in initiating the binding between the algae and the particles

(Bhattacharya et al., 2010). Moreover, exposure of *Scenedesmus obliquus* to 70 nm PS nanoparticles inhibited growth and reduced the concentration of chlorophyll in the cells (Besseling et al., 2014). Exposure of the green freshwater algae *Pseudokirchneriella subcapitata* to nonfunctionalized, carboxylated, and aminated PS nanoparticles of different size ranges (50–500, 110, and 20 nm) showed that natural and positively charged particles had a stronger adsorption to the cell wall of *P. subcapitata* than negatively charged particles. The results suggest that the binding affinity is a function of both interparticle and particle-cell wall interactions, which are influenced by the medium hardness and particle concentration. Moreover, medium conditions and material properties were also found to be important factors for nanoparticle adsorption to algal cell walls (Nolte et al., 2017). In a study of the effects from exposure to 23 nm PS, nanoparticles on excretion of exopolymeric substances from three marine phytoplankton species, *Amphora* sp., *Ankistrodesmus angustus*, and *Phaeodactylum tricornutum* (Chen et al., 2011) showed that the nanoparticles induced significant acceleration in exopolymeric substance excretion in *Amphora* sp., and after 72 h, the interactions reached equilibrium and microscopic gels of 4–6 μm were formed. The response by the other two species was less pronounced (Chen et al., 2011).

More studies on nanoplastic toxicity in algae need to be carried out to complement the sparse research currently available primarily on how algae respond to PS nanoparticles. Nonetheless, the results from these few studies imply that positively charged PS nanoparticles have more pronounced effects than negatively charged PS nanoparticles. Furthermore, the uptake of nanoplastics by algae is dependent on particle charge and surface functionalization but ultimately appears to cause physiological distress upon entry in the algal metabolism.

13.6.2 Nanoplastic Effects on Filter Feeders

The next step in the food chain is the filter feeding zooplankton which filter and concentrate suspended matter and food particles such as algal cells from water. Filter feeders play an important role in cleaning water and they, together with phytoplankton, serve as natural points of entry into the food web for nanoparticles. Filter feeders can ingest nanoparticles from their environment, but more likely, they will ingest particles that are aggregated with other particles (Ward and Kach, 2009). However, it is not known if the aggregated particles have the same physiological impact as the primary particles (Andrady, 2011). The literature provides many studies on the effects of nanoparticles on filter feeders, but only a few address plastic nanoparticles which are therefore highlighted in the following sections.

D. magna, a freshwater invertebrate commonly found in lakes and ponds all over the world, is one of the most commonly used organisms in toxicity tests. They can ingest particles in the range of 20 nm to 70 μm from the surrounding water (Zhu et al., 2008; Rosenkranz et al., 2009). With their filtering apparatus, they filter up to 18.5 mL of water per hour per mg dry wt *Daphnia* (Burns, 1969) and are therefore able to replace 3.1% of their body weight every 10 min (Stobbart et al., 1977). Out of the literature provided on the topic of effects of nanoplastics on *D. magna*, only one study investigates PMMA (poly(methyl methacrylate)) while several have investigated PS as a source of nanoplastics. Blessing et al. describe experiments in which *D. magna* were exposed to a mixture of 70 nm nanoparticles and algae (*S. obliquus*) (1) immediately, (2) after 5 days of nanoparticle and algal mixing, (3) following filtering of algae, and (4) instead of algae, kairomones from the fish predator perch (*Perca fluviatilis*) were added with nanoplastics. The mortality rate of *D. magna* was six times higher for the preexposed algae (2) compared with fresh algae (1) (Besseling et al., 2014), which is likely due to the fact that the preexposed algae had absorbed a higher amount of particles and thereby the uptake by *Daphnia* was higher. Moreover, the aging process may enhance the transfer of styrene monomers from the particles to the algae and thereby increase the bioavailability of styrene. *Daphnia*, receiving the fish kairomones, assumed to be stressful for the *Daphnia*, had a lower reproduction rate and a more pronounced reduction in body size. The authors explained the reduction in body size as a change in survival strategy, with and without the presence of a predator (Besseling et al., 2014).

Rist et al. (2017) exposed *D. magna* to 100 nm and to 2 μm fluorescent PS particles under four different conditions. Under condition (1), *Daphnia* was exposed to particles during a 24 h ingestion phase and a 24 h egestion phase in clean medium. For this condition, the total particle mass per animal (body burden of plastic particles) was measured during the ingestion phase and during the egestion phase. Condition (2) entailed the same scenario described in condition (1), however, with the addition of algae food (*Raphidcelis subcapitata*) during the ingestion and egestion phases for separate *Daphnia* groups. In condition (3), in addition to the parameters described in condition (2), algal cell densities were noted four times within the measurement period and body burden was measured. Finally, in condition (4), *Daphnia* were exposed for 21 days, and nanoparticle-induced effects on growth and reproduction, as well as on body burden, were assessed. From their experiments, the authors concluded that the 2 μm particles ingested and egested a higher mass concentration of the particles than the 100 nm particles. Furthermore, when food was present, the body burdens decreased. The 100 nm particles decreased the filtration rate whereas the larger particles did not. Taken all parameters into account, the 100 nm particles were potentially the most hazardous to *Daphnia* in this study (Rist et al., 2017).

D. magna have also been exposed to 20 and 1000 nm fluorescent carboxylated PS nanoparticles for four different exposure times (Rosenkranz et al., 2009). The uptake was higher for the 1000 nm particles when calculating the uptake rate based on mass of the particles. However, when calculating the uptake based on surface area, the uptake rate of the 20 nm particles was higher (Rosenkranz et al., 2009). These results are in agreement with the study performed by Rist et al. (2017) that compared two differently sized particles to conclude that uptakes were higher for larger particles when considering mass. However, both studies also registered a higher uptake for smaller particles when considering surface area, which might be an important conclusion when considering toxicity of nanoparticles because of their special properties.

In studies conducted by Ma et al. (2016), 1-day-old and 10-day-old *D. magna* were exposed to different dosages and exposure regimens of 50 nm and micrometer-sized ^{14}C-Phe labeled PS particles. EC_{50} values around 15 mg/L were observed for the 50 nm particles within the dosage range of 1–50 mg/L, whereas no immobilization was observed for the larger particles with concentrations up to 100 mg/L. Moreover, the 50 nm particles also caused physical damage to the *Daphnia* body in, for example, the thoracopods (essential to swimming and creating water currents for filter feeding), whereas the larger particles were excreted following ingestion into the intestinal tract. The increased damage on the *Daphnia* body can partly be explained by a stronger electrostatic interaction between the nanoparticles and daphnids. Toxic impurities, such as styrene monomers or surfactants, arising after synthesis are likely responsible for the reported high toxicity values. The bioaccumulation of ^{14}C-Phe increased over time within the 14 day exposure period. Additionally, at the end of the exposure trials, the levels of ^{14}C-Phe were three times higher than in the control group (Ma et al., 2016).

In a 48 h acute toxicity assessment undertaken by Casado et al. (2013), *D. magna* were exposed to 55 and 110 nm fluorescently labeled polyethylenimine PS particles. The 55 and 110 nm particles had EC_{50} values of 0.77 and 0.66 μg/mL, respectively. Because the 55 nm particles contain a higher degree of surface functionalization owing to the possible core size effect, their half maximal effective concentration values were higher than that for the 110 nm particles (Casado et al., 2013).

In another study, *D. magna* and *Corophium volutator* were exposed to fluorescently labeled and nonlabeled sets of 86 nm PMMA and 125 nm PMMA-PSMA (poly(methyl methacrylate-*co*-stearyl methacrylate)) (Booth et al., 2016). The *Daphnia* were exposed to three environmentally realistic concentrations (0.01, 0.1, and 1.0 mg/L) and to one extremely high concentration (1000 mg/L) of each plastic particle type. Immobilization, uptake and depuration studies took place during the 48 h period of nanoparticle exposure. The PMMA particles did not cause any significant mortality in any of the concentrations but the PMMA-PSMA induced a significant toxicity at most concentrations. During the uptake study, an increased level of fluorescence in the gut was registered already after 24 h, which then increased with time. Following the relocation of the exposed animals to clean media for 24 h, fluorescence was not observed in the gut, and the animals appeared as healthy as the control group. The *C. volutator* was exposed to the particles with the same concentrations as used for *Daphnia*, but the high dose was reduced to 500 mg/L. First, the authors added particles to filtered seawater to avoid colloids and aggregation of the particles. When the particles had precipitated to the sediment, which occurred after 5–6 h, the test animals were added. After 10 days of exposure, the animals were moved into beakers with clean sediments, and the viability and sublethal effects were investigated. No significant effects were found (Booth et al., 2016).

The available studies where *D. magna* were exposed to nanoplastics verify that surface area is a more important factor than mass in determining toxicity. Both Rist et al. (2017) and Rosenkranz et al. (2009) showed a higher uptake for the smaller sized particles when comparing surface area. Moreover, all studies show toxicity of organisms exposed to nanoplastic particles or particles very close to the nanometer definition, that is, particles with a size of 125 and 110 nm (Casado et al., 2013; Booth et al., 2016). The presence of algal food has also been shown to influence toxicity (Besseling et al., 2014; Rist et al., 2017). All these studies, nonetheless, exposed the *Daphnia* in a media, according to the OECD or US EPA recommendations, which is not entirely reflective of the natural aquatic environment. Although these results demonstrate important insight on the topic, they may not mirror effects occurring in a natural aquatic systems. The reason is that, for example, the media is less complex than the natural aquatic environment considering that it does not contain secreted proteins or biomolecules. Proteins can, depending on identity and concentration, destabilize or stabilize nanoparticles (Cukalevski et al., 2015) and thereby affect their uptake and toxicity. Hence, extensive research efforts must be placed for conducting these toxicity experiments in more ecologically relevant aquatic environments.

Only one study has used conditioned water, that is, water that has been prefiltered by *D. magna* as media. *Daphnia* were exposed to 90 nm carboxylic acid and amino-functionalized PS nanoparticles, and the toxicity, uptake, and removal were studied (Nasser and Lynch, 2016). A corona made up of proteins and biomolecules in the media formed around the particles that started aggregating after 6 h. However, when comparing the media without proteins and biomolecules present, the particle size remained unaffected. The EC_{50} values for both types of particles decreased over time in the conditioned water and had a lower value compared with the pure media, that is, no conditioned water. However, the removal rate was lower in the preconditioned exposure, which can be seen as a secondary effect of nanoplastics. As expected and in agreement with other studies, the positively charged particles were more toxic than the negatively charged particles (Nasser and Lynch, 2016).

Bivalvia is a class of marine and freshwater filter-feeding molluscs such as oysters, mussels, and clams; all of which contain two component calcium carbonate-consisting shells. They are a prey to many predators, including humans. The particle capture rate of bivalves, which is approximately 90%, decreases asymptotically with decreasing particle size (Clausen and Riisgard, 1996). The risk for them to ingest nonaggregated nanoparticles is therefore low but increases for aggregated nanoparticles in the environment. Assessing the amount of excreted pseudofeces can imply the biological effects of nonaggregated and aggregated particles. This method is used, for example, on *Mytilus edulis* that ingests food particles and then excretes particles with low nutritive value in the form of pseudofeces (Clausen and Riisgard, 1996).

Cole et al. (2011) exposed Pacific oyster (*Crassostrea gigas*) larvae to surface-modified PS particles and investigated the impact the particles had on larval feeding and growth. First, they used fluorescently labeled PS nanoparticles with a size distribution of 70 nm to 20 μm to investigate the particle size range *C. gigas* could ingest. The animals were exposed to PS and algae. To elucidate the relationship between nanoparticle surface properties and ingestion rates, three different PS surface chemical compositions were tested: unmodified, aminated, and carboxylated. The authors found no differences in size of oyster larvae exposed to any particles during 8 days of exposure. Although the smallest (70 nm) particles could not be recorded and thereby were excluded from the study, the second smallest size class of nanoparticles (160 nm) were ingested by the oysters at a higher frequency than the larger, micrometer-sized, particles (Cole and Galloway, 2015).

Ward and Kach (2009) studied the ingestion and egestion rates of mussels (*M. edulis*) and oysters (*Crassostrea virginica*) exposed to 100 nm PS nanoparticles, aggregated 100 nm PS, and 10 μm PS particles. The ingestion rates were highest for the aggregated particles in both species. Moreover, mussels had a higher ingestion rate of the aggregated particles than the oysters. In the beginning of the exposure, the egestion rate was higher for the larger particles (10 μm) compared with the aggregated ones. However, the egestion rate for the aggregated particles increased with time. As expected, it was easier for the animals to ingest aggregated particles than nanoparticles, but a higher gut retention time was seen for the animals fed with nanoparticles. Ward and Kach (2009) also measured the produced amount of pseudofeces and found that the mussels and oysters that had received nanoparticles produced less pseudofeces than the animals fed with larger particles, suggesting that nanoparticles can accumulate in the animals.

The blue mussel, *M. edulis*, was indirectly and directly exposed to 30 nm PS nanoparticles with concentrations ranging from 0.1 to 0.3 g/L by Wegener et al. The indirect exposure resulted from the mixing of algal food (*Pavlova lutheri*) with particles. In both cases, the nanoparticles aggregated as soon as they came into contact with seawater, but 15 out of 16 mussels produced pseudofeces, whereas there was no production of pseudofeces in the control group. The measured size of the particles, when pseudofeces was produced, was less than 1 μm. This suggests that the mussels identified the particles as low-nutritive food and that the particles most likely were absorbed to the gills. The filtering activity decreased when nanoparticles were present depending on particle concentration. However, both results suggest that mussels were able to detect the presence of particles and thereby filtered less and produced more pseudofeces (Wegner et al., 2012).

Studies have also been performed on filtering rotifers, such as *Brachionus manjavacas*, exposed to different sizes of fluorescent PS particles. For example, Snell and Hicks (2011) exposed *B. manjavacas* to PP particles of the sizes—38, 83, 217, 546, and 2980 nm—and studied immediate and long-term health effects and how size and concentration affected the uptake, distribution, reproduction rate, feeding behavior, and offspring fitness. Both newborn neonates and reproductive, ovigerous females were exposed to particles during 2 h or 2 days. A size-dependent decrease in reproduction rate was registered for the 37 nm particles, whereas the larger particles did not affect the reproduction rate. All particles larger than 83 nm remained in the stomach and intestine until defecation. However, the 37 nm particles passed through the intestinal wall and entered the tissue and were also passed from mother to the extruded egg. Moreover, the feeding rate was particle-size-dependent, and since the chemical compounds of the particles are nontoxic in bulk sizes, the toxicity of the 37 nm particles was likely related to particle size (Snell and Hicks, 2011).

A considerable amount of nanoparticles present in the aquatic environment will likely occur as aggregates. However, both Ward and Kach (2009) and Wegner et al. (2012) showed results implying the important features of nanosize effects, despite the fact that the nanoparticles were aggregated in their studies. Also, Ward and Kach (2009) registered reduced production of pseudofeces in *M. edulis* when exposed to nanoparticles, whereas Wegner et al. (2012) showed that production of pseudofeces increased with nanoparticle exposure.

13.6.3 Nanoplastic Effects on Top Consumers

There are many different food chains in both marine and freshwater that consist of, for example, phytoplankton, zooplankton, planktivorous and piscivorous fish (Brönmark and Hansson, 2005). Food chains often cause biomagnification of ingested food particles, even if nanoparticle exposure is confined to one organism. A particle can, for example, be nontoxic to one trophic level such as *D. magna*, but higher up, the effect might be transferred and enhanced (Cedervall et al., 2012; Mattsson et al., 2015a).

Few studies have exposed higher trophic levels, such as fish, to nanoparticles. Nonetheless, the response of Crucian Carp (*Carassius carassius*) to PS nanoparticles through a food chain, from algae (*Scenedesmus* sp.) through zooplankton (*D. magna*), has been studied (Cedervall et al., 2012; Mattsson et al., 2015a). The experiments were designed as 3-day cycles where particles were initially added to algae that were allowed to ingest the particles for 24 h. On day 2, *D. magna* were allowed to feed on the algae for 24 h. Both studies addressed the effects on *Daphnia* during the exposure and found no behavioral impact or toxicity. On day three, the *Daphnia* were collected and washed with water to avoid any potential contamination from free nanoparticles before they were given to fish. After 2 months of exposure, the fish feeding behavior was recorded and analyzed. Behavioral changes, such as feeding time, activity, and shoaling behavior were reported. Moreover, a disturbed fat metabolism, increased ethanol concentration in the liver, and increased levels of inosine/adenosine and lysine in muscles of fish that had received nanoparticles through the food chain were reported (Mattsson et al., 2015a; Cedervall et al., 2012).

Another study on fish was performed by Manabe et al. (2011) in which embryos and larvae of the fish species medaka (*Oryzias latipes*) were subjected to four different types of fluorescent PS nanoparticles: nonfunctionalized 50 and 500 nm and carboxylated c-50 and c-500 nm. In this study, mortality and uptake of nanoparticles of 3 h-old embryos exposed for 165 h were assessed. The excretion rate was measured for embryos exposed from the age of 3 h to 3 days, and after the exposure, they were moved to embryo culture medium for 4 days. Yolk-sac larvae were exposed to particles until they reached 50 days post hatching, to investigate mortality in two different sizes of tanks: one large plastic container where the larvae were together and one small cup where they were alone. The uptake was higher for the smaller particles (50 and c-50 nm) than for the larger ones. The highest uptake was found for the 50 nm particles and the lowest for the c-500 nm particles. The measured excretion rate was lower for the smaller particles and lowest for the c-50 nm. The larvae exposed in a large tank showed no significant difference in survival rate. However, the larvae in the small cups exposed to smaller particles, 50 and c-50 nm, had lower survival rates in comparison with the control group. These findings show that nanoparticles are easier to ingest and harder to excrete than larger particles. Additionally, this study demonstrates that group rearing decreases mortality rate (Manabe et al., 2011).

Sea urchin embryos (*Paracentrotus lividus*) were subjected to 40 nm negatively charged carboxy-modified PS and 50 nm positively charged amino-modified PS in studies conducted by Della Torre et al. (2014). The sea urchins were exposed to the seawater-induced aggregate forms of the positively and negatively charged particles with sizes of less than 100 and 1000 nm, respectively. The aggregated negatively charged particles showed no toxicity and accumulated inside the digestive tract of the embryos, whereas the nanometer-sized positively charged particles showed a toxicity, which increased with time. Moreover, the positively charged particles caused severe developmental defects, and the particles were more dispersed inside the embryo. These effects were shown to be dependent on the difference in surface charge and aggregation in sea water (Della Torre et al., 2014).

Kashiwada (2006) also exposed the fish species medaka (*O. latipes*) to fluorescent latex particles in the size range of 40 nm to 42 μm. Four different experiments were performed. The first three all investigated the adsorption and accumulation in medaka eggs but from three different perspectives. The first experiment looked at not only the adsorption and accumulation but also the distribution of nanoparticles in posthatch larvae. In the second experiment, the particle size dependence was analyzed, and in the third experiment, the effects of salinity on adsorption and accumulation in eggs were investigated in addition to the aggregation of the nanoparticles in the solution. The fourth experiment looked at the distribution of nanoparticles in the blood, as well as other organs in adult medaka. The eggs were exposed for 3 days, whereas the adults were exposed for 7 days. No mortality was registered in any of the experiments. Fluorescence was detected at a higher amount in whole eggs and egg envelopes (chorion), as well as in oil droplets, compared with the yolk area. The particles were also shown to accumulate in the oil droplets. The highest adsorption and accumulation value was found for the 474 nm particles. The nanoparticle aggregation depended on the salinity where the higher salinity led to more aggregation. Particles were registered in several organs, including the gills, kidney, liver, intestine, ovaries, tissue, and brain. The highest dose was found in the gills, and the author argued that the particles entered through the gills and then reached the bloodstream and moved further to other organs. The most surprising finding was that both the tissue and the brain contained nanoparticles. Although these results were not significant, they may indicate the possibility for particles to pass biological barriers (Kashiwada, 2006), a notion strengthened by a study on crucian carp, detecting nanoparticles in brain tissue of the fish (Mattsson et al., 2017).

Nanoplastic is harmful to the exposed top consumers according to these five studies in part by the fact that they travel through food chains (Cedervall et al., 2012; Mattsson et al., 2015b). Smaller particles (i.e., particles <100 nm), especially those that are positively charged, showed higher toxicity profiles in zooplankton owing to their higher uptake and slower excretion rate (Manabe et al., 2011; Nasser and Lynch, 2016). However, these studies are not entirely reflective of what would occur in a natural aquatic environment since they were conducted with high nanoparticle concentrations and only specific types of particles were used. Nonetheless, these results provide a glimpse to troubling concepts such as the ability of nanoplastics to pass biological barriers and accumulate in organs depending on their size. Additionally, even if most nanoplastic particles in the aquatic environment are in their aggregated or coated forms there can be mechanisms within the organisms that will separate the particles and allow them to penetrate biological barriers although such effects are yet to be demonstrated.

13.7 SUMMARY

Nanoplastic particles are found in the aquatic environment as primary particles or, more likely, as secondary particles derived from degraded plastic material. These nanoparticles will interact with their surroundings although their exact fate is, at present, difficult to predict considering that nanoparticles are a recent environmental problem. However, nanoparticles do interact with living organisms as either free nanoplastic particles or as aggregates. The effects these interactions have are not fully elucidated, but there are studies showing that nanoparticles have considerable effects on biota and on the ecosystem function in natural systems (Kashiwada, 2006; Ward and Kach, 2009; Bhattacharya et al., 2010; Manabe et al., 2011; Snell and Hicks, 2011; Cedervall et al., 2012; Besseling et al., 2014; Mattsson et al., 2015a; Booth et al., 2016; Greven et al., 2016; Rist et al., 2017). Most of the laboratory studies have used high concentrations of one type of plastic particles (Kashiwada, 2006; Rosenkranz et al., 2009; Ward and Kach, 2009; Chen et al., 2011; Snell and Hicks, 2011; Cedervall et al., 2012; Wegner et al., 2012; Besseling et al., 2014; Cole and Galloway, 2015; Mattsson et al., 2015a; Nasser and Lynch, 2016; Rist et al., 2017) or as homoaggregated particles (Ward and Kach, 2009; Bhattacharya et al., 2010), which may not reflect the scenario occurring in natural aquatic environments. However, such studies provide insights into target mechanisms and assist in understanding which metabolic features are affected. Moreover, nanoparticles can accumulate in specific regions such as tissues (Kashiwada, 2006), and therefore, some organisms may be exposed to concentrations higher than the mean expected concentration. The fact that many organisms can take up these particles highlights the need for more research regarding potential toxic effects of exposure to nanoplastics. There is also a need for method development for both collections of nanoparticles in the aquatic environment and identification of the collected particles. In a broader context, the multitude of studies identifying considerable effects of nanoplastics should serve as a signal for decision-makers to consider both the production of nanomaterials and the breakdown of plastic consumer products as potential environmental problems. However, the strong scientific engagement toward understanding ecological effects of nanoplastic exposure, together with findings suggesting that some particles are less reactive than others, may provide insights into producing less potent nanomaterials.

REFERENCES

Alivisatos, A.P., 1996. Semiconductor clusters, nanocrystals, and quantum dots. Science 271, 933–937.

Andrady, A.L., 2011. Microplastics in the marine environment. Mar. Pollut. Bull. 62, 1596–1605.

Auffan, M., Rose, J., Bottero, J.Y., Lowry, G.V., Jolivet, J.P., Wiesner, M.R., 2009. Towards a definition of inorganic nanoparticles from an environmental, health and safety perspective. Nat. Nanotechnol. 4, 634–641.

Besseling, E., Wang, B., Lurling, M., Koelmans, A.A., 2014. Nanoplastic affects growth of *S. obliquus* and reproduction of *D. magna*. Environ. Sci. Technol. 48, 12336–12343.

Besseling, E., Quik, J.T., Sun, M., Koelmans, A.A., 2017. Fate of nano- and microplastic in freshwater systems: a modeling study. Environ. Pollut. 220, 540–548.

Bhattacharya, P., Lin, S.J., Turner, J.P., Ke, P.C., 2010. Physical adsorption of charged plastic nanoparticles affects algal photosynthesis. J. Phys. Chem. C 114, 16556–16561.

Biswas, P., Wu, C.Y., 2005. Nanoparticles and the environment. J. Air Waste Manage. Assoc. 55, 708–746.

Booth, A.M., Hansen, B.H., Frenzel, M., Johnsen, H., Altin, D., 2016. Uptake and toxicity of methylmethacrylate-based nanoplastic particles in aquatic organisms. Environ. Toxicol. Chem. 35, 1641–1649.

Brönmark, C., Hansson, L.-A., 2005. The Biology of Lakes and Ponds, second ed. Oxford University Press, Oxford.

Brown, D.M., Wilson, M.R., MacNee, W., Stone, V., Donaldson, K., 2001. Size-dependent proinflammatory effects of ultrafine polystyrene particles: a role for surface area and oxidative stress in the enhanced activity of ultrafines. Toxicol. Appl. Pharmacol. 175, 191–199.

Burns, C.W., 1969. Relation between filtering rate, temperature, and body size in 4 species of *Daphnia*. Limnol. Oceanogr. 14, 693.

Carr, S.A., Liu, J., Tesoro, A.G., 2016. Transport and fate of microplastic particles in wastewater treatment plants. Water Res. 91, 174–182.

Casado, M.P., Macken, A., Byrne, H.J., 2013. Ecotoxicological assessment of silica and polystyrene nanoparticles assessed by a multitrophic test battery. Environ. Int. 51, 97–105.

Cedervall, T., Lynch, I., Lindman, S., Berggard, T., Thulin, E., Nilsson, H., Dawson, K.A., Linse, S., 2007. Understanding the nanoparticle-protein corona using methods to quantify exchange rates and affinities of proteins for nanoparticles. Proc. Natl. Acad. Sci. U. S. A. 104, 2050–2055.

Cedervall, T., Hansson, L.A., Lard, M., Frohm, B., Linse, S., 2012. Food chain transport of nanoparticles affects behaviour and fat metabolism in fish. PLoS One 7, e32254.

Chen, C.S., Anaya, J.M., Zhang, S.J., Spurgin, J., Chuang, C.Y., Xu, C., Miao, A.J., Chen, E.Y.T., Schwehr, K.A., Jiang, Y.L., Quigg, A., Santschi, P.H., Chin, W.C., 2011. Effects of engineered nanoparticles on the assembly of exopolymeric substances from phytoplankton. PLoS One 6.

Chow, J.C., Watson, J.G., Savage, N., Solomon, C.J., Cheng, Y.S., McMurry, P.H., Corey, L.M., Bruce, G.M., Pleus, R.C., Biswas, P., Wu, C.Y., 2005. Nanoparticles and the environment. J. Air Waste Manage. Assoc. 55, 1411–1417.

Clausen, I., Riisgard, H.U., 1996. Growth, filtration and respiration in the mussel *Mytilus edulis*: no evidence for physiological regulation of the filter-pump to nutritional needs. Mar. Ecol. Prog. Ser. 141, 37–45.

Cole, M., Galloway, T.S., 2015. Ingestion of nanoplastics and microplastics by Pacific oyster larvae. Environ. Sci. Technol. 49, 14625–14632.

Cole, M., Lindeque, P., Halsband, C., Galloway, T.S., 2011. Microplastics as contaminants in the marine environment: a review. Mar. Pollut. Bull. 62, 2588–2597.

Colton Jr., J.B., Burns, B.R., Knapp, F.D., 1974. Plastic particles in surface waters of the northwestern Atlantic. Science 185, 491–497.

Cukalevski, R., Ferreira, S.A., Dunning, C.J., Berggard, T., Cedervall, T., 2015. IgG and fibrinogen driven nanoparticle aggregation. Nano Res. 8, 2733–2743.

Dale, A.L., Casman, E.A., Lowry, G.V., Lead, J.R., Viparelli, E., Baalousha, M., 2015. Modeling nanomaterial environmental fate in aquatic systems. Environ. Sci. Technol. 49, 2587–2593.

Della Torre, C., Bergami, E., Salvati, A., Faleri, C., Cirino, P., Dawson, K.A., Corsi, I., 2014. Accumulation and embryotoxicity of polystyrene nanoparticles at early stage of development of sea urchin embryos *Paracentrotus lividus*. Environ. Sci. Technol. 48, 12302–12311.

Dell'Orco, D., Lundqvist, M., Oslakovic, C., Cedervall, T., Linse, S., 2010. Modeling the time evolution of the nanoparticle-protein corona in a body fluid. PLoS One 5,e10949.

Derraik, J.G., 2002. The pollution of the marine environment by plastic debris: a review. Mar. Pollut. Bull. 44, 842–852.

Gallego-Urrea, J.A., Tuoriniemi, J., Pallander, T., Hassellov, M., 2010. Measurements of nanoparticle number concentrations and size distributions in contrasting aquatic environments using nanoparticle tracking analysis. Environ. Chem. 7, 67–81.

Gigault, J., Pedrono, B., Maxit, B., Ter Halle, A., 2016. Marine plastic litter: the unanalyzed nano-fraction. Environ. Sci. Nano 3, 346–350.

Greven, A.C., Merk, T., Karagöz, F., Mohr, K., Klapper, M., Jovanović, B., Palić, D., 2016. Polycarbonate and polystyrene nanoplastic particles act as stressors to the innate immune system of fathead minnow (Pimephalespromelas). Environ. Toxicol. Chem. 35, 3093–3100.

Haruta, M., 2004. Gold as a novel catalyst in the 21st century: preparation, working mechanism and applications. Gold Bull. 37, 27–36.

Herzke, D., Anker-Nilssen, T., Nost, T.H., Gotsch, A., Christensen-Dalsgaard, S., Langset, M., Fangel, K., Koelmans, A.A., 2016. Negligible impact of ingested microplastics on tissue concentrations of persistent organic pollutants in northern fulmars off coastal Norway. Environ. Sci. Technol. 50, 1924–1933.

Hidalgo-Ruz, V., Gutow, L., Thompson, R.C., Thiel, M., 2012. Microplastics in the marine environment: a review of the methods used for identification and quantification. Environ. Sci. Technol. 46, 3060–3075.

Karlsson, H.L., Gustafsson, J., Cronholm, P., Moller, L., 2009. Size-dependent toxicity of metal oxide particles—a comparison between nano- and micrometer size. Toxicol. Lett. 188, 112–118.

Kashiwada, S., 2006. Distribution of nanoparticles in the see-through medaka (*Oryzias latipes*). Environ. Health Perspect. 114, 1697–1702.

Koelmans, A.A., Bakir, A., Burton, G.A., Janssen, C.R., 2016. Microplastic as a vector for chemicals in the aquatic environment: critical review and model-supported reinterpretation of empirical studies. Environ. Sci. Technol. 50, 3315–3326.

Lambert, S., Wagner, M., 2016. Characterisation of nanoplastics during the degradation of polystyrene. Chemosphere 145, 265–268.

Lambert, S., Sinclair, C.J., Bradley, E.L., Boxall, A.B.A., 2013. Effects of environmental conditions on latex degradation in aquatic systems. Sci. Total Environ. 447, 225–234.

Lorenz, C., Von Goetz, N., Scheringer, M., Wormuth, M., Hungerbuhler, K., 2011. Potential exposure of German consumers to engineered nanoparticles in cosmetics and personal care products. Nanotoxicology 5, 12–29.

Lowry, G.V., Gregory, K.B., Apte, S.C., Lead, J.R., 2012. Transformations of nanomaterials in the environment. Environ. Sci. Technol. 46, 6893–6899.

Lundqvist, M., Stigler, J., Cedervall, T., Berggard, T., Flanagan, M.B., Lynch, I., Elia, G., Dawson, K., 2011. The evolution of the protein corona around nanoparticles: a test study. ACS Nano 5, 7503–7509.

Lynch, I., Dawson, K.A., Linse, S., 2006. Detecting cryptic epitopes created by nanoparticles. Sci. STKE. 2006. pe14.

Ma, Y., Huang, A., Cao, S., Sun, F., Wang, L., Guo, H., Ji, R., 2016. Effects of nanoplastics and microplastics on toxicity, bioaccumulation, and environmental fate of phenanthrene in fresh water. Environ. Pollut. 219, 166–173.

Manabe, M., Tatarazako, N., Kinoshita, M., 2011. Uptake, excretion and toxicity of nano-sized latex particles on medaka (*Oryzias latipes*) embryos and larvae. Aquat. Toxicol. 105, 576–581.

Mato, Y., Isobe, T., Takada, H., Kanehiro, H., Ohtake, C., Kaminuma, T., 2001. Plastic resin pellets as a transport medium for toxic chemicals in the marine environment. Environ. Sci. Technol. 35, 318–324.

Mattsson, K., Ekvall, M.T., Hansson, L.A., Linse, S., Malmendal, A., Cedervall, T., 2015a. Altered behavior, physiology, and metabolism in fish exposed to polystyrene nanoparticles. Environ. Sci. Technol. 49, 553–561.

Mattsson, K., Hansson, L.A., Cedervall, T., 2015b. Nano-plastics in the aquatic environment. Environ. Sci. Process. Impacts 17, 1712–1721.

Mattsson, K., Johnson, E.V., Malmendal, A., Linse, S., Hansson, L.A., Cedervall, T., 2017. Brain damage and behavioural disorders in fish induced by plastic nanoparticles delivered through the food chain. Sci. Rep. 7, 11452.

Moore, C.J., 2008. Synthetic polymers in the marine environment: a rapidly increasing, long-term threat. Environ. Res. 108, 131–139.

Nasser, F., Lynch, I., 2016. Secreted protein eco-corona mediates uptake and impacts of polystyrene nanoparticles on *Daphnia magna*. J. Proteome 137, 45–51.

Nolte, T.M., Hartmann, N.B., Kleijn, J.M., Garnaes, J., van de Meent, D., Jan Hendriks, A., Baun, A., 2017. The toxicity of plastic nanoparticles to green algae as influenced by surface modification, medium hardness and cellular adsorption. Aquat. Toxicol. 183, 11–20.

Nowack, B., Bucheli, T.D., 2007. Occurrence, behavior and effects of nanoparticles in the environment. Environ. Pollut. 150, 5–22.

Oberdorster, G., Oberdorster, E., Oberdorster, J., 2005. Nanotoxicology: an emerging discipline evolving from studies of ultrafine particles. Environ. Health Perspect. 113, 823–839.

O'Brine, T., Thompson, R.C., 2010. Degradation of plastic carrier bags in the marine environment. Mar. Pollut. Bull. 60, 2279–2283.

PlasticsEurope, 2015. Plastics—The Facts 2014/2015 an Analysis of European Plastics Production, Demand and Waste Data. Plastics Europe, Association of Plastic Manufactures, Brussels, p. 34.

Pruter, A.T., 1987. Sources, quantities and distribution of persistent plastics in the marine-environment. Mar. Pollut. Bull. 18, 305–310.

Rist, S., Baun, A., Hartmann, N.B., 2017. Ingestion of micro- and nanoplastics in *Daphnia magna*—quantification of body burdens and assessment of feeding rates and reproduction. Environ. Pollut. 228, 398–407.

Rosenkranz, P., Chaudhry, Q., Stone, V., Fernandes, T.F., 2009. A comparison of nanoparticle and fine particle uptake by *Daphnia magna*. Environ. Toxicol. Chem. 28, 2142–2149.

Ryan, P.G., Connell, A.D., Gardner, B.D., 1988. Plastic ingestion and Pcbs in seabirds—is there a relationship. Mar. Pollut. Bull. 19, 174–176.

Ryan, P.G., Moore, C.J., van Franeker, J.A., Moloney, C.L., 2009. Monitoring the abundance of plastic debris in the marine environment. Philos. Trans. R. Soc. Lond. Ser. B Biol. Sci. 364, 1999–2012.

Sharma, S., Chatterjee, S., 2017. Microplastic pollution, a threat to marine ecosystem and human health: a short review. Environ. Sci. Pollut. Res. Int. 24, 21530–21547.
Simonet, B.M., Valcarcel, M., 2009. Monitoring nanoparticles in the environment. Anal. Bioanal. Chem. 393, 17–21.
Singh, B., Sharma, N., 2008. Mechanistic implications of plastic degradation. Polym. Degrad. Stabil. 93, 561–584.
Snell, T.W., Hicks, D.G., 2011. Assessing toxicity of nanoparticles using *Brachionus manjavacas* (Rotifera). Environ. Toxicol. 26, 146–152.
Stobbart, R.H., Keating, J., Earl, R., 1977. Study of sodium uptake by water flea *Daphnia-magna*. Comp. Biochem. Phys. A 58, 299–309.
Talvitie, J., Mikola, A., Setala, O., Heinonen, M., Koistinen, A., 2017. How well is microlitter purified from wastewater?—A detailed study on the stepwise removal of microlitter in a tertiary level wastewater treatment plant. Water Res. 109, 164–172.
Tenzer, S., Docter, D., Kuharev, J., Musyanovych, A., Fetz, V., Hecht, R., Schlenk, F., Fischer, D., Kiouptsi, K., Reinhardt, C., Landfester, K., Schild, H., Maskos, M., Knauer, S.K., Stauber, R.H., 2013. Rapid formation of plasma protein corona critically affects nanoparticle pathophysiology. Nat. Nanotechnol. 8. 772–U1000.
Ter Halle, A., Ladirat, L., Gendre, X., Goudouneche, D., Pusineri, C., Routaboul, C., Tenailleau, C., Duployer, B., Perez, E., 2016. Understanding the fragmentation pattern of marine plastic debris. Environ. Sci. Technol. 50, 5668–5675.
Thompson, R., Moore, C., Andrady, A., Gregory, M., Takada, H., Weisberg, S., 2005. New directions in plastic debris. Science 310, 1117.
UNEP, 2015. Biodegradable Plastics and Marine Litter. Misconceptions, concerns and impacts on marine environments. United Nations Environment Programme (UNEP).
von Moos, N., Burkhardt-Holm, P., Kohler, A., 2012. Uptake and effects of microplastics on cells and tissue of the blue mussel *Mytilus edulis* L. after an experimental exposure. Environ. Sci. Technol. 46, 11327–11335.
Ward, J.E., Kach, D.J., 2009. Marine aggregates facilitate ingestion of nanoparticles by suspension-feeding bivalves. Mar. Environ. Res. 68, 137–142.
Wegner, A., Besseling, E., Foekema, E.M., Kamermans, P., Koelmans, A.A., 2012. Effects of nanopolystyrene on the feeding behavior of the blue mussel (*Mytilus edulis* L.). Environ. Toxicol. Chem. 31, 2490–2497.
Zettler, E.R., Mincer, T.J., Amaral-Zettler, L.A., 2013. Life in the "plastisphere": microbial communities on plastic marine debris. Environ. Sci. Technol. 47, 7137–7146.
Zhu, H., Han, J., Xiao, J.Q., Jin, Y., 2008. Uptake, translocation, and accumulation of manufactured iron oxide nanoparticles by pumpkin plants. J. Environ. Monitor. 10, 713–717.
Ziccardi, L.M., Edgington, A., Hentz, K., Kulacki, K.J., Kane Driscoll, S., 2016. Microplastics as vectors for bioaccumulation of hydrophobic organic chemicals in the marine environment: a state-of-the-science review. Environ. Toxicol. Chem. 35, 1667–1676.
Zoroddu, M.A., Medici, S., Ledda, A., Nurchi, V.M., Lachowicz, J.I., Peana, M., 2014. Toxicity of nanoparticles. Curr. Med. Chem. 21, 3837–3853.

INDEX

Note: Page numbers followed by *f* indicate figures, *t* indicate tables, and *b* indicate boxes.

G

H

I

K

L

N

O

P

R

S

T

U

V

W

Z